Sustainable Agriculture and Advances of Remote Sensing
Volume 1: In Earth Observation

Sustainable Agriculture and Advances of Remote Sensing
Volume 1: In Earth Observation

Editors

Dimitrios S. Paraforos
Anselme Muzirafuti
Giovanni Randazzo
Stefania Lanza

MDPI • Basel • Beijing • Wuhan • Barcelona • Belgrade • Manchester • Tokyo • Cluj • Tianjin

Editors
Dimitrios S. Paraforos
Hochschule Geisenheim University
Germany

Anselme Muzirafuti
University of Messina
Italy

Giovanni Randazzo
University of Messina
Italy

Stefania Lanza
University of Messina
Italy

Editorial Office
MDPI
St. Alban-Anlage 66
4052 Basel, Switzerland

This is a reprint of articles from the Special Issue published online in the open access journal *Applied Sciences* (ISSN 2076-3417) (available at: https://www.mdpi.com/journal/applsci/special_issues/Agriculture).

For citation purposes, cite each article independently as indicated on the article page online and as indicated below:

LastName, A.A.; LastName, B.B.; LastName, C.C. Article Title. *Journal Name* **Year**, *Volume Number*, Page Range.

Volume 1
ISBN 978-3-0365-5337-5 (Hbk)
ISBN 978-3-0365-5338-2 (PDF)

Volume 1-2
ISBN 978-3-0365-5335-1 (Hbk)
ISBN 978-3-0365-5336-8 (PDF)

Cover image courtesy of Anselme Muzirafuti

Contents

About the Editors

Dimitrios S. Paraforos

Prof. Dr. Dimitrios S. Paraforos is serving as a Deputy Professor of Agricultural Engineering in Special Crops at the Hochschule Geisenheim University (Department of Technology, Von-Lade-Str. 1, D-65366 Geisenheim, Germany). He holds an MSc from the University of Thessaly (Greece) and, in 2016, obtained a PhD from the University of Hohenheim, both in Agricultural Engineering. His research focus is on precision farming, digital technologies in agriculture, and, more generally, on control systems, robotics, and automation applied to agriculture; in addition, he has obtained industry experience working as an automation engineer in the food industry. He is the author of more than 53 papers indexed by Scopus, with an important contribution in the field of sensors and ISOBUS technologies for the enhancement of agricultural practices. Currently, he is coordinating the ERA-NET ICT-Agri II European project iFAROS on developing methods for increasing the efficiency and precision of site-specific fertilizer application.

Anselme Muzirafuti

Dr. Anselme Muzirafuti, born in Rwanda, is an assistant professor at the University of Messina (Department of Mathematics, Computer Sciences, Physics and Earth Sciences, Via F. Stagno d'Alcontres, 31–98166 Messina, Italy). He holds a master's degree in Applied Geophysics and Geology Engineering obtained in 2015 from University of Moulay Ismail, Meknes (Morocco), and a PhD in Hydrogeophysics obtained in 2021 from the same University. Since 2009, he received multi-excellence scholarships for his higher education studies from different governments, including the Government of Rwanda, the Government of the Kingdom of Morocco, and the European Union. In 2016, 2018 and 2019, he participated in major conferences on climate change, sustainability and geoscience, namely, the 22nd Conference of Parties held in Marrakech (Morocco); the 24th International Sustainable Development Research Society Conference (action for a sustainable world: from theory to practice), held in Messina (Italy); and the 2019 European Geoscience Union General Assembly, held in Vienna (Austria). His research interest has been focused on the use of Structural Geology, Geomatics and Geophysics for sustainable management of territories. He worked on different projects, in Morocco and in Italy, related to geomorphological mapping and surveys using images acquired by satellites and drones. He recently worked as analysist of satellite images in the BESS project "(Pocket Beaches management and Remote Monitoring Systems)—Program Interreg VA Italia Malta 2014–2020". The results of his works have been presented at international conferences and published in international journals.

Giovanni Randazzo

Prof. Dr. Giovanni Randazzo is an associate professor of Coastal Geomorphology and Environmental Geology at the University of Messina (Department of Mathematics, Computer Sciences, Physics and Earth Sciences, Via F. Stagno d'Alcontres, 31–98166 Messina, Italy). He holds a PhD in Marine Environment and Resources obtained from University of Messina (Italy). Since 1987, his research interest has been focused on the study of the coastal area, on its management and protection. During these 30 years, he collaborated with the Smithsonian Institution of Washington D.C. in the study of the Nile Delta; with the Thai Geological Service in the study of the east coast of the local peninsula; and with ENEA (Italian National Agency for New Technologies, Energy and Sustainable Economic Development), he participated in the X Italian expedition in Antarctica. He has collaborated in the environmental assessment impact of various public works (especially in the coastal area), and he participated in the drafting of the Territorial Landscape Plan of the Province of Messina (Sicily, Italy). In recent years, he has actively participated in the debate on the emergence of waste, writing scientific articles, intervening in the local press, and participating in various debates, where he presented a scheme of management of the emergency alternative to those not acting for the Sicilian Region. In 2013, he founded Geologis s.r.l., a branch of the University of Messina, active in the field of territory surveys using aerial and marine drones equipped with RGB cameras, LiDAR sensors, and thermal imaging cameras. On behalf of the European Union, he has coordinated at national level and/or as a local unit several projects related to coastal management and territorial security. Since December 2017, he has been the lead partner of the Pocket Beaches management and Remote Monitoring Systems (BESS) project as part of the Interreg Italy—Malta Program. He is the author of more than 120 scientific publications.

Stefania Lanza

Dr. Stefania Lanza is an administrator of Geologis s.r.l, an Academic Spin-Off of the University of Messina (Via F. Stagno d'Alcontres, 31–98166 Messina, Italy). In 2007, she obtained a PhD in Geology from the University of Messina (Italy) with a thesis on "The risk assessment of coastal areas: from planning to monitoring". She worked on different projects related to sedimentology and geomorphology mapping. In 2008, she took the final exam of the master course with a thesis entitled: "Coastal monitoring of the coast of Badalona (Spain) contribution to the 2007–2008 survey campaign" supervised by Prof. Jordi Serra of the Autonomous University of Barcelona. In 2013, she co-founded Geologis s.r.l., active in the field of territory surveys using aerial and marine drones equipped with RGB cameras, LiDAR sensors, and thermal imaging cameras. She recently worked as Coordinator of the Geomorphological and Sedimentological Activities of the Project in the context of BESS project "Pocket Beach Management & Remote Surveillance System—Program Interreg VA Italia Malta 2014–2020". She is currently working as coordinator of field activities in the context of the "BIOBLU project—Robotic Bioremediation for Coastal Debris in Blue Flag Beach and in a Maritime Protected Area—Interreg V-A Italy—Malta 2014–2020 program".

Preface to "Sustainable Agriculture and Advances of Remote Sensing Volume 1: In Earth Observation"

This Special Issue on "Sustainable Agriculture and Advances of Remote Sensing" falls within the scope of current efforts to mitigate and adapt to the changing climate. It has been launched with the aim of collecting and promoting recent scientific studies proposing and evaluating advances in remote sensing technology and agricultural engineering leading to sustainable agriculture. It is mainly addressed to the policy makers, entrepreneurs and academicians engaged in the fight against climate change, in zero hunger initiatives, in natural resource management and in environment protection research. A special thanks is addressed to the authors who submitted their manuscripts to contribute to these initiatives.

Dimitrios S. Paraforos, Anselme Muzirafuti, Giovanni Randazzo, and Stefania Lanza
Editors

applied sciences

MDPI

Article

Integration of Sentinel 1 and Sentinel 2 Satellite Images for Crop Mapping

Shilan Felegari [1], Alireza Sharifi [2], Kamran Moravej [1], Muhammad Amin [3], Ahmad Golchin [1], Anselme Muzirafuti [4], Aqil Tariq [5] and Na Zhao [6,*]

1 Department of Soil Science, Faculty of Agriculture, University of Zanjan, Zanjan 45371-38791, Iran; shilan.felegari@znu.ac.ir (S.F.); kmoravej@znu.ac.ir (K.M.); agolchin2011@yahoo.com (A.G.)
2 Department of Surveying Engineering, Faculty of Civil Engineering, Shahid Rajaee Teacher Training University, Tehran 16788-15811, Iran; a_sharifi@sru.ac.ir
3 Institute of Geo-Information & Earth Observation, PMAS Arid Agriculture University Rawalpindi, Rawalpindi 46300, Pakistan; m.amin@uaar.edu.pk
4 Interreg Italia-Malta-Progetto: Pocket Beach Management and Remote Surveillance System, University of Messina, Via F. Stagno d'Alcontres, 31-98166 Messina, Italy; a.muzirafuti@edu.umi.ac.ma
5 State Key Laboratory of Information Engineering in Surveying Mapping and Remote Sensing, Wuhan University, Wuhan 430079, China; aqiltariq@whu.edu.cn
6 State Key Laboratory of Resources and Environmental Information System, Institute of Geographic Sciences and Natural Resources Research, Chinese Academy of Sciences, Beijing 100101, China
* Correspondence: zhaon@lreis.ac.cn

Abstract: Crop identification is key to global food security. Due to the large scale of crop estimation, the science of remote sensing was able to do well in this field. The purpose of this study is to study the shortcomings and strengths of combined radar data and optical images to identify the type of crops in Tarom region (Iran). For this purpose, Sentinel 1 and Sentinel 2 images were used to create a map in the study area. The Sentinel 1 data came from Google Earth Engine's (GEE) Level-1 Ground Range Detected (GRD) Interferometric Wide Swath (IW) product. Sentinel 1 radar observations were projected onto a standard 10-m grid in GRD output. The Sen2Cor method was used to mask for clouds and cloud shadows, and the Sentinel 2 Level-1C data was sourced from the Copernicus Open Access Hub. To estimate the purpose of classification, stochastic forest classification method was used to predict classification accuracy. Using seven types of crops, the classification map of the 2020 growth season in Tarom was prepared using 10-day Sentinel 2 smooth mosaic NDVI and 12-day Sentinel 1 back mosaic. Kappa coefficient of 0.75 and a maximum accuracy of 85% were reported in this study. To achieve maximum classification accuracy, it is recommended to use a combination of radar and optical data, as this combination increases the chances of examining the details compared to the single-sensor classification method and achieves more reliable information.

Keywords: Sentinel 1 and 2; Copernicus Sentinels; crop classification; food security; agricultural monitoring; remote sensing; data analysis; SAR; random forest

Citation: Felegari, S.; Sharifi, A.; Moravej, K.; Amin, M.; Golchin, A.; Muzirafuti, A.; Tariq, A.; Zhao, N. Integration of Sentinel 1 and Sentinel 2 Satellite Images for Crop Mapping. *Appl. Sci.* **2021**, *11*, 10104. https://doi.org/10.3390/app112110104

Academic Editor: Amerigo Capria

Received: 29 August 2021
Accepted: 26 October 2021
Published: 28 October 2021

1. Introduction

To ensure food security, each region must produce high-consumption agricultural crops on time and in sufficient quantities [1]. Plant inventories of the crop season as one of the important components of agricultural statistics and estimation of crop fertility [2], as well as recognizing the region's capabilities for production, information about the type of crop, depending on the existing conditions, is one of the main preconditions for controlling anomalies benefit the agricultural and insurance industries as a private sector, as well as the public sector. Remote sensing [3] is one of the more advanced methods for mapping crops from various regions. The most common method for classifying crops in remote sensing is with optical images. With advances in remote sensing and spatial, temporal, and spectral separations, classification results became more professional [4]. Sentinel 2A was launched into orbit on 23 June 2015 as part of the Copernicus Sentinels mission. Sentinel 2 is made up of two satellites, Sentinel 2A and Sentinel 2B, which are very similar to each other. Sentinel 2 has a revisit time of 5 days, rather than 10 days, because of these 2 satellites (2A and 2B). Short-infrared, near-infrared, and visible wavelengths are among the electromagnetic spectrum parties of 13 bands of Sentinel 2's multispectral device.

Sentinel 2 was used in a variety of research fields for product classification due to its unique and advanced specifications [5]. However, the presence of clouds is one of the drawbacks of using optical sensors. Because light rays cannot penetrate the cloud, there is a gap in visual images due to the presence of clouds and cloud shadows, and this lack of influence is a significant problem in classification and monitoring of crops. The technique of combining multiple sensors is used to solve the problem of the cloud and its shadow; this technique can effectively use different parts of the electromagnetic spectrum [6]. The size of cloud particles, for example, is smaller than the wavelength of microwave radiation in the C band, allowing it to influence the cloud. With radar sensors, satellites emit energy and measure its reflection, allowing them to benefit from various parts of the electromagnetic spectrum due to these sensors. A synthetic aperture radar (SAR) [7] can be defined as a system that uses tool movement to achieve acceptable ground resolution. Despite the fact that SAR data from space is now widely available to the public, it used to necessitate special procedures [8]. Following the launch of the Sentinel 1 mission, SAR data became freely available for a limited time [9]. Sentinel 1A and Sentinel 1B satellites have a six-day recurring frequency. Because of the overlap and combination of ascending and descending orbits, this period is repeated every two days for Sentinel 1 in Europe. Because SAR images determine plant structure and moisture content, and visual images specify vegetation biophysical processes, the combination of optical and SAR images provides a source of supplementary data.

Figure 1 shows the radar backscatter and NDVI (Normal Difference Vegetation Index) profiles from the Sentinel 1 and Sentinel 2 satellites. Radar data will be used to determine the structural development of the wheat plant, and it will show a significant decrease in VV during the vertical increase stage of the plant stem. Useful information is used to examine crops that exist in SAR backscatter amplitudes [10] to achieve classification, especially for rice and forest mapping. The merging of data from optical and radar references, as well as the development of software' ability to perform classification methods, is what makes SAR data so important in integrated land classification [11].

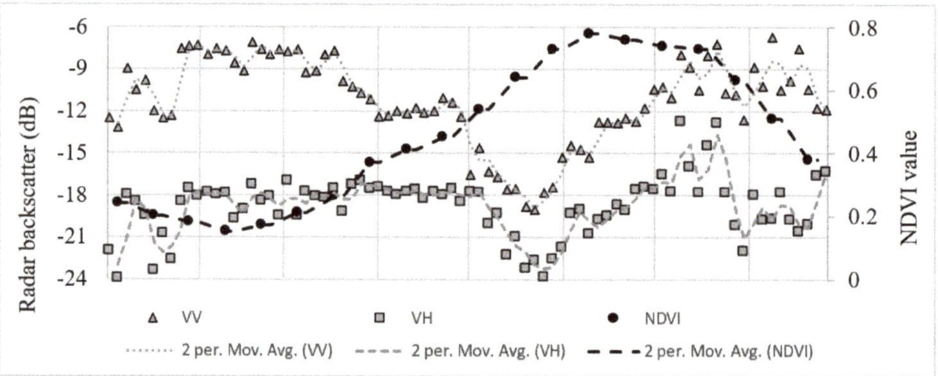

Figure 1. Example of profiles of (upper panel) Sentinel 2 normalized difference vegetation index (NDVI) and (lower panel) Sigma^0 VV and VH backscatter intensities for a winter wheat field.

McNairn et al. [12] reported successful results by integrating optical and SAR images to provide annual crop inventories. Soria–Ruiz et al. [13] applied radar and optical imagery in cloudy areas of Mexico to provide acceptable accuracy for land use classification

Inglada et al. [14] shared the use of high-resolution optical image and SAR time series; using Landsat 8 and Sentinel 1 combined data to improve early detection of crop type, they proposed the integration of Sentinel 2 images for initial crop identification. The results of a recent study using Sentinel 1 and Sentinel 2 data to assess groundwater and identify irrigation crops in southern India, that mainly use Sentinel 1 data, showed that when used in the monsoon season, they have a good ability to identify a variety of irrigated crops [15]. Torbick et al. [6] used real-time close-up images of the Sentinel 1, Sentinel 2, and Landsat 8 by combining intermediate-resolution ground observations to map seasonal crop types in the United States.

Joshi et al. [16] in a study of 112 different land use areas to investigate the integration of optical and radar data concluded that optical and radar data as complementary data are also effective in determining the details of land use map with high accuracy. Aiming to evaluate different methods of integrating optical and multipolar radar data for land mapping in Brazil, Pereira et al. [17] concluded that radar information improves user accuracy, while the polarization data of HH (horizontal transmission and reception) more than horizontal polarization (HV) (horizontal transmission and vertical reception) leads to the differentiation of different land use classes, but the integration of radar and optical data had the best statistical results for land mapping. Zhou et al. [18] used SAR images, optical images and the integration of both data types to evaluate the possibility of winter wheat mapping. The classification map was performed using a combination of Sentinel 1 information and optical images using a random forest method. The best results (F1 = 98%) were obtained by combining SAR and optical images for winter wheat mapping. Campus-Taberner et al. [19] used a multitemporal algorithm to combine Sentinel 2 and Landsat 8 data. Their results showed that there is a high consistency between ground estimates and measurements, and a high correlation and accuracy ((RMSE < 0.83, RMSEm < 23.6% and RMSEr < 16.6%)) as a result of the performance of Sentinel 2 and Landsat 8 images were reported.

As mentioned above, many studies examined the performance of the optical and radar image combination method to identify the type of crop. So far, these studies were limited to the following:

i. Combine Sentinel 1 and Sentinel 2 time series data;
ii. Classification of the majority of strategic national crops;
iii. Provide crop classification data with acceptable accuracy.

The aim of this research is to scrutinize the deficiency and strength of the combined radar data and optical images to identify the type of crops. This research was conducted in 2020 in the (Iranian) region and, to achieve the purpose of researching data and images of time series, Sentinel 1 and Sentinel 2 were used to answer the following questions with Tarom help of the obtained results: (i) how can acceptable classification accuracy be achieved given the changes in plant growth during the growing season? (ii) What is the contribution of each data set used in this study in estimating the research objectives? (iii) How can the accuracy of the information be measured for acceptable classification?

2. Materials and Methods

2.1. Study Area

This research was carried out in Tarom City (Zanjan Province), Iran, which has a wide range of climates (Figure 2). The semiarid cold climate, which occupies about 34% of the city area and, unlike the cold and humid climate, has the lowest location, is the driest. The region's lowest point is 300 m above sea level, and the region's highest point is 2700 m in the northeastern mountainous areas. Tarom receives 450 mm of annual rainfall on average, ranging from 200 mm in the lowlands to 1050 mm in the northern highlands. The average annual temperature is 17.3 degrees Celsius, with lows of 11 degrees Celsius and highs of 45 degrees Celsius. Autumn and spring are the rainiest seasons.

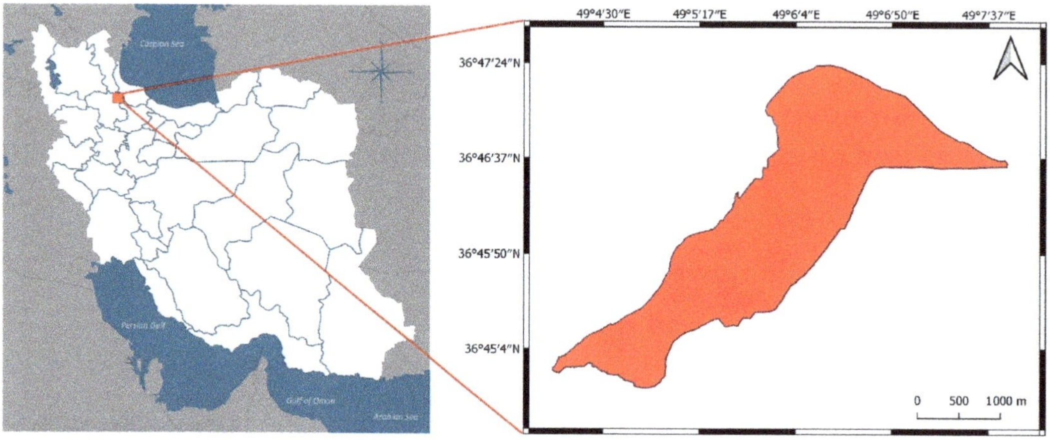

Figure 2. Location of the study area.

2.2. Field Data

Separately from agricultural and horticultural crops, the area under cultivation and the type of crop harvested were extracted from statistics from the Ministry of Jihad Agriculture and the Agricultural Jihad Organization of Zanjan region, as well as face-to-face interviews with Jihad Agricultural experts. More than 92,000 hectares of agrarian land will be covered by the agricultural sector with the cultivation of 68 types of crops and 21 types of horticultural crops in the region, and the volume of runoff produced by the region amounting to 2.2 billion cubic meters, which was constructed by several large and small dams or is under construction soon.

2.3. Sentinel 1 Data

The Sentinel 1 data came from Google Earth Engine's (GEE) Level 1 Ground Range Detected (GRD) Interferometric Wide Swath (IW) product [20]. Sentinel 1 radar observations were projected onto a standard 10-m grid in GRD output. GEE preprocessed the data with the Sentinel 1 toolbox. Thermal noise reduction, radiometric calibration, and terrain correction were all part of the preprocessing. Sentinel 1's key characteristics were the VV and VH polarized backscatter readings (in decibels, dB). The orientation of the transmitted radar beam has a significant impact on backscatter. Because of the considerably varied viewing orientations of the ascending and descending satellite overpasses, these were split and treated as supplementary observations. We used an improved Lee filter (Lee, 1981) with a damping value of 1 and a kernel size of 7 × 7 to minimize radar speckle in the pictures (Figure 3). To avoid the influence of changes in the angle of incidence on the return values, we employed two methods:

Ignore any observations with incidence angles less than 32° or greater than 42° because their geometries differed too much from the average incidence angle in our region of 37°. According to Equation (1), the remaining backscatter values observed at the angle of incidence θ are converted to backscatter values viewed at a reference angle θ_{ref} (1).

$$\sigma^0_{\theta_{ref}} = \frac{\sigma^0_\theta cos^2\left(\theta_{ref}\right)}{cos^2(\theta)} \quad (1)$$

In this equation, σ^0_θ is the measured incidence angle θ backscatter intensity, and $\sigma^0_{\theta re}$ is the predicted backscatter intensity under a reference angle θ_{ref} of 37 °C. This simplified adjustment was based on Lambert's law of optics and assumed scattering processes. According to Lambert's law, an ideal Lambert's reflector reflects the quantity of light equal to the cosine of the angle of incidence of the radiation source in any direction; hence, according to this rule and for our categorization, the earth's surface is not an ideal Lambert's reflector. After the severe incidence angle was covered, two to five statements were recorded for each location over 12 days, resulting in a 12-day return mosaic from the combined visits of Sentinel 1A and Sentinel 1B. Because each satellite has a 12-month repeat visit, after the severe incidence angle was covered, two to five statements were recorded for each location over 12 days. All recorded backscatter values inside the 12-day window for each pixel were transformed from their dB values to the original values, averaged, and converted back to the 12-day mosaic backscatter values in dB.

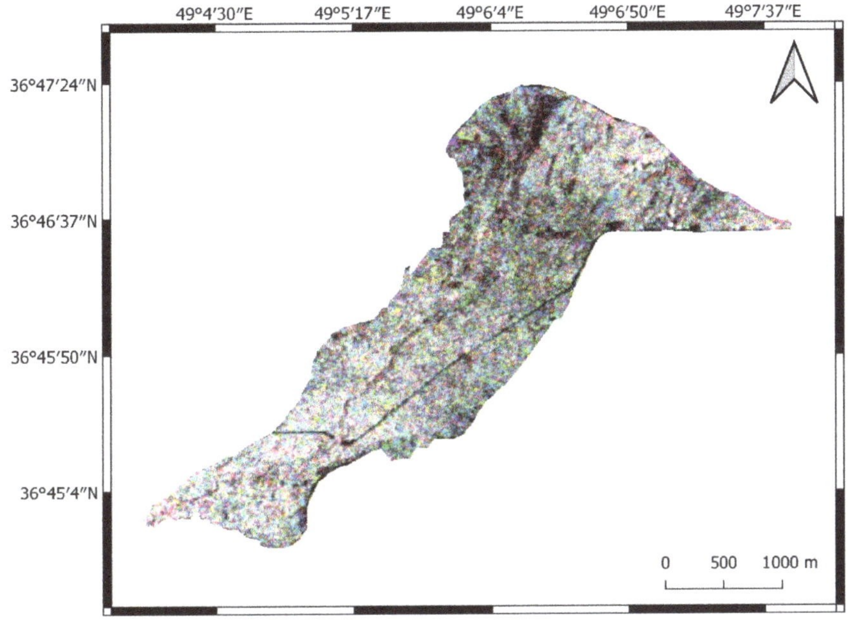

Figure 3. Sentinel 1 VH backscatter mosaics from 12 days in RGB composite. Dates are 1–13 March 2020 (red), 17–29 June 2020 (green), and 16–28 August 2020 (blue).

2.4. Sentinel 2 Data

The Sen2Cor method [21] was used to mask for clouds and cloud shadows the Sentinel 2 Level-1C data from the Copernicus Open Access Hub, and then the iCOR atmospheric correction scheme [22] was used to atmospherically correct the data. An extra geometric adjustment based on manually determined ground control points was performed on Sentinel 2 scenes that were badly coregistered (i.e., had a multitemporal coregistration error of >0.5 pixels). At a spatial resolution of 10-m, we utilized the NDVI value derived from the red (B4) and near-infrared (B8) bands [23]. The NDVI measure was chosen as a typical optical descriptor because of its past effectiveness in crop categorization studies [20]. While NDVI was reported to saturate during the most productive parts of the growing season [24], utilizing this index in time series rather than single-date images was proven to overcome this problem. The categorization algorithm used these NDVI values as input. However, crop categorization over a wide area encompassing many image tiles is complicated by frequent and uniform cloud cover. An improved version of a pixel-wise weighted least-squares smoothing of the NDVI data over time [25] was used to eliminate cloud blockage. Between 1 March 2020, and 31 August 2020, smoothed NDVI images were created at 10-day intervals.

2.5. Classification of Hierarchical Random Forest

The decision tree is one of the most effective tools for estimating target variables or classifying patterns. A decision tree divides the input space into sections and assigns a response value to each section [25]. In simple terms, the average of the target values related to the patterns in each area can be used to determine the answer in regression problems, or, in other words, the responsibilities assigned to each area based on the average of the target values corresponds to the learning patterns in each area. RF is a new development method for decision trees that uses grounded rules to combine the predictions of several single algorithms. To create each tree, a different set of existing patterns is chosen, with each fixed design being replaced. The total number of available ways [26] will be used to determine the size of this chosen category. Because it performs better in research with extensive input data and various features, and estimates the purposes required for mapping, the RF algorithm is much more efficient than other classification models, such as neural networks. We use a unique process called bootstrapping in the random forest method. Each tree in this method represents one of the training samples that is chosen at random, and sub-branches in each of these trees are treated as a random set of input features. As can be seen in Figure 4, the classification method was broken down into two stages. The first stage involved creating classes, determining water and forest crop classes, and the second stage involved classifying the crops studied in this study. The appropriate network parameters are determined by the search in the random forest method. The minimum sample size required for a leaf node, the minimum sample size required to divide a node, the impurity criterion, and the number of trees are examples of these parameters.

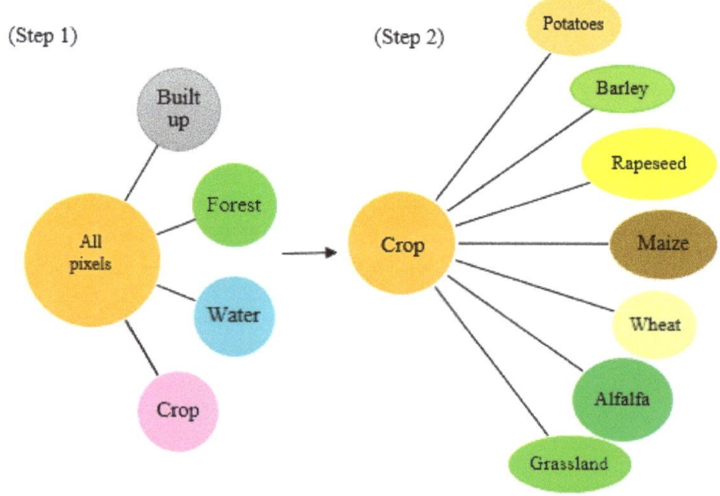

Figure 4. Schematic overview of two-step hierarchical classification procedure.

The RF algorithm falls under the ensemble learning methods, in which multiple decision trees (forming a random forest) are built during training, after which the mode of the predicted classes of the individual trees forms the output class of the forest. The RF classifier usually outperforms simple decision trees due to less over-fitting. The random forest is constructed using a bootstrapping technique in which each tree is fitted based on a random subset of training samples with replacement, while at each split in this tree, a random subset of the input features is also selected. The classification method was divided into two different stages, the first stage including classes made, and water and forest with crop classes were determined, and in the second stage, the classification of crops studied in this research was done. In the random forest method, the appropriate parameters of the network are determined by the search. These parameters can be defined as the minimum sample size needed for a leaf node, the minimum sample size required to divide a node, the impurity criterion, and the number of trees [27].

2.6. Calibration and Validation Data

The data in the database were randomly divided into a set of 80% validity and 20% calibration, the purpose of which was to examine the percentage of data calibration and validation, and it was also considered that the final value in this range is not much different from the actual amount of data. For water-based, human-made, and forest-created classes, manual and training sets are generated, and the 20–80 division rule is used to capture all data. Subsets were considered in plots over 2 ha. In this study, both validation methods, calibration, and classification were based on pixels, and the buffer operation was not performed on validation. Due to the omission of small strings in training and different field sizes, a marked difference was observed between the pixel ratio and the initial verification and calibration ratio with the independent validation sample (Table 1).

Table 1. Calibration and validation parcels and pixels per class.

Code	Class	Calibration Parcels	Validation Parcels	Calibration Pixels	Validation Pixels
1	Potatoes	613	5522	121,884	3,867,261
2	Barley	175	1590	31,189	1,333,238
3	Rapeseed	10	77	1520	50,159
4	Maize	1894	17,063	293,880	15,614,296
5	Wheat	848	7652	171,864	5,700,540
6	Alfalfa	352	3179	73,059	1,913,304
7	Grassland	2614	23,549	322,425	21,200,420
-	Total	7567	67,481	1,426,520	56,515,322

2.7. Classification Schemes

The main goal of this study was to determine the stock of individual and composite optical and SAR pictures in terms of classification accuracy. Furthermore, objective insight into the evolution of classification accuracy throughout the course of the growing season gives important information about the predicted accuracy of a classification during a certain phase of the growth season. One of the essential goals of this research is to compare SAR and optical images alone and the combination of these two in the classification process, and during the growing season we will find out the accuracy of classification with the help of these two images, and we can even comment on the accuracy of the classification with the help of these images in a certain period. According to Table 2, 18 classification designs were determined. Sentinel 1 SAR images were used only for the first six designs, Sentinel 2 NDVI images for the second six designs, and Sentinel 1 and Sentinel 2 composite images for the third six designs. In all 18 classification schemes, the performance estimators were OA classification and Kappa Cohen (K) agreement coefficient. The following equation represents the OA calculation:

$$\mathrm{OA} = \frac{\sum correct\ predictions}{total\ number\ of\ predictions} \tag{2}$$

In the Equation (2), the predictions are presented for all validation examples, the expected and real cases are comparable, and the following equation representing the K calculation is also used:

$$\mathrm{K} = \frac{p_0 - p_e}{1 - p_e} \tag{3}$$

where p_0 is the relative observed agreement among raters, and p_e is the hypothetical probability of chance agreement.

Table 2. Overall accuracy (OA) and kappa coefficient (k) of various categorization plans.

	Sentinel 2						Sentinel 1							
#	**March**	**April**	**May June**		**July**	**August**	**March**	**April**	**May June**		**July**	**August**	**OA**	**κ**
1							X						0.44	0.33
2							X	X					0.60	0.52
3							X	X	X				0.72	0.61
4							X	X	X	X			0.73	0.70
5							X	X	X	X	X		0.72	0.64
6							X	X	X	X	X	X	0.78	0.70
7	X												0.42	0.30
8	X	X											0.56	0.42
9	X	X	X										0.70	0.57
10	X	X	X	X									0.73	0.65
11	X	X	X	X	X								0.74	0.72
12	X	X	X	X	X	X							0.80	0.73
13	X						X						0.55	0.42
14	X	X					X	X					0.67	0.57
15	X	X	X				X	X	X				0.74	0.67
16	X	X	X	X			X	X	X	X			0.81	0.74
17	X	X	X	X	X		X	X	X	X	X		0.83	0.80
18	X	X	X	X	X	X	X	X	X	X	X	X	0.84	0.79

2.8. Classification Accuracy

Estimating the purpose of the classification, the random forest classification method is also used to predict the classification accuracy, and the average probability of the expected class of trees in the forest about the possibilities of the predicted class is an input sample. The probability of a winning class is defined based on the classification certainty for a particular instance. More reliable classification and strong agreement between different trees indicate high accuracy, but disagreement between trees reduces the likelihood of predicting a reliable classification. The prediction result will be shown at the pixel- or sample-level [28].

3. Result

The trend of changes in the two variables kappa and OA in the classification methods in this study is shown in the following table (Table 2). The table shows that as the number of images used as input values grows, so do the values of the two variables kappa and OA. Controlling the degree of resolution of the categories and the differences between the types of crops can be used to investigate this increase. The differences between Sentinel 1 and Sentinel 2 and optical classification versus SAR classification can be analyzed by examining the results.

Sentinel 1 classification performed better than Sentinel 2 classification in March. Moreover, the optical-only classification performed better than the SAR-only classification throughout the growing season (kappa of 0.69 vs. 0.67, OA of 77 percent vs. 75 percent). According to these findings, different crops in early growth have different characteristics that can be detected and compared with the light spectrum, but this conclusion can only be applied to the crops studied in this study, such as winter cereals. Optical and radar signatures for crops grown in April and May, such as potatoes and corn, reveal the type of management method used and reflect the winter plant cover, which is difficult to distinguish between optical and radar studies. A combination of Sentinel 1 and Sentinel 2 images performs better than using a single sensor for classification. The maximum accuracy obtained in the last days of July was 81 percent, which could not be increased by combining with August images. Figure 5 depicts the final classification as of the last day of August 2020. For this purpose, no filtration was used, and the image's recognizability is the result of the classification packages in the crop area's landscape. Due to the similarity between alfalfa and potatoes at the start of the growing season, all crops were classified as potatoes at first, but with time and growth evolution, this error was eliminated in the last days of

August, and the distinction between alfalfa and potato was evident in the crop classification.

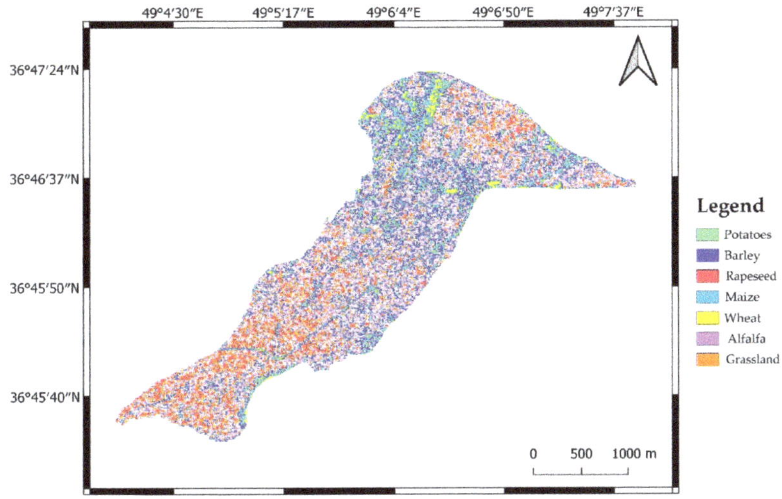

Figure 5. Final categorization result based on Sentinel 1 and 2 inputs through August 2020.

Figure 6 shows the results, where certain within-field zones in the center field, which is part of the validation dataset, were incorrectly classified as alfalfa in the early mapping stage (taken here in June 2020) but were correctly identified as potato by the end of August. Figure 6 was specified to eliminate this ambiguity with two letters (a) and (b). At the beginning of June 2020, the whole crop was identified as alfalfa in the study area due to the phenological similarities between the two crops of alfalfa and potatoes and the lack of development of plant growth, but at the end of August, due to full plant growth and obvious phenological differences, it was possible to distinguish between the two crops (purple represents alfalfa, and green represents potatoes).

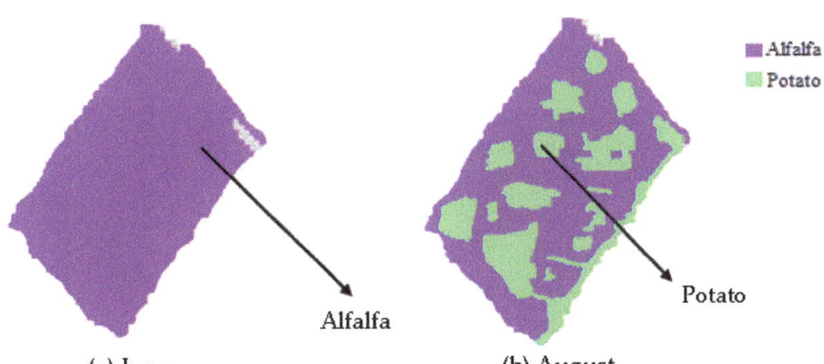

Figure 6. Zones in middle field were misclassified as alfalfa in June (**a**) but were correctly labeled as potato in August (**b**).

Figure 7 depicts the classification result's dependability. Because the side pixels are due to the integration with the signals of the surrounding terrain, the data have low validity along the considered boundaries. Figure 7 shows the results for the high data invalidity at the pixel boundary, which is close to the central packet. Wheats were the crops grown in this package, but the data uncertainty can be seen in the central portion of the packages, which can be attributed to the inconsistency created in the pixels' background.

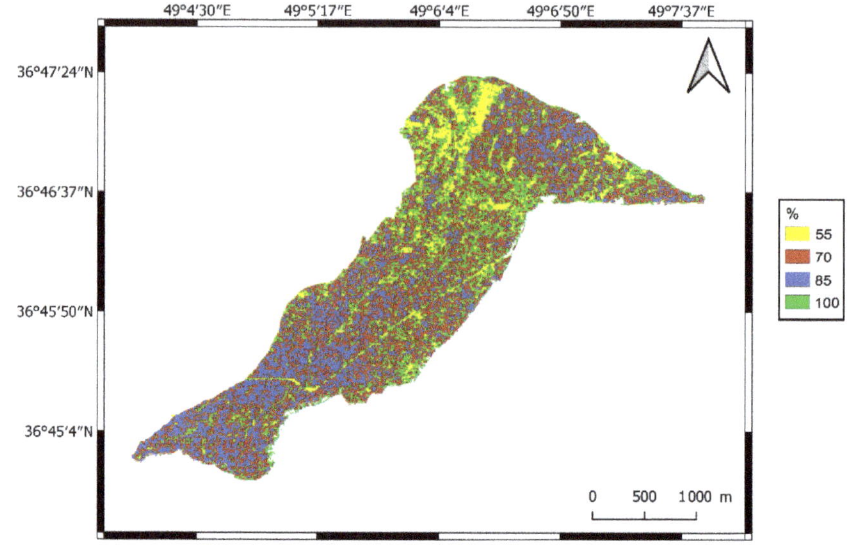

Figure 7. Classification confidence defined as random forest predicted class probability of majority class for each pixel at end of August 2020.

To ensure classification performance, which is a function of classification reliability, the method of quantifying classification accuracy was used. To accomplish this, all samples will be used, all data will be validated, the RF classifier will be used, and all samples will be averaged at a certain level. In terms of Gini, this function calculates the characteristics of the input data. The significance of the two radar and optical input sources' characteristics must be examined in the data [29]. We use the Gini significance feature as a reduction of impurities when using the random forest classifier, which is used for all forest trees. Gini's personality is extremely valuable. Its high importance indicates that it plays a key role in the forecasting process; on the other hand, if the feature's importance is low, it means that, according to Sentinel 1 sensor data, this information was limited for prediction before May, despite being consistent with plant structure. Early April and May play a critical role in the longitudinal development of winter crops and the ability to distinguish summer crops from one another. Plant development shows a difference in their NDVI values over time and during the growing season. Changes in the amount of NDVI will cause plants to differentiate due to differences in their phenological structure. This distinction is particularly noticeable between July and August, when the crops are distinguishable from one another due to growth and development.

4. Discussion

If we want to correctly classify crops in a region, we need to concentrate on the uniformity of all available inputs. If the data is related to Sentinel 1 data, 12-day return mosaics can be created. It is crucial to be cautious when it comes to reducing the impact of angles on the output data. We need to use a filter to get high-quality and high-resolution radar images. The Lee filter is one of the most comprehensive filters for this situation. The sharpness of radar images is reduced due to blurry effects. Twelve-day return mosaics and time series can be used to compensate for this flaw, and this annoying effect can be easily removed or reduced using this technique. However, because the classification was not the main criterion in terms of time, this disorder will not cause a primary problem, according to the method used in this study. Quegan et al. [30] used a special time filter in their study that could be considered a new method, but using their method was not a priority in this study. For Sentinel 2 data, this method was used to smooth out all cloud effects in NDV images, and 10-day NDVI mosaics were considered without cloud and fog coverage. When all predicting classes in one step lead to a 1.5 percent increase in the OA index in Tarom for mapping crops using random forest hierarchical and time-series model inputs, the results showed that the method of two-step hierarchical use of the method is nonhierarchical.

The first classification step was completed in August, with an OA index of around 84 percent due to differences in radar and optical effects for three different crops (forest, construction, water). The random forest method can be used in the second stage to identify more specialized differences between classes. During rejection, significant increases in both kappa and OA were reported, though by the end of May, summer crops (potatoes and corn) had grown significantly, but winter crops had also grown significantly. Increased classification accuracy will result from increased awareness of crop phenological development stages. Previous research has demonstrated that the use of optical time series improves the principles of classification by allowing for the separation of crops [31]. Separation of crops, such as cereals and vegetable crops, as well as winter and summer crops, is strongly recommended to improve the quality of classification work, despite the difficulties that such separation poses due to the similarity of crops such as winter wheat and winter barley. To solve the problem, it is suggested that winter crops be classified alongside summer crops. It is difficult to separate grasslands from winter crops (cereals) and winter crops from summer crops. Making a classification error between grassland and winter crops can be explained by the fact that both of these crops grow well in the months of April and May, and the green mass is visible on both crops, making it difficult to distinguish between them. Furthermore, vegetable crops such as potatoes and corn are very similar to one another until the end of plant growth, making it difficult to separate them by the end of April in practice. This can be thought of as a drawback to using remote sensing for such purposes.

Given the high values of the two kappa and OA indices during the study period (growth season), the items obtained as a result become more citational over time, but the results cannot be obtained in a specific time frame. As a result, it is important to remember that we will have to wait a certain amount of time to get the desired and required results from the crops for classification, because the variables in this study are crops that require time, and a set of observations takes time to evolve. One of the study's most important findings is that the difference in the OA variable for the classification process at the start of the growing season using two sensors, Sentinel 1 and Sentinel 2, revealed that this index (OA) was higher in studies using Sentinel 1. The difference between the use of Sentinel 1 and Sentinel 2 sensors will be determined after 30 days of the growing season, so that the data as predictor variables from these two sensors will be significantly different. The characteristics of the input data can be attributed to the greater validity of radar data for the crop classification process. The values of the OA variable were 35 percent lower when the analyses were done solely with VH backscatter, which was even lower than when the analyses were done with Sentinel 2 NDVI. On the basis of this evidence, it is impossible to say with certainty that the radar results differ from the optical classification of the first months of the growing season.

One of the issues with classifications that rely solely on optical methods is the use of a predictor variable, which is frequently the NDVI predictor. Previous research [31] focused on this factor, but it was later determined that NDVI could not provide a complete view of Sentinel 2 optical images, so it is recommended that more bands be used to achieve a more comprehensive view of optical images [32]. The method described in Section 2.5 does not apply to the NDVI index or single spectral bands. Although the use of different optical indices is recommended for future research, the focus of this study is on the use of a valid vegetation index (10-m resolution NDVI). The use of radar data, such as interferometry coherence stacks [33], can improve the validity of the classification process. Future research should consider whether classification based on radar data is more accurate than classification based on optical data. In the growing season, this case can answer the research hypothesis that radar data is more efficient than cloud position.

This study found that two periods are more important for classification when usin the NDVI predictor variable. Winter crops grew the most longitudinally in May and Apri and it was during this time that summer crops began to grow, making it possible to classif bare soil and other plants with summer crops. Due to the clear sky and lack of cloud during this time, optical images can be obtained in greater detail. August and July are th two most important months for making more use of sunlight. The winter grain harves season and greater access to summer crops in the summer are two other factors that aid i the sorting process. According to Zhou et al. [18], the VV index takes precedence over th HV index when classifying crops, but we found no difference in the classification proces between the two variables in this study.

NDVI is the most important predictor when considering the characteristics of Sentinel and Sentinel 2, as well as their ability to estimate forecasts. The most important crop classif cation predictions are made using a combination of radar and optical features. The accurac of classification in areas close to the center is greater than the accuracy of classificatio at pixel borders when it comes to crop classification at the pixel level. The data are a qualitative, indicating a close relationship between classification accuracy and classificatio reliability, but statistical probability cannot be used to estimate the percentage of reliabilit and statistics was ineffective in this study when it came to crop classification.

The results of this study showed that using Sentinel 2 optical data as a supplement t Sentinel 1 SAR data provided comprehensive information on plant structure [34]. Furthe more, by using time-series images rather than single images, the problem of determinin the date of the images is eliminated [35]. High-accuracy results were reported in severa studies that used a combined method of optical and radar images. The reason for this i that in some studies, the classification performed was related to the level [36], or filtratio was used in the classification [31], but in this study, the unfiltered method of classificatio in pixels was used in addition to a small amount of OA. It is difficult to group similar crop together, but if we look into several specific classes of different crops, the accuracy of cro classification will improve. This obvious difference in the structure of plants will facilitat classification if the plants are classified in terms of structure and family [37,38], such a the classification of horticultural and agricultural crops [39], paddy and grassland. Th findings of our study are in line with those of previous studies. Data and their propertie can be prioritized over pixels in such studies, allowing the data from the two sensor Sentinel 1 and Sentinel 2 to be used as input. Closed space can also be used with th classification method, which has the advantage of lowering the signal-to-noise ratio. W can look into the differences between radar and optical data, spectroscopic interferenc optical image SWIR bands, and plant phenological differences as one of the input factor for classification in the future.

5. Conclusions

The need to feed the population is prioritized as a result of population growth. Ident fication of high-consumption crops and their alternative varieties is one of the main issue in future planning to meet the nutritional needs of large-scale crop cultivation. Supplyin the cultivated crop's fertilizer needs, awareness of water needs, and early detection o anomalies are all critical in the second stage. These issues were largely solved by nev technologies such as remote sensing and the use of satellite imagery. When compared t that of traditional classification methods, the Copernicus program develops the potentia for classification through the simultaneous use of multiple sensors. The classificatio accuracy was said to improve with the combination of radar data and optical signals i previous studies because one of the benefits of combining these data is that the cloud effec is reduced. We used 10-day Sentinel 2 smoothed NDVI mosaics and 12-day Sentinel 1 bac mosaics to create a classification map for the 2020 growing season in Tarom. The randon forest method was used for classification purposes. This study reported a kappa coefficien of 0.75 and a maximum accuracy of 85 percent. It is recommended to use a combinatio of radar and optical data to achieve maximum classification accuracy, as this combinatio will increase the chances of examining the details compared to a single-sensor classification method and provide more reliable information. This study found that combining optical and radar data was the most important factor in predicting the final classification, and that using optical data alone produced acceptable results. Finally, because the level of reliability is low in areas such as closed border areas, it is much more difficult to predict the classification than to give a general conclusion using such methods. It is also necessary to compare the results of various sensors to better assess their ability to classify crops and to assess the potential of various sensors for such research.

Author Contributions: Conceptualization, A.S.; methodology, A.S.; validation, S.F., A.G., K.M.; formal analysis, A.T., M.A.; writing—original draft preparation, A.S. and S.F.; writing—review and editing, A.M.; visualization, A.M.; supervision, A.S.; funding acquisition, N.Z. All authors have read and agreed to the published version of the manuscript.

Funding: This research was funded by the National Natural Science Foundation of China, grant number 42071374.

Institutional Review Board Statement: Not applicable.

Informed Consent Statement: Not applicable.

Data Availability Statement: The data that support the findings of this study are available from the corresponding author, upon reasonable request.

Acknowledgments: The authors would like to thank Andia Sharifi for her helpful assistance.

Conflicts of Interest: The authors declare no conflict of interest.

References

1. Jain, M.; Balwinder-Singh; Rao, P.; Srivastava, A.K.; Poonia, S.; Blesh, J.; Azzari, G.; McDonald, A.J.; Lobell, D.B. The impact of agricultural interventions can be doubled by using satellite data. *Nat. Sustain.* **2019**, *2*, 931–934. [CrossRef]
2. Sharifi, A. Estimation of biophysical parameters in wheat crops in Golestan province using ultra-high resolution images. *Remote Sens. Lett.* **2018**, *9*, 559–568. [CrossRef]
3. Kosari, A.; Sharifi, A.; Ahmadi, A.; Khoshsima, M. Remote sensing satellite's attitude control system: Rapid performance sizing for passive scan imaging mode. *Aircr. Eng. Aerosp. Technol.* **2020**, *92*, 1073–1083. [CrossRef]
4. Wu, F.; Wu, B.; Zhang, M.; Zeng, H.; Tian, F. Identification of crop type in crowdsourced road view photos with deep convolutional neural network. *Sensors* **2021**, *21*, 1165. [CrossRef]
5. Van Tricht, K.; Gobin, A.; Gilliams, S.; Piccard, I. Synergistic use of radar sentinel 1 and optical sentinel 2 imagery for crop mapping: A case study for Belgium. *Remote Sens.* **2018**, *10*, 1642. [CrossRef]
6. Ghaderizadeh, S.; Abbasi-Moghadam, D.; Sharifi, A.; Zhao, N.; Tariq, A. Hyperspectral Image Classification Using a Hybrid 3D-2D Convolutional Neural Networks. *IEEE J. Sel. Top. Appl. Earth Obs. Remote Sens.* **2021**, *14*, 7570–7588. [CrossRef]
7. Sharifi, A.; Amini, J.; Sumantyo, J.T.S.; Tateishi, R. Speckle reduction of PolSAR images in forest regions using fast ICA algorithm. *J. Indian Soc. Remote Sens.* **2015**, *43*, 339–346. [CrossRef]
8. McNairn, H.; Brisco, B. The application of C-band polarimetric SAR for agriculture: A review. *Can. J. Remote Sens.* **2004**, *30*, 525–542. [CrossRef]
9. Baillarin, S.J.; Meygret, A.; Dechoz, C.; Petrucci, B.; Lacherade, S.; Tremas, T.; Isola, C.; Martimort, P.; Spoto, F. Sentinel 2 level 1 products and image processing performances. In Proceedings of the 2012 IEEE International Geoscience and Remote Sensing Symposium, Munich, Germany, 22–27 July 2012; pp. 7003–7006. [CrossRef]
10. Sharifi, A. Development of a method for flood detection based on Sentinel 1 images and classifier algorithms. *Water Environ. J.* **2020**, *35*, 924–929. [CrossRef]
11. Uddin, K.; Matin, M.A.; Meyer, F.J. Operational flood mapping using multi-temporal Sentinel 1 SAR images: A case study from Bangladesh. *Remote Sens.* **2019**, *11*, 1581. [CrossRef]
12. McNairn, H.; Champagne, C.; Shang, J.; Holmstrom, D.; Reichert, G. Integration of optical and Synthetic Aperture Radar (SAR) imagery for delivering operational annual crop inventories. *ISPRS J. Photogramm. Remote Sens.* **2009**, *64*, 434–449. [CrossRef]
13. Soria-Ruiz, J.; Fernandez-Ordoñez, Y.; Woodhouse, I.H. Land-cover classification using radar and optical images: A case study in Central Mexico. *Int. J. Remote Sens.* **2010**, *31*, 3291–3305. [CrossRef]
14. Inglada, J.; Vincent, A.; Arias, M.; Marais-Sicre, C. Improved early crop type identification by joint use of high temporal resolution sar and optical image time series. *Remote Sens.* **2016**, *8*, 362. [CrossRef]
15. Ferrant, S.; Selles, A.; Le Page, M.; Herrault, P.A.; Pelletier, C.; Al-Bitar, A.; Mermoz, S.; Gascoin, S.; Bouvet, A.; Saqalli, M.; et al. Detection of irrigated crops from Sentinel 1 and Sentinel 2 data to estimate seasonal groundwater use in South India. *Remote Sens.* **2017**, *9*, 1119. [CrossRef]

16. Joshi, N.; Baumann, M.; Ehammer, A.; Fensholt, R.; Grogan, K.; Hostert, P.; Jepsen, M.R.; Kuemmerle, T.; Meyfroidt, P.; Mitchard, E.T.A.; et al. A review of the application of optical and radar remote sensing data fusion to land use mapping and monitoring. *Remote Sens.* **2016**, *8*, 70. [CrossRef]
17. De Oliveira Pereira, L.; Da Costa Freitas, C.; Anna, S.J.S.S.; Lu, D.; Moran, E.F. Optical and radar data integration for land use and land cover mapping in the Brazilian Amazon. *GIScience Remote Sens.* **2013**, *50*, 301–321. [CrossRef]
18. Zhou, T.; Pan, J.; Zhang, P.; Wei, S.; Han, T. Mapping winter wheat with multi-temporal SAR and optical images in an urban agricultural region. *Sensors* **2017**, *17*, 1210. [CrossRef] [PubMed]
19. Campos-Taberner, M.; García-Haro, F.J.; Camps-Valls, G.; Grau-Muedra, G.; Nutini, F.; Busetto, L.; Katsantonis, D.; Stavrakoudis, D.; Minakou, C.; Gatti, L.; et al. Exploitation of SAR and optical sentinel data to detect rice crop and estimate seasonal dynamics of leaf area index. *Remote Sens.* **2017**, *9*, 248. [CrossRef]
20. Bellón, B.; Bégué, A.; Seen, D.L.; de Almeida, C.A.; Simões, M. A remote sensing approach for regional-scale mapping of agricultural land-use systems based on NDVI time series. *Remote Sens.* **2017**, *9*, 600. [CrossRef]
21. De Keukelaere, L.; Sterckx, S.; Adriaensen, S.; Knaeps, E.; Reusen, I.; Giardino, C.; Bresciani, M.; Hunter, P.; Neil, C.; Van der Zande, D.; et al. Atmospheric correction of Landsat-8/OLI and Sentinel 2/MSI data using iCOR algorithm: Validation for coastal and inland waters. *Eur. J. Remote Sens.* **2018**, *51*, 525–542. [CrossRef]
22. Zheng, B.; Myint, S.W.; Thenkabail, P.S.; Aggarwal, R.M. A support vector machine to identify irrigated crop types using time-series Landsat NDVI data. *Int. J. Appl. Earth Obs. Geoinf.* **2015**, *34*, 103–112. [CrossRef]
23. Rouse, W.; Haas, R.H.; Schell, J.A.; Deering, D.W. Monitoring vegetation systems in the Great Plains with ERTS. *NASA Spec. Publ.* **1974**, *351*, 309.
24. Cai, Y.; Guan, K.; Peng, J.; Wang, S.; Seifert, C.; Wardlow, B.; Li, Z. A high-performance and in-season classification system of field-level crop types using time-series Landsat data and a machine learning approach. *Remote Sens. Environ.* **2018**, *210*, 35–47. [CrossRef]
25. Estel, S.; Kuemmerle, T.; Alcántara, C.; Levers, C.; Prishchepov, A.; Hostert, P. Mapping farmland abandonment and recultivation across Europe using MODIS NDVI time series. *Remote Sens. Environ.* **2015**, *163*, 312–325. [CrossRef]
26. Ouyang, F.; Su, W.; Zhang, Y.; Liu, X.; Su, J.; Zhang, Q.; Men, X.; Ju, Q.; Ge, F. Ecological control service of the predatory natural enemy and its maintaining mechanism in rotation-intercropping ecosystem via wheat-maize-cotton. *Agric. Ecosyst. Environ.* **2020**, *301*, 107024. [CrossRef]
27. Belgiu, M.; Drăgu, L. Random forest in remote sensing: A review of applications and future directions. *ISPRS J. Photogramm. Remote Sens.* **2016**, *114*, 24–31. [CrossRef]
28. Sharifi, A.; Amini, J.; Tateishi, R. Estimation of forest biomass using multivariate relevance vector regression. *Photogramm. Eng. Remote Sens.* **2016**, *82*, 41–49. [CrossRef]
29. Snapp, S.; Bezner Kerr, R.; Ota, V.; Kane, D.; Shumba, L.; Dakishoni, L. Unpacking a crop diversity hotspot: Farmer practice and preferences in Northern Malawi. *Int. J. Agric. Sustain.* **2019**, *17*, 172–188. [CrossRef]
30. Quegan, S.; Toan, T.L.; Yu, J.J.; Ribbes, F.; Floury, N. Multitemporal ERS SAR analysis applied to forest mapping. *IEEE Trans. Geosci. Remote Sens.* **2000**, *38*, 741–753. [CrossRef]
31. Yan, S.; Shi, K.; Li, Y.; Liu, J.; Zhao, H. Integration of satellite remote sensing data in underground coal fire detection: A case study of the Fukang region, Xinjiang, China. *Front. Earth Sci.* **2020**, *14*, 1–12. [CrossRef]
32. Francini, S.; McRoberts, R.E.; Giannetti, F.; Mencucci, M.; Marchetti, M.; Scarascia Mugnozza, G.; Chirici, G. Near-real time forest change detection using PlanetScope imagery. *Eur. J. Remote Sens.* **2020**, *53*, 233–244. [CrossRef]
33. Shanmugapriya, S.; Haldar, D.; Danodia, A. Optimal datasets suitability for pearl millet (Bajra) discrimination using multiparametric SAR data. *Geocarto Int.* **2020**, *35*, 1814–1831. [CrossRef]
34. Moreau, D.; Pointurier, O.; Nicolardot, B.; Villerd, J.; Colbach, N. In which cropping systems can residual weeds reduce nitrate leaching and soil erosion? *Eur. J. Agron.* **2020**, *119*, 126015. [CrossRef]
35. Saleem, M.; Pervaiz, Z.H.; Contreras, J.; Lindenberger, J.H.; Hupp, B.M.; Chen, D.; Zhang, Q.; Wang, C.; Iqbal, J.; Twigg, P. Cover crop diversity improves multiple soil properties via altering root architectural traits. *Rhizosphere* **2020**, *16*, 100248. [CrossRef]
36. Muoni, T.; Koomson, E.; Öborn, I.; Marohn, C.; Watson, C.A.; Bergkvist, G.; Barnes, A.; Cadisch, G.; Duncan, A. Reducing soil erosion in smallholder farming systems in east Africa through the introduction of different crop types. *Exp. Agric.* **2019**, *56* 183–195. [CrossRef]
37. Ahmad, A.; Ahmad, S.R.; Gilani, H.; Tariq, A.; Zhao, N.; Aslam, R.W.; Mumtaz, F. A synthesis of spatial forest assessment studies using remote sensing data and techniques in Pakistan. *Forests* **2021**, *12*, 1211. [CrossRef]
38. Burke, M.; Lobell, D.B. Satellite-based assessment of yield variation and its determinants in smallholder African systems. *Proc Natl. Acad. Sci. USA* **2017**, *114*, 2189–2194. [CrossRef] [PubMed]
39. Li, G.; Messina, J.P.; Peter, B.G.; Snapp, S.S. Mapping land suitability for agriculture in Malawi. *Land Degrad. Dev.* **2017**, *28* 2001–2016. [CrossRef]

applied sciences

MDPI

Article

Adaptive Metaheuristic-Based Methods for Autonomous Robot Path Planning: Sustainable Agricultural Applications

Farzad Kiani [1,*,†], Amir Seyyedabbasi [1,2], Sajjad Nematzadeh [3], Fuat Candan [4], Taner Çevik [3], Fateme Aysin Anka [5,‡], Giovanni Randazzo [6], Stefania Lanza [7] and Anselme Muzirafuti [6,*]

1 Software Engineering Department, Faculty of Engineering and Natural Science, Istinye University, Istanbul 34396, Turkey; amir.seyyedabbasi@gmail.com
2 Computer Engineering Department, Faculty of Engineering and Architecture, Beykent University, Istanbul 34398, Turkey
3 Computer Engineering Department, Faculty of Engineering and Architecture, Nisantasi University, Istanbul 34398, Turkey; nematzadeh.sajjad@gmail.com (S.N.); taner.cevik@nisantasi.edu.tr (T.Ç.)
4 Computer Engineering Department, Faculty of Engineering and Natural Science, Istanbul Sabahattin Zaim University, Istanbul 34303, Turkey; fuat.candan@std.izu.edu.tr
5 Political Science and Public Administration Department, Faculty of Economics, Administrative and Social Sciences, Istinye University, Istanbul 34396, Turkey; fatemeh.dehghan@istinye.edu.tr
6 Dipartimento di Scienze Matematiche e Informatiche, Scienze Fisiche e Scienze della Terra, Università degli Studi di Messina, Via F. Stagno d'Alcontres, 31-98166 Messina, Italy; giovanni.randazzo@unime.it
7 GeoloGIS s.r.l., Dipartimento di Scienze Matematiche e Informatiche, Scienze Fisiche e Scienze della Terra, Università degli di Messina, Via F. Stagno d'Alcontres, 31-98166 Messina, Italy; stefania.lanza@unime.it
* Correspondence: farzad.kiyani@gmail.com (F.K.); anselme.muzirafuti@unime.it (A.M.); Tel.: +90-850-283-6000 (F.K.)
† Other publish name is Ferzat Anka.
‡ Other publish name is Fatemeh Dehghan Khangahi.

Citation: Kiani, F.; Seyyedabbasi, A.; Nematzadeh, S.; Candan, F.; Çevik, T.; Anka, F.A.; Randazzo, G.; Lanza, S.; Muzirafuti, A. Adaptive Metaheuristic-Based Methods for Autonomous Robot Path Planning: Sustainable Agricultural Applications. *Appl. Sci.* **2022**, *12*, 943. https://doi.org/10.3390/app12030943

Academic Editor: José Miguel Molina Martínez

Received: 28 December 2021
Accepted: 10 January 2022
Published: 18 January 2022

Abstract: The increasing need for food in recent years means that environmental protection and sustainable agriculture are necessary. For this, smart agricultural systems and autonomous robots have become widespread. One of the most significant and persistent problems related to robots is 3D path planning, which is an NP-hard problem, for mobile robots. In this paper, efficient methods are proposed by two metaheuristic algorithms (Incremental Gray Wolf Optimization (I-GWO) and Expanded Gray Wolf Optimization (Ex-GWO)). The proposed methods try to find collision-free optimal paths between two points for robots without human intervention in an acceptable time with the lowest process costs and efficient use of resources in large-scale and crowded farmlands. Thanks to the methods proposed in this study, various tasks such as tracking crops can be performed efficiently by autonomous robots. The simulations are carried out using three methods, and the obtained results are compared with each other and analyzed. The relevant results show that in the proposed methods, the mobile robots avoid the obstacles successfully and obtain the optimal path cost from source to destination. According to the simulation results, the proposed method based on the Ex-GWO algorithm has a better success rate of 55.56% in optimal path cost.

Keywords: autonomous robots; remote sensing; smart agriculture; climate change; environmental protection; drone; photogrammetry; path planning; internet of things; environmental monitoring

1. Introduction

In recent years, environmental protection and sustainability have become fundamental needs. Environmental sustainability is the conservation of natural resources and meeting the needs of future generations to avoid potential hazards, and for this purpose, it is vital to interact with the planet responsibly. In this situation, it is necessary to provide future generations with a lifestyle at least an equal in quality to the current generations, and in this direction, it is necessary to use existing natural resources efficiently [1]. In recent times, one of the most popular areas of sustainability is agriculture. In the last few years, researchers have made traditional agriculture more efficient and functional with new technologies, concepts, and methods within the scope of smart agriculture. In this context, sustainable agriculture can be achieved, and resources such as human and natural resources will be used more efficiently. On the other hand, with the prediction that the world population will reach 9 billion people by 2050, agricultural products should be increased by 70% [2]. Currently, the food industry is responsible for 30% of the world's energy consumed and 22% of greenhouse gas emissions. In addition, if a product variety is not suitable for certain regional conditions and the planning in planting and harvesting is wrong, it causes the overconsumption of resources, crop culling, and consequent food shortages. These problems may even cause forced migration in some regions [3]. Therefore, the agricultural sector has to address serious issues such as climate change issues, limited arable land, and increasing demand for freshwater. In this regard, it is essential that the development policies of states for agriculture are in a sustainable framework [4].

The role of smart systems in sustainable agriculture is increasing day by day. In this direction, many technological methods are used and recommended. One of them is to use autonomous robots' technology, but in an environment with many autonomous robots and obstacles, one of the most critical tasks is to transfer these robots safely between two points without them colliding with each other or with obstacles. For an autonomous robot, the problem of searching for a safe path from a source to a destination is called path planning [5,6]. This issue can be addressed using various new technologies (e.g., Wireless Sensor Networks (WSNs) and Internet of Things (IoT)) that have a wide range of applications [7–9] since they can be designed with heterogeneous or homogeneous devices in distributed, central, or Peer-to-Peer (P2P) architectures. One of the application areas of these technologies which has become popular in recent years is agriculture [10–13]. This field has a wide range of smart applications and systems from the cultivation of agricultural products to their logistics [14–17]. Although there are many agricultural studies in the literature, the design of smart and autonomous devices and applications that use effective and efficient resources have not been developed. One of them is the proposal of efficient 3D path planning algorithms for mobile devices used in large-scale farmlands, which has many obstacles.

It is important to consider the environment in three dimensions in order for it to be applicable to real-world applications and projects in complex environments. Furthermore, when it comes to mobile robots, three-dimensional movements and areas seem more acceptable. In real application areas, considering the resources of mobile robots, such as energy, finding the optimal path is important. Optimal path planning means that the shortest path length, where the selected path is as far as possible from obstacles, must be smooth without sharp turns and must consider motion constraints. Finding an optimal 3D path planning is a Non-deterministic Polynomial-time (NP-hard) problem [5,6]. This makes metaheuristic algorithms a good choice for designing a solution to such a problem. Considering that large-scale environments in 3D environments increase the applicability of this study in real applications, as such, one of the fundamental problems related to robots from past to present is 3D path planning for aerial robots. This problem can become even more complex in large-scale agricultural areas with many obstacles.

In this study, we focused on Gray Wolf Optimizer (GWO)-based algorithms to solve the mentioned problem. In general, GWO-based algorithms have a balanced behavior transition between discovery and use phases because they use the hierarchical group working mechanism of wolves, and they also use a minimum number of control mathematical parameters. In this way, the chance of finding the optimal solution in a short time is high; in addition, the use of resources is also efficient. On the other hand, a GWO-based method was proposed in [18] for solving the mentioned problem, and they proved it was better than other metaheuristic-based algorithms. In this study, two methods, inspired by Incremental Gray Wolf Optimization (I-GWO) [19] and Expanded Gray Wolf Optimization (Ex-GWO) [19], are proposed to address the above issue. The classical GWO algorithm can behave more stably in normal situations (for a somewhat standard environment without many obstacles). The Ex-GWO-based path planning method may be performed more successfully in larger and more crowded environments with larger populations and iterations, and the I-GWO-based path planning method may give good results in medium and smaller, less populated environments. However, the I-GWO is faster than other algorithms.

These methods can be applied in different and diverse agricultural application areas and thus can be useful work for farming and smart agriculture. This paper presents optimized, reliable, and shortest pathfinding mechanisms for smart agricultural robots (e.g., autonomous tractors and agricultural drones) that track crops on large-scale farmlands without the need for the intervention of any human using distributed IoT [20] and WSN technologies. Thanks to the algorithms proposed in this study, efficient resource consumption and product growth rate can be achieved with low risk and cost. On the other hand, avoiding obstacles in the path planning of agricultural areas is more complex than in other path planning areas because of a dense population of objects that can serve as obstacles such as trees, plants, and buildings. As mentioned above, the most critical problem these mobile robots face is the efficient use of resources such as energy, so this issue is given importance in this paper. In other words, the management of resources with minimal loss is the aim of the paper. In addition, a smooth and efficient pathfinding mechanism is very important for robots; because of this, the system must showcase a sustainable performance. Therefore, the method used with the mobile robots must deliver them to the destination point using the best path. To achieve all these purposes, two different algorithms based on metaheuristic algorithms are presented for each autonomous mobile robot. Indeed, the proposed methods find collision-free optimal paths in an acceptable time with the lowest process costs in different environments containing various obstacles. In this study, it is assumed that there are many obstacles in agricultural land in order to ensure that environmental conditions are realistic. Therefore, the proposed algorithms are simulated and evaluated in a similar environment. The mobile robots in this farmland try to find the optimal paths while bypassing possible obstacles in the farmland with our proposed methods. In addition, in a developed application by the authors for farmers, these employed robots can be monitored and controlled.

In Section 2 of this paper, the literature studies are presented. The proposed algorithms and their related applications are explained in Section 3. In Section 4, the simulation results and performances of each method are evaluated. The last section of the paper includes the conclusions and possible further studies.

2. Literature Review

2.1. Unmanned Aerial Robots' Applications in Agriculture

IoT and similar technologies such as WSN, which have become popular in recent years, are used to meet the needs in the agriculture fields. Along with the IoT, the widespread use of autonomous robots such as Unmanned Aerial Robots (UAVs) increases productivity in agriculture. In recent years, studies related to this subject have gained acceleration [21–24]. In [25], the authors used UAVs to detect possible drainage pipes. Often, farmers need to repair or construct drain lines to efficiently remove water from soil. Therefore, in this study, they wanted to increase resource consumption and productivity in agriculture by focusing on this issue. In [26], the combined application of UAV and Unmanned Ground Robot (UGV) was proposed to monitor and manage crops. The authors proposed a system that can periodically monitor the condition of crops, capture multiple images of them, and determine the state of the crops. In addition to many UAV-based studies and products, recently, the concepts of IoT and autonomous robots have begun to be presented together. In this way, the data detected by the UAVs or each autonomous robot reach the place where they need to be sent instantly, the necessary actions can be taken on this data, and it can quickly provide a decision mechanism to the farmer or other technological devices. For example, in [27], the authors presented a farm monitoring system via UAV, IoT, and Long-Range Wide Area Network (LoRAWAN) technologies for efficient resource management and data delivery. In this regard, they monitored water quality. In [28], the authors proposed a new model to minimize the post-disaster inspection cost to serve a disaster-affected area. In this study, battery charging costs, service costs, drone hovering, turning, acceleration, cruise, and deceleration costs were considered. In this regard, the authors used two heuristics (*not meta-heuristics*) algorithms, but it was not possible to avoid the fundamental problems of heuristics [19]. In [29], the study aimed to deliver to a number of customers by UAVs, namely drones. Here, it focused on three issues. One was the launch points of the drones, the second was the launch points of the customers, and the third was the distance between the customer and the drone. The proposed method goal was to minimize the total operational cost, including an explicit calculation of the energy consumption of the drone as a function of the drone speed.

The most common role of drones in agriculture is to assess and monitor crops. For this, remote sensing is carried out, but this task is not enough when agricultural applications become more widespread. For this, autonomous mobile robots such as drones and other UAVs with technologically different features are designed for various agricultural purposes. In [30], the authors used satellite images to crop mapping. They used the remote sensing feature and utilized advantages of combined radar data and optical images to identify the type of crops. The authors claim that this combination provides an increased chance of examining details and provides more reliable information compared to a single-sensor classification method. We can generally categorize UAV/drone-based agricultural applications into three categories: Monitoring Applications, (b) Spraying Applications, and (c) Multi-robots Applications. In the first category, crops are tracked, and certain appropriate information and vegetation indices are extracted. For this, it is necessary to provide the imaging data that are processed later. Thus, we can identify problem areas in the crop that suffer from various diseases and pests. The data received by UAVs sensors can be characterized based on their spectral, spatial, and temporal properties. The selection of suitable sensors and data depends on the nature of their applications. There are many studies in the literature related to this [31–33]. Most studies in the second category have focused on applications that can spray pesticides and fertilizers in appropriate and correct amounts. Most of the papers reviewed install a spray device and take into account various conditions that can affect this process, such as weather [34–36]. We should not forget that these agricultural chemical products can cause various problems such as environmental disasters and human diseases such as cancer. Currently, most of the existing studies in the literature generally focus on a single autonomous, mobile robot performing a monitoring operation. For example, in some cases such as large crops, a single mobile device (e.g., UAV) cannot complete the monitoring process as it is characterized by limited power sources (limited battery). On the contrary, a multi-robot application can overcome this difficulty by dividing the area into multiple sub-areas corresponding to the number of UAVs/drones [37–39]. In addition, different purposes and applications are carried out on a single drone. However, the need for more than one mobile robot to work is increasing day by day. In particular, parallel processing is very important in terms of performance and process speed. In this regard, one of the most important issues is that these autonomous mobile robots can work together as soon as possible and use fewer resources without colliding with each other. The situation becomes even more difficult, especially in large-scale agricultural land, which consists of various barriers. Thus, the problem of path planning seems to be quite important, and an efficient mechanism can be used in many various agricultural applications; it can also be coded and embedded with different hardware devices. Therefore, in the next subsection, the topic of three-dimensional path planning in the literature is discussed.

2.2. Path Planning in Agricultural Applications

It is very important that autonomous robots used in smart agriculture perform their duties efficiently and that resources are used efficiently. In this regard, a vital issue is that these robots do their tasks with the most optimum mechanism. Therefore, it is necessary to focus on the NP-hard type of 3D path planning problem. A general classification of 3D path planning consists of four types, as shown in Figure 1. These types are sampling-based algorithms [40], node-based algorithms [41], mathematical-based algorithms [42], and nature-based algorithms [43]. The methods in the first three categories suffer from high time complexity and local minima trap, especially when mobile robots face multiple constraints when planning a path. Metaheuristic algorithms, a set of nature-inspired algorithms, are the fourth category in this taxonomy that imitate natural, biological, interactive behaviors or physical events [44,45]. These methods try to find an almost optimal path by eliminating the process of creating complex environment models based on stochastic approaches. The stochastic approaches can be efficient and fast in solving large and complex optimization problems, especially in non-differentiable, multi-objective, and multimodal problems [20,46].

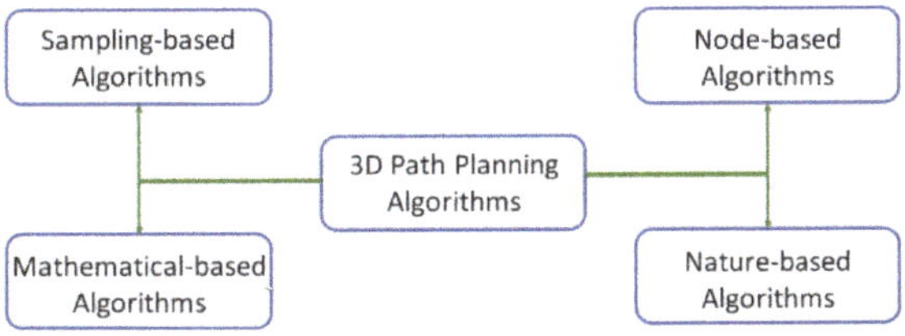

Figure 1. 3D path planning algorithms taxonomy [5,47].

Finding the best shortest path entails some problems such as the existence of many possible obstacles in its route. In addition, this path should be smooth without sharp turns and must consider movement restrictions. These problems may be even more cumbersome when considering large land areas and similar agricultural environments. Solution techniques in path planning algorithms for mobile autonomous mobile robots may include a visibility graph [48], probable road maps [49], and random exploring algorithms [50]. However, judging from the results of numerous studies in the literature, metaheuristic methods may be better overall [51–53]. Metaheuristic methods try to find an almost optimal path by eliminating the process of creating complex environment models based on stochastic approaches. These methods are among the most appropriate approaches to solve unifying and nonlinear global optimization problems [54]. Worth mentioning here is the No-Free-Lunch (NFL) [55] theorem. It asserts that there is no specific metaheuristic algorithm that provides the best solution for every optimization problem. This means that if one algorithm can solve a kind of problem effectively, then it may not be effective to solve another kind of problem. As such, there is a considerable demand to develop new metaheuristic algorithms that can be used in various problems.

As previously stated, the path planning problem has become popular in recent years and the metaheuristic algorithms can be the most appropriate solution for it, but in the literature, there are not many works that study agricultural lands for various purposes. Many agricultural studies in the literature have focused on issues such as the farmer's income from harvest, the variety of land use, the type and amount of employment, labor productivity, biodiversity indices based on landscape ecological measures, and soil erosion [56–58]. In the literature, although there are some studies on path planning in agricultures [59,60], they have generally not focused either on 3D path planning or on the problem of having many obstacles in the real environment farmlands and how to detect them.

In [61], the authors addressed the coverage path problem in a particular region with many known obstacles for mobile robots in agriculture. The study proposed a practical method, considering the geometry properties and obstacles of the area. It used an obstacle avoidance mechanism to find a coverage path for agricultural drones. However, optimal pathfinding and its usability in a 3D space were not taken into account. Additionally, the complexity time and space of their proposed method are not efficient in comparison with metaheuristic-based algorithms. In [59], the authors showed the simulation results of an algorithm designed to autonomously perform the path planning process for UAVs in agricultural lands. The purpose of this study was to provide the appropriate conditions to automate the process and carry out further audit activities. The algorithm considers photogrammetric parameters such as ground sample distance (GSD) and overlap between photos. For this, image processing techniques were used. In [62], the authors used autonomously acting ground robots for various agricultural applications. They researched different applications for path planning techniques to various agricultural contexts and applied land coverage and point-to-point navigation techniques. They used the D* to find the optimal path in a partial environment. However, this method is not very efficient since it uses a node-based algorithm (D* algorithm), and it is also designed for 2D areas [63]. As mentioned before, among the 3D path planning methods, metaheuristics may be the most efficient method. In [64], the authors proposed a custom model to navigate semi-autonomous agricultural robots with trailer. However, the geometry features were considered in 2D. In addition, the authors did not focus on finding the optimal path. Therefore, mobile devices moving on the non-optimal path map may not be successful in using their resources efficiently.

In [65], the authors proposed a path planning method inspired by the Ant Colony Optimization (ACO) algorithm to multipoint measurements in potato ridge cultivation. However, the related method did not perform successfully in finding optimal paths and is also useful for 2D areas. It may be unlikely to be implemented on real robots due to the fact that they did not focus on the recognition of obstacles and the avoidance of mechanisms of them. In [57], three local search metaheuristic algorithms, which were simulated by annealing and tabu search references, were used to calculate annual crop planning with a new irrigation mechanism. The objective function of this study was to maximize the gross benefits associated with the allocation of crops. The authors claimed that the tabu search method gave the best results in comparisons. In [66], an evolutionary algorithm was used for a complex strategic land use problem based on the management of a farming system. This study aimed to pursue a multi-purpose strategy that fulfilled spatial constraints in the 50-year planning management of the farm. Although the study is comprehensive, the metaheuristic method used and proposed may not be a very performant and efficient solution.

3. Materials and Methods

With the increase in the world population, the need for agricultural and food products has also increased. At the same time, the importance and need for smart agricultural systems and methods have also increased. Therefore, it is very important to plan optimal paths without harming objects (barriers) such as plants and trees in agricultural areas. Thanks to the methods proposed in this study, various tasks such as tracking crops in large farmlands can be performed efficiently by autonomous robots. Accordingly, it is necessary to find the optimal path between two points for robots without human intervention. Therefore, in this paper, two adaptive 3D path planning methods were presented for autonomous agricultural UAVs to find collision-free optimal paths in an acceptable time with the lowest process costs in different environments, containing various obstacles. These methods were developed, inspired by two metaheuristic algorithms (I-GWO and Ex-GWO). In addition, many obstacles were assumed to be present in the field in order to prove that the proposed methods are functional, and robots had to find their paths in relation to these obstacles. In addition, this study also used a mechanism for obstacle management. The studies in the literature either do not mention how to detect and prevent obstacles or they used the features of an existing device and did not suggest an algorithm or technique [20,64]. This mechanism can be embedded in various sensors and IoT devices.

3.1. *Definitions*

Before describing the proposed algorithms, the problem must be defined. The main purpose of 3D path planning is to find an optimum (or nearly optimal) path between the source (start) and the destination (target) stations. The path planning function is defined as outlined in Equation (1).

$$f(source,\ destination) \rightarrow Trajectory \tag{1}$$

Source and destination denote the relative coordinates of the source and destination positions on the map. Each path has a cost during movement from source to destination. There are different parameters used to define a cost between two points. In most studies, the cost is considered as the consumption of energy, Euclidean distance, and velocity [20,21]. For example, the position matrix determines how many stations robots travel from the source to destination. This matrix is defined using Equation (2).

$$Positions = [p_1\ , p_2, p_3, p_4 \ldots p_D] \tag{2}$$

where pi represents the position coordinates of each station that our robot takes on the map. In order to find an optimized trajectory, the proposed algorithms try to minimize cost (length of trajectory in our experiment). The cost of the trajectories is calculated using Equation (3), where i and j denote the current and next stations.

$$Cost_{(i,j)} = \sum_{i=s}^{j=D} distance_{i,j} + CurrentPower_i \tag{3}$$

Based on Equation (3), the cost of each founded path is obtained by the sum of distances between tuples from source to destination. Drones can be blended with metaheuristics so they can carry out their mission efficiently. In this regard, not only the distance parameters of the drones but also the remaining power amounts of drones are taken into account in the fitness function, which is defined to be more realistic. Therefore, the result from the metaheuristic algorithm can be used in real environments when applied to mobile robots. Random and optimized trajectories are used for UAV movement from source to destination, as shown in Figure 2. Here, the UAV moved through different stations. In path planning methods, usually, either the robots randomly move or costly processes are undertaken in finding an optimized path, but in this study, we tried to find the most efficient optimized path. This process is performed gradually between both stations. In this way, the UAV tries to find a path between two points. To optimize randomly created paths and to find the best possible trajectory, a method is proposed in this section with a minimum computational cost. Thanks to this method, robots can also actively avoid obstacles. In the final phase, the sum of all tuples' costs is calculated, and the cost of the path is obtained. The purpose of this study was to find the best path between the start and target stations of each UAV.

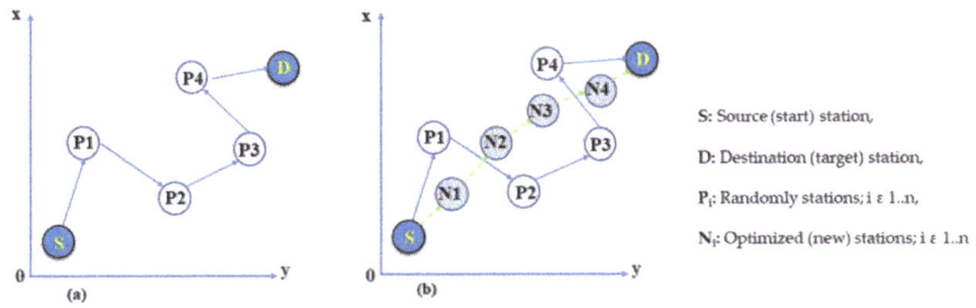

Figure 2. The randomly created and optimized trajectory. (**a**) Randomly Created Path; (**b**) Optimized Created Path.

Typically, the first step in path planning is to represent the workspace as a map. In the maps, many obstacles were used to make the mobile robots' tasks of finding the path realistic and complex. The challenge was to avoid various obstacles and to reach the position of the destination. In this study, a large-scale map was prepared to evaluate the proposed algorithms. The boundary of this map is shown in Figure 3a. In addition, three mobile robots with different start and destination stations were used, and their three-dimension points are given in Figure 3b. In this paper, it was assumed that the number of obstacles was quite high in order to make our proposed methods applicable in real areas. The number of obstacles was considered to be 150. Therefore, the coordination of some of the obstacles is presented in Figure 3c, and the full list is presented in Supplementary File 1. The problem of avoiding and managing obstacles is one of the most important aspects of path planning. The used mechanism includes two main steps and algorithms that take place sequentially, which were inspired by [47].

a) 3D map boundary

Map	Start boundary	End Boundary
Crowded Large map	(0, 0, 0)	(1000, 1000, 1000)

b) 3D map UAV initial and final positions

	UAV 1	UAV 2	UAV 3
Initial	(0, 0, 0)	(0, 500, 0)	(0, 1000, 0)
Final	(1000, 1000, 1000)	(1000, 500, 1000)	(1000, 0, 1000)

c) 3D obstacles for relevant map

Obstacles number	Crowded Large map	Obstacles number	Crowded Large map	Obstacles number	Crowded Large map
1	(10, 890, 0) - (40, 920, 15)	11	(180, 900, 0) - (210, 930, 15)	21	(60, 700, 0) - (90, 730, 15)
2	(30, 930, 0) - (60, 960, 15)	12	(200, 860, 0) - (230, 890, 15)	22	(110, 680, 0) - (140, 710, 15)
3	(70, 940, 0) - (100, 970, 15)	13	(160, 840, 0) - (190, 870, 15)	23	(160, 650, 0) - (190, 680, 15)
4	(160, 940, 0) - (190, 970, 15)	14	(210, 820, 0) - (240, 850, 15)	24	(50, 630, 0) - (80, 660, 15)
5	(110, 930, 0) - (140, 960, 15)	15	(170, 800, 0) - (200, 830, 15)	25	(30, 590, 0) - (60, 620, 15)
6	(60, 890, 0) - (90, 920, 15)	16	(130, 880, 0) - (160, 810, 15)	26	(80, 590, 0) - (110, 620, 15)
7	(30, 850, 0) - (60, 880, 15)	17	(120, 820, 0) - (150, 850, 15)	27	(120, 590, 0) - (150, 620, 15)
8	(80, 850, 0) - (110, 880, 15)	18	(90, 750, 0) - (120, 780, 15)	28	(200, 600, 0) - (230, 630, 15)
9	(100, 890, 0) - (130, 920, 15)	19	(80, 800, 0 - 110, 830, 15)	29	(160, 580, 0) - (190, 610, 15)
10	(140, 880, 0) - (170, 910, 15)	20	(40, 810, 0) - (70, 840, 15)	30	(220, 560, 0) - (250, 590, 15)

Figure 3. Land map (**a**), UAVs' positions (**b**) and obstacles coordinates (**c**).

3.2. GWO-Based Path Planning

In this paper, the path planning for autonomous agricultural robots was realized using the proposed method, inspired by Incremental Gray Wolf Optimization (I-GWO) and Expanded Gray Wolf Optimization (Ex-GWO) algorithms. These algorithms are inspired by gray wolves in nature. The natural behaviors of gray wolves such as encircling, hunting, and attacking prey have been modeled mathematically. Encircling in the I-GWO and Ex-GWO are calculated based on Equations (4) and (5). The hunting and attacking mechanism in the I-GWO can be obtained by Equations (9)–(11), and in the Ex-GWO, this is based on Equations (12)–(14). There are four types of wolves in each pack; alpha, beta, delta, and omega wolves. Each wolf has different responsibilities in the pack. Alpha, beta, and delta wolves are involved in encircling the prey, and omega wolves update their own positions based on them to attack the prey. The I-GWO algorithm is based on leader wolf's behavior. Other wolves in the pack update their own positions based on all the wolves selected before themselves. This may result in these wolves being present in similar regions. Thus, they only search for prey (solution) in a particular and similar area, which may be a missing point. The nth wolf in the pack updates its own position based on the $n-1$ wolf before it. This algorithm is completely dependent on the alpha wolf. In the I-GWO algorithm, all relative operations are addressed according to Equations (4)–(11), where t indicates the current iteration, T demonstrates maximum iteration number, $\vec{X}$ indicates the position vector of a wolf, and $\vec{X_p}$ is the position vector of the prey. Additionally, D is a vector that depends on the location of the target.

$$\vec{D} = \left|\vec{C} \cdot \vec{X_p} - \vec{X_t}\right| \tag{4}$$

$$\vec{X}(t+1) = \vec{X_t} - \vec{A} \cdot \vec{D} \tag{5}$$

$$\vec{A} = 2\vec{a} \cdot \vec{r_1} - \vec{a} \tag{6}$$

$$\vec{C} = 2 \cdot \vec{r_2} \tag{7}$$

$$\vec{a} = 2\left(1 - \frac{t^2}{T^2}\right) \tag{8}$$

$$\vec{D_\alpha} = \left|\vec{C_\alpha} \cdot \vec{X_\alpha} - \vec{X}\right| \tag{9}$$

$$\vec{X_\alpha} = \vec{X_\alpha} - \vec{A_\alpha} \cdot \vec{D_\alpha} \tag{10}$$

$$\vec{X_n}(t+1) = \frac{1}{n-1} \sum_{i=1}^{n-1} X_i(t);\ n = 2,\ 3,\ \ldots m \tag{11}$$

Another metaheuristic algorithm (Ex-GWO) is based on the first three hierarchies of the wolves (alpha, beta, and delta) in a pack. The fourth level of the wolves in a pack update their positions based on the leading three wolves. Generally, the nth wolf updates its own position relative to the prey according to the previous and the first three wolves (Equations (12)–(14)). In the Ex-GWO algorithm, the attacking mechanism is used to avoid the prey from escaping.

$$\vec{D_\alpha} = \left|\vec{C_1} \cdot \vec{X_\alpha} - \vec{X}\right|, \vec{D_\beta} = \left|\vec{C_2} \cdot \vec{X_\beta} - \vec{X}\right|, \vec{D_\delta} = \left|\vec{C_3} \cdot \vec{X_\delta} - \vec{X}\right| \tag{12}$$

$$\vec{X_1} = \vec{X_\alpha} - \vec{A_1} \cdot \vec{D_\alpha}, \vec{X_2} = \vec{X_\beta} - \vec{A_2} \cdot \vec{D_\beta}, \vec{X_3} = \vec{X_\delta} - \vec{A_3} \cdot \vec{D_\delta} \tag{13}$$

$$\vec{X_n}(t+1) = \frac{1}{n-1} \sum_{i=1}^{n-1} X_i(t);\ n = 4,\ 5,\ \ldots\ m \tag{14}$$

It is assumed that the coefficient vectors $\vec{A}$ and $\vec{C}$ lead to encircle the prey. The parameter $\vec{a}$ decreases from 2 to 0 relative to the iteration number. It is used to improve the convergence speed of the algorithm. These parameters control the tradeoff between exploration and exploitation phases. It is used to get closer to the solution range. $\vec{r_1}$ and $\vec{r_2}$ are the random vectors in a range of [0, 1]. In every algorithm, the leader encircles the prey, then hunts the prey, and finally attacks the prey based on the $\vec{A}$ value. If $\left|\vec{A}\right| < 1$, the wolf is attacking the prey; otherwise, it is busy trying to find prey (solution). Figure 4 depicts the working of the proposed algorithms, considering exploration and exploitation phases. Thanks to these features, the proposed 3D path planning methods were able to act in a balance between the two phases and try to find the most appropriate path without falling into any local optima trap.

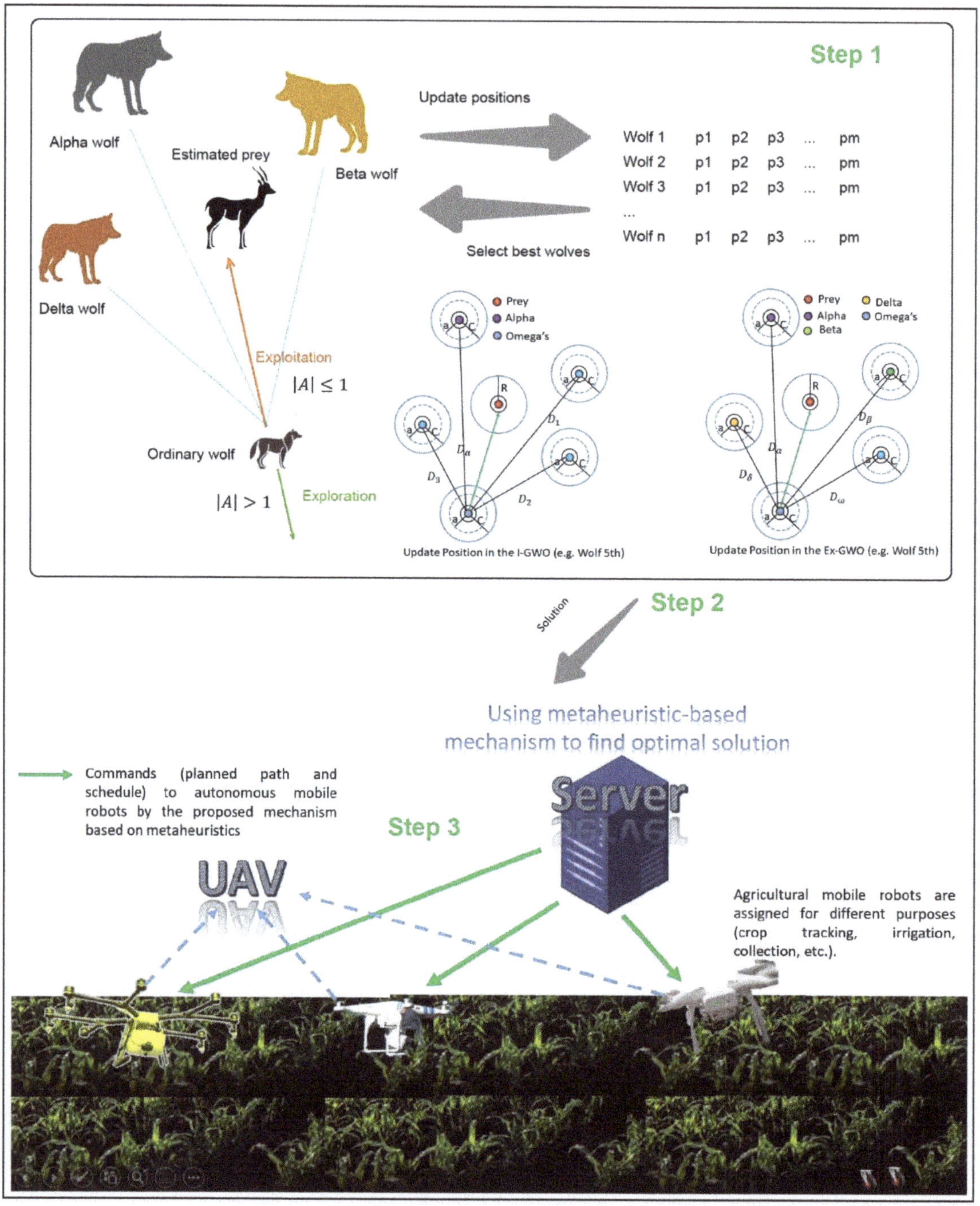

Figure 4. Working mechanism of the proposed method in UAV-based agricultural applications.

3.3. Working Mechanism of the Method

One of the most commonly applied methods of 3D path planning is to provide a robot with a defined number of static stations and to allow an algorithm to discover the most appropriate path. These types of algorithms are easier to apply mathematically, but generally, their time and space complexity is relatively higher. Here, a pool of stations is assumed so that these stations can be created randomly. Since the station selection in our methods is based on metaheuristic algorithms, it works appropriately with fewer parameters, and therefore, it can work efficiently by consuming resources in an acceptable time. Mathematically, this pool has been described in the structure of the 3xn matrix. The elements of search space represent distances between stations. Each station in the pool is a possible position that a mobile robot can choose as the next station. This pool is used to control the mobile robot's movement in the area. In addition, by using the information of this pool, it may be possible to help to avoid obstacles. The station selection process used in our methods is presented in Algorithm 1. In this study, the number and positions of stations (mobile robot stopovers) and obstacles are predefined similar to other studies in the literature [5,6,15,20,53]. On the other hand, obstacle avoidance is one of the many challenges that exist in the path planning problem. In this study, a method was used to avoid the collision of the UAVs with obstacles (objects or other robots), which benefits from geometric and calculus-based formulae. It was inspired by [47].

Algorithm 1. Pseudocode of station selection

```
State is array of candidate stations
w = distance obtained from metaheuristics      //Equations (10) and (13)
d = The list of distances
For each station (i) in pool
     di = distance between current and next stations + distance between next and destination stations
End For
MinDist = Min(d) //Min function indicates minimum distance in the list
if (MinDist < w)
          Select station with minimum distance as next station
Else
          Select station by metaheuristics as next station
End if
```

Primarily, the proposed methods initialize the random position matrix. Each row of the position matrix defines the path, and the columns represent the number of steps in the path to the destination. These number of stations are denoted as p. The (x_n^m, y_n^m, z_n^m) presents a coordinate of each station, where m is the aforementioned index of stations and n is the number of search agents in each method (Figure 5a). The search agents are the configuration parameter of the metaheuristic algorithms. Then, for each metaheuristic algorithm, a search space, based on the position matrix, is initialized. The search space is shown in Figure 5b, which represents the distance between tuples. In this table, each row represents a path length. Each element of the row shows the distance between two points as $d_{(i,j)}^n$, where i is the current state and j is the previous state. Furthermore, n is in the number of search agents. In addition, in the proposed methods, the path cost based on a fitness function that was presented in Equation (3) is calculated.

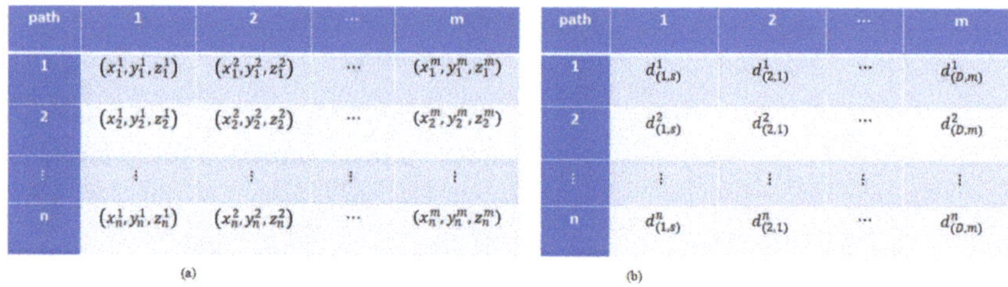

Figure 5. The working mechanism of the method. (**a**) The position matrix of each path; (**b**) the search space that represents distance between tuples.

In the next step, the proposed methods calculate the distance between each tuple for each station in the pool. In this case, we have a distance cost (d) between the current station and candidate next stations. The d includes two values: first is the distance between the current and next states, and the second is the distance between next and destination states. However, the metaheuristic algorithms find the best solution for the next station of each current station. If the distance of the possible next stations is smaller than the obtained value from metaheuristic algorithms (w), the relevant station with the minimum value is selected as the elected next station. Otherwise, the UAV chooses the achieved solution of the metaheuristic algorithms as the next station (Algorithm 1). The proposed method's aim is to reduce the cost of each path and try to find the optimal path with minimum cost for multi-UAVs. In this study, three UAVs were used that had dissimilar start (source) and final (destination) stations. The results obtained from this method are explained in the analysis and results section. The pseudocode and flowchart of the proposed path planning can be found in Algorithm 2 and Figure 6.

Algorithm 2. Pseudocode of proposed path planning

```
Initialize the grey wolf population X_i (i = 1, 2, . . . .., n)
Initialize A, C and a        //Equations(6)–(8)
Initialize positions matrix and search space
Calculate fitness of each agent        //Equation (3)
While (t < Max number of iterations)
    For each search agent
        Update the position of current search agent        //Algorithm 1.
    End For
    Update a, A and C
    Calculate the fitness of all search agents
    Update position        //Equation (11) or Equation (14)
    Update the search space matrix
    t = t +1
End While
```

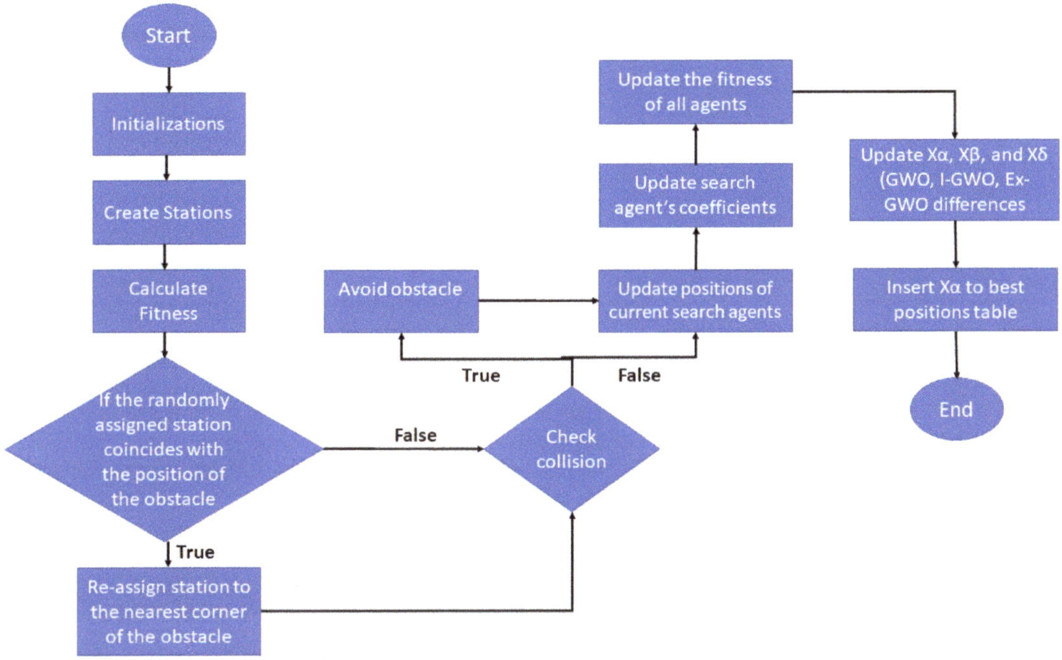

Figure 6. Proposed path planning method flowchart.

3.4. Other Possible Features: Applicability in Farmlands

Based on the functionality of farmland, farmers can analyze the data to increase productivity before the agricultural year begins. Most farmers fertilize their farmland based on the experimental information. Modern agriculture tries to use the source efficiently and encourages farmers to use new technologies in cultivation to increase productivity along with their income. While data that have been collected are stored in a light server to serve the clients, peer-to-peer communication can be held between monitoring devices with the robots via the Global System for Mobile communications (GSM). In precision agriculture, farmers are able to increase productivity by using the previous data analysis. As the connections are bidirectional, exchanging urgent commands such as changing tasks, terminating current tasks, and more can be performed. A tiny unit of computers of robots provides a mid-layer infrastructure to receive commands and to respond to requests. Thanks to the proposed algorithms, the farmer is able to lunch a UAV with predetermined states to monitor and control their land. Farmers can track the whole of their agricultural land and their crops remotely, and they can also meet their needs such as irrigation and harvesting using the related autonomous UAV and agricultural robot robots on an optimal path and minimum costs. In addition, the proposed methods can be used to find optimal routes for multiple UAVs at the same time in parallel or concurrently. In this case, each UAV perceives the other UAV as an obstacle andso, the relevant UAV can continue its mission without colliding with our obstacle management mechanism. In addition, thanks to this mechanism, it will be possible for the proposed methods to work successfully in dynamic or uncertain environments.

4. Results and Discussion

This section presents the performance of the proposed methods, which is analyzed and compared with the GWO-based 3D path planning method [20]. The authors used the Gray Wolf optimizer (GWO) algorithm to find an optimal path with minimal cost in 3D environments. According to the results of the study, in path planning, the GWO-based method is better than Dijkstra, A*, D*, and several famous metaheuristic-based methods. They proved that the GWO-based method presents a more balanced and better performance in similar problems. In addition, GWO-based algorithms are sought after in many research and application areas due to their balanced behavior amongst various metaheuristic algorithms [19,20]. Therefore, we selected this method for the comparison of results and performance. The implantation and analysis presentation was performed in Java and MATLAB. The algorithms proposed in this study were performed on a Core i7-5500 U 2.4 processor with 8GB of RAM.

4.1. Simulation Setting

In the simulation, large-scale, agricultural land was considered. The size of the environment was 1000 m $*$ 1000 m $*$ 1000 m. In addition, 150 obstacles were also placed in this area. Three UAVs with different start and endpoints were considered in each simulation of the used methods. The map boundaries, UAVs, and obstacle positions were assumed based on Figure 3. Each used algorithm was run 15 times. Furthermore, simulation parameters are presented in Table 1. The best, worst, and average costs (distance traveled in meters), execution times and complexity, and finally, convergence curve analysis for each UAV in each algorithm was applied by different population sizes and iteration numbers.

Table 1. The simulation parameters.

Parameter	Value
Population size	30, 50, 100
Maximum iteration	50, 100, 200
Farmland Square	1000 m $*$ 1000 m $*$ 1000 m
r_1, r_2	Rand [0,1]
a	linearly decreased from 2 to 0 over
A	[−2a, 2a]
C	Rand [0, 2]

4.2. Analysis and Evaluation (Cost of Paths)

In this section, the proposed path planning methods are analyzed based on the cost function, introduced in Equation (3). The results obtained are presented in Table 2. The starting and ending points of UAVs are assumed to be different from each other. In this table, the costs for these autonomous robots were obtained from a set of various populations and iterations. This process was evaluated for all algorithms used. According to the results obtained, the Ex-GWO-based method achieved the best result compared to other used path planning methods. The Ex-GWO-based method gave the best solution in five of the assumed nine scenarios and ranked first among the three methods with a 55.56% success rate. In the ranking, the I-GWO-based method was second with 38.88% and the GWO-based method was third with 5.56%. These results are presented in Table 3.

Table 2. Simulation results for each path planning algorithm on crowded large-scale map.

Algorithm	Pop	Iter	UAV_1 (Cost-m)			UAV_2 (Cost-m)			UAV_3 (Cost-m)			Overall Simulation Time (s)
			Best	*Ave*	*Worst*	*Best*	*Ave*	*Worst*	*Best*	*Ave*	*Worst*	
GWO	30	50	**2199**	2645	3385	2019	2328	2566	2148	2611	3526	**8.416**
I-GWO	30	50	**2199**	2736	3290	2087	2378	2579	**2107**	2446	2723	11.747
EX-GWO	30	50	2250	2649	3205	**1996**	2314	2655	2156	2713	4195	9.910
GWO	50	100	2145	2573	3295	1955	2234	2497	2095	2526	3469	**15.040**
I-GWO	50	100	2181	2558	3119	1993	2321	2520	**2090**	2361	2662	17.654
EX-GWO	50	100	**2128**	2621	3121	**1918**	2238	2593	2102	2638	4139	16.686
GWO	100	200	2070	2522	3202	1870	2136	2403	2026	2453	3417	**23.378**
I-GWO	100	200	**2045**	2537	3047	1936	2223	2466	2016	2261	2588	26.659
EX-GWO	100	200	2088	2479	2983	**1858**	2186	2528	**2007**	2568	4065	24.794

* The best values are bold.

Table 3. Ranking summary of metaheuristic algorithms in cost parameter.

Algorithm	Success Rate (Percent)	Rank
GWO	5.56	3
I-GWO	38.88	2
Ex-GWO	**55.56**	**1**

Based on the obtained results, it is determined that the Ex-GWO-based method exhibits good performance in large-scale and complex farmland with a high number of obstacles. This is because the Ex-GWO-based method finds the best solution according to the alpha, beta, delta wolves, and whole pack. The wolves use the whole pack's location knowledge to update their positions, so for experiments with larger population sizes and more iterations, the Ex-GWO-based method has a better chance of reaching the best solution. Therefore, the wolves in the pack minimize the escape paths of the hunt (prey), and hence, the prey can be caught faster. The fact that this mechanism can be better than other methods can be seen more clearly in large and crowded environments. On the other hand, another method proposed in this study, the I-GWO-based path planning method, outperforms good results in smaller and less populated environments. The basic update process in this method is very dependent on the alpha setup. Therefore, the speed of growth and the selection of the right places for the first wolf is of great importance. In this method, there is the possibility of finding problem solutions (prey) much faster in fewer iterations. For this reason, our proposed methods may be the most appropriate choices in various real-life application areas of mobile autonomous systems such as the use of UAVs for different and varied purposes and environments. In general, the GWO-based method has good performance in medium, small-sized, and not too crowded environments. In fact, the usage capacity of it may be between our two methods, but its success rate is not considered good according to the results. Briefly, the Ex-GWO-based path planning method may be performed more successfully in larger and crowded environments with larger populations and more iterations, and the I-GWO-based path planning method may give good results in medium and smaller, less populated environments.

Additionally, in Figure 7, the movements of UAVs for each algorithm are also shown on the defined map based on the obtained simulation results in various population sizes and iterations. In this figure, the circles show the start state of each UAV, and the star symbols show the destination state of each UAV.

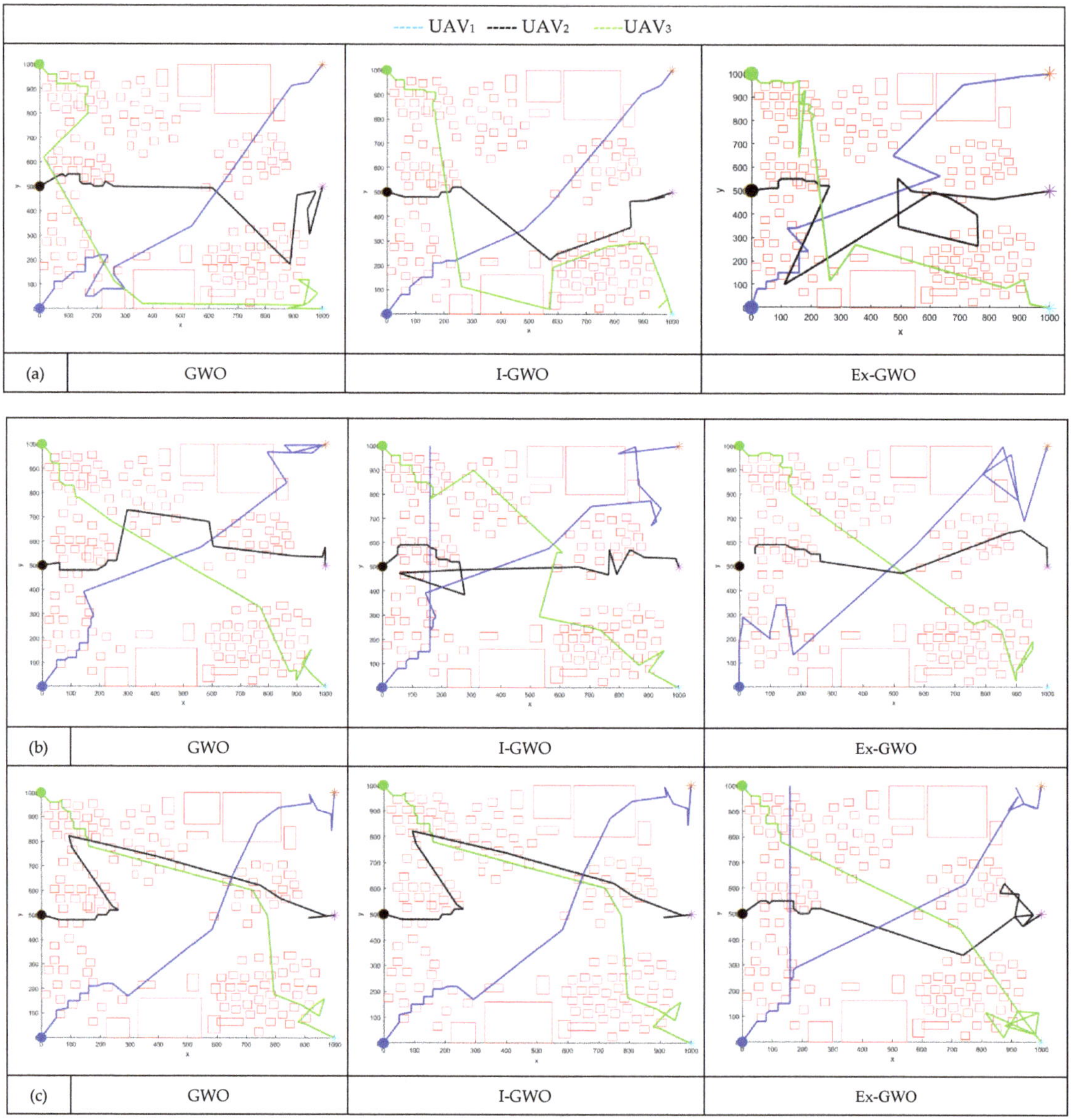

Figure 7. The movement of UAVs on the generated paths in each method. (**a**) population of 30, 50 iterations; (**b**) population of 50, 100 iterations; (**c**) population of 100, 200 iterations.

4.3. Analysis and Evaluation (Taken Times and Complexity)

The execution times of the proposed methods are also taken into consideration. The best execution time analyses for each method are presented in Figure 8 for various population sizes and iterations. The GWO has the best overall-time performance, while Ex-GWO and I-GWO rank second and third, respectively. Indeed, the GWO-based path planning for three UAVs in the parallel periods consumes the minimum time to reach its destination. The reason for this may be due to the fact that it depends on the three first wolves. In the I-GWO, the incorrect position or the wrong movement of the alpha wolf can move the whole pack away from the target or cause them to catch the prey late. At the Ex-GWO, each pack member has more roles and contributions than the other two methods, which means that this algorithm may need a longer execution time. However, the Ex-GWO may not be the worst in terms of time, as seen with its better performance in crowded environments. The results show that in the crowded map scenarios with many barriers, the I-GWO-based method does not perform well with regard to the overall time and optimum path cost. The fact that the GWO had the best overall time does not mean that the other two methods were bad, because these two methods could be concluded in an acceptable time. In addition, the time complexity analysis of the proposed methods is $O(n^2)$.

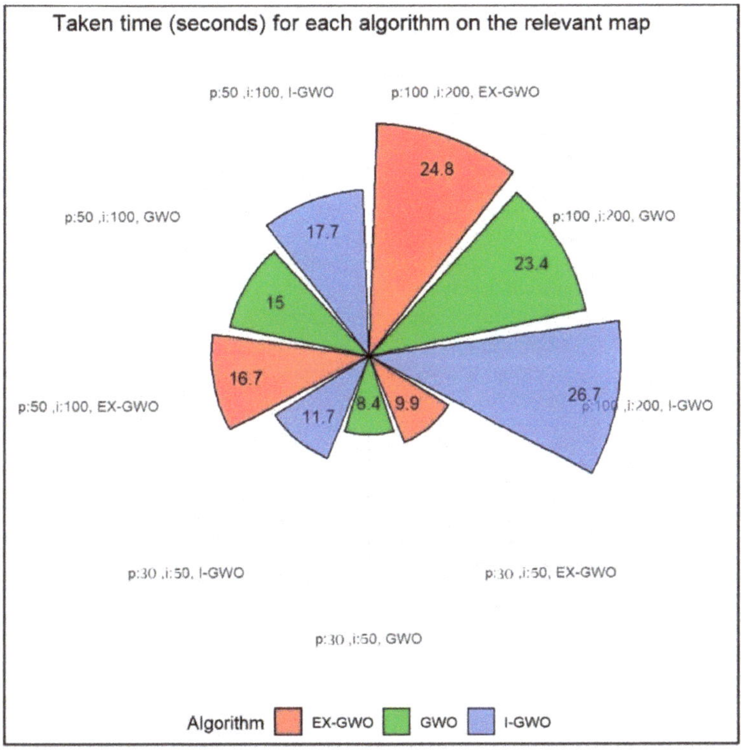

Figure 8. Taken time for each method on the relevant map.

4.4. Analysis and Evaluation (Convergence Curve)

Figure 9 presents the convergence curve of each proposed path planning algorithm. As mentioned before, the number of obstacles and the boundary sizes of the map are listed in Figure 3. The three metaheuristic algorithms used have different structures in the exploration and exploitation phases. Figure 9 illustrates the convergence curve of each method with various iterations and population sizes. In the I-GWO algorithm, the transition from exploration to the exploitation phase is faster than the other two metaheuristic algorithms (GWO and Ex-GWO). As a result of the observations, it was concluded that 50 iterations were enough to analyze of convergence rate because the results achieved did not display remarkable differences [20].

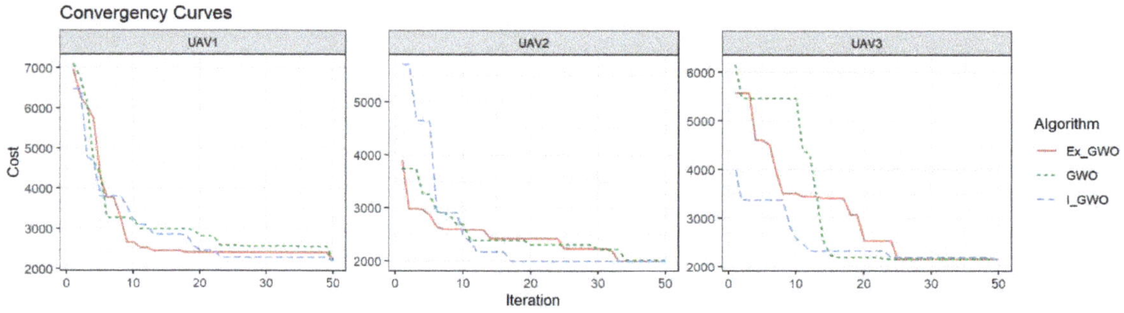

Figure 9. Convergence analysis for each UAV on the relevant map in population of 30 and 50 iterations.

In Figure 10, the statistical results of the path planning methods with the boxplot graph are presented. Boxplots are a standard method for displaying data distribution based on statistical indicators such as minimum, first quartile (Q1), median, third quartile (Q3), and maximum. This diagram also provides information regarding the existence of outlier data. In addition, the symmetry ratio in the data can be analyzed from this graph. The values were obtained from three metaheuristic algorithms with a population size of 30 and 50 iterations after 15 runs. The box plot graph analysis describes the maximum and minimum values of the obtained best cost and the frequency of the values. The x-axis of each figure indicates the name of the respective algorithm, whereas the y-axis indicates the average of best cost obtained. From Figure 10, it can be observed that the results obtained using the Ex-GWO algorithm are near to the best solution, whilst the algorithm tries to find the best solution. As well as this, after initial iterations in the exploitation phase the Ex-GWO obtained results near to the best cost.

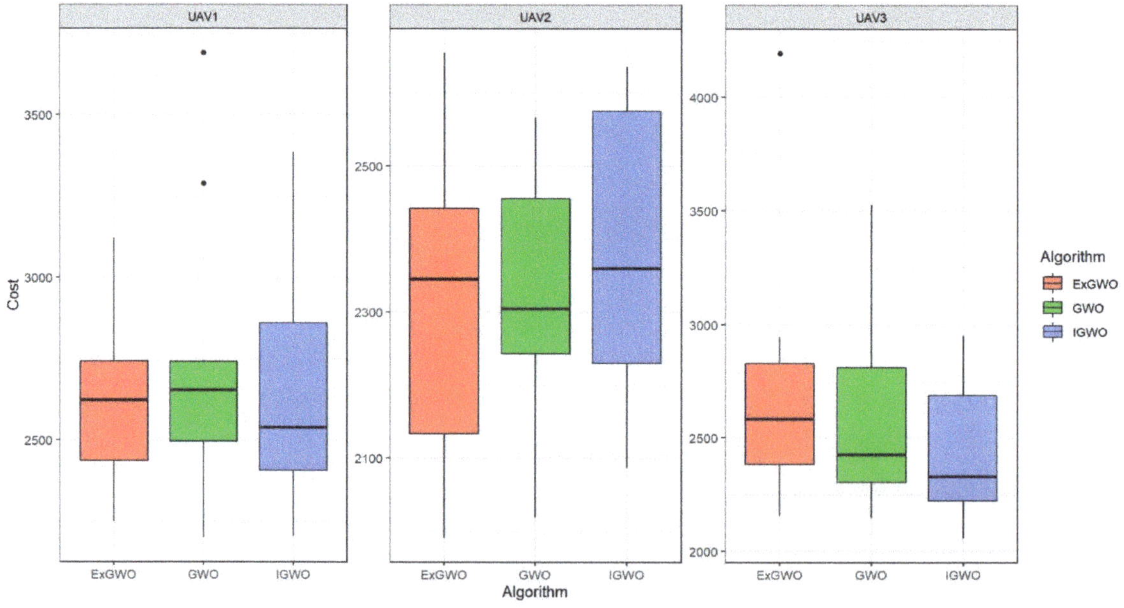

Figure 10. Boxplot graph analysis for each UAV in population of 30 and 50 iterations.

Figure 11 illustrates the distributions of costs in 15 runs. While the UAV_2 has an almost uniform distribution, both UAV_1 and UAV_3 have lower cost densities. Note that obstacles are employed as a marker for these metaheuristic algorithms, and paths with an appropriate number of obstacles help to improve the performances. For this purpose, a Student's t-test was applied for all combinations of UAVs. The *p*-value of UAV_1 and UAV_2 for all combinations of algorithms, population, and iterations count in 15 runs is 9.3×10^{-12}, the *p*-value of UAV_1, and UAV_3 is 0.201, and the *p*-value of UAV_2 and UAV_3 is 9.8×10^{-6}. Generally, *p*-values less than 0.05 are accepted for hypothesis rejection. The null hypothesis is that all UAV s do not have a meaningful relationship. The significant difference between UAV_2 with both UAV_1 and UAV_3 is proved. Therefore, the null hypothesis is rejected.

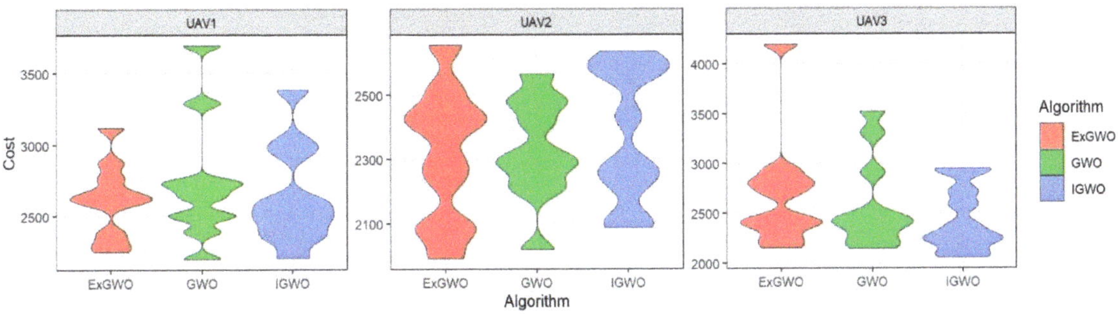

Figure 11. Distributions of costs in 15 runs for each UAV in population of 30 and 50 iterations.

5. Conclusions

The focus of the paper was to solve the NP-hard problem of efficient crop harvesting by finding the most suitable and optimal paths for UAVs. This study presented adaptive 3D path planning methods using metaheuristic algorithms (I-GWO and Ex-GWO) for autonomous agricultural UAVs. Therefore, in this study, maximum profit was achieved by consuming the least energy by harvesting the most crops in the shortest possible time. In addition, the use of resources such as human and natural resources was carried out efficiently by creating sustainable and smart agriculture. In other words, the method allows farmers to monitor crop variability and stress conditions continuously and harvest the best crops, resulting in efficient resource consumption and an increase in profits. The proposed methods tried to find the best solution in an acceptable time without falling into any local optima trap. The proposed method's aim was to reduce the cost of each path and try to find the optimal path with the minimum cost for multi-UAVs. In addition, this study also proposed a mechanism for obstacle management. In this study, a large-scale farmland map with many various obstacles was considered. From the results, it can be concluded that in terms of the minimum execution time parameter, the GWO-based method did the best, whereas in finding the optimal path with the minimum cost, the Ex-GWO-based method was better. The proposed method based on the Ex-GWO attained a 55.56% success rate, the I-GWO, and the GWO-based method attained 38.88% and 5.56% success rates in optimal path costs, respectively. In addition, in the analysis of convergence curve behavior for metaheuristic algorithms, the proposed I-GWO-based method was observed to offer the best solution. Thanks to the algorithms proposed in this study, efficient resource consumption and product growth rate can be achieved with low risk and cost. They can also be used in real agricultural applications. In addition, the consideration and installation of specific mechanical and hardware devices to mobile robots and farmland can play an important role. In this regard, information about the mission environment can be gathered by other types of sensors (e.g., laser spot), which are mounted on mobile UAVs. These sensors can provide information about the shape, size, and location of an obstacle. Using sensory information, robots may advance towards a target without colliding with an obstacle or coming under enemy radars. On the other hand, various sensor devices are used to collect information on parameters such as humidity, temperature, etc. in agricultural land for different applications. Briefly, since the map information and the starting and destination points of each mobile device are certain, the developed method can be easily embedded on these devices. Thanks to the obstacle and object detection feature used of the method, the supposed parameters in the defined fitness function, and the station selection feature on the map, method can be applicable in the real world as well. It can also be even more useful when combined with special equipment used in farmland and mobile devices.

In future studies, we would like to explain our roadmap below with a focus on smart and sustainable agriculture. Alongside mobile robots, it will focus on a method that tracks and harvests crops in large-scale farmland with Internet of Vehicles (IoVs). In such a scenario, the mobile robots would only be tasked with monitoring the farmland. In this case, a blended mechanism with image processing methods will be presented. It will then use the results from these autonomous robots as an input matrix for the IoVs. Since a complex and NP-hard type of problem will arise here, metaheuristic-based algorithms will again come into play. In this regard, hybrid or new algorithms will be presented. On the other hand, the 3D path planning methods proposed in this study can be applied to IoT systems such as smart cities, industries, and agriculture in hybrid form with machine learning algorithms such as reinforcement-learning- or game-theory-based algorithms.

Supplementary Materials: The supporting information about maps can be downloaded at: https://www.mdpi.com/article/10.3390/app12030943/s1.

Author Contributions: Conceptualization, F.K., S.N. and F.A.A.; methodology, F.K., T.Ç., F.C. and A.S.; software, A.S. and S.N.; validation, A.S., S.N., G.R., S.L., A.M.; formal analysis, A.S., F.C. and S.N.; investigation, A.S., F.A.A. and F.C.; resources, F.K., F.C., T.Ç., A.M. and A.S.; data curation, A.S. and S.N.; writing—original draft preparation, F.K., A.S., F.C., G.R., S.L. and A.M.; writing—review and editing, F.K., S.N., T.Ç., F.A.A., G.R., S.L. and A.M.; visualization, A.S., F.C., G.R., S.L. and A.M.; supervision, F.K., A.M. and S.N.; project administration, F.K.; All authors have read and agreed to the published version of the manuscript.

Funding: This research received no external funding.

Institutional Review Board Statement: Not applicable.

Informed Consent Statement: Not applicable.

Data Availability Statement: Data will be made available from the corresponding author on reasonable request.

Conflicts of Interest: The authors declare that they have no conflict of interest.

References

1. United Nations Environment Programme. Sustainability. Available online: https://www.unep.org/about-un-environment/sustainability (accessed on 29 September 2021).
2. Sylvester, G. *E-Agriculture in Action: Drones for Agriculture*; Food and Agriculture Organization of the United Nations and International Telecommunication Union: Bangkok, Thailand, 2018; pp. 11–22.
3. Dehghan Khangahi, F. Ecological Problems and Social Mobilization: The Case of Urmia Lake. Ph.D. Thesis, Istanbul University, Istanbul, Turkey, 2020.
4. Dehghan Khangahi, F.; Kiani, F. Social Mobilization and Migration Predictions by Machine Learning Methods: A study case on Lake Urmia. *Int. J. Innov. Technol. Explor. Eng.* **2021**, *10*, 123–127. [CrossRef]
5. Yang, L.; Qi, J.; Song, D.; Xiao, J.; Han, J.; Xia, Y. Survey of Robot 3D Path Planning Algorithms. *J. Control Sci. Eng.* **2016**, *2016*, 7426913. [CrossRef]
6. Yang, L.; Qi, J.; Xiao, J.; Yong, X. A literature review of UAV 3D path planning. In Proceeding of the 11th World Congress on Intelligent Control and Automation, Shenyang, China, 29 June–4 July 2014; pp. 2376–2381. [CrossRef]
7. Agrawal, H.; Dhall, R.; Iyer, K.S.S.; Chetlapalli, V. An improved energy efficient system for IoT enabled precision agriculture *J. Ambient. Intell. Humaniz. Comput.* **2020**, *11*, 2337–2348. [CrossRef]
8. Kiani, F.; Nematzadehmiandoab, S.; Seyyedabbasi, A. Designing a dynamic protocol for real-time Industrial Internet of Things-based applications by efficient management of system resources. *Adv. Mech. Eng.* **2019**, *11*, 1–23. [CrossRef]
9. Kiani, F.; Seyyedabbasi, A.; Nematzadeh, S. Improving the performance of hierarchical wireless sensor networks using the metaheuristic algorithms: Efficient cluster head selection. *Sens. Rev.* **2021**, *41*, 368–381. [CrossRef]
10. Tzounis, A.; Katsoulas, N.; Bartzanas, T.; Kittas, C. Internet of Things in agriculture, recent advances and future challenges *Biosyst. Eng.* **2017**, *164*, 31–48. [CrossRef]

11. Nayyar, A.; Nguyen, B.-L.; Nguyen, N.G. The Internet of Drone Things (IoDT): Future Envision of Smart Drones. In *First International Conference on Sustainable Technologies for Computational Intelligence*; Advances in Intelligent Systems and Computing; Springer: Singapore, 2020; Volume 1045, pp. 563–580.
12. Anastasios, L.; Thomas, L.; Panagiotis, S.; Michalis, Z.; George, L. Towards smart farming: Systems, frameworks and exploitation of multiple sources. *Comput. Netw.* **2020**, *172*, 107147.
13. Verónica, S.; Francisco, R. From Smart Farming towards Agriculture 5.0: A Review on Crop Data Management. *Agronomy* **2020**, *10*, 207.
14. Kiani, F.; Seyyedabbasi, A. Wireless Sensor Network and Internet of Things in Precision Agriculture. *Int. J. Adv. Comput. Sci. Appl.* **2018**, *9*, 99–103. [CrossRef]
15. Benos, L.; Bechar, A.; Bochtis, D. Safety and ergonomics in human-robot interactive agricultural operations. *Biosyst. Eng.* **2020**, *200*, 55–72. [CrossRef]
16. Panagiotis, R.; Panagiotis, S.; Thomas, L.; Ioannis, M. A compilation of UAV applications for precision agriculture. *Comput. Netw.* **2020**, *172*, 107148.
17. Kadeghe, G.; Wesley, M.; Edward, M.; Glen, C. An Extensive Review of Mobile Agricultural Robotics for Field Operations: Focus on Cotton Harvesting. *AgriEngineering* **2020**, *2*, 150–174.
18. Dewangan, R.K.; Shukla, A.; Godfrey, W.W. Three dimensional path planning using Grey wolf optimizer for UAVs. *Appl. Intell.* **2019**, *49*, 2201–2217. [CrossRef]
19. Seyyedabbasi, A.; Kiani, F. I-GWO and Ex-GWO: Improved algorithms of the Grey Wolf Optimizer to solve global optimization problems. *Eng. Comput.* **2021**, *37*, 509–532. [CrossRef]
20. Seyyedabbasi, A.; Kiani, F. MAP-ACO: An efficient protocol for multi-agent pathfinding in real-time WSN and decentralized IoT systems. *Microprocess. Microsyst.* **2020**, *79*, 103325. [CrossRef]
21. Chung, H.M.; Maharjan, S.; Zhang, Y.; Eliassen, F.; Strunz, K. Placement and Routing Optimization for Automated Inspection with Unmanned Aerial Robots: A Study in Offshore Wind Farm. *IEEE Trans. Ind. Inform.* **2020**, *17*, 3032–3043. [CrossRef]
22. del Cerro, J.; Cruz Ulloa, C.; Barrientos, A.; de León Rivas, J. Unmanned Aerial Robots in Agriculture: A Survey. *Agronomy* **2021**, *11*, 203. [CrossRef]
23. Macrina, G.; Di Puglia Pugliese, L.; Guerriero, F.; Laporte, G. Drone-aided routing: A literature review. *Transp. Res. Part C Emerg. Technol.* **2020**, *120*, 102762. [CrossRef]
24. Baik, H.; Valenzuela, J. An optimization drone routing model for inspecting wind farms. *Soft Comput.* **2021**, *25*, 2483–2498. [CrossRef]
25. Allred, B.; Eash, N.; Freeland, R.; Martinez, L.; Wishart, D. Effective and efficient agricultural drainage pipe mapping with UAS thermal infrared imagery: A case study. *Agric. Water Manag.* **2018**, *197*, 132–137. [CrossRef]
26. Vasudevan, A.; Kumar, D.A.; Bhuvaneswari, N.S. Precision farming using unmanned aerial and ground robots. In Proceedings of the IEEE Technological Innovations in ICT for Agriculture and Rural Development (TIAR), Chennai, India, 15–16 July 2016; pp. 146–150.
27. Behjati, M.; Noh, A.M.; Alobaidy, H.; Zulkifley, M.; Nordin, R.; Abdullah, N. LoRa Communications as an Enabler for Internet of Drones towards Large-Scale Livestock Monitoring in Rural Farms. *Sensors* **2021**, *21*, 5044. [CrossRef] [PubMed]
28. Chowdhury, S.; Shahvari, O.; Marufuzzaman, M.; Li, X.; Bian, L. Drone routing and optimization for post-disaster inspection. *Comput. Ind. Eng.* **2021**, *159*, 107495. [CrossRef]
29. Dukkanci, O.; Kara, B.Y.; Bektaş, T. Minimizing energy and cost in range-limited drone deliveries with speed optimization. *Transp. Res. Part C Emerg. Technol.* **2021**, *125*, 102985. [CrossRef]
30. Felegari, S.; Sharifi, A.; Moravej, K.; Amin, M.; Golchin, A.; Muzirafuti, A.; Tariq, A.; Zhao, N. Integration of Sentinel 1 and Sentinel 2 Satellite Images for Crop Mapping. *Appl. Sci.* **2021**, *11*, 10104. [CrossRef]
31. Matrice, D. Matrice 100. 100, 2016, 2017. Available online: https://www.dji.com/product/matrice100 (accessed on 14 November 2021).
32. Papachristos, C.; Kamel, M.; Popović, M.; Khattak, S.; Bircher, A.; Oleynikova, H.; Dang, T.; Mascarich, F.; Alexis, K.; Siegwart, R. Autonomous Exploration and Inspection Path Planning for Aerial Robots Using the Robot Operating System. In *Studies in Computational Intelligence*; Koubaa, A., Ed.; Springer Science and Business Media LLC: Toledo, Spain, 2018; pp. 67–111.
33. Christiansen, M.P.; Laursen, M.S.; Jørgensen, R.N.; Skovsen, S.; Gislum, R. Designing and Testing a UAV Mapping System for Agricultural Field Surveying. *Sensors* **2017**, *17*, 2703. [CrossRef] [PubMed]
34. Faical, B.S.; Pessin, G.; Filho, G.P.; Carvalho, A.C.; Furquim, G.; Ueyama, J. Fine-Tuning of UAV Control Rules for Spraying Pesticides on Crop Fields. In Proceedings of the 2014 IEEE 26th International Conference on Tools with Artificial Intelligence, Limassol, Cyprus, 10–12 November 2014; pp. 527–533.
35. Dai, B.; He, Y.; Gu, F.; Yang, L.; Han, J.; Xu, W. A vision-based autonomous aerial spray system for precision agriculture. In Proceedings of the 2017 IEEE International Conference on Robotics and Biomimetics (ROBIO), Macao, China, 5–8 December 2017; pp. 507–513.
36. Klauser, F.; Pauschinger, D. Entrepreneurs of the air: Sprayer drones as mediators of volumetric agriculture. *J. Rural. Stud.* **2021**, *84*, 55–62. [CrossRef]
37. Skobelev, P.; Budaev, D.; Gusev, N.; Voschuk, G. Designing Multi-agent Swarm of UAV for Precise Agriculture. In *International Conference on Practical Applications of Agents and Multi-Agent Systems*; Javier, B.J., Ed.; Springer: Toledo, Spain, 2018; pp. 47–59.

38. Ju, C.; Son, H.I. Multiple UAV Systems for Agricultural Applications: Control, Implementation, and Evaluation. *Electronics* **2018**, *7*, 162. [CrossRef]
39. Li, X.; Zhao, Y.; Zhang, J.; Dong, Y. A Hybrid PSO Algorithm Based Flight Path Optimization for Multiple Agricultural UAVs. In Proceedings of the 2016 IEEE 28th International Conference on Tools with Artificial Intelligence (ICTAI), San Jose, CA, USA, 6–8 November 2016; pp. 691–697.
40. Noreen, I.; Khan, A.; Habib, Z. Optimal Path Planning using RRT*-Adjustable Bounds. *Intell. Serv. Robot* **2018**, *11*, 41–52. [CrossRef]
41. Guruji, A.K.; Agarwal, H.; Parsediya, D. Time-efficient A* Algorithm for Robot Path Planning. *Procedia Technol.* **2016**, *23*, 144–149. [CrossRef]
42. Jason, G.; Xin, M.; Liu, F.; Ying, W.; Ren, H. Mathematical Modeling and Intelligent Algorithm for Multi-Robot Path Planning. *Math. Probl. Eng.* **2017**, *2017*, 1465158.
43. Li, J.; Yang, S.X.; Xu, Z. A Survey on Robot Path Planning using Bio-inspired Algorithms. In Proceedings of the 2019 IEEE International Conference on Robotics and Biomimetics (ROBIO), Dali, China, 6–8 December 2019; pp. 2111–2116.
44. Choudhury, N.; Mandal, R.; Kar, S.K. Bioinspired robot path planning using PointBug algorithm. In Proceedings of the 2016 International Conference on Electrical, Electronics, and Optimization Techniques (ICEEOT), Chennai, India, 3–5 March 2016; pp. 2638–2643.
45. Wu, L.; Li, Y.; Liu, J. Based on Improved Bio-Inspired Model for Path Planning by Multi-AUV. In Proceedings of the Proceedings of the 2018 International Conference on Electronics and Electrical Engineering Technology, Tianjin, China, 19–21 September 2018; pp. 128–134.
46. Seyyedabbasi, A.; Aliyev, R.; Kiani, F.; Gulle, M.U.; Basyildiz, H.; Shah, M.A. Hybrid algorithms based on combining reinforcement learning and metaheuristic methods to solve global optimization problems. *Knowl.-Based Syst.* **2021**, *223*, 107044. [CrossRef]
47. Kiani, F.; Seyyedabbasi, A.; Aliyev, R.; Gulle, M.U.; Basyildiz, H.; Shah, M.A. Adapted-RRT: Novel hybrid method to solve three-dimensional path planning problem using sampling and metaheuristic-based algorithms. *Neural Comput. Appl.* **2021**, *33*, 15569–15599. [CrossRef]
48. Flemming, S.; la Anders, C.H.; Morten, B. Configuration Space and Visibility Graph Generation from Geometric Workspaces for UAVs. In Proceedings of the AAIA, Guidance, Navigation, and Control Conference, Portland, OR, USA, 8–11 August 2011; pp. 1–12.
49. Bera, T.; Bhat, M.S.; Ghose, D. Analysis of Obstacle based Probabilistic RoadMap Method using Geometric Probability. *IFAC Proc. Vol.* **2014**, *47*, 462–469. [CrossRef]
50. Kavraki, L.; Svestka, P.; Latombe, J.-C.; Overmars, M. Probabilistic roadmaps for path planning in high-dimensional configuration spaces. *IEEE Trans. Robot. Autom.* **1996**, *12*, 566–580. [CrossRef]
51. Kiani, F.; Seyyedabbasi, A.; Aliyev, R.; Shah, M.A.; Gulle, M. 3D Path Planning Method for Multi-UAVs Inspired by Grey Wolf Algorithms. *J. Internet Technol.* **2021**, *22*, 1–12.
52. Pandey, P.; Shukla, A.; Tiwari, R. Three-dimensional path planning for unmanned aerial robots using glowworm swarm optimization algorithm. *Int. J. Syst. Assur. Eng. Manag.* **2018**, *9*, 836–852. [CrossRef]
53. Qu, G.; Gai, W.; Zhong, M.; Zhang, J. A novel reinforcement learning based grey wolf optimizer algorithm for unmanned aerial robots (UAVs) path planning. *Appl. Soft Comput. J.* **2020**, *89*, 1–12. [CrossRef]
54. Wolpert, D.H.; Macready, W.G. A comparison of deterministic and probabilistic optimization algorithms for nonsmooth simulation-based optimization. *Build. Environ.* **2004**, *39*, 989–999. [CrossRef]
55. Wolpert, H.; William, G. No free lunch theorems for optimization. *IEEE Trans. Evol. Comput.* **1997**, *1*, 67–82. [CrossRef]
56. Cantelli, L.; Bonaccorso, F.; Longo, D.; Melita, C.D.; Schillaci, G.; Muscato, G. A Small Versatile Electrical Robot for Autonomous Spraying in Agriculture. *AgriEngineering* **2019**, *1*, 29. [CrossRef]
57. Chetty, S.; Adewumi, A.O. Three New Stochastic Local Search Metaheuristics for the Annual Crop Planning Problem Based on a New Irrigation Scheme. *J. Appl. Math.* **2013**, *2013*, 158538. [CrossRef]
58. Memmah, M.-M.; Lescourret, F.; Yao, X.; Lavigne, C. Metaheuristics for agricultural land use optimization. A review. *Agron. Sustain. Dev.* **2015**, *35*, 975–998. [CrossRef]
59. Palomino-Suarez, D.; Pérez-Ruiz, A. Towards Automatic UAV Path Planning in Agriculture Oversight Activities. In *LACAR 2019: Advances in Automation and Robotics Research*; Springer: Cali, Colombia, 2019; pp. 22–30.
60. Zeng, J.; Ju, R.; Qin, L.; Hu, Y.; Yin, Q.; Hu, C. Navigation in Unknown Dynamic Environments Based on Deep Reinforcement Learning. *Sensors* **2019**, *19*, 3837. [CrossRef] [PubMed]
61. Wang, K.; Meng, Z.; Wang, L.; Wu, Z.; Wu, Z. Practical Obstacle Avoidance Path Planning for Agriculture UAVs. In *IEA/AIE 2019: Wotawa F. Advances and Trends in Artificial Intelligence. From Theory to Practice*; Springer: Graz, Austria, 2019; pp. 196–203.
62. Juman, M.A.; Wong, Y.W.; Rajkumar, R.K.; H'Ng, C.Y. An integrated path planning system for a robot designed for oil palm plantations. In Proceedings of the TENCON 2017—2017 IEEE Region 10 Conference, Penang, Malaysia, 5–8 November 2017; pp. 1048–1053.
63. Santos, L.C.; Santos, F.N.; Pires, E.S.; Valente, A.; Costa, P.; Magalhães, S. Path Planning for ground robots in agriculture: A short review. In Proceedings of the 2020 IEEE International Conference on Autonomous Robot Systems and Competitions (ICARSC), Azores, Portugal, 15–17 April 2020; pp. 61–66.

64. Pichler-Scheder, M.; Ritter, R.; Lindinger, C.; Amerstorfer, R.; Edelbauer, R. Path planning for semi-autonomous agricultural robots. In *Reinventing Mechatronics*; Yan, X.T., Ed.; Springer: Cham, Switzerland, 2020; pp. 35–53.
65. Mai, T.; Shao, S.; Yun, Z. The Path Planning of Agricultural AGV in Potato Ridge Cultivation. *Ann. Adv. Agric. Sci.* **2019**, 3, 32003. [CrossRef]
66. Chikumbo, O.; Goodman, E.; Deb, K. Approximating a multidimensional Pareto front for a land use management problem: A modified MOEA with an epigenetic silencing metaphor. In Proceedings of the 2012 IEEE Congress on Evolutionary Computation (CEC), Brisbane, Australia, 10–15 June 2012; pp. 1–9.

applied sciences

MDPI

Article

WebGIS Implementation for Dynamic Mapping and Visualization of Coastal Geospatial Data: A Case Study of BESS Project

Giovanni Randazzo [1,2,3], Franco Italiano [1,4], Anton Micallef [1,5], Agostino Tomasello [1,6], Federica Paola Cassetti [1,6], Anthony Zammit [1,7], Sebastiano D'Amico [5], Oliver Saliba [1,7], Maria Cascio [1], Franco Cavallaro [1], Antonio Crupi [1], Marco Fontana [1], Francesco Gregorio [1], Stefania Lanza [1,3], Emanuele Colica [1,5] and Anselme Muzirafuti [1,8,*]

1 Interreg Italia–Malta–Progetto: Pocket Beach Management & Remote Surveillance System, University of Messina, Via F. Stagno d'Alcontres, 31-98166 Messina, Italy; giovanni.randazzo@unime.it (G.R.); francesco.italiano@ingv.it (F.I.); anton.micallef@um.edu.mt (A.M.); agostino.tomasello@unipa.it (A.T.); federicapaola.cassetti@unipa.it (F.P.C.); anthony.b.zammit@gov.mt (A.Z.); oliversaliba90@gmail.com (O.S.); maria.cascio86@gmail.com (M.C.); frcavall@tin.it (F.C.); antoniocrupi5@hotmail.it (A.C.); marco.fontana92@yahoo.it (M.F.); gregoriofrancesco92@gmail.com (F.G.); stefania.lanza@unime.it (S.L.); emanuele.colica@umi.edu.mt (E.C.)
2 Dipartimento di Scienze Matematiche e Informatiche, Scienze Fisiche e Scienze della Terra, Università degli Studi di Messina, Via F. Stagno d'Alcontres, 31-98166 Messina, Italy
3 GeoloGISs.r.l. Spin Off, Via F. Stagno d'Alcontres, 31-98166 Messina, Italy
4 Istituto Nazionale di Geofisica e Vulcanologia, Osservatorio Etneo-Sezione di Catania, 95125 Catania, Italy
5 Euro-Mediterranean Centre on Insular Coastal Dynamics (ICoD), Institute of Earth Systems, University of Malta, MSD-2080 Msida, Malta; sebastiano.damico@um.edu.mt
6 Department of Earth and Marine Science (DiSTeM), University of Palermo, Viale delle Scienze Ed. 16, 90128 Palermo, Italy
7 Ministry for Gozo St. Francis Square, VCT-1335 Victoria, Malta
8 Laboratory of Geoengineering and Environment, Department of Geology, Faculty of Sciences, University of Moulay Ismail, BP. 11201 Zitoune, Meknes 50000, Morocco
* Correspondence: a.muzirafuti@edu.umi.ac.ma or muzansel@gmail.com

Citation: Randazzo, G.; Italiano, F.; Micallef, A.; Tomasello, A.; Cassetti, F.P.; Zammit, A.; D'Amico, S.; Saliba, O.; Cascio, M.; Cavallaro, F.; et al. WebGIS Implementation for Dynamic Mapping and Visualization of Coastal Geospatial Data: A Case Study of BESS Project. *Appl. Sci.* **2021**, *11*, 8233. https://doi.org/10.3390/app11178233

Academic Editor: Hyung-Sup Jung

Received: 5 August 2021
Accepted: 2 September 2021
Published: 5 September 2021

Abstract: Within an E.U.-funded project, BESS (Pocket Beach Management and Remote Surveillance System), the notion of a geographic information system is an indispensable tool for managing the dynamics of georeferenced data and information for any form of territorial planning. This notion was further explored with the creation of a WebGIS portal that will allow local and regional stakeholders/authorities obtain an easy remote access tool to monitor the status of pocket beaches (PB) in the Maltese Archipelago and Sicily. In this paper, we provide a methodological approach for the implementation of a WebGIS necessary for very detailed dynamic mapping and visualization of geospatial coastal data; the description of the dataset necessary for the monitoring of coastal areas, especially the PBs; and a demonstration of a case study for the PBs of Sicily and Malta by using the methodology and the dataset used during the BESS project. Detailed steps involved in the creation of the WebGIS are presented. These include data preparation, data storage, and data publication and transformation into geo-services. With the help of different Open Geospatial Consortium protocols, the WebGIS displays different layers of information for 134 PBs including orthophotos, sedimentological/geomorphological beach characteristics, shoreline evolution, geometric and morphological parameters, shallow water bathymetry, and photographs of pocket beaches. The WebGIS allows not only for identifying, evaluating, and directing potential solutions to present and arising issues, but also enables public access and involvement. It reflects a platform for future local and regional coastal zone monitoring and management, by promoting public/private involvement in addressing coastal issues and providing local public administrations with an improved technology to monitor coastal changes and help better plan suitable interventions.

Keywords: geographic information system (GIS); pocket beaches; coastal management; Interreg; climate change; remote sensing; drone; Sicily; Malta; Gozo; Comino

1. Introduction

The numerous fields of application of geographic information systems (GISs) have now become an effective and irreplaceable element in the study of anthropogenic activities and natural phenomena, thanks to sophisticated technologies and the growing interest generated by the science of geographic information [1]. Geospatial data gathered on natural phenomena are required to construct maps and they are increasingly being used in GISs. These are systems that can capture, store, analyze, manage, and share data that are linked to geographical locations [2]. Over the years, GISs have become more dynamic, flexible, and accessible to users [3]. Once a GIS project has been prepared, web publishers can create and publish interactive web pages characterized by a high level of customization in the form of a WebGIS. It allows the information, including access to the maps published online, to be available to end users, such as citizens, tourists, and regional administrations, by using a common web browser by connecting from remote Internet locations [4]. The WebGIS combines two powerful technologies: GIS and the Internet, providing connectivity at the global level [5]; the result of this synergy results in greater ease in finding data, sharing analytical tools, and reaching a larger number of users [6,7]. Opdam [7] has argued that communication between science and society constitutes a relevant tool to optimize any planning and management. Veenendaal et al. [8] discussed the development of web mapping and presented a timeline of major web mapping events starting just after the creation of the World Wide Web with the publication of online maps in 1993 to the development of real-time services in 2017.

Examples of WebGISs have been developed in various fields, including radon risk management [9] and a risk assessment system for heavy metal pollution [10], georeferenced bibliographies [11], wastewater treatments [12] and water environment monitoring and management systems [13], planning and emergency phases in case of floods [14], transport infrastructure management [15], management of abandoned mines [16], civic education on peace and conflict [17], and as a landslide early warning system [18].

In the coastal field, examples of WebGISs have been utilized for visualizing coastal flooding vulnerability and planning for resiliency [19] and as a support to the management of coastal areas [20–22].

Application and interface servers have been used in WebGISs already consolidated by important public administrations, such as, for example, the Province of Belluno [23], Arpa Puglia [24], and the Metropolitan City of Venice [25].

The WebGIS approach can be defined as a set of geographic information services for the internet, based on a network that uses different forms of internet access to provide geographic information, analytical tools, and different GIS services [26]. The WebGIS is a GIS that uses web technology to communicate between the web application server and the end user client [19,27,28]. While incredibly powerful, the adoption of desktop GIS software has often lagged, due to several reasons, such as the expense of site licenses and higher-end computer hardware and the complexity of GIS software requiring high levels of training and expertise. With WebGISs, users do not need to purchase and install expensive GIS software to access and work with maps and databases [29]. Also, users do not need to become experts in sophisticated GIS applications, since the functionality is made available through a regular web browser and an integrated viewer with a simple, user-friendly interface. Otherwise, GIS tools and data are often beyond the reach of ordinary citizens with an interest in a particular place-based decision problems [19].

Implementation of WebGISs applied on pocket beaches (PBs) of Sicily, Malta, Gozo, and Comino represents one of the outputs of the Pocket Beach Management and Remote Surveillance System project (BESS), co-financed by the European Union (European Regional Development Fund, within the Operative Program Italy—Malta 2014–2020), coordinated by the Department of Mathematical and Computer Sciences, Physical Sciences, and Earth Sciences (MIFT) of the University of Messina (UNIME). The Maltese partners included the Ministry for Gozo and University of Malta (represented by the Euro-Mediterranean Center on Insular Coastal Dynamics (ICoD)); the Sicilian partners included the Department of

Earth Sciences (DiSTeM) of the University of Palermo (UNIPA) and the National Institute of Geophysics and Volcanology (INGV).The aim of the BESS project was to achieve, following a large number of studies, a coastal monitoring platform that contained geological and morphological data, biological and sedimentological analysis, bathymetric information, and aerial photogrammetric and anemometric data. This platform was developed to allow the management of these input data within a WebGIS to study the evolution of 134 PBs in Sicily and Malta.

PBs represent coastal features characterized by their evolving shape, their geomorphological influence on sedimentary input, and their ecological value. These enclosed beaches are small beaches set between headlands that diffract or refract incoming waves [30–32]. The term "pocket beach" describes a beach controlled by a geological structure or a human structure, such as a groin or jetty [33–35] and they are common along rocky coasts throughout the world [36,37].

Detailed studies of PBs have previously been conducted [35,37–42], PBs are widespread throughout the entire coastline [42–44] and the characteristics of their exceptional natural landscape make them very attractive to tourists. Natural and man-made PBs are frequent elements of the Sicilian and Maltese coast and are often the most attractive segments of rocky coasts, forming a hub for tourist activity.

The main contributions of this paper are to provide:

- A methodological approach for the implementation of a WebGIS necessary for a very detailed dynamic mapping and visualization of geospatial coastal data.
- The description of the dataset necessary for the monitoring of coastal areas, especially the PBs.
- A demonstration of a case study for the PBs of Sicily and Malta by using the methodology and the dataset used during the BESS project.

In this paper, the process followed during the implementation of the BESS WebGIS is presented. The WebGIS contains the dataset obtained from a series of monitoring procedures carried out with a holistic approach during the BESS project. The project was adopted in the aim of reaching a turning point in terms of management of the geological and naturalistic heritage represented by the PBs, by considering, above all, implications of the well-being of society that arise from the protection of highly valued tourist sites. This paper describes how WebGIS technology was employed for geospatial data representation and dynamic mapping of the PBs. It further demonstrates the process of the implementation of a WebGIS, which is considered to be useful in terms of mapping, monitoring, and sensitization of coastal geomorphological peculiarities, namely the PBs.

2. Materials and Methods

2.1. Study Area

The islands of Malta and Sicily are characterized by distinctive environmental and climatic similarities, as they are both located in the center of the Mediterranean Sea (Figure 1). They have, in fact, many geological similarities since they form a spur on the northern edge of the African continental plate that includes Malta, SE Sicily, the Pelagian Islands, eastern Tunisia, and the northwestern Libyan shelf [45].

During the last one million years, following eustatic variations, both islands have alternately been physically linked to each other and to the Italian peninsula [45–47]. In particular, during the Last Glacial Maximum they were physically connected by an isthmus and formed a territorial unicum, located in the center of the Mediterranean, acquiring the shape of the three-legged island [48].

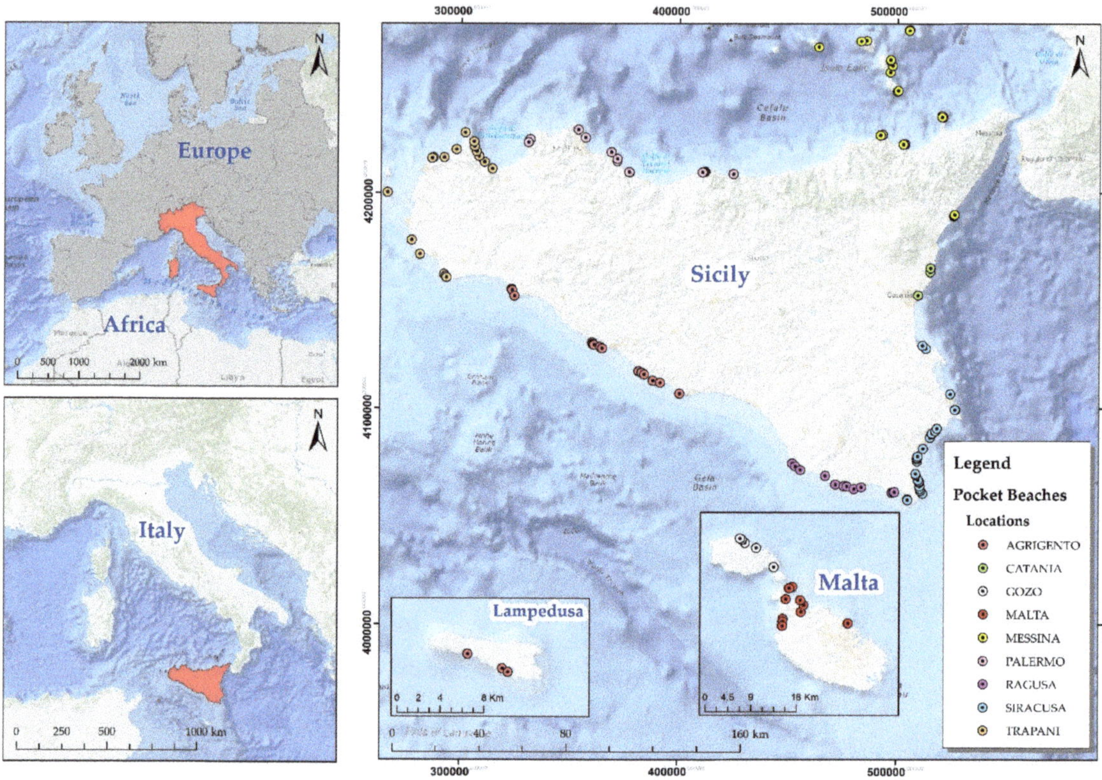

Figure 1. Geographical framework of the study area (map adapted from ESRI, GEBCO, NOAA, National Geographic, Garmin, HERE, Geonames.org, and other contributors).

Despite this, while Sicily shows a great geodiversity, due to the geologically complex history dating from the Paleozoic to Pleistocene periods, involving metamorphic, magmatic, and sedimentary outcrops, the geology of the Maltese Islands are predominantly composed of marine sedimentary rocks, mainly limestone with subsidiary marls and clays [49,50], deposited in shallow marine conditions between the late Oligocene and Miocene age [49], with sporadic occurrences of quaternary deposits in some areas (Figure 2).

Sicily is the second largest island in the Mediterranean Sea, with a 1623 km coastal length, divided into 26% of rocky coasts and 74% of sandy and/or pebbly beaches [51,52]; among them 110 PBs consist of both natural and artificial features, having different dimensions and various planform geometries, differently exposed to incident wave energy and with limited sediment sourcing by drainage basins.

The Maltese Archipelago consists of the three main islands of Malta, Gozo, and Comino and a number of other minor islands and rocks [53,54]. The entire coastline measures 272 km [55], with the predominant shore type being rocky (90.5%); sand and shingle shores only comprise 2.4% of the remaining coastline [56]. The coastal morphology of the Maltese Islands has been largely determined by tectonic activity, primarily in the Holocene period, which has been uninterrupted to present day [57].

The PBs of the Maltese islands (22 PBs) are exclusively small (the largest, Ghadira, is only 1 km long), flanked between rocky headlands or anthropogenic infrastructures, and there is limited sediment exchange between beaches located at various spatial scales [58,59]

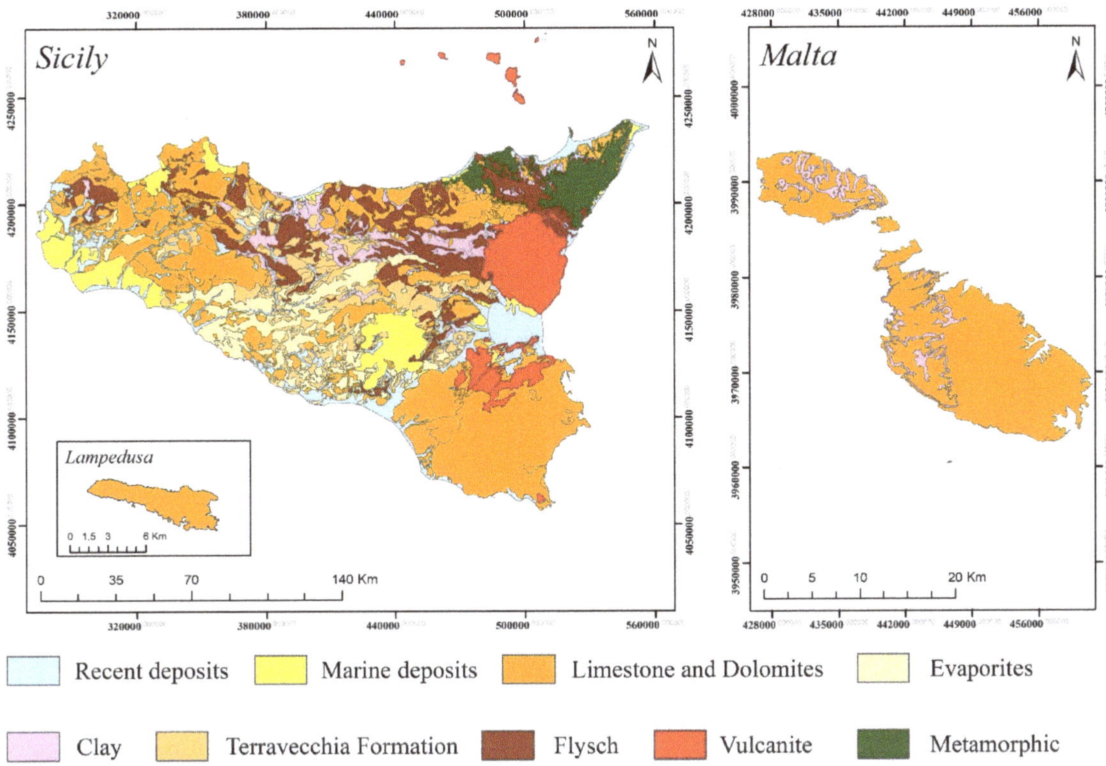

Figure 2. Lithological framework of Sicily and the Maltese Archipelago. (Revised from Geologic Map of Sicily—Lentini and Carbone, 2014).

2.2. Dataset Preparation and Data Description

The data loaded into the BESS GIS are diverse (Table A1) and concern all the parameters that allow a detailed geo-characterization of the PBs present in Sicily, Malta, Gozo, and Comino. A synthesis of these data fed the WebGIS, opening the project results to a wide stakeholder audience.

Two flight surveys were conducted using different DJI (Mavic 2, 210 RTK, 600 Pro) drones during the Spring–Summer season of 2019 and Autumn–Winter season of 2019/2020. These types of drones were chosen due to their size, flight time, flight safety, stability, and the payload of sensors they can carry.

DJI Mavic 2 is a small UAV equipped with an excellent Hasselblad camera, with a one-inch sensor. It has a collapsible frame and two powerful landing lights, plus collision sensors on all sides. This drone combines the high technology of stability, flight safety, and miniaturization of electronics. DJI 210 RTK is a large UAV equipped with a flight terminator and parachute. It has upper, lower, and front anti-collision sensors and a sophisticated system (DJI Airsense) for signaling proximity to traditional aircrafts. The drone is waterproof and is equipped with an RTK (real-time kinematic) system. DJI 600 Pro is a large UAV equipped with a flight terminator and parachute. It has 3 GPS and 3 inertial attitude and altitude systems (IMU). It is powered by 6 batteries, can carry a maximum of 5 kg payload, and has a flight range of about 30 min without any payload.

The acquired images were processed with Pix4D mapper software to build very-high-resolution orthophotos of 1.6 cm spatial resolution. In addition, satellite images and historical maps were also used. Based on these drone orthophotos, satellite images, historical maps, data related to geomorphological-sedimentological, geometric parameters, lithology, bathymetry, land-use/land-cover mapping of Sicilian and Maltese PBs were extracted.

The drone orthophotos were geo-referenced through on-ground placement of markers whose locations were acquired using GPS RTK (GNSS TOPCON Hiper HR composed of base + rover) [60,61]. Following Bowman et al. [42], Storlazzi and Field [62], and Narcross et al. [63], the orthophotos were also used to define the geometry and morphological characteristics of PBs.

The next step, considering the various ground control points (GCPs), was to build the digital surface models (DSMs) of the PBs and then derive the slopes expressed as a percentage with linear interpolation (to express the details at a better resolution), finally classified into 4 categories (0–5–10–15%). Beaches that were difficult to reach, such as nature reserves and private beaches or those too close to airports or military installations, were analyzed using both free (Sentinel-2 with 10 m pixel spatial resolution) and commercial (WorldView-2, 50 cm pixel spatial resolution) satellite images.

The satellite images were also used to derive the bathymetry [64–66], processed by the ENVI software and then reclassified by QGIS version 3.4 (Open Source Geospatial Foundation, Chicago, USA). The reclassification was necessary to differentiate each beach because of different degrees of resolution, depending on multiple factors, such as presence of *Posidonia oceanica*, clouds, sun, and quality of previous bathymetric data. Using this bathymetry, submerged beach closure depth was calculated, following Lisi et al. [67], and the mapping of the *Posidonia oceanica* was performed for the whole number of PBs following Tomasello et al. [68]; Ventura et al. [69] and Rende et al. [70].

Beach sediment samples were collected along 3 shore-perpendicular transects on beaches with a shoreline length > 300 m (and along 1 transect for < 300 m). During the first field survey, sediments at backshore, shoreline, and −1 m were collected; this was restricted to a shoreline sample in a second survey. The samples were then transported to the laboratory and subjected to particle size analysis.

Land use/land cover [71] and lithology were surveyed from different sources (maps, drones, and satellite); a buffer of 0.5 km for land use and 1 km for lithology was used.

An additional project output involved the establishment of a Wi-Fi-connected remote surveillance network of ten sites, with a central control room feeding a specific sector of the GIS. Each site was equipped with one or more cameras and with an anemometer powered by solar panels. In three of these sites, the Italian National Institute of Geophysics and Volcanology (INGV) placed a monitoring system composed of accelerometric/velocimetric devices aimed at detecting the microseismicity induced by sea waves in three selected PBs.

The data ordered within the GIS were uploaded within the WebGIS, dividing them into two macro areas: raster and vector file format. Raster images are those with variable resolution, depending on the object they show and the degree of detail they intend to represent, including orthophotos, satellite images, and bathymetry. Vectors are instead points, lines, and polygons, imported by GIS, which contain an attribute table with all the element details within the shapefile—sediment samples, in situ photo collections, remote surveillance systems (points), geometric and morphological parameters (points, lines, and polygons) and lithology, land use, *Posidonia oceanica* (polygons). The coordinate system chosen was WGS84 33N (EPSG: 32633). The data uploaded to WebGIS amounted to about 200 GB.

2.3. Methodology

The WebGIS platform allows the display of geospatial data on the website with Open Geospatial Consortium (OGC) protocols such Web Map Service (WMS), Web Feature Service (WFS) and Web Feature Service with transactions (WFS-T), and Web Coverage Service (WCS) and allows displays on a web page both from PC and smartphone. The creation of the maps of the layer components, the organization, and the preparation of the data was carried out through the QGIS Desktop program version 3.4 and, subsequently, following various operations including the management of the scales, basic levels, metadata, preparation for QGIS server, and additional function settings, these were displayed within the Lizmap web client application (Figure 3). The various geographic data, once obtained and stored in GIS, were published and transformed into geo-services into WebGIS. A geo-service allows the consultation, the processing, and the return of geographic data through the internet. OGC defines the common aspects of the OpenGIS Web Service (OWS) and provides a software interface through which other applications' clients can access and use geospatial data located remotely. This communication is based on extensible markup language (XML) through the HTTP communication protocol; so that the service that is made available is independent to the platform and operating system. The publication of geo-services can take place in two different ways:

- WebGIS: publication of geographic information data through an interactive web page that does not require the use of specific GIS software for their use.
- Geo web service: a particular architecture that allows the return of geographic data located remotely through specific protocols, following a request made either through a simple web browser or through specific client applications.

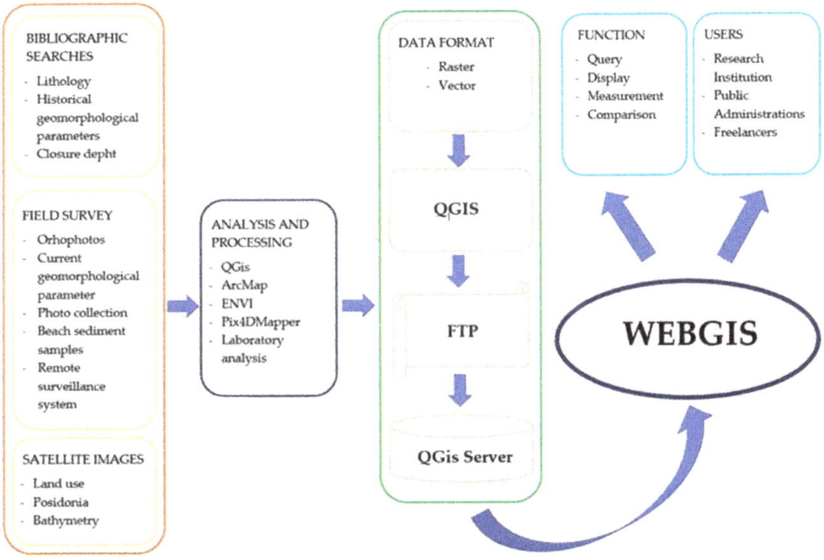

Figure 3. Chart flow methodology of BESS WebGIS implementation.

Both these approaches were used for the BESS project to allow data visualization both remotely (GIS desktop) and via WebGIS. The OWS services used for the usability of the data were WMS, for raster data, and WFS, for vector data. The database was made usable as a geo-service through the QGIS Server application and was subsequently made accessible online through the Lizmap client interface. This application was chosen because it uses the same libraries as the QGIS Desktop application and, therefore, it allows maps with complex graphical representations to be published with the same characteristics as the Desktop application, while keeping all the parameters defined in QGIS desktop unchanged. The main advantage of QGIS Server is undoubtedly the integration with QGIS Desktop itself; it also does not require specific skills in the field of publishing web services and significantly shortens processing times. Once QGIS Server has created the standard geo-services (WMS/WFS), the Lizmap client interface can publish data online. This interface was chosen as it can be configured from a plugin (Lizmap) within QGIS Desktop and does not require special knowledge of programming languages. While in the present study, a dataset with data of different spatial resolutions was used, the BESS WebGIS allows the visualization of information at the scale of 1:2257. This is in accordance with the Open

Street Map basemap included. However, this scale can be improved and reach 1:1128 by disactivating the underlying basemap.

3. Results

3.1. WebGIS Visualization and Mapping

By accessing the WebGIS platform via the following link (http://51.38.247.246/mylizmap/lizmap/www/index.php/view/map/?repository=bess&project=bess (accessed on 8 February 2021)), a screen will appear where it will be possible to manage the functions of the service (Figure 4); the left part represents the toolbar and the various layers present, while the central and right part is dedicated to navigation. While the default language is in Italian, a user can choose the name of the various layers and then put it in any convenient language.

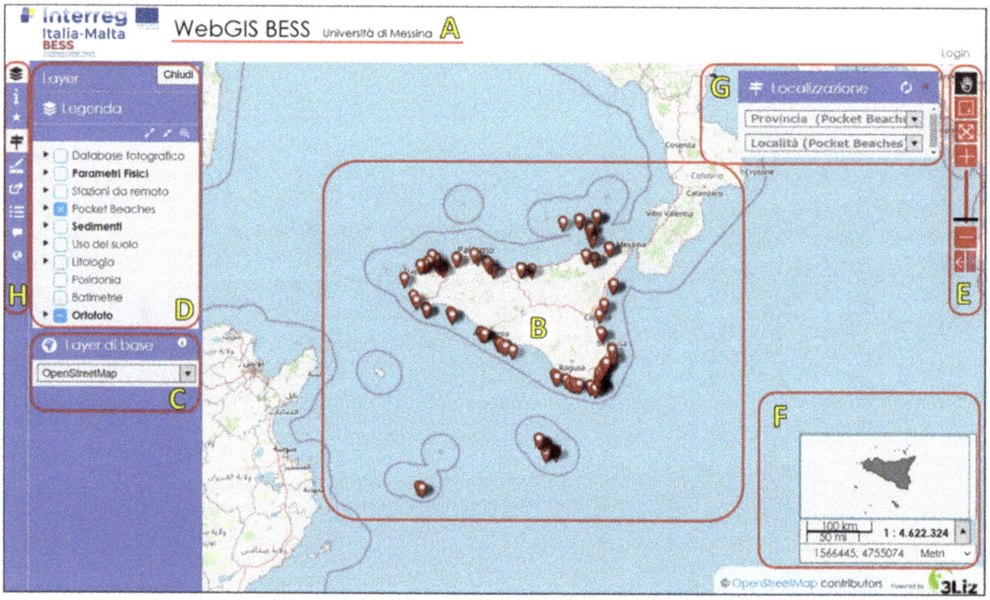

Figure 4. WebGIS home screen with yellow letters indicating the main default functions.

The defaults functions and features indicated in the WebGIS home screen (Figure 5) are illustrated below:

A. Title of the web window reflecting the container of all the information relating to the project.

B. Initial map with the default scale set

C. Tool that allows for activating or deactivating of a basemap layer from the map background. In this study, OpenStreetMap was used due to the amount of information it contains.

D. The legend lists the various data layers (raster and vector) in order of overlap, inserted within the map; both single and multi-layers may be accessed. Some layers can be grouped according to the composition of the project. By clicking on the small triangle ▸ on the left of each layer, the relative symbol will appear, opening the data related to the information layer. By selecting a specific layer, it will be possible to obtain relevant information, like its spatial extension, or to modify, with various levels, its degree of opacity. Once any layer is activated, it will appear in the map according to a presentation style previously created in QGIS. By clicking with the left mouse button on any vector element, a popup window will appear on the right containing the information of the objects positioned in that particular point (Figure 5).

E. This tool allows navigation of the map through various buttons:

(1). Pan—the horizontal shift (the hand sign symbol) allows movement around the map by clicking and holding the left mouse button. It is also possible to use the mouse wheel, both forwards and backwards, thus activating the modification of the scale, keeping the mouse cursor in the center of the map;

(2). Zoom window (red plus symbol), which allows a specific zoom by dragging the pointer in order to draw a rectangle, which then defines the area of interest to be scaled;

(3). Zoom to initial map extension's symbol allows zoom of the starting map;

(4). Zoom in—zoom out's symbol allows selection of the zoom level through a scale bar;

(5). Previous and next zoom: these two indicators allow one to scroll through the zoom history.

F. The panel displaying the map overview uses a blue square highlight to indicate the current surface shown on the map. Furthermore, a scale bar and coordinates of the cursor point position based on the reference system set are also shown inside this panel, in this case UTM WGS84 33N. It is possible to change the map view by moving the blue square on the overview. In the lower right corner, there is also the exact geographical position. Finally, a small drop-down menu also allows choosing the unit of measurement (meters/degrees/degrees and minutes/degrees, minutes, and seconds).

G. The location tool allows a specific position to be reached, based on any categorized item of the project, in this case, an individual PB under study. These actions are initially carried out by identifying the area in which the various PBs are grouped, for example, the Sicilian Provinces (Messina, Palermo, Catania, Syracuse, Ragusa, Agrigento, and Trapani), Malta, Gozo, and Comino. Once this parameter has been identified, the beaches discriminated for that area will appear in the second window; by selecting the latter, it will appear on the map to the same extent as the selected geometric object (Figure 6). The activated layer corresponds to the orthophoto.

H. Toolbar

1. Layer symbol activates or deactivates the layer management panel and its legend (D)

2. Information allows obtaining information relating to the map description, set properties, contacts of the person responsible for the published data, and other features.

3. Star tool allows selecting and filtering the geometries of a single layer into the map using various tools, and, subsequently, displays in the attributes table.

4. Localization symbol activates or deactivates the localization tool (G).

5. By clicking Ruler tool it is possible to choose the type of the activated measure (length, area, and perimeter).
6. Permalink allows users to obtain an external link that will allow them to save the map in the exact extension, projection, and scale of that moment.
7. Table attribute symbol activates the attribute tables of the vector layers present in the project, with the possibility to sort and select the fields, as well as to filter them in order to make comparisons.
8. Popup (message) symbol represents the pop-up window for the information of the selected data.
9. Atlas (globe) symbol represents the atlas, or a sequential display of a given layer. In this case the single PBs layer was chosen for display. It is also possible to automatically reproduce the sequence of the beaches using the "play" button.

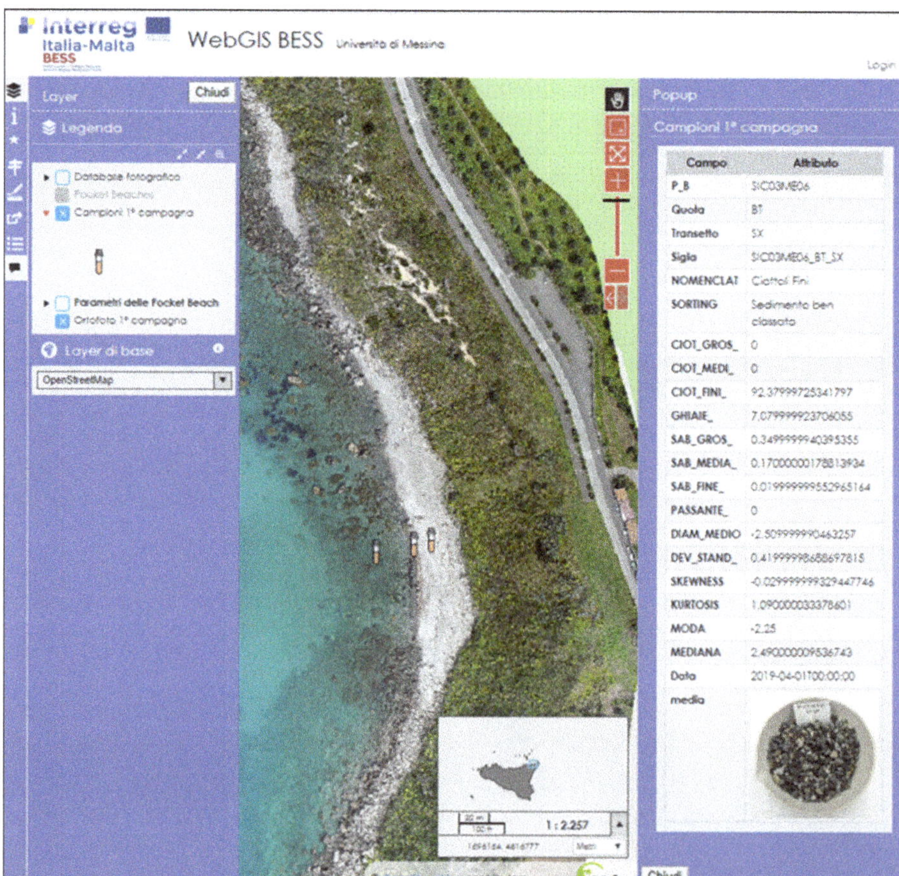

Figure 5. Display of the platform with the opening of a popup relative to the grain size information about the shoreline sample of Capo Milazzo's PB.

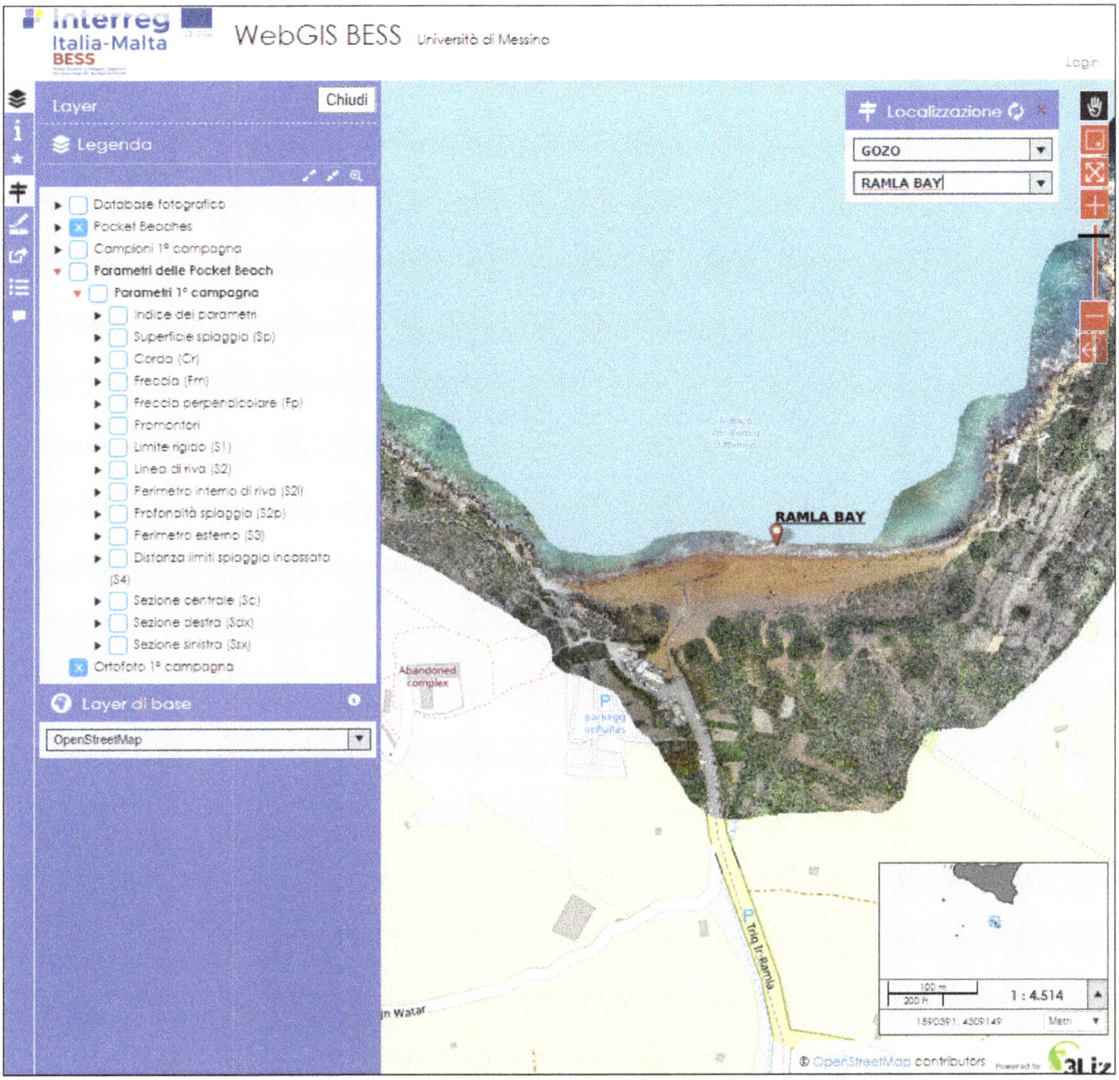

Figure 6. Visualization on the WebGIS platform of the PB of Ramla Bay (Gozo).

Starting from these instruments WebGIS can be used to investigate the different layers of the project, crossing queries and comparing information. This platform offers the possibility of turning on and off layers of different or the same areas, as well as of different time intervals; this, in both cases, shows, in detail, the characteristics of the PBs and their evolution over time.

3.2. Examples of Comparison

Regarding the orthophotos, there is a possibility to view and overlap them in order, to visually notice the differences, in particular by playing on the opacity of one of the two to compare them more clearly. To make the comparison between orthophotos more effective, it is also possible to activate the layers relating to the respective shorelines (Figure 7). For example, certain geomorphological differences have also been digitized and shown within the layers concerning the geomorphological parameters, which can be interrogated to view their characteristics.

(a) (b)

Figure 7. The figure shows a comparison of the pocket beach of Rais Gerbi (Palermo) in two different time intervals through the management of opacity: (**a**) Spring-Summer Season; (**b**) overlap of Autumn-Winter season on Spring-Summer season. The respective shorelines are highlighted in orange.

Among the various combination that can be made by switching different layers on or off, it may be useful to combine lithology, samples and photo collections. Thanks to this overlap, it is possible to query the map in order to notice an affinities or differences at the mesoscale. The following example (Figure 8) shows how, by clicking on the photographic icon, the popups relating to both the latter and the underlying lithology layer will open simultaneously. In Figure 9, on the other hand, by clicking on the nearby shoreline sample, the popup will show its characteristics.

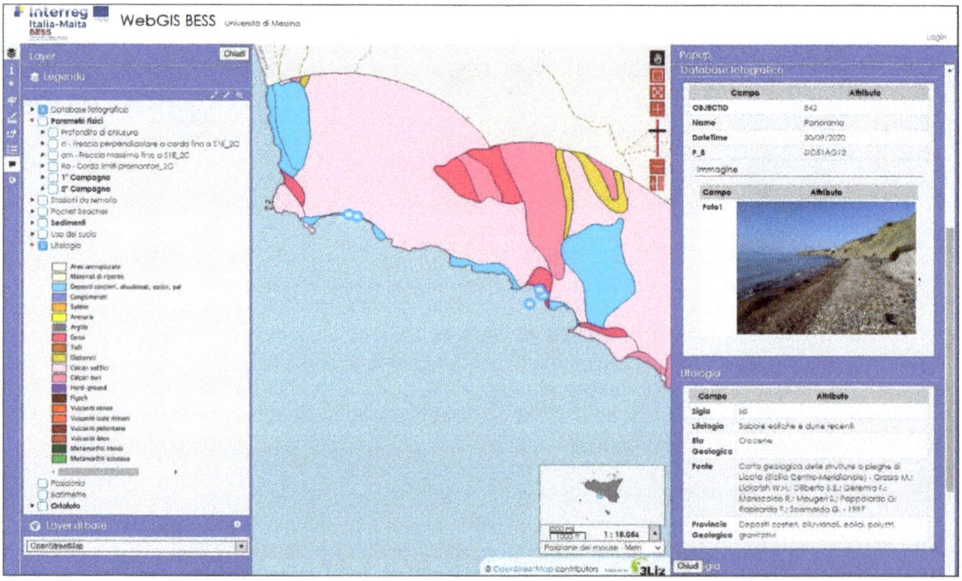

Figure 8. WebGIS platform that shows the overlap between samples, lithology, and photo collection of the pocket beach of Palma di Montechiaro (Agrigento). On the right, the popup between the last two obtained by clicking on the photo collection.

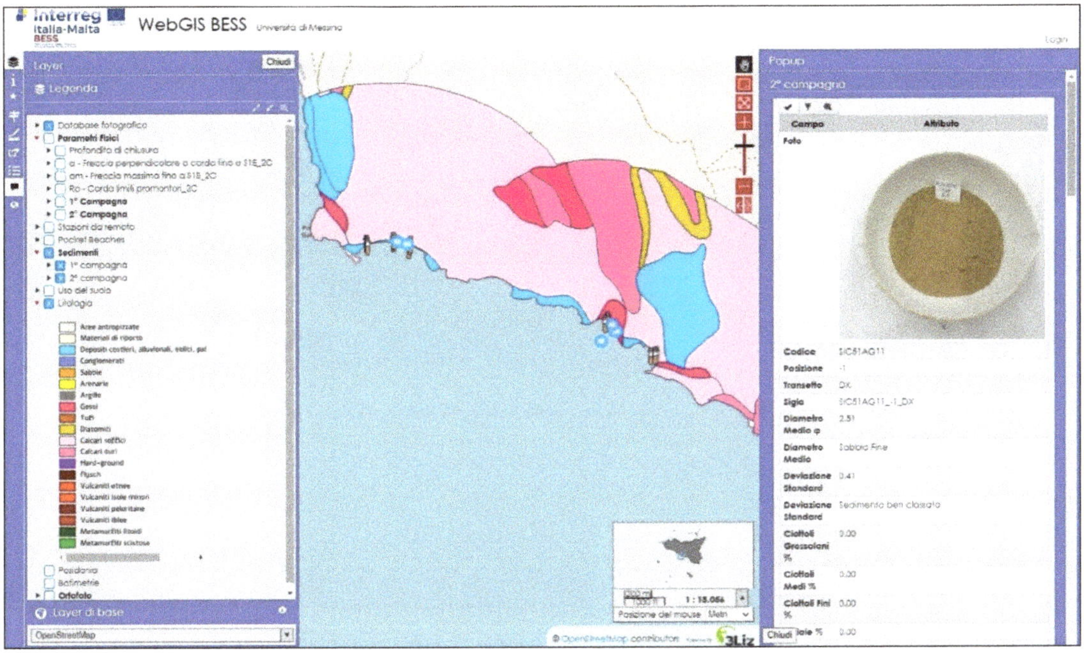

Figure 9. WebGIS platform that shows the overlap between samples, lithology, and photo collection of the pocket beach of Palma di Montechiaro (Agrigento). On the right, the pop-up of the shoreline sample obtained by clicking on the photo collection.

It is evident, through these simple steps, how it is possible to notice the similarities between the lithological characteristics of the outcrop, in this case metamorphic, and the respective photographs of the study area and the nearby beach sediments.

Another comparison can be made with regard to the sedimentological analysis, in order to learn the differences between the grain size and other statistical parameters (Figure 10). To compare samples related to the same vector layer, the selection tool should be used to click the geometries to study. By clicking the filter button, the attribute table of the selected geometries may be viewed.

It may be useful to compare the surfaces of the beaches of successive surveys. Figure 11 shows how, by activating the respective layers and clicking on them in the map, the user can view both the visual and area differences by observing the popup.

One of the layers present in the WebGIS consists of the mapping of *Posidonia oceanica* proximal to the PBs; through this platform, it is possible to observe its distribution according to the various bathymetric zones. In Figure 12, for example, it emerged that *Posidonia oceanica* develops mainly between 5 and 20 m deep.

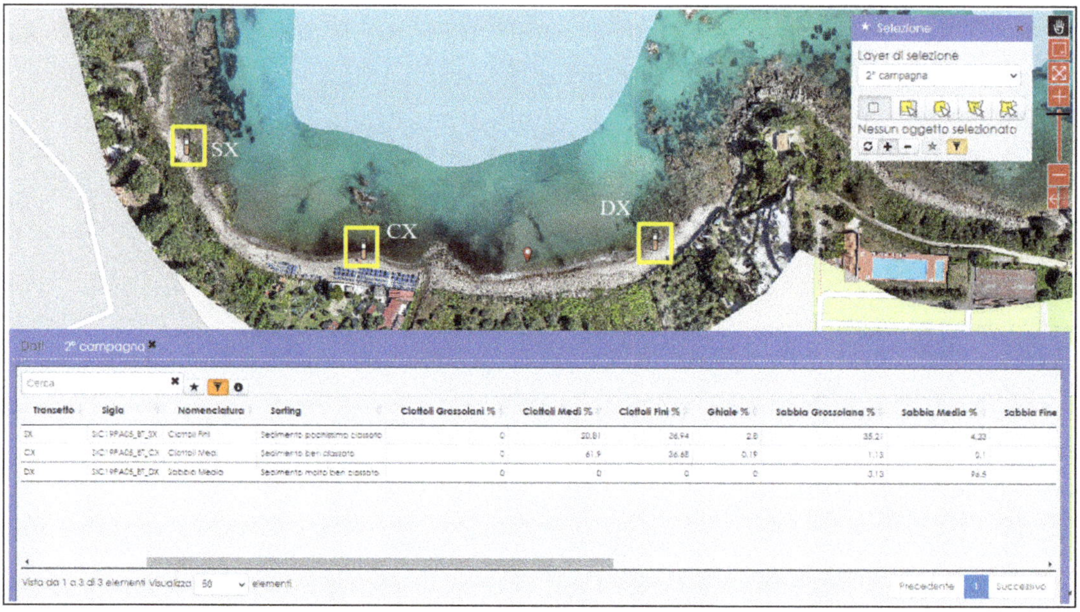

Figure 10. Example of comparison between beach sediments for three different transepts taken along the PB of Cefalù (Palermo).

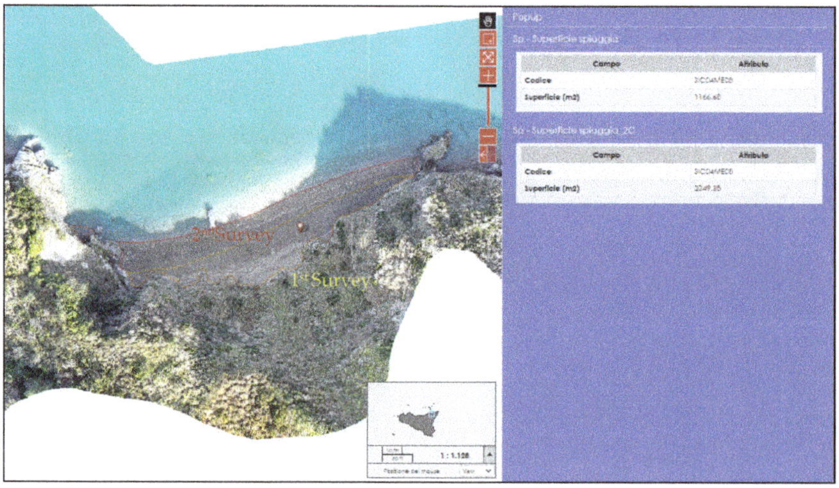

Figure 11. Example of comparison between beach surfaces of two successive surveys made in the PB of Tindari (Messina).

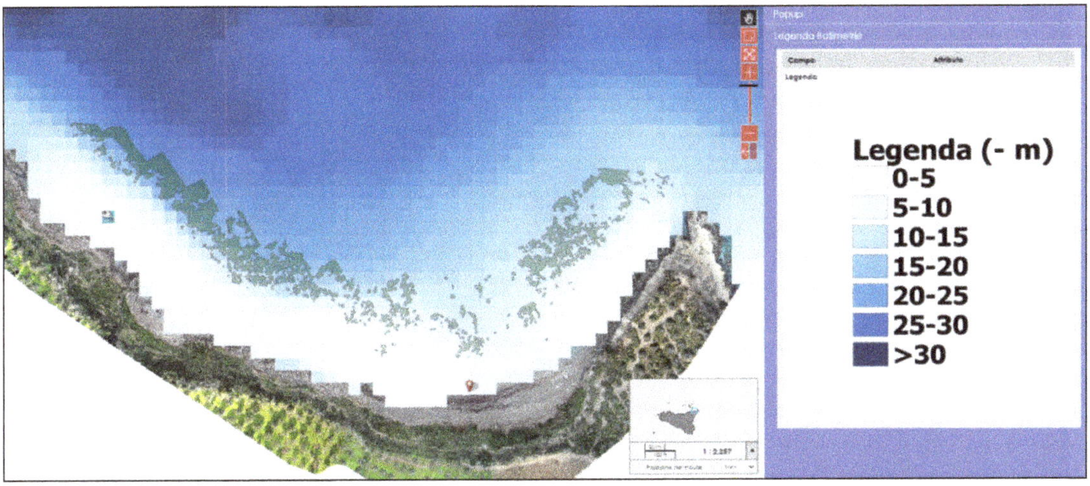

Figure 12. Distribution of *Posidonia oceanica* according to the various bathymetric zones on the PB of Capo Milazzo (Messina).

4. Discussion

While many of the beaches in the Maltese islands are subject to some form of management of varying degrees, (the most established being primarily based on management indicated in the achievement of the Blue Flag beach quality status), there is still a need for an overall holistic management policy and strategy. In Sicily, there is a lack of coastal planning, where coastal management has been focused on local or sectorial erosional rates, which had previously caused severe damage to the natural and archaeological cultural heritage [52]. A very large amount of public funds has been expended on coastal protection without prior consideration of coastal planning to better understand the needs of each individual beach and the interaction of the specific interventions with natural system processes.

The importance of developing (within the BESS project) a monitoring system supported by a surveillance network that specifically targets pocket beaches, is supported by a widespread concern that climate change may be disrupting the stability of otherwise natural systems [72]. Pocket beaches have been defined as ones that have, in their evolution, achieved a sediment balance or a state of equilibrium; they have also been referred to as 'sediment tight' systems. However, with an increasing influence of climate change on natural processes that influence either the arrival of sediments to a pocket beach (such as that via precipitation influenced watersheds) or the potential loss of sediments (such as the wave climate influencing a pocket beach embayment), it is paramount that such potential changes are clearly identified and monitored.

Considering the multiplicity of parameters involved in the study of PBs, and considering their small size, these micro-beaches can be considered as a sentinel system in which, changes in the shape or grain size of the deposit or in the composition of the fauna and flora, could give an indication of the global trend imposed by the effects of climate change. A geospatial database is a fundamental key for the creation of the GIS of the BESS project. All geospatial information acquired at different locations highlighted naturalistic and ecological details of particular interest. Then, they allowed modelling of the beach evolution and the definition of hazard levels in each PB. Such a work on a specific aspect of the much wider coast is clearly highly useful in contributing to a more holistic coastal zone management approach. The relevance of this work is also borne out by the acceptance by the E.U.'s Interreg Italia–Malta funding mechanism to support this initiative aimed at developing a WebGIS for mapping and visualization of coastal data.

In fact, the GIS plays an important and useful role in managing a large amount of parametric data describing PBs. It is, thus, necessary to create a comprehensive data management system that allows cross-referencing of information related to different fields interacting in the driving of coastal dynamics processes. WebGIS is an online-distributed system that allows users to access geographic information data by processing services available through a web browser; it has a strong interactivity and dynamics. Compared with traditional GIS software (or online GIS mapping), WebGIS has a wider application range, more timely data updates, lower construction costs, and higher security. Many large internet companies, such as Google, Baidu, and Tencent, recently developed online maps that are based on WebGIS technology for public use [10]. The work described in this paper illustrates the strong potential of such a WebGIS in the manipulation and presentation of a wide set of territorial and marine datasets that may be utilized for more effective understanding and management purposes. Additionally, the WebGIS developed by the BESS project is a particularly useful tool to allow a wider user audience to better appreciate pocket beaches not just visually but also in terms of their functioning and overall complexity; the WebGIS system presents technical data in a manner that will, however, allow both expert and non-expert use, facilitating different levels of manipulation and understanding. The raising of such awareness in the general public has been shown to initiate a process that transforms their initial curiosity into a role of controller and stimulus/driver. On a wider scope, not limited to the PBs, the utility of such a system as WebGIS is much greater.

The lack of interest that is often demonstrated on public items may be countered by increasing interaction with such amenities. The BESS project describes 134 PBs as a territorial database, the evolution of which will hence forth become more visible through the developed website and, in particular, a remote surveillance system; this will allow public monitoring of the evolution of these systems, not only in the territorial sense, but also in terms of the actions taken by local politicians and stakeholders.

The results demonstrate that through the development of the BESS WebGIS, (dedicated only to PBs), it is possible to clearly describe all the geometric and morphological characteristics supported by geological, sedimentological, and bathymetric data. This simple and immediate online geo-database allows a variety of users to access the platform to view and draw as much information as possible. Unlike Desktop GIS, the use of which is limited to a personal computer that contains all the data of a specific project, a WebGIS is accessible remotely via a server, with no restrictions on the place or number of users who access it. WebGIS has the advantage of combining the potential of GIS with the usefulness of the Internet in showing interactive maps and spatial data, as well as any subsequent updates; this clearly makes WebGIS the cheapest and easiest way to distribute geospatial data. The results highlighted how simple and immediate it can be to view the results of related field campaigns once uploaded to a WebGIS platform.

The numerous examples listed demonstrate the effectiveness of WebGIS in comparing the various levels present, which helps to better understand its scientific value and to encourage research in this field. While the knowledge of dynamic evolution of a given layer describes the temporal trend, it is when different types of data are crossed that the potential of the WebGIS grows and multiplies the research ideas. While the BESS WebGIS is undoubtedly useful for those involved in research in this field, its usefulness can be extended to tourism, civil protection, and land-use purposes. Additionally, the BESS WebGIS portal allows:

(1) The creation of a management model extended to a sufficiently large area.
(2) Ensuring that planning is functional to the defense of the natural asset and its conscious and sustainable use.
(3) Making the model implementable continuously, wherever possible involving local administrations, and available to public and private users, using state of the art technologies.

Through the developed WebGIS it will be possible to create a management system for these specific territorial niches, based on an active monitoring plan, at low cost, and with a high technological component. This monitoring–management model is in line with the scenarios of future management plans, which will provide for a detailed mapping of the environments concerned, so that protection systems can be created, not necessarily those that are structural, but rather systems based on good continuous management practice exploiting the intrinsic resilience of the system rather than weakening it. The challenge lies precisely in identifying the coastal management systems that make it possible to enhance the intrinsic characteristics of resilience of the beaches. Such systems will allow users to counter the effects of the changes that occur with the sea level rise, as well as with the worsening of marine weather conditions.

GIS has been improving rapidly over the years, but the use and technology development of the Internet still remain the key factors in the development of WebGIS. Only in this way, through a complete involvement of users, will it be possible to aim for an important information and awareness campaign linked to PBs. This will allow for the PBs, ecological pearls, to be monitored and preserved.

5. Conclusions

A WebGIS portal including various pocket-beach-related data and models in different temporal and spatial scales was created to present the results of the Interreg project, BESS, to a wider group of people. It is an effective way to share information about Sicilian and Maltese PBs, acquired by the project, and to compare coastal data with each other. The development process suggested that the ease of use, combined with a negligible realization cost, provides an opportunity for replicability and scalability in other geographical and administrative contexts, and also for different purposes. It also proposes that the barriers that limit end-user targets may be addressed though this mechanism, allowing a much wider audience to interact with geographic information data, requiring only an internet service for access; in this manner, not only experts but also those with less informatics skills and less high-performing computers, such as the general public, may interact with this database. The current BESS WebGIS was presented not as a point of arrival, but as a starting point: a platform that must be implemented for the overall management of all coastal data necessary for constant active monitoring. This project suggests that data management will be useful, above all, in predictive terms of the coastal morphological trend, as an effective tool that will support future planning, no longer local, but increasingly on a regional scale. To assist in addressing the project's development toward these directions, a video surveillance system was created, allowing a network of mayors and stakeholders involved with pocket beaches in their territories to become more directly involved with the project and its extensive data. Bringing together various public and private entities in the same direction increases the possibility of giving greater emphasis to coastal problems and identifying solutions. The portal and its further development at a regional level should improve beach-user choices by providing useful data, such as traffic conditions, services, and the quality of each location. The portal should also facilitate the process of initiating Integrated Coastal Zone Management at the local level. This is particularly so in Sicily, where the results of the project and the structure and methodology followed in the BESS project has already been integrated within the Regional Plan Against the Coastal Erosion, which is intended to become the official instrument to manage the coastal area in Sicily.

In addition, various entities will be given the opportunity to actively participate in a process of continuous study and monitoring. End users will have the possibility to view a familiar basemap namely OpenStreetMap, to help them explore the portal by following the specific guidelines. Local public administrations will be able to see the effects of changes in the coastline and, where necessary, plan interventions aimed at the conservation of the coast, which represents a heritage of great importance from a naturalistic and tourist point of view. Probably, defining GIS is a more complex challenge than naming it because the field itself is continually evolving and morphing. It is important to note that the results of the BESS project have been used for implementing the Regional Plan Against the Coastal Erosion (PRCEC with the Italian acronym) and they also contributed to scientific discussions about coastal area management as well.

Author Contributions: Conceptualization, G.R., A.M. (Anselme Muzirafuti) and S.L.; methodology, M.C., F.G., A.M. (Anselme Muzirafuti), and M.F.; software, M.C., F.G., A.M. (Anselme Muzirafuti) and M.F.; validation, A.M. (Anton Micallef), S.D., A.C., A.Z., O.S., F.I., A.T. and F.C.; formal analysis, A.M. (Anton Micallef), G.R., A.M. (Anselme Muzirafuti) and S.L.; investigation, M.C., F.G., E.C., A.M. (Anselme Muzirafuti) and M.F.; resources, F.G. and M.F.; data curation, M.C., F.G., A.M. (Anselme Muzirafuti) and M.F.; writing—original draft preparation, G.R., A.M. (Anselme Muzirafuti), F.G. and M.F.; writing—review and editing, G.R., A.M. (Anton Micallef) and A.M. (Anselme Muzirafuti); visualization, M.C., S.L. and F.P.C.; supervision, G.R.; project administration, A.Z.; funding acquisition, A.Z. All authors have read and agreed to the published version of the manuscript.

Funding: A.Z. is supported by the European Regional Development Fund (INTERREG Italia–Malta) for the Project BESS: Pocket Beach Management and Remote Surveillance System.

Institutional Review Board Statement: Not applicable.

Data Availability Statement: "BESS project website" at http://bess.pa.ingv.it/?lang=it, accessed on 5 August 2021.

Acknowledgments: This work was partially funded by the European Regional Development Fund (INTERREG Italia–Malta) under the Project BESS: Pocket Beach Management and Remote Surveillance System.

Conflicts of Interest: The authors declare no conflict of interest. The funders had no role in the design of the study; in the collection, analyses, or interpretation of data; in the writing of the manuscript; or in the decision to publish the results.

Appendix A

Table A1. Summary table of the layers/data present in the GIS.

Layer	Description
Orthophotos and satellite images	• Acquisition of orthophotos through APR. (two field surveys: Spring–Summer 2019, Autumn–Winter 2019/2020) • Sentinel-2 and Worldview-2 satellite images (Spring–Summer 2019)
Beach sediment samples	Sedimentological parameters (Folk and Ward [73]).
Geometric and geomorphological parameters	• Ro—length of the control line/headland spacing/entrance of embayment • a—bay indentation ("depth" of the bay) • am—bay indentation max (m) • S1—Length of embayed shoreline (m) • S1E—External length of embayed shoreline (m) • S2—Length of embayed beach (m) • S2e—Shoreline internal perimeter • S2p—Maximum beach width • S3—linear distance between the edges of the beach • Sp—Beach surface • CD—closure depth
Lithology	Within a 1 km buffer from the shoreline
Land use	Extracted from Sentinel-2 images for 500 m surface from the shoreline.
Bathymetry	By satellite images (the degree of precision is related to the quality images)
Posidonia oceanica	Mapping of the shallow water *Posidonia oceanica*, extracted from the clearest orthophotos
In situ photo collection	Pictures of the beaches from different point of view
Remote surveillance system	Two types of monitoring systems: photographic/climatic stations and stations with accelerometer and velocimetric devices

References

1. Caiaffa, E. *Un Percorso Attraverso Concetti e Nozioni Fondamentali Per Addentrarsi Nel Vasto Mondo Della Scienza Della Informazione Geografica*; ENEA: Roma, Italy, 2006; p. 1922.
2. Bhat, M.A.; Shah, R.M.; Ahmad, B. Cloud computing: A solution to geographical information systems (GIS). *Inter. J. Comp. Sci. Eng.* **2011**, *3*, 594–600.
3. Dragicevic, S. The potential of Web-based GIS. *J. Geogr. Syst.* **2004**, *6*, 79–81. [CrossRef]
4. Barricelli, B.R.; Maiellaro, N.; Padula, M.; Scala, P.L. Dal GIS al WebGIS in Ambiente Open Source. *L'edilizia* **2009**, *23*, 94–97.

5. De Castro, A.F.; Amaro, V.E.; Grigio, A.M.; Cavalcante, R.G. Modeling and development of a WebGis for environmental monitoring of coastal areas that are influenced by the oil industry. *J. Coast. Res.* **2011**, *64*, 1643–1647.
6. Tang, W.; Selwood, J. *Connecting Our World: GIS Web Services*; ESRI Press: Hong Kong, 2003.
7. Opdam, P. Learning science from practice. *Landsc. Ecol.* **2010**, *25*, 821–823. [CrossRef]
8. Veenendaal, B.; Brovelli, M.A.; Li, S. Review of Web Mapping: Eras, Trends and Directions. *ISPRS Int. J. Geo-Inf.* **2017**, *6*, 317. [CrossRef]
9. Lopes, S.I.; Moreira, P.M.; Cruz, A.M.; Martins, P.; Pereira, F.; Curado, A. RnMonitor: A WebGIS-based platform for expedite in situ deployment of IoT edge devices and effective Radon Risk Management. In Proceedings of the 2019 IEEE International Smart Cities Conference (ISC2), Casablanca, Morocco, 14–17 October 2019; pp. 451–457.
10. Pan, S.; Wang, K.; Wang, L.; Wang, Z.; Han, Y. Risk Assessment System Based on WebGIS for Heavy Metal Pollution in Farmland Soils in China. *Sustainability* **2017**, *9*, 1846. [CrossRef]
11. Howell, R.G.; Petersen, S.L.; Balzotti, C.S.; Rogers, P.C.; Jackson, M.W.; Hedrich, A.E. Using WebGIS to Develop a Spatial Bibliography for Organizing, Mapping, and Disseminating Research Information: A Case Study of Quaking Aspen. *Angelands* **2019**, *41*, 244–247. [CrossRef]
12. Seng, B.; Liang, H.; Zhao, Y.; Tang, Y. Design and Implementation of Visualization System for Wastewater Treatment in Dianchi Lake Based on WebGIS. *E3S Web Conf.* **2019**, *118*, 04032. [CrossRef]
13. Luo, K.; Zhang, T.; Liu, Y. Design and Implementation of the Water Environment Monitoring and Management System Based on WebGIS for the Four-Lake Basin of Hubei Province. *E3S Web Conf.* **2019**, *136*, 06036. [CrossRef]
14. Villani, G.; Nanni, S.; Tomei, F.; Pasetti, S.; Mangiaracina, R.; Agnetti, A.; Leoni, P.; Folegani, M.; Mazzini, G.; Botarelli, L.; et al. The RainBO Platform for Enhancing Urban Resilience to Floods: An Efficient Tool for Planning and Emergency Phases. *Climate* **2019**, *7*, 145. [CrossRef]
15. Santoso, M.I.; Gumelar, R.G.; Irawan, B. Development of the WebGIS application for transport infrastructure management in the city of Serang. *IOP Conf. Ser. Mater. Sci. Eng.* **2019**, *673*, 012072. [CrossRef]
16. Stanković, R.; Vulović, N.; Lilić, N.; Obradović, I.; Tošović, R.; Pešić-Georgiadis, M. A WebGIS Decision Support System for Management of Abandoned Mines. *Energies* **2016**, *9*, 567. [CrossRef]
17. Wirkus, L. An Open Source WebGIS Application for Civic Education on Peace and Conflict. *ISPRS Int. J. Geo-Inf.* **2015**, *4*, 1013–1032. [CrossRef]
18. Ahmed, B.; Rahman, M.S.; Islam, R.; Sammonds, P.; Zhou, C.; Uddin, K.; Al-Hussaini, T.M. Developing a Dynamic Web-GIS Based Landslide Early Warning System for the Chittagong Metropolitan Area, Bangladesh. *ISPRS Int. J. Geo-Inf.* **2018**, *7*, 485. [CrossRef]
19. Lathrop, R.; Auermuller, L.; Trimble, J.; Bognar, J. The Application of WebGIS Tools for Visualizing Coastal Flooding Vulnerability and Planning for Resiliency: The New Jersey Experience. *ISPRS Int. J. Geo-Inf.* **2014**, *3*, 408–429. [CrossRef]
20. Oliveira, A.; Jesus, G.; Gomes, G.L.; Rogeiro, J.; Azevedo, A.; Rodrigues, M.; Fortunato, A.B.; Dias, D.M.; Tomas, L.M.; Vaz, L.; et al. An interactive WebGIS observatory platform for enhanced support of integrated coastal management. *J. Coast. Res.* **2014**, *70*, 507–512. [CrossRef]
21. Gourmelon, F.; Le Guyader, D.; Fontenelle, G. A dynamic GIS an Efficient Tool for Integrated Coastal zone Management. *ISPR Int. J. Geo-Inf.* **2014**, *3*, 391–407. [CrossRef]
22. Hansen, H.S.; Fuglsang, M. An Operational Web-Based Indicator System for Integrated Coastal Zone Management. *ISPRS Int. J. Geo-Inf.* **2014**, *3*, 326–344. [CrossRef]
23. Available online: https://webgis.provincia.belluno.it/ (accessed on 5 August 2021).
24. Available online: http://www.webgis.arpa.puglia.it/ (accessed on 5 August 2021).
25. Available online: https://webgis.cittametropolitana.ve.it/lm/ (accessed on 5 August 2021).
26. Peng, Z.R.; Tsou, M. *Internet GIS: Distributed Geographic Information Services for the Internet and Wireless Networks*; John Wiles and Sons Inc.: Hoboken, NJ, USA, 2003.
27. Haklay, M.; Singleton, A.; Parker, C. Web Mapping 2.0: The Neogeography of the GeoWeb. *Geogr. Compass* **2008**, *2*, 2011–2039.
28. Fu, P.; Sun, J. *Webgis Principles and Application*; ESRI Press: Redlands, CA, USA, 2010.
29. Zhang, X.; Yao, J.; Zhu, C.; Li, Y.; Ding, X. A WebGis for Spatial Data Processing, Analysis and Distribution for Coastal Investigation and Assessment (mission 908) in Jiangsu, China. *The International Archives of the Photogrammetry, Remote Sensing and Spatial Information Sciences*; Part B4 Beijing, 2008, XXXVII.B. Available online: https://isprs.org/proceedings/XXXVII/congress/4_pdf/140.pdf (accessed on 5 August 2021).
30. Silverster, R.; Tsuchiya, Y.; Shibano, T. Zeta Bays, Pocket Beaches and Headland Control. In Proceedings of the 17th International Conference on Coastal Engineering, Sydney, Australia, 23–28 March 1980; pp. 1306–1319.
31. Komar, P.D. Rhythmic Shoreline Features and Their Origins. In *Mega-Geomorphology*; Gardner, R., Scoging, H., Eds.; Clarendon Press: Oxford, UK, 1983; pp. 92–112.
32. Uda, T.; Serizawa, M.; San-Nami, T.; Furuike, K. Shoreline changes of a pocket beach caused by elongationof harbor breackwater and their prediction. *Trans. Jpn. Geomorphol. Union* **2002**, *23*, 395–413.
33. Hsu, J.R.C.; Yu, M.J.; Lee, F.C.; Benedet, L. Static Bay beach concept for scientists and engineers: A review. *Coast. Eng.* **2010**, *57*, 76–91. [CrossRef]
34. Short, A.D. Role of geological inheritance in Australian beach morphodynamics. *Coast. Eng.* **2010**, *57*, 92–97. [CrossRef]

35. Pranzini, E.; Rosas, V.; Jackson, N.L.; Nordstrom, K.F. Beach changes from sediment delivered by streams to pocketbeaches during a major flood. *Geomorphology* **2013**, *199*, 36–47. [CrossRef]
36. Short, A.D.; Masselik, G. Embayed and Structurally Controlled Beaches. In *Handbook of Beach and Shoreface Morphodynamics;* Short, A.D., Ed.; John Willey & Sons: New York, NY, USA, 1999; pp. 230–249.
37. Bowman, D.; Rosas, V.; Pranzini, E. Pocket beaches of Elba Island (Italy)—Planview geometry, depth of closure and sediment dispersal. *Estuar. Coast. Shelf Sci.* **2014**, *138*, 37–46. [CrossRef]
38. Hsu, J.R.C.; Evans, C. Parabolic bay shapes and applications. *Proc. Inst. Civil Eng.* **1989**, *2*, 557–570. [CrossRef]
39. Silvester, R.; Hsu, J.R.C. *Coastal Stabilization: Innovative Concepts*; Prentice-Hall: Englewood Cliffs, NJ, USA, 1993; p. 578.
40. Klein, A.H.F.; Filho, L.B.; Schumacher, D.H. Short-term beach rotation processes in distinct headland bay beach systems. *J. Coast. Res.* **2002**, *18*, 442–458.
41. Hsu, J.; Benedet, L.; Klein, A.H.; Raabe, A.; Tsai, C.-P.; Hsu, T. Appreciation of static bay beach concept for coastal management and protection. *J. Coast. Res.* **2008**, *24*, 198–215. [CrossRef]
42. Bowman, D.; Guillen, J.; Lopez, L.; Pellegrino, V. Planview Geometry and morphological characteristics of pocket beaches on the Catalan coast (Spain). *Geomorphology* **2009**, *108*, 191–199. [CrossRef]
43. Brignone, M.; Schiaffino, C.F.; Isla, F.I.; Ferrari, M. A system for beach video-monitoring: Beachkeeper plus. *Comput. Geosci.* **2012**, *49*, 53–61. [CrossRef]
44. Simeone, S.; Palombo, L.; De Falco, G. Morphodynamics of a Nontidal Embayed Beach: The Case Study of Is Arutas (Western Mediterranean). *J. Coast. Res.* **2013**, *29*, 63–71. [CrossRef]
45. Cappadonia, C.; Coratza, P.; Agnesi, V.; Soldati, M. Malta and Sicily Joined by Geoheritage Enhancement and Geotourism within the Framework of Land Management and Development. *Geosciences* **2018**, *8*, 253. [CrossRef]
46. Furlani, S.; Antonioli, F.; Biolchi, S.; Gambin, T.; Gauci, R.; Lo Presti, V.; Sulli, A. Holocene sea level change in Malta. *Quat. Int.* **2013**, *288*, 146–157.
47. Agnesi, V.; Angileri, S.E.; Cappadonia, C.; Coratza, P.; Costanzo, D.; Soldati, M.; Tonelli, C. Geositi nel paesaggio mediterraneo: Confronto tra aree costiere maltesi e siciliane. *Cult. Territ. Ling.* **2014**, *4*, 1–12.
48. Randazzo, G. Il canale di Sicilia. In *La Costa d'Italia (a cura di S.Ginesu)*; da Carlo Delfino, E., Ed.; Tavole Nazionali 7: Sassari, Italy, 2011; pp. 79–84.
49. Pedley, M.; Clarke, M.H. *Limestone Isles in a Crystal Sea: The Geology of the Maltese Islands*; Publishers Enterprises Group: San Gwann, Malta, 2002.
50. Baldassini, N.; Di Stefano, A. Stratigraphic features of the Maltese Archipelago: A synthesis. *Nat. Hazards* **2017**, *86*, 203–231. [CrossRef]
51. Lanza, S.; Randazzo, G. Improvements to a Coastal Management Plan in Sicily (Italy): New approaches to borrow sediment management. *J. Coast. Res.* **2011**, *64*, 1357–1361.
52. Randazzo, G.; Lanza, S. Regional Plan against Coastal Erosion: A Conceptual Model for Sicily. *Land* **2020**, *9*, 307. [CrossRef]
53. House, M.R.; Dunham, K.C.; Wigglesworth, J.C. Geology and Structure of the Maltese Islands. In *Malta, A Background for Development*; Bowen, J.H., Dewdney, J.C., Fisher, W.B., Eds.; Durham University Press: Durham, UK, 1961; pp. 24–33.
54. Ransley, N. *A Geography of the Maltese Islands*; St. Aloysius College: B'Kara, Malta, 1974; p. 60.
55. Axiak, V.; Gauci, V.; Mallia, A.; Mallia, E.A.; Schembri, P.J.; Vella, A.J. State of the Environment Summary Report for Malta 1998. Environ. Prot. Dept., Min. for the Environment, Floriana (Malta), 1999, p. 130. Available online: https://era.org.mt/wp-content/uploads/2019/05/SOEReport1998.pdf (accessed on 5 August 2021).
56. Gauci, M.J.; Deidun, A.; Schembri, P.J. Faunistic diversity of Maltese pocket sandy and shingle beaches: Are these of conservation value? *Oceanologia* **2005**, *47*, 219–241.
57. Drago, A. Physical Oceanography in the Coastal Waters of the Maltese Islands. In *Intergovernmental Oceanographic Commission–International Commission on the Scientific Exploration of the Mediterranean Sea (IOC-ICSEM) Workshop on Coastal Erosion;* Foundation for International Studies: Valletta, Malta, 1991; pp. 1–5.
58. Micallef, A. Monitoring Beach Stability on the Mediterranean Island of Malta. In *Proceedings of the Sixth International Conference on the Mediterranean Coastal Environments, MedCoast'03, Ravenna, Italy, 7–11 October 2003;* Ozhan, E., Ed.; Middle East Technical University: Ankara, Turkey, 2003; pp. 1357–1368.
59. Deidun, A.; Schembri, P.J. Long or short? Investigating the effect of beach length and other environmental parameters on macrofaunal assemblages of Maltese pocket beaches. *Estuar. Coast. Shelf Sci.* **2008**, *79*, 17–23.
60. Muzirafuti, A.; Cascio, M.; Lanza, S. UAV Photogrammetry-Based Mapping of Sicilian Pocket Beaches: First Approach. In *Planet Care from Space;* Dessena, M.A., Melis, M.T., Eds.; AIT Series, Trends in Earth Observation; Italian Society of Remote Sensing: Cagliari, Italy, 2021.
61. Muzirafuti, A.; Cascio, M.; Lanza, S. UAV Photogrammetry-based Mapping the Pocket Beach of Isola Bella, Taormina (Northeastern Sicily). In Proceedings of the 2021 IEEE International Workshop on Metrology for the Sea (MetroSea 2021), Reggio Calabria, Italy, 4–6 October 2021.
62. Storlazzi, C.D.; Field, M.E. Sediment distribution and transport along a rocky embayed coast: Monterey Peninsula and Carmel Bay, California. *Mar. Geol.* **2000**, *170*, 289–316. [CrossRef]
63. Norcross, Z.M.; Fletcher, C.H.; Merrifield, M. Annual and interannual changes on a reef-fringed pocket beach: Kailua Bay, Hawaii. *Mar. Geol.* **2002**, *190*, 553–580. [CrossRef]

64. Muzirafuti, A.; Barreca, G.; Crupi, A.; Faina, G.; Paltrinieri, D.; Lanza, S.; Randazzo, G. The Contribution of Multispectral Satellite Image to Shallow Water Bathymetry Mapping on the Coast of Misano Adriatico, Italy. *J. Mar. Sci. Eng.* **2020**, *8*, 126. [CrossRef]
65. Muzirafuti, A.; Crupi, A.; Lanza, S.; Barreca, G.; Randazzo, G. Shallow Water Bathymetry by Satellite Image: A case Study on the Coast of San Vito Lo Capo Peninsula, Northwestern Sicily, Italy. In Proceedings of the IMEKO TC-19 International Workshop on Metrology for the Sea, Genoa, Italy, 3–5 October 2019.
66. Randazzo, G.; Barreca, G.; Cascio, M.; Crupi, A.; Fontana, M.; Gregorio, F.; Lanza, S.; Muzirafuti, A. Analysis of Very High Spatial Resolution Images for Automatic Shoreline Extraction and Satellite-Derived Bathymetry Mapping. *Geosciences* **2020**, *10*, 172. [CrossRef]
67. Lisi, I.; Bruschi, A.; Del Gizzo, M.; Archina, M.; Barbano, A.; Corsini, S. Le unità fisiografiche e le profondità di chiusura della costa italiana. *L'ACQUA* **2010**, *2*, 35–52.
68. Tomasello, A.; Cassetti, F.P.; Savona, A.; Pampalone, V.; Pirrotta, M.; Calvo, S.; Signa, G.; Andolina, C.; Mazzola, A.; Vizzini, S.; et al. The use of very high resolution images for studying Posidonia oceanica reefs. *Vie Milieu* **2020**, *70*, 25–35.
69. Ventura, D.; Bonifazi, A.; Gravina, M.F.; Belluscio, A.; Ardizzone, G. Mapping and Classification of Ecologically Sensitive Marine Habitats Using Unmanned Aerial Vehicle (UAV) Imagery and Object-Based Image Analysis (OBIA). *Remote Sens.* **2018**, *10*, 1331. [CrossRef]
70. Rende, S.F.; Bosman, A.; Di Mento, R.; Bruno, F.; Lagudi, A.; Irving, A.D.; Dattola, L.; Giambattista, L.D.; Lanera, P.; Proietti, R.; et al. Ultra-High-Resolution Mapping of *Posidoniaoceanica* (L.) Delile Meadows through Acoustic, Optical Data and Object-based Image Classification. *J. Mar. Sci. Eng.* **2020**, *8*, 647. [CrossRef]
71. Randazzo, G.; Cascio, M.; Fontana, M.; Gregorio, F.; Lanza, S.; Muzirafuti, A. Mapping of Sicilian Pocket Beaches Land Use/Land Cover with Sentinel-2 Imagery: A Case Study of Messina Province. *Land* **2021**, *10*, 678. [CrossRef]
72. Petrişor, A.-I.; Hamma, W.; Nguyen, H.D.; Randazzo, G.; Muzirafuti, A.; Stan, M.-I.; Tran, V.T.; Aştefănoaiei, R.; Bui, Q.-T.; Vintilă, D.-F.; et al. Degradation of Coastlines under the Pressure of Urbanization and Tourism: Evidence on the Change of Land Systems from Europe, Asia and Africa. *Land* **2020**, *9*, 275. [CrossRef]
73. Folk, R.L.; Ward, W.C. Brazos River Bar: A study in the significance of grain size parameters. *J. Sedim. Petro.* **1957**, *27*, 3–26. [CrossRef]

applied sciences

MDPI

Review

IoT-Enabled Smart Agriculture: Architecture, Applications, and Challenges

Vu Khanh Quy [1], Nguyen Van Hau [1], Dang Van Anh [1], Nguyen Minh Quy [1], Nguyen Tien Ban [2], Stefania Lanza [3], Giovanni Randazzo [4] and Anselme Muzirafuti [4,*]

1 Faculty of Information Technology, Hung Yen University of Technology and Education, Hung Yen 160000, Vietnam; quyvk@utehy.edu.vn (V.K.Q.); haunv@utehy.edu.vn (N.V.H.); dangvananh@utehy.edu.vn (D.V.A.); minhquy@utehy.edu.vn (N.M.Q.)
2 Faculty of Telecommunication 1, Posts and Telecommunications Institute of Technology, Hanoi 100000, Vietnam; bannt@ptit.edu.vn
3 GeoloGIS s.r.l., Dipartimento di Scienze Matematiche e Informatiche, Scienze Fisiche e Scienze della Terra, Università degli Studi di Messina, Via F. Stagno d'Alcontres, 31-98166 Messina, Italy; stefania.lanza@unime.it
4 Dipartimento di Scienze Matematiche e Informatiche, Scienze Fisiche e Scienze della Terra, Università degli Studi di Messina, Via F. Stagno d'Alcontres, 31-98166 Messina, Italy; giovanni.randazzo@unime.it
* Correspondence: anselme.muzirafuti@unime.it

Abstract: The growth of the global population coupled with a decline in natural resources, farmland, and the increase in unpredictable environmental conditions leads to food security is becoming a major concern for all nations worldwide. These problems are motivators that are driving the agricultural industry to transition to smart agriculture with the application of the Internet of Things (IoT) and big data solutions to improve operational efficiency and productivity. The IoT integrates a series of existing state-of-the-art solutions and technologies, such as wireless sensor networks, cognitive radio ad hoc networks, cloud computing, big data, and end-user applications. This study presents a survey of IoT solutions and demonstrates how IoT can be integrated into the smart agriculture sector. To achieve this objective, we discuss the vision of IoT-enabled smart agriculture ecosystems by evaluating their architecture (IoT devices, communication technologies, big data storage, and processing), their applications, and research timeline. In addition, we discuss trends and opportunities of IoT applications for smart agriculture and also indicate the open issues and challenges of IoT application in smart agriculture. We hope that the findings of this study will constitute important guidelines in research and promotion of IoT solutions aiming to improve the productivity and quality of the agriculture sector as well as facilitating the transition towards a future sustainable environment with an agroecological approach.

Keywords: sustainable agriculture; food security; green technologies; Internet of Things; natural resources; sustainable environment; IoT ecosystem

Citation: Quy, V.K.; Hau, N.V.; Anh, D.V.; Quy, N.M.; Ban, N.T.; Lanza, S.; Randazzo, G.; Muzirafuti, A. IoT-Enabled Smart Agriculture: Architecture, Applications, and Challenges. *Appl. Sci.* **2022**, *12*, 3396. https://doi.org/10.3390/app12073396

Academic Editor: Manuel Armada

Received: 7 March 2022
Accepted: 25 March 2022
Published: 27 March 2022

1. Introduction

In order to meet the current global needs of humanity, new solutions and technologies are constantly being proposed and implemented. This has led to the advent of the Internet of Things (IoT) [1,2]. IoT is defined as the network of all objects that are embedded within devices, sensors, machines, software and people through the Internet environment to communicate, exchange information and interact in order to provide a comprehensive solution between the real world and the virtual world [3]. In recent years, IoT has been applied in a series of domains, such as smart homes [4,5], smart cities [6,7], smart energy [8,9], autonomous vehicles [10,11], smart agriculture [12–15], campus management [16,17], healthcare [18,19], and logistics [20,21]. Series of other IoT applications have been described by Shafique et al. [22]. An illustration of rich and diverse IoT applications for smart agriculture is provided in Figure 1.

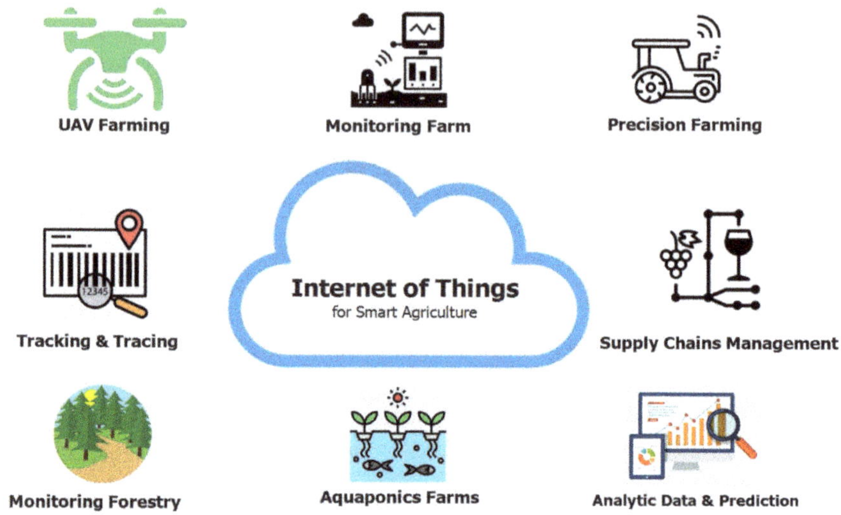

Figure 1. An illustration of IoT applications for smart agriculture.

According to the United Nations' (UN 2019) statistics, the world population is estimated to grow to 10 billion by 2050 [23]. As a consequence, the requirements of agricultural products are continually increasing. However, farmlands are declining, natural resources are increasingly depleted, and the rise of unpredictable nature challenges, such as global warming, salinization, and flooding, make food security the most concerning problem for all nations worldwide.

In recent years, with the aim of increasing agricultural production, new solutions and technologies have been introduced in the agriculture sector [24]. An emerging trend is the application of the IoT and big data. A significant number of studies have been focused on research, experiments, and applications [25,26]. According to the Cisco forecast, over 500 billion IoT devices will be connected to the Internet by 2030 [27]. The use of IoT and big data will enable smart agriculture and is expected to enhance efficiency and productivity [28].

Over the years, wireless sensor networks (WSN) have been strongly applied in the agricultural sector, building the foundation for developing smart agriculture [29]. The unique characteristics of WSN, such as the ability to self-organize, self-configure, self-establish, and self-recover, make it suitable for smart agriculture [30]. The sensor device consists of a radio frequency (RF) transceiver, sensor, microcontroller, and battery power [31]. The WSN focuses on applications such as environmental monitoring, machine control automation, and traceability [32–35].

Along with the development of science and technology, the urgent requirement for breakthrough solutions and technologies aiming at improving productivity and efficiency in the agriculture sector has led to adoption of the IoT. The primary motivation for their applications is the breakthrough progress of smart agriculture and its inevitable role as the future of smart and sustainable environment management. IoT integrates a series of existing solutions and technologies, such as WSN, cognitive radio, ad hoc networks, cloud computing, and end-user applications [36]. In the smart agricultural sector, automation solutions and technologies, mechanical machines, knowledge, decision-making tools, services, and software are integrated seamlessly to help farmers improve productivity, product quality, and profitability [37].

In this work, a comprehensive survey of IoT applications for smart agriculture is conducted. An analysis of 135 relevant works published between 2017 and 2022 was conducted. Firstly, relevant 550 papers published in the period of (2017–2022) were retrieved from major scientific databases, namely IEEE Xplore Digital Library, Science Direct, MDPI, and Springer, by using keywords such as IoT-enabled smart agriculture, smart agriculture, Internet of Things, aquaponics, monitoring forestry based on IoT, tracking and tracing, smart precision farming, greenhouse production, Sigfox, LoRa, Wi-Fi, LoRaWAN, and IoT ecosystems. In the next step, we excluded papers that were published in low-repute conferences and journals, and then we conducted the content analysis for the obtained paper. Finally, 135 papers were selected for the preparation of the present work.

In addition, we analyzed and discussed the benefits and challenges, open issues, trends, and opportunities of IoT in the smart agriculture sector. This work is organized as follows: Section 1 introduces our work, and in Section 2, we present an IoT ecosystem architecture for smart agriculture that consists of three main components: IoT devices, communication technology, and data storage and big data processes. Section 3 presents the IoT applications in agriculture, including (1) monitoring, (2) tracking and traceability, (3) precision agriculture, and (4) greenhouses. Section 4 introduces some open issues and future research challenges of IoT for smart agriculture. Issues are discussed for two main directions: business and technology. In Section 5, we present the main conclusions of this work.

2. IoT Ecosystem Architecture for Smart Agriculture

In this section, we present a common framework of an IoT ecosystem for smart agriculture based on three main components, including (1) IoT devices, (2) communication technologies, and (3) data process and storage solutions. An illustration of the IoT ecosystem for smart agriculture is presented in Figure 2.

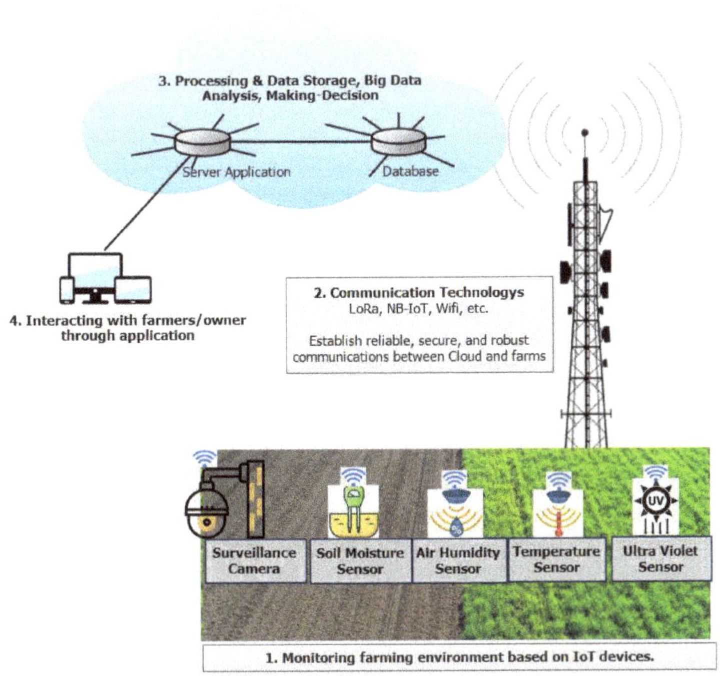

Figure 2. An illustration of IoT ecosystems' architecture for smart agriculture.

2.1. IoT Devices

The common architecture of an IoT device consists of sensors to collect information from the environment, actuators based on wired or wireless connections, and an embedded system that has a processor, memory, communication modules, input–output interfaces, and battery power [38,39]. The common architecture of a typical IoT device for smart agriculture is shown in Figure 3.

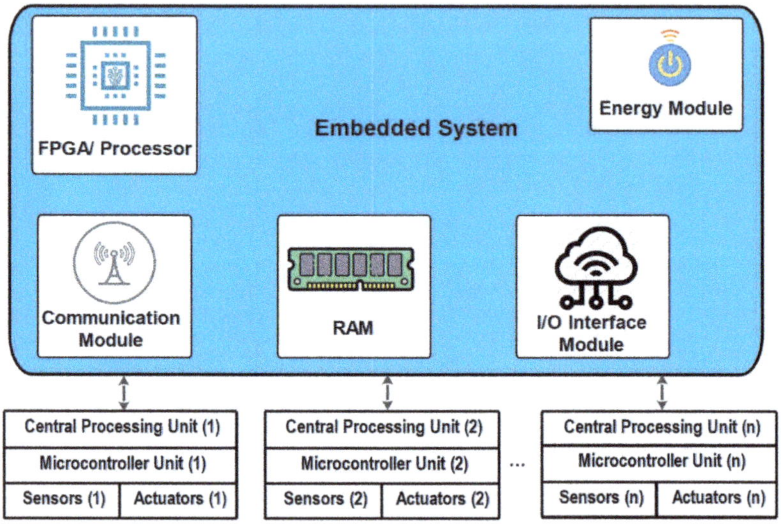

Figure 3. An illustration of the common architecture of an IoT device.

Embedded systems are programmable interactive modules, namely FPGAs (field programmable gate arrays). Sensor devices are specially designed to operate in open environments, in nature, in soil, water, and air to measure and collect environmental parameters that affect production, such as soil nutrients, humidity, temperature, etc. Smart farming solutions are agricultural operations that are often deployed on large farmlands, outdoors, so the devices that support solutions need some unique characteristics, such as the ability to withstand the effects of weather, humidity, and temperature instability throughout their service lifecycle. Some of their main features, as shown in Figure 4, make IoT devices suitable for smart agriculture solutions [40–42].

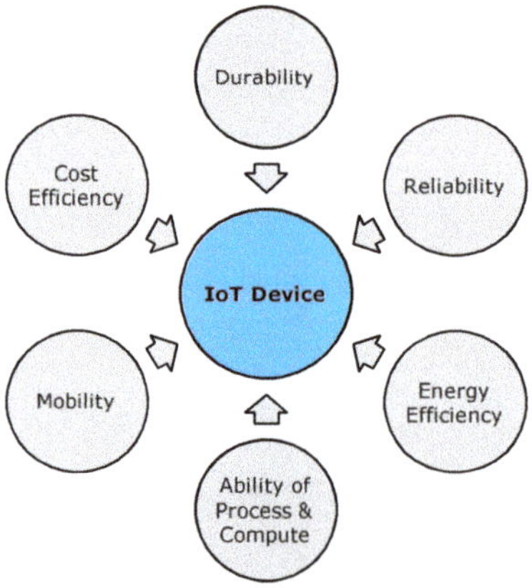

Figure 4. The main characteristics of IoT devices.

Depending on the required operation, there are several typical sensors applied in the smart agriculture sector. Sensors can be divided into several typical categories, such as (1) location sensor, (2) optical sensor, (3) mechanical sensor, (4) electrochemical sensor, and

(5) air flow sensor. These sensors are used to collect information such as air temperature, soil temperature, air humidity, soil moisture, leaf moisture, precipitation, wind speed, wind direction, and solar radiation, and barometric pressure [21,24,36].

2.2. Communication Technology

The survey of communication technologies for IoT [43,44] indicated that to integrate IoT into the smart agriculture sector, communication technologies must progressively improve the evolution of IoT devices. They play an important role in the development of IoT systems. The existing communication solutions can be classified as: protocol, spectrum, and topology.

Protocols: many wireless communication protocols have been proposed for the smart agriculture sector. Based on these protocols, devices in a smart agricultural system can interact, exchange information, and make decisions to monitor and control farming conditions and improve yields and production efficiency. The typical, low-power communication protocol numbers commonly used in smart agriculture can be divided into short-range and long-range categories based on the communication range.

- Short-range: NFMI (near-field magnetic induction) [45], Bluetooth [46], ZigBee [47], terahertz (Z-Wave) [48,49], and RFID [50].
- Long-range: LoRa [51], Sigfox [52], and NB-IoT (Narrowband IoT) [53].

Table 1 presents some typical communication technologies for the smart agriculture sector. The values in Table 1 indicate that short-range communication technologies have a transmission distance of less than 20 (m), high energy efficiency, and low data rate. These protocols are often employed in sensor networks, while long-range communication technologies have transmission distances of up to several tens of kilometres, consume more energy, and are installed for backhaul device-to-device communications. A diverse survey of low-power communication technologies for IoT that presents solutions, challenges, and some open issues is described by Sundaram et al. [54].

Table 1. Some typical communication technologies for smart agriculture.

Type	Spectrum	Transmission Distance	Type of Network	Frequency	Data Rate
802.11a/b/g/n/ac	Unlicensed	100 m	WLAN	2.4–5 GHz	2–700 Mbps
802.11ah	Unlicensed	1000 m	WLAN	Several Sub-GHz	78 Mbps
802.11p	Licensed	1 km	WLAN	5.9 GHz	3–27 Mbps
802.11af	Licensed	1 km	WLAN	54–790	25–550 Mbps
SigFox	Licensed	Rural: 50 km Urban:10 km	LPWA	Zwave	100–600 bps
LoRaWAN	Licensed	20 km	LPWA	Several Sub-GHz	0.3–100 kbps
NB-IoT	Licensed	35 km	LPWA	Zwave	250 kbps
LTE-3GPP	Licensed	5 km	WWAN	1.4 MHz	200 kbps
EC-GPRS	Licensed	5 m	WWAN	GSM bands	240 kbps
WiMAX	Hybrid	50–80 km	WWAN	Several Sub-GHz	70 Mbps
Bluetooth	Unlicensed	100 m	WPAN	2.4 GHz	2–26 Mbps
ZigBee	Unlicensed	1 km	WHAN	2.4 GHz	250 kbps
Z-Wave	Unlicensed	100 m	WHAN	900 MHz	100 kbps
6LoWPAN	Unlicensed	30 m	WHAN	Zwave	250 kbps
NFC	Unlicensed	20 cm	D2D	13.56 MHz	424 kbps

Spectrum: Each radio device uses certain frequency bands for communication. The FCC (Federal Communications Commission) has defined unlicensed spectrum bands for unlicensed operations in scientific, industrial, and medical purposes [55]. These spectrum bands are often applied for low-power levels and short-range applications. Consequently, a series of common technologies for the smart agriculture sector, from wireless machine control and UAVs to communication technologies such as Wi-Fi and Bluetooth, use unlicensed spectrum bands [56]. However, the use of unlicensed spectra faces several challenges, such as the quality of service guarantee, the cost of setting up the initial infrastructure, and the interference generated by the huge number of IoT devices [57,58].

A licensed spectrum usually is allocated to mobile networks. It provides more efficient network traffic, more reliability, enhances the quality of service (QoS), offers security, provides extensive coverage, and involves lower initialization infrastructure costs for users. However, the use of licensed spectrum bands has faced some limitations, such as high data transmission costs and the low energy efficiency of IoT devices [59].

Several recent studies have demonstrated the efficiency of unlicensed spectrum bands in the mm wave range. It uses extremely low power but provides large transmission distances and high data rates [60,61]. One limitation of the mm wave spectrum is that the data rate is strongly affected by weather conditions, especially rain [62].

Topology: The establishment of the communication spectrum band and operation protocol of IoT devices depends on the structure that deploys IoT devices for smart agriculture applications. Network structures for smart agriculture usually have two main types of nodes: sensor and backhaul nodes [63]. The common characteristics of IoT sensor nodes are short communication distance, low data rate, and high energy efficiency. In contrast, IoT backhaul nodes often require large transmission distances, high throughput, and data rates. Therefore, based on the role of each IoT network node, the sensor node or backhaul node selects and installs appropriate communication technologies [64]. Figure 5 presents a typical low-power network topology designed for measuring and monitoring factors in a smart farm. The system includes:

(1) IoT sensor nodes collect information from the farming environment, such as soil moisture, air humidity, temperature, nutrient ingredients of soil, pest images, and water quality, then transmit collected data to IoT backhaul devices. Depending on the operation purpose and installation location, IoT sensor nodes can be installed as RFDs (reduced-function devices), which only communicate with FFDs (full-function devices). These nodes cannot communicate with the other RFDs, aiming to save energy and decrease investment costs.

(2) IoT backhaul nodes, besides having the role of an IoT sensor node, also play a role as intermediate nodes to receive information from other IoT nodes and transmit it to the control centre. IoT backhaul nodes are often installed as an FFD device, which can connect to other FDD and RFD devices.

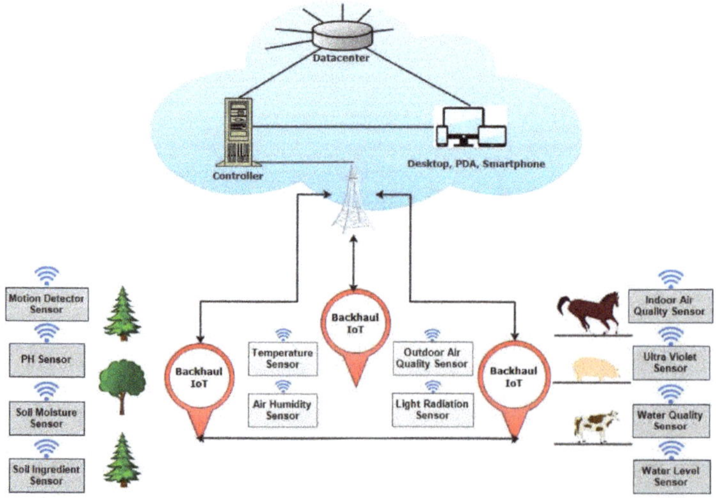

Figure 5. An illustration of the common IoT-based smart agriculture topology.

2.3. Data Analytics and Storage Solutions

In the smart agriculture domain, besides the main problems of sensing, collecting data, and controlling devices to respond to the real farming environment, data storage and processing are also important problems and face some challenges [26,28]. In reality, the number of collected data is huge, and traditional data storage, organization, and processing solutions are not feasible. Therefore, big data processing solutions need to be researched and applied for smart agriculture [65,66].

The complexity of data storage and processing is due to the unique characteristics of the smart agriculture field, including unstructured data and various formats, such as text, images, audio and video, economic figures, and market information. Recent solutions and technologies have introduced the use of cloud platforms for storage and data analytics, which are collected from farms [36,67]. In addition, cloud-assisted big data analytic solutions, such as edge computing [68] or fog computing [69], are also proposed to reduce latency and costs and support QoS.

The survey results demonstrate that, in recent years, many management information systems for smart agriculture have been proposed [70–72]. Nowadays, possible solutions have been developed and commercialized, providing solutions and services for farmers to manage farms and fields, aiming to increase productivity, reduce human labour, and enhance farming efficiency, as follows:

- *OnFarm* [73]: It is part of the SWIM family, which specializes in providing solutions and technology for smart agriculture. *OnFarm* is a technology platform that allows farmers to manage and use data in the simplest way. It is also a comprehensive solution for managing, using, and controlling water on smart farms.
- *Farmobile* [74]: It is a commercialized online platform to manage smart farms that allows farmers, traders, scientists, and insurance companies to operate and communicate centrally on an online platform.
- *Silent Herdsman Platform* [75]: It is a platform that allows monitoring of the activities of cow colonies and predicting their milk production.
- *CropX* [76]: It is a platform that will enable monitor and control nutrients of farming soil based on a sensor system and big data analytic solutions.
- *FarmX* [77]: It is an all-in-one platform for tree crops. *FarmX* provides a series of diverse farming management solutions, including irrigation, fertilizer, cropland management systems, environmental monitoring, and crop production forecasting.
- *Easyfarm* [78]: It is a platform that provides software to help farmers manage and account for figures of farms. *Easyfarm* provides visual figures, including input and output supplies management, production forecasting, and market connectivity, to help farmers fully manage their farms.
- *KAA* [79]: It is a cloud-based IoT platform that aims to provide comprehensive, end-to-end solutions for farmers, including: (1) connecting and managing IoT devices in farms or fields; and (2) monitoring and controlling behaviours of devices based on data analysis results.
- *Farmlogs* [80]: It is a platform that provides tools and solutions for: (1) automating the production cost calculation process; (2) managing the day-to-day activities of the farm in real-time; and (3) supporting marketing and increasing sales of products.

3. Typical Applications of IoT in Smart Agriculture

In recent years, a series of IoT applications for agriculture have been introduced. According to survey results, we divided these applications into categories based on their purpose, including monitoring, tracking and traceability, and greenhouse production. The detailed results are presented in the following subsection.

3.1. Monitoring

In the agriculture sector, factors affecting the farming and production process can be monitored and collected, such as soil moisture, air humidity, temperature, pH level, etc.

These factors depend on the considered agricultural sector. Some smart agricultural sectors are applying the following monitoring solutions:

Crop Farming: In this sector, some vital factors that affect the farming process and production efficiency include air temperature, precipitation, air humidity, soil moisture, salinity, solar radiation, pest status, soil nutrient ingredients, etc. In [81], the authors designed an IoT device called FarmFox. This device allows real-time collection and analysis of the composition of the farming soil and transmits the information to farmers/owners via the Internet. The results demonstrate the health of the soil is monitored in real time to provide timely recommendations to farmers aiming to increase productivity and farming efficiency.

Furthermore, in [82], the authors proposed an IoT device to allow intelligent control of temperature and humidity factors, called a weather radar. This device will automatically turn on the warning mode using the light signal and send messages to the farmer when the temperature or humidity exceeds a pre-installed threshold. In [83], the authors introduced an IoT system based on Web GIS to monitor pest status and provide early warnings. In addition, this study also proposes a predictive model based on monitoring the habitat of pests and diseases. The efficiency of the proposed system was indicated, based on the predicted figures of the locust epidemic, to have a high accuracy rate (over 87%) in 2019 (China).

Monitoring information, such as soil condition, moisture, and temperature, and the prediction of natural factors, such as rainfall and weather, support the control of growing conditions of crops, helping farmers plan and make irrigation decisions to optimize production and reduce labour costs. In addition, the collected data, combined with big data processing technology, can provide recommendations for implementing preventive and remedial solutions against pests and diseases in farming.

Aquaponics: It is an integration of aquaculture and hydroponics. Aquaponics is a farming technique where fish waste becomes a source of nutrients needed by plants. One of the most important issues in such farms is constantly monitoring water quality, water level, temperature, salinity, pH, sunlight, etc. [84]. According to this research direction, in [85], the authors designed an IoT system to monitor the temperature and pH value of water for aquaponics farms. Moreover, this system is also equipped with a control system of water metrics to keep the fish habitat stable and an automatic fish feeding function to increase the productivity of the fish. The results show that the IoT system had stable operation and provided real-time monitoring parameters. The authors of [86] designed an aquaponics farm for households/urban areas based on IoT. This system recommends the proper ratio of fish and plants.

Consequently, the system decreases feed consumption as well as reduces carbon emissions into the environment. The primary purpose of this proposal aims to balance the self-sustaining ability of the aquaponics system. The experimental results demonstrate the number of fish decreases from 30 to 15, and the number of plants increases from 20 to 30, but the crop production will increase by more than 50%. A detailed and diverse survey of the IoT systems and devices for control and monitoring of aquaponics farms is introduced in [87]. Based on the obtained data, monitoring can improve the production of fish and plants through the control, supplementation, and regulation of nutritional ingredients in the water. The collected data were also used to automate the management of aquaponics farms to reduce labour costs.

Forestry: Humans depend on forests for survival. Moreover, forests play a vital role in the carbon cycle and provide a habitat for more than two-thirds of animal species in the world. Forests also have the effect of protecting watersheds, limiting floods, and mitigating climate change. The main factors that need to be monitored in a forest include soil ingredients, air temperature, humidity, and concentration of several different gases, such as oxygen, methane, ammonia, and hydrogen sulphide. A series of forest control systems and solutions are presented in [88,89] based on IoT and big data analytics.

In [90], the authors designed a peatland forest environmental monitoring system. This forest area plays a very vital role in the rainforest ecosystem of Brunei. However, the peatland forest type is very burnt. This work designed an IoT system to monitor environmental conditions, such as temperature, humidity, wind direction, barometric pressure, and manage possible disasters. For the purpose of enhancing feasibility, IoT devices use the solar-powered system and communicate with the monitoring centre based on the LoRa network. In [91], the authors proposed a solution to control forest changes and vitality by using high-resolution RapidEye satellite imagery. This solution has been deployed commercially in several states in Germany and has detected leaf diseases in a pine forest affected by pests. Survey results indicate that monitoring in forestry focuses on providing early warning systems against forest fires, pest control, or deforestation.

Livestock Farming: It is defined as the process of raising domesticated animals, such as cows, pigs, sheep, and goats, chickens, etc., in an agricultural environment to obtain traction, serve production, and obtain products such as meat, eggs, milk, fur, leather, etc. In this area, the factors to be monitored depend on the type and number of farming animals [92]. In [93], the authors designed a support system for the diagnosis, prevention, and treatment of diseases for livestock called VetLink. This system can provide recommendations for animal health for farmers in rural areas where it is difficult to access veterinary doctors immediately. In [94], the authors proposed a noncontact temperature measurement system and monitoring of animals to ensure early detection of diseases and animal health. This system can be used for remote monitoring of animal health and timely anomaly detection. In [95], the authors introduced a monitoring system for large-scale pig farms based on IoT. The specific solution is to attach an IC tag on each pig to monitor the behaviour of each pig, such as their period of feeding and resting and exercise. Data from sensors are collected and combined with data analytics solutions that can make recommendations for pig health.

The monitoring data of water, feed, and animal health for livestock in the farming process helps farmers set up livestock plans, reduce labour costs, and enhance production efficiency. While a series of solutions has been provided for monitoring large-scale farms, their application in small and medium-sized farms is very limited, especially in developing countries. This can be attributed to the high cost and the lack of knowledge needed to set up, manage, and operate IoT systems. Therefore, effective and low-cost solutions for agricultural IoT have much potential.

3.2. Tracking and Tracing

In order to meet the needs of consumers and increase profit value, in the future, farms need to demonstrate that products offered to the market are clean products and can be tracked and traced conveniently, thereby enhancing the trust of consumers in product safety and health-related issues. In order to solve this problem, a series of tracking- and tracing-based problems for the smart agricultural sector has been proposed, specifically as follows:

In [96], the authors designed an information system that allows tracking and tracing of agricultural products and foods such as dairy and vegetables, called SISTABENE. This system helps suppliers track the production process and errors arising in the supply chain, and helps end-users trace the origin of food. In [97], the authors proposed a food supply chain traceability system based on blockchain technology. It helps to track and trace agri-food supply chains' production process and trace the origin of agricultural products. This solution has been employed at Shanwei Lvfengyuan Modern Agricultural Development Co., Ltd. (Shanwei, China). Although there are still limitations, the results demonstrate that this solution has successfully supported the tracing of food and agricultural products through QR codes, improving product quality and ensuring the clear traceability of products. In [98,99], the authors proposed smart agricultural solutions to tracking and tracing agricultural products, thereby allowing consumers to know the product's entire history. These solutions enable tracking and tracing some of the data collected along the supply chain, ensuring that consumers and other stakeholders can identify products' origin, location, and history.

3.3. Smart Precision Farming

The advent of the GPS (global positioning system) has created breakthrough advances in many fields of science and technology. The GPS provides the most important parameters for locating a device, such as location and time. GPS systems have been successfully deployed in many fields, such as smartphones, vehicles, and IoT ecosystems. However, GPS is only good support for outdoor systems and the sky. Meanwhile, the demand for the locating and navigating systems in the home and on the streets of smart cities is growing rapidly. Aiming to solve this problem, an advanced global navigation satellite system (GNSS) is being deployed [100]. Based on GPS and GNSS systems, suitable farming maps have been established for fields and farms. As a result, agricultural machinery and equipment can be operated autonomously [101]. Figure 6 presents an illustration of the typical cloud-assisted, IoT-based precision agriculture platform.

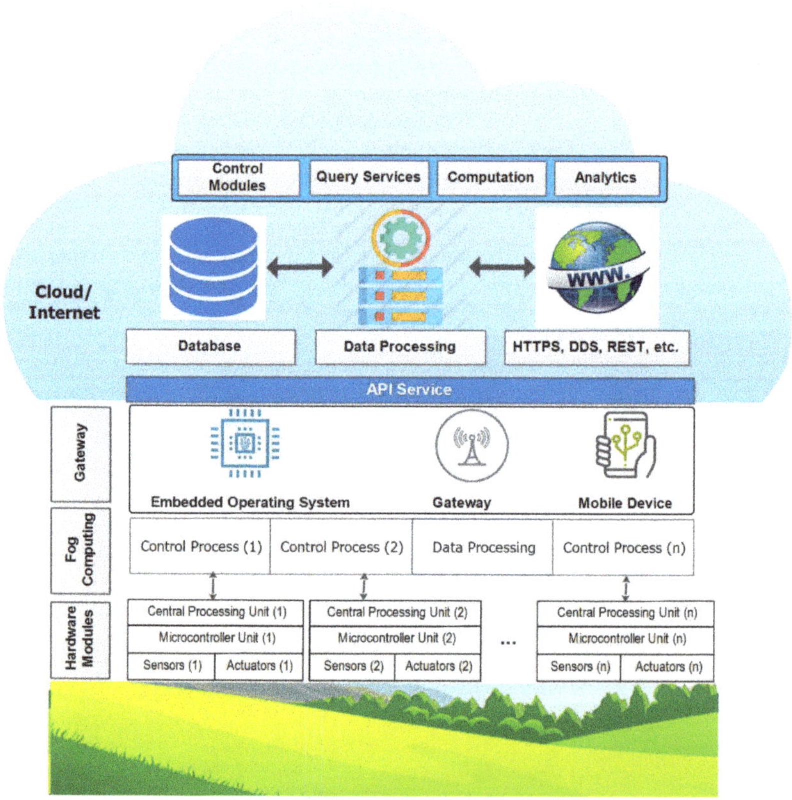

Figure 6. Cloud-assisted IoT-based precision agriculture platform.

In smart precision farming, one of the most important applications is the use of drones in monitoring and farming activities. Some common farming tasks using UAVs include spraying pesticides, fertilizing, sowing seeds, evaluating and mapping, and monitoring crop growth. In [102], the authors presented a detailed survey of drone applications for smart agriculture, including applications, control technology, and future trends of the UAV application for smart agriculture. In [103], the authors designed an automatic agricultural product classification system based on camera systems, image processing algorithms, and mechanical actuators. The experimental results for agriculture products such as oranges and tomatoes present a classification success rate of over 95%, and the sorting time for each product is less than 1(s). This solution can be adapted and applied to the classification of different agricultural products. In [104], the authors proposed a solution to estimate grape production. The proposed solution combines an RGB-D camera mounted on a mobile robot platform and size estimation algorithm for a bunch of grapes. The experimental results present an average error in the range of [2.8–3.5] (cm). The results demonstrate this solution is a feasible method for evaluating the productivity of large-scale grape farms.

The survey results show that smart precision agricultural equipment, such as irrigation systems, unmanned aerial vehicles (UAV), and smart agricultural equipment, etc., are configurable in an autonomous-control mode based on certain conditions or can be controlled remotely by the farmer via the Internet [105,106]. Smart precision farming helps to improve productivity and production efficiency and is suitable for large-scale farms. Nowadays, suppliers of precision agricultural equipment have IoT modules built into their machines, allowing machines to operate autonomously and remotely via the Internet [107].

3.4. Greenhouse Production

A greenhouse consists of walls and a roof, which are usually made from transparent materials, such as plastic or glass. In a greenhouse, plants are grown in a controlled environment, including controlling for moisture, nutrient ingredients of the soil, light, temperature, etc. Consequently, greenhouse technology makes it possible for humans to grow any plant, at any time, by providing suitable environmental conditions [108]. Figure 7 illustrates a smart agriculture IoT system for monitoring greenhouse farming factors based on IoT ecosystems.

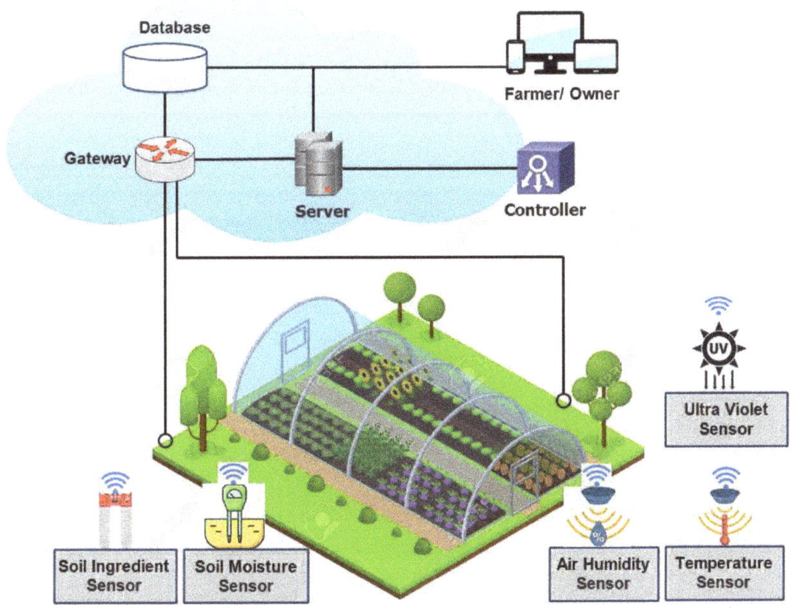

Figure 7. An illustration of IoT application for monitoring farming conditions in a greenhouse.

In [109], the authors introduced an IoT-based greenhouse environmental monitoring system for multipoint monitoring in large greenhouses. Instead of using multiple sensors at different locations, this solution involves a drive system that allows the sensor system to move to different locations in the greenhouse. The experimental results show that the proposed system can effectively monitor multiple points in large greenhouses. In [110], the authors introduced an energy-saving temperature control technology for smart greenhouses. This study proposed two intelligent control methods: active disturbance rejection control and fuzzy active disturbance rejection control. The experimental results demonstrate that the proposed technology saves over 15% of the total energy consumption of the greenhouse. In [111], the authors designed an intelligent IoT system to monitor and control greenhouse temperature for energy efficiency and improve crop productivity. The experimental results for the Kingdom of Saudi Arabia, where daytime temperatures can be above 50 °C, demonstrate the efficiency of the proposed solution, including saving energy and predicting the rate of plant growth.

Recent studies indicated that solutions integrating IoT, big data processing, and artificial intelligence could be applied in greenhouses to reduce labour and energy efficiency. Moreover, it also provides direct connections between the greenhouse farms and the customer [112–115].

4. Challenges and Open Research Directions

The survey results indicate that IoT components for the smart agriculture sector, including hardware and software, have been focused on research and achieved many breakthrough results. Several IoT solutions have been deployed on large-scale farms/fields. However, the widespread deployment of IoT in the agricultural sector still presents some challenges. We have present two main problems: economic efficiency and technical problems. We consider these issues coupled with policies that will drive the integration of IoT technologies in agriculture.

4.1. Economic Efficiency

In agricultural economics, one of the most important characteristics is a low rate of profit of an investment project, which presents many risks from natural conditions. The benefit–cost of a new technology seeking deployment in agriculture should be carefully calculated to ensure a trade-off between the cost of technology implementation and the profit potential. Therefore, we discuss the economic aspects related to IoT implementation in smart agriculture.

There are several types of costs related to the implementation of IoT in agriculture. We divided them into categories, including (1) the system initialization cost and (2) the system operating cost. The system initialization cost includes hardware purchases (IoT devices, gateways, base station infrastructure). The system operating cost includes service registration cost and the cost of labour to manage IoT devices. Furthermore, additional operating costs include incurred costs from energy consumption, maintenance, data exchange among IoT devices, gateways and cloud servers. According to the opinion of Turgut and Boloni [116], the successful deployment of the IoT technologies will only happen if the customer benefits (customers need to know the benefits and potential) that IoT systems provide exceed their physical value and privacy costs. The businesses participating in the IoT domain will profit and achieve success. We can describe this process using these two conditions, as follows:

$$Success\ of\ IoT\ Applications = \begin{cases} V_{service} > C_{pri} + C_h^{user} + C_{pay},\ Farmer\ Benefits & (1) \\ V_{info} + R_{pay} > C_h^{business},\ Businesses\ Benefits & (2) \end{cases}$$

where

$V_{service}$ is the expected value received by the IoT service users.
C_{pri} is the cost of the loss of privacy.
C_h^{user} is the equipment and hardware costs the user pays.
C_{pay} is the payment for the service fee.
V_{info} is the received information value.
R_{pay} is the received direct payment.
$C_h^{business}$ is the share of the hardware and maintenance costs of the business.

According to the opinion of the service user (farmers or the owner of the farm), Equation (1) shows that the perceived value of the service for the user ($V_{service}$) must be higher than the total of costs, including: the cost of the loss of privacy (C_{pri}), the equipment and hardware costs the user pays (C_h^{user}), and the payments for the service fee (C_{pay}), while the opinion of the service provider, as shown in Equation (2), shows that the received information value (V_{info}) and the received direct payments (R_{pay}) must be higher than the share of the hardware and maintenance costs of the business ($C_h^{business}$).

There is still a gap between service providers and service users (farmers or the owner of the farm), leading to the slow deployment of IoT applications in smart agriculture. In terms of the economic aspect, the analyzed results show that the need for a support policy from regulatory agencies and governments to allow service providers and service users to use IoT-based smart agriculture applications in their infancy can be met. As discussed in [117], to promote smart agriculture, the European Union has issued supportive economic policies, the so-called the European CAP (Common Agricultural Policy), whose annual budget amounts to approximately EUR 59 billion and is paid for by the nations of the EU.

In our view, to be able to apply IoT in the field of smart agriculture, service costs (C_{pay}) and the operating and system initialization cost of IoT (C_h^{user}) needs to constantly be improved and optimized to reduce the cost of the IoT services for farmers. In addition, IoT businesses (service providers) also need to maximize the value of information obtained (V_{info}) to improve the profitability of the service providers.

In reality, service providers may commercially exploit the information received (V_{info}) in the period of providing services for farms, aiming to encourage the deployment of IoT applications in smart agriculture. Nowadays, several IoT platform providers allow free registration and use of services with some limitation conditions regarding services' functionality and ability processing; the number of connected IoT devices; and the number of data stored while premium functions and services charge users a fee.

In addition, one of the significant factors slowing down IoT adoption in agriculture is farmers' knowledge and ability to use IoT devices. In developed countries, this issue can be easily solved due to the accessibility of new technologies of farmers. Otherwise, in developing countries, where the majority of farmers in rural areas have very limited access to advanced technologies, this issue is a significant challenge [118,119].

4.2. Technical Problems

Interference: Deploying a huge number of IoT devices for smart agriculture can cause interference to different network systems, especially some IoT networks using short spectrum bands such as ZigBee, Wi-Fi, Sigfox, and LoRa (See Table 1). Interference can degrade system performance as well as reduce the reliability of IoT ecosystems. IoT networks that use cognitive technology to reuse unlicensed spectra increase the cost of the device. In our opinion, the advent of the 6G network [120] will allow a huge number of devices to connect to the Internet with an extremely high access speed and extremely large bandwidth. The full interference problem of IoT networks will be solved.

Security and Privacy: One of the most important problems of applying IoT in smart agriculture is the security problem, including the protection of data and systems from attacks on the Internet. In regard to system security, IoT devices' limited capacity and ability led to complex encryption algorithms that are impossible to implement on IoT devices. As a result, IoT systems can be attacked using the Internet to gain system control rights; IoT gateways are also attacked via denial of service [121–123]. In addition, cloud servers can be attacked by data spoofing to perform unauthorized tasks that affect the autonomous farming processes of farms. Cloud infrastructures can also be controlled by attackers [124,125]. Several issues of detailed IoT data privacy and security measures have been discussed in [126–128]. According to *Neshenko* et al., the IoT data security issue is one of the biggest problems slowing down IoT adoption in smart agriculture [129].

Regarding data security, the obtained information from IoT systems in farms is collected, processed, and commercially exploited by service providers to varying degrees.

Therefore, one of the most important problems of policies regards the validity and legal status of farm data [130]. In reality, these data are of great value when aggregated and analyzed for large-scale agricultural activities. Consequently, without policies, the data privacy and security of farms can affect the competitive advantage of farmers/farm owners. In our opinion, using cryptography coupled with access keys is a possible solution to solve this problem. Keys could be made available based on a regional user group and to those who contributed to the database. For further complex cases, secure multiparty computation can be used, where the homomorphic encryption method [131,132], or this method combined with the blockchain [133], can be applied for the purpose of balancing privacy and data utility.

In our opinion, the security problems of IoT systems will be an exciting research topic and garner attention for both academia and industry research. An in-depth survey of threats and solutions to improve robustness, trust, and privacy for future IoT systems is presented in [134].

Reliability: Most IoT devices are expected to be deployed outdoors (in fields and farms). Harsh work environments lead to the rapid degradation of IoT devices' quality and can lead to unexpected manufacturer failures. The mechanical safety of IoT devices and systems must be ensured so they can withstand extremes of weather, such as temperature, humidity, rainstorms, and floods [135]. In our opinion, new materials and technologies need to continue to be studied to improve the durability of devices.

The open problems and challenges discussed in this section indicate that for IoT to be widely deployed in the smart agriculture sector, there are still many issues to be solved. Service providers need to reduce the service costs, more effectively exploiting the information collected from the farm. On the other hand, farmers need to improve their skills to be able to apply IoT solutions on their farm to enhance productivity and farming efficiency. Researchers need to continually study and propose optimal solutions and technologies to ensure IoT systems' privacy and security and improve the durability of IoT devices. These are really major challenges and exciting research topics in the future so IoT can be widely applied in the smart agriculture sector.

5. Conclusions

In this study, we presented an overview of IoT and big data for the smart agriculture sector. Several issues related to promoting IoT deployment in the agriculture sector have been discussed in detail. Survey results indicate that many studies have been performed to apply IoT for smart agriculture, aiming to enhance productivity, reduce human labour, and improve production efficiency. The benefits of applying IoT and big data in agriculture were discussed. In addition, we also pointed out the challenges we need to overcome to be able to accelerate the deployment of IoT in smart agriculture. However, there are still some challenges that need to be addressed for IoT solutions to be affordable for the majority of farmers, including small- and medium-scale farm owners. In addition, security technologies need to be continuously improved, but in our opinion, the application of IoT solutions for smart agriculture is inevitable and will enhance productivity, provide clean and green foods, support food traceability, reduce human labour, and improve production efficiency. On the other hand, this survey also points out some interesting research directions for security and communication technologies for IoT. We think that these will be very exciting research directions in the future.

Author Contributions: Conceptualization, V.K.Q., N.V.H., and A.M.; methodology, V.K.Q. and A.M.; validation, N.M.Q., D.V.A., A.M. and N.T.B.; resources, A.M.; writing—original draft preparation, V.K.Q., N.V.H., D.V.A., N.M.Q. and A.M.; writing—review and editing, V.K.Q., G.R., S.L. and A.M.; visualization, V.K.Q., N.V.H., D.V.A., N.M.Q., N.T.B., G.R., S.L. and A.M.; supervision, A.M. All authors have read and agreed to the published version of the manuscript.

Funding: This research received no external funding.

Institutional Review Board Statement: Not applicable.

Informed Consent Statement: Not applicable.

Conflicts of Interest: The authors declare that they have no conflict of interest.

References

1. Quy, V.K.; Nam, V.H.; Linh, D.M.; Ngoc, L.A.; Gwanggil, J. Wireless Communication Technologies for IoT in 5G: Vision, Applications, and Challenges. *Wirel. Commun. Mob. Comput.* **2022**, *2022*, 3229294. [CrossRef]
2. Sinche, S.; Raposo, D.; Armando, N.; Rodrigues, A.; Boavida, F.; Pereira, V.; Silva, J.S. A Survey of IoT Management Protocols and Frameworks. *IEEE Commun. Surv. Tutor.* **2020**, *22*, 1168–1190. [CrossRef]
3. Elijah, O.; Rahman, T.A.; Orikumhi, I.; Leow, C.Y.; Hindia, M.N. An Overview of Internet of Things (IoT) and Data Analytics in Agriculture: Benefits and Challenges. *IEEE Internet Things J.* **2018**, *5*, 3758–3773. [CrossRef]
4. Li, W.; Logenthiran, T.; Phan, V.; Woo, W.L. A Novel Smart Energy Theft System (SETS) for IoT-based Smart Home. *IEEE Internet Things J.* **2019**, *6*, 5531–5539. [CrossRef]
5. Shin, D.; Yun, K.; Kim, J.; Astillo, P.V.; Kim, J.-N.; You, I. A Security Protocol for Route Optimization in DMM-Based Smart Home IoT Networks. *IEEE Access* **2019**, *7*, 142531–142550. [CrossRef]
6. An, J.G.; Le Gall, F.; Kim, J.; Yun, J.; Hwang, J.; Bauer, M.; Zhao, M.; Song, J.S. Toward Global IoT-Enabled Smart Cities Interworking Using Adaptive Semantic Adapter. *IEEE Internet Things J.* **2019**, *6*, 5753–5765. [CrossRef]
7. Cirillo, F.; Gomez, D.; Diez, L.; Maestro, I.E.; Gilbert, T.B.J.; Akhavan, R. Smart City IoT Services Creation through Large-Scale Collaboration. *IEEE Internet Things J.* **2020**, *7*, 5267–5275. [CrossRef]
8. Ammad, M.; Shah, M.A.; Islam, S.U.; Maple, C.; Alaulamie, A.A.; Rodrigues, J.J.P.C.; Mussadiq, S.; Tariq, U. A Novel Fog-Based Multi-Level Energy-Efficient Framework for IoT-Enabled Smart Environments. *IEEE Access* **2020**, *8*, 150010–150026. [CrossRef]
9. Metallidou, C.K.; Psannis, K.E.; Egyptiadou, E.A. Energy Efficiency in Smart Buildings: IoT Approaches. *IEEE Access* **2020**, *8*, 63679–63699. [CrossRef]
10. Quy, V.K.; Nam, V.H.; Linh, D.M.; Ban, N.T.; Han, N.D. Communication Solutions for Vehicle Ad-hoc Network in Smart Cities Environment: A Comprehensive Survey. *Wirel. Pers. Commun.* **2022**, *122*, 2791–2815. [CrossRef]
11. Kiani, F.; Seyyedabbasi, A.; Nematzadeh, S.; Candan, F.; Çevik, T.; Anka, F.A.; Randazzo, G.; Lanza, S.; Muzirafuti, A. Adaptive Metaheuristic-Based Methods for Autonomous Robot Path Planning: Sustainable Ag-ricultural Applications. *Appl. Sci.* **2022**, *12*, 943. [CrossRef]
12. Patle, K.S.; Saini, R.; Kumar, A.; Palaparthy, V.S. Field Evaluation of Smart Sensor System for Plant Disease Prediction Using LSTM Network. *IEEE Sens. J.* **2022**, *22*, 3715–3725. [CrossRef]
13. Vangala, A.; Das, A.K.; Kumar, N.; Alazab, M. Smart Secure Sensing for IoT-Based Agriculture: Blockchain Perspective. *IEEE Sens. J.* **2020**, *21*, 17591–17607. [CrossRef]
14. Citoni, B.; Fioranelli, F.; Imran, M.A.; Abbasi, Q.H. Internet of Things and LoRaWAN-Enabled Future Smart Farming. *IEEE Internet Things Mag.* **2019**, *2*, 14–19. [CrossRef]
15. Kumar, R.; Mishra, R.; Gupta, H.P.; Dutta, T. Smart Sensing for Agriculture: Applications, Advancements, and Challenges. *IEEE Consum. Electron. Mag.* **2021**, *10*, 51–56. [CrossRef]
16. Chang, Y.; Lai, Y. Campus Edge Computing Network Based on IoT Street Lighting Nodes. *IEEE Syst. J.* **2020**, *14*, 164–171. [CrossRef]
17. Sutjarittham, T.; Habibi Gharakheili, H.; Kanhere, S.S.; Sivaraman, V. Experiences with IoT and AI in a Smart Campus for Optimizing Classroom Usage. *IEEE Internet Things J.* **2019**, *6*, 7595–7607. [CrossRef]
18. Rani, S.; Ahmed, S.H.; Shah, S.C. Smart Health: A Novel Paradigm to Control the Chickungunya Virus. *IEEE Internet Things J.* **2019**, *6*, 1306–1311. [CrossRef]
19. Zhou, Z.; Yu, H.; Shi, H. Human Activity Recognition Based on Improved Bayesian Convolution Network to Analyze Health Care Data Using Wearable IoT Device. *IEEE Access* **2020**, *8*, 86411–86418. [CrossRef]
20. Humayun, M.; Jhanjhi, N.; Hamid, B.; Ahmed, G. Emerging Smart Logistics and Transportation Using IoT and Blockchain. *IEEE Internet Things Mag.* **2020**, *3*, 58–62. [CrossRef]
21. Song, Y.; Yu, F.R.; Zhou, L.; Yang, X.; He, Z. Applications of the Internet of Things (IoT) in Smart Logistics: A Comprehensive Survey. *IEEE Internet Things J.* **2021**, *8*, 4250–4274. [CrossRef]
22. Shafique, K.; Khawaja, B.A.; Sabir, F.; Qazi, S.; Mustaqim, M. Internet of Things (IoT) for Next-Generation Smart Systems: A Review of Current Challenges, Future Trends & Prospects for Emerging 5G-IoT Scenarios. *IEEE Access* **2020**, *8*, 23022–23040. [CrossRef]
23. Available online: https://www.un.org/development/desa/en/news/population/world-population-prospects-2019.html (accessed on 7 May 2021).
24. Yang, X.; Shu, L.; Chen, J.; Ferrag, M.A.; Wu, J.; Nurellari, E.; Huang, K. A Survey on Smart Agriculture: Development Modes, Technologies, and Security and Privacy Challenges. *IEEE/CAA J. Autom. Sin.* **2021**, *8*, 273–302. [CrossRef]
25. Ayaz, M.; Ammad-Uddin, M.; Sharif, Z.; Mansour, A.; Aggoune, E.-H.M. Internet-of-Things (IoT)-Based Smart Agriculture: Toward Making the Fields Talk. *IEEE Access* **2019**, *7*, 129551–129583. [CrossRef]
26. Alfred, R.; Obit, J.H.; Chin, C.P.-Y.; Haviluddin, H.; Lim, Y. Towards Paddy Rice Smart Farming: A Review on Big Data, Machine Learning, and Rice Production Tasks. *IEEE Access* **2021**, *9*, 50358–50380. [CrossRef]

27. Zikria, Y.B.; Ali, R.; Afzal, M.K.; Kim, S.W. Next-Generation Internet of Things (IoT): Opportunities, Challenges, and Solutions. *Sensors* **2021**, *21*, 1174. [CrossRef]
28. Kour, V.P.; Arora, S. Recent Developments of the Internet of Things in Agriculture: A Survey. *IEEE Access* **2020**, *8*, 129924–129957. [CrossRef]
29. Saad, A.; Benyamina, A.E.H.; Gamatié, A. Water Management in Agriculture: A Survey on Current Challenges and Technological Solutions. *IEEE Access* **2020**, *8*, 38082–38097. [CrossRef]
30. Tyagi, S.K.S.; Mukherjee, A.; Pokhrel, S.R.; Hiran, K.K. An Intelligent and Optimal Resource Allocation Approach in Sensor Networks for Smart Agri-IoT. *IEEE Sens. J.* **2020**, *21*, 17439–17446. [CrossRef]
31. Li, X.; Pu, T.; Li, L.; Ao, J. Enhanced Sensitivity of GaN-Based Temperature Sensor by Using the Series Schottky Barrier Diode Structure. *IEEE Electron Device Lett.* **2020**, *41*, 601–604. [CrossRef]
32. Gopalakrishnan, S.; Waimin, J.; Raghunathan, N.; Bagchi, S.; Shakouri, A.; Rahimi, R. Battery-Less Wireless Chipless Sensor Tag for Subsoil Moisture Monitoring. *IEEE Sens. J.* **2021**, *21*, 6071–6082. [CrossRef]
33. Udutalapally, V.; Mohanty, S.P.; Pallagani, V.; Khandelwal, V. sCrop: A Novel Device for Sustainable Automatic Disease Prediction, Crop Selection, and Irrigation in Internet-of-Agro-Things for Smart Agriculture. *IEEE Sens. J.* **2020**, *21*, 17525–17538. [CrossRef]
34. Spachos, P.; Gregori, S. Integration of Wireless Sensor Networks and Smart UAVs for Precision Viticulture. *IEEE Internet Comput.* **2019**, *23*, 8–16. [CrossRef]
35. Abdelnour, A.; Buchin, F.; Kaddour, D.; Tedjini, S. Improved Traceability Solution Based on UHF RFID for Cheese Production Sector. *IEEE J. Radio Freq. Identif.* **2018**, *2*, 68–72. [CrossRef]
36. Friha, O.; Ferrag, M.A.; Shu, L.; Maglaras, L.; Wang, X. Internet of Things for the Future of Smart Agriculture: A Comprehensive Survey of Emerging Technologies. *IEEE/CAA J. Autom. Sin.* **2021**, *8*, 718–752. [CrossRef]
37. Farooq, M.S.; Riaz, S.; Abid, A.; Abid, K.; Naeem, M.A. A Survey on the Role of IoT in Agriculture for the Implementation of Smart Farming. *IEEE Access* **2019**, *7*, 156237–156271. [CrossRef]
38. Mishra, D.; Zema, N.R.; Natalizio, E. A High-End IoT Devices Framework to Foster Beyond-Connectivity Capabilities in 5G/B5G Architecture. *IEEE Commun. Mag.* **2021**, *59*, 55–61. [CrossRef]
39. Javed, F.; Afzal, M.K.; Sharif, M.; Kim, B. Internet of Things (IoT) Operating Systems Support, Networking Technologies, Applications, and Challenges: A Comparative Review. *IEEE Commun. Surv. Tutor.* **2018**, *20*, 2062–2100. [CrossRef]
40. Poyen, F.B.; Ghosh, A.; Kundu, P.; Hazra, S.; Sengupta, N. Prototype Model Design of Automatic Irrigation Controller. *IEEE Trans. Instrum. Meas.* **2021**, *70*, 9502217. [CrossRef]
41. Wang, Y.; Rajib, S.M.S.M.; Collins, C.; Grieve, B. Low-Cost Turbidity Sensor for Low-Power Wireless Monitoring of Fresh-Water Courses. *IEEE Sens. J.* **2018**, *18*, 4689–4696. [CrossRef]
42. El-Basioni, B.M.M.; El-Kader, S.M.A. Laying the Foundations for an IoT Reference Architecture for Agricultural Application Domain. *IEEE Access* **2020**, *8*, 190194–190230. [CrossRef]
43. Alam, M.M.; Malik, H.; Khan, M.I.; Pardy, T.; Kuusik, A.; Le Moullec, Y. A Survey on the Roles of Communication Technologies in IoT-Based Personalized Healthcare Applications. *IEEE Access* **2018**, *6*, 36611–36631. [CrossRef]
44. Chettri, L.; Bera, R. A Comprehensive Survey on Internet of Things (IoT) Toward 5G Wireless Systems. *IEEE Internet Things J.* **2020**, *7*, 16–32. [CrossRef]
45. Pal, A.; Kant, K. NFMI: Connectivity for Short-Range IoT Applications. *Computer* **2019**, *52*, 63–67. [CrossRef]
46. Collotta, M.; Pau, G.; Talty, T.; Tonguz, O.K. Bluetooth 5: A Concrete Step Forward toward the IoT. *IEEE Commun. Mag.* **2018**, *56*, 125–131. [CrossRef]
47. Bacco, M.; Berton, A.; Gotta, A.; Caviglione, L. IEEE 802.15.4 Air-Ground UAV Communications in Smart Farming Scenarios. *IEEE Commun. Lett.* **2018**, *22*, 1910–1913. [CrossRef]
48. Gente, R.; Busch, S.F.; Stubling, E.-M.; Schneider, L.M.; Hirschmann, C.B.; Balzer, J.C.; Koch, M. Quality Control of Sugar Beet Seeds With THz Time-Domain Spectroscopy. *IEEE Trans. Terahertz Sci. Technol.* **2016**, *6*, 754–756. [CrossRef]
49. Afsharinejad, A.; Davy, A.; Naftaly, M. Variability of Terahertz Transmission Measured in Live Plant Leaves. *IEEE Geosci. Remote Sens. Lett.* **2017**, *14*, 636–638. [CrossRef]
50. Wang, X.; Zhang, J.; Yu, Z.; Mao, S.; Periaswamy, S.C.G.; Patton, J. On Remote Temperature Sensing Using Commercial UHF RFID Tags. *IEEE Internet Things J.* **2019**, *6*, 10715–10727. [CrossRef]
51. Ali, S.; Glass, T.; Parr, B.; Potgieter, J.; Alam, F. Low Cost Sensor with IoT LoRaWAN Connectivity and Machine Learning-Based Calibration for Air Pollution Monitoring. *IEEE Trans. Instrum. Meas.* **2021**, *70*, 5500511. [CrossRef]
52. Joris, L.; Dupont, F.; Laurent, P.; Bellier, P.; Stoukatch, S.; Redouté, J. An Autonomous Sigfox Wireless Sensor Node for Environmental Monitoring. *IEEE Sens. Lett.* **2019**, *3*, 5500604. [CrossRef]
53. Popli, S.; Jha, R.K.; Jain, S. A Survey on Energy Efficient Narrowband Internet of Things (NBIoT): Architecture, Application & Challenges. *IEEE Access* **2019**, *7*, 16739–16776. [CrossRef]
54. Sundaram, J.P.S.; Du, W.; Zhao, Z. A Survey on LoRa Networking: Research Problems, Current Solutions, and Open Issues. *IEEE Commun. Surv. Tutor.* **2020**, *22*, 371–388. [CrossRef]
55. Xing, C.; Li, F. Unlicensed Spectrum-Sharing Mechanism Based on Wi-Fi Security Requirements Implemented Using Device to Device Communication Technology. *IEEE Access* **2020**, *8*, 135025–135036. [CrossRef]
56. Jiang, X.; Zhang, H.; Yi, E.A.B.; Raghunathan, N.; Mousoulis, C.; Chaterji, S.; Peroulis, D.; Shakouri, A.; Bagchi, S. Hybrid Low-Power Wide-Area Mesh Network for IoT Applications. *IEEE Internet Things J.* **2021**, *8*, 901–915. [CrossRef]

57. Lagen, S.; Giupponi, L.; Goyal, S.; Patriciello, N.; Bojovic, B.; Demir, A.; Beluri, M. New Radio Beam-Based Access to Unlicensed Spectrum: Design Challenges and Solutions. *IEEE Commun. Surv. Tutor.* **2020**, *22*, 8–37. [CrossRef]
58. Shah, S.W.H.; Mian, A.N.; Crowcroft, J. Statistical QoS Guarantees for Licensed-Unlicensed Spectrum Interoperable D2D Communication. *IEEE Access* **2020**, *8*, 27277–27290. [CrossRef]
59. Lu, X.; Petrov, V.; Moltchanov, D.; Andreev, S.; Mahmoodi, T.; Dohler, M. 5G-U: Conceptualizing Integrated Utilization of Licensed and Unlicensed Spectrum for Future IoT. *IEEE Commun. Mag.* **2019**, *57*, 92–98. [CrossRef]
60. Razavieh, A.; Chen, Y.; Ethirajan, T.; Gu, M.; Cimino, S.; Shimizu, T.; Hassan, M.K.; Morshed, T.; Singh, J.; Zheng, W.; et al. Extremely-Low Threshold Voltage FinFET for 5G mmWave Applications. *IEEE J. Electron Devices Soc.* **2021**, *9*, 165–169. [CrossRef]
61. Patriciello, N.; Lagén, S.; Bojović, B.; Giupponi, L. NR-U and IEEE 802.11 Technologies Coexistence in Unlicensed mmWave Spectrum: Models and Evaluation. *IEEE Access* **2020**, *8*, 71254–71271. [CrossRef]
62. Mezzavilla, M.; Polese, M.; Zanella, A.; Dhananjay, A.; Rangan, S.; Kessler, C.; Rappaport, T.S.; Zorzi, M. Public Safety Communications above 6 GHz: Challenges and Opportunities. *IEEE Access* **2018**, *6*, 316–329. [CrossRef]
63. Kassim, M.R.M. IoT Applications in Smart Agriculture: Issues and Challenges. In Proceedings of the IEEE Conference on Open Systems (ICOS), Kota Kinabalu, Malaysia, 17–19 November 2020; pp. 19–24. [CrossRef]
64. Boursianis, A.D.; Papadopoulou, M.S.; Gotsis, A.; Wan, S.; Sarigiannidis, P.; Nikolaidis, S.; Goudos, S.K. Smart Irrigation System for Precision Agriculture—The AREThOU5A IoT Platform. *IEEE Sens. J.* **2020**, *21*, 17539–17547. [CrossRef]
65. López, I.D.; Figueroa, A.; Corrales, J.C. Multi-Dimensional Data Preparation: A Process to Support Vulnerability Analysis and Climate Change Adaptation. *IEEE Access* **2020**, *8*, 87228–87242. [CrossRef]
66. Bhat, S.A.; Huang, N.F. Big Data and AI Revolution in Precision Agriculture: Survey and Challenges. *IEEE Access* **2021**, *9*, 110209–110222. [CrossRef]
67. Roopaei, M.; Rad, P.; Choo, K.R. Cloud of Things in Smart Agriculture: Intelligent Irrigation Monitoring by Thermal Imaging. *IEEE Cloud Comput.* **2017**, *4*, 10–15. [CrossRef]
68. Zhang, X.; Cao, Z.; Dong, W. Overview of Edge Computing in the Agricultural Internet of Things: Key Technologies, Applications, Challenges. *IEEE Access* **2020**, *8*, 141748–141761. [CrossRef]
69. Quy, V.K.; Hau, N.V.; Anh, D.V.; Ngoc, L.A. Smart healthcare IoT applications based on fog computing: Architecture, applications and challenges. *Complex Intell. Syst.* **2021**. [CrossRef]
70. Kopishynska, O.; Utkin, Y.; Galych, O.; Marenych, M.; Sliusar, I. Main Aspects of the Creation of Managing Information System at the Implementation of Precision Farming. In Proceedings of the 2020 IEEE 11th International Conference on Dependable Systems, Services and Technologies (DESSERT), Kyiv, Ukraine, 14–18 May 2020; pp. 404–410. [CrossRef]
71. Soeparno, H.; Perbangsa, A.S.; Pardamean, B. Best Practices of Agricultural Information System in the Context of Knowledge and Innovation. In Proceedings of the 2018 International Conference on Information Management and Technology (ICIMTech), Jakarta, Indonesia, 3–5 September 2018; pp. 489–494. [CrossRef]
72. Zhang, F.; Cao, N. Application and Research Progress of Geographic Information System (GIS) in Agriculture. In Proceedings of the 2019 8th International Conference on Agro-Geoinformatics (Agro-Geoinformatics), Istanbul, Turkey, 16–19 July 2019; pp. 1–5. [CrossRef]
73. Available online: http://www.onfarm.com (accessed on 7 May 2021).
74. Available online: https://www.farmobile.com (accessed on 7 May 2021).
75. Available online: https://www.thoughtworks.com/clients/silent-herdsman (accessed on 7 May 2021).
76. Available online: https://cropx.com (accessed on 7 May 2021).
77. Available online: https://www.farmx.co (accessed on 7 May 2021).
78. Available online: http://www.easyfarm.com (accessed on 7 May 2021).
79. Available online: https://www.kaaproject.org (accessed on 7 May 2021).
80. Available online: https://farmlogs.com (accessed on 7 May 2021).
81. Sengupta, A.; Debnath, B.; Das, A.; De, D. FarmFox: A Quad-Sensor based IoT box for Precision Agriculture. *IEEE Consum. Electron. Mag.* **2021**, *10*, 63–68. [CrossRef]
82. Kaiyi, L.; Hengyuan, K.; Huansheng, M.; Fan, Z. Design of a New Generation of Weather Radar Intelligent Temperature and Humidity Monitoring System Based on ZigBee. In Proceedings of the 2019 International Conference on Meteorology Observations (ICMO), Chengdu, China, 28–31 December 2019; pp. 1–3. [CrossRef]
83. Dong, Y.; Xu, F.; Liu, L.; Du, X.; Ren, B.; Guo, A.; Geng, Y.; Ruan, C.; Ye, H.; Huang, W.; et al. Automatic System for Crop Pest and Disease Dynamic Monitoring and Early Forecasting. *IEEE J. Sel. Top. Appl. Earth Obs. Remote Sens.* **2020**, *13*, 4410–4418. [CrossRef]
84. Ghandar, A.; Ahmed, A.; Zulfiqar, S.; Hua, Z.; Hanai, M.; Theodoropoulos, G. A Decision Support System for Urban Agriculture Using Digital Twin: A Case Study with Aquaponics. *IEEE Access* **2021**, *9*, 35691–35708. [CrossRef]
85. Pasha, A.K.; Mulyana, E.; Hidayat, C.; Ramdhani, M.A.; Kurahman, O.T.; Adhipradana, M. System Design of Controlling and Monitoring on Aquaponic Based on Internet of Things. In Proceedings of the 2018 4th International Conference on Wireless and Telematics (ICWT), Nusa Dua, Bali, Indonesia, 12–13 July 2018; pp. 1–5. [CrossRef]
86. Lee, C.; Jhang, J. System Design for Internet of Things Assisted Urban Aquaponics Farming. In Proceedings of the 2019 IEEE 8th Global Conference on Consumer Electronics (GCCE), Osaka, Japan, 15–18 October 2019; pp. 986–987. [CrossRef]
87. Wei, Y.; Li, W.; An, D.; Li, D.; Jiao, Y.; Wei, Q. Equipment and Intelligent Control System in Aquaponics: A Review. *IEEE Access* **2019**, *7*, 169306–169326. [CrossRef]

88. Marcu, A.-E.; Suciu, G.; Olteanu, E.; Miu, D.; Drosu, A.; Marcu, I. IoT System for Forest Monitoring. In Proceedings of the 42nd International Conference on Telecommunications and Signal Processing (TSP), Budapest, Hungary, 1–3 July 2019; pp. 629–632. [CrossRef]
89. Zou, W.; Jing, W.; Chen, G.; Lu, Y.; Song, H. A Survey of Big Data Analytics for Smart Forestry. *IEEE Access* **2019**, *7*, 46621–46636. [CrossRef]
90. Essa, S.; Petra, R.; Uddin, M.R.; Suhaili, W.S.H.; Ilmi, N.I. IoT-Based Environmental Monitoring System for Brunei Peat Swamp Forest. In Proceedings of the 2020 International Conference on Computer Science and Its Application in Agriculture (ICOSICA), Bogor, Indonesia, 16–17 September 2020; pp. 1–5. [CrossRef]
91. Marx, A.; Tetteh, G.O. A Forest Vitality and Change Monitoring Tool Based on RapidEye Imagery. *IEEE Geosci. Remote Sens. Lett.* **2017**, *14*, 801–805. [CrossRef]
92. Chaudhry, A.A.; Mumtaz, R.; Hassan Zaidi, S.M.; Tahir, M.A.; Muzammil School, S.H. Internet of Things (IoT) and Machine Learning (ML) enabled Livestock Monitoring. In Proceedings of the 2020 IEEE 17th International Conference on Smart Communities: Improving Quality of Life Using ICT, IoT and AI (HONET), Charlotte, NC, USA, 14–16 December 2020; pp. 151–155. [CrossRef]
93. Yang, Y.; Ren, R.; Johnson, P.M. VetLink: A Livestock Disease-Management System. *IEEE Potentials* **2020**, *39*, 28–34. [CrossRef]
94. Ma, S.; Yao, Q.; Masuda, T.; Higaki, S.; Yoshioka, K.; Arai, S.; Takamatsu, S.; Itoh, T. Development of Noncontact Body Temperature Monitoring and Prediction System for Livestock Cattle. *IEEE Sens. J.* **2021**, *21*, 9367–9376. [CrossRef]
95. Lee, G.; Kim, M.; Koroki, K.; Ishimoto, A.; Sakamoto, S.H.; Ieiri, S. Wireless IC Tag Based Monitoring System for Individual Pigs in Pig Farm. In Proceedings of the 2019 IEEE 1st Global Conference on Life Sciences and Technologies (LifeTech), Osaka, Japan, 12–14 March 2019; pp. 168–170. [CrossRef]
96. Tradigo, G.; Vizza, P.; Veltri, P.; Lambardi, P.; Caligiuri, F.M.; Caligiuri, G.; Guzzi, P.H. SISTABENE: An information system for the traceability of agricultural food production. In Proceedings of the 2019 IEEE International Conference on Bioinformatics and Biomedicine (BIBM), San Diego, CA, USA, 18–21 November 2019; pp. 2304–2309. [CrossRef]
97. Wang, L.; Xu, L.; Zheng, Z.; Liu, S.; Li, X.; Cao, L.; Li, J.; Sun, C. Smart Contract-Based Agricultural Food Supply Chain Traceability. *IEEE Access* **2021**, *9*, 9296–9307. [CrossRef]
98. Cui, P.; Dixon, J.; Guin, U.; Dimase, D. A Blockchain-Based Framework for Supply Chain Provenance. *IEEE Access* **2019**, *7*, 157113–157125. [CrossRef]
99. Abdulhussein, A.B.; Hadi, A.K.; Ilyas, M. Design a Tracing System for a Seed Supply Chain Based on Blockchain. In Proceedings of the 2020 3rd International Conference on Engineering Technology and its Applications (IICETA), Najaf, Iraq, 6–7 September 2020; pp. 209–214. [CrossRef]
100. Kong, S.; López-Salcedo, J.A.; Wu, Y.; Kim, E. IEEE Access Special Section Editorial: GNSS, Localization, and Navigation Technologies. *IEEE Access* **2019**, *7*, 131649–131652. [CrossRef]
101. Alatise, M.B.; Hancke, G.P. A Review on Challenges of Autonomous Mobile Robot and Sensor Fusion Methods. *IEEE Access* **2020**, *8*, 39830–39846. [CrossRef]
102. Kim, J.; Kim, S.; Ju, C.; Son, H.I. Unmanned Aerial Vehicles in Agriculture: A Review of Perspective of Platform, Control, and Applications. *IEEE Access* **2019**, *7*, 105100–105115. [CrossRef]
103. Zhou, K.; Meng, Z.; He, M.; Hou, J.; Li, T. Design and Test of a Sorting Device Based on Machine Vision. *IEEE Access* **2020**, *8*, 27178–27187. [CrossRef]
104. Kurtser, P.; Ringdahl, O.; Rotstein, N.; Berenstein, R.; Edan, Y. In-Field Grape Cluster Size Assessment for Vine Yield Estimation Using a Mobile Robot and a Consumer Level RGB-D Camera. *IEEE Robot. Autom. Lett.* **2020**, *5*, 2031–2038. [CrossRef]
105. Zhou, X.; Zhou, J. Data-Driven Driving State Control for Unmanned Agricultural Logistics Vehicle. *IEEE Access* **2020**, *8*, 65530–65543. [CrossRef]
106. Favenza, A.; Imam, R.; Dovis, F.; Pini, M. Detecting water using UAV-based GNSS-Reflectometry data and Artificial Intelligence. In Proceedings of the 2019 IEEE International Workshop on Metrology for Agriculture and Forestry (MetroAgriFor), Portici, Italy, 24–26 October 2019; pp. 7–12. [CrossRef]
107. Available online: http://www.claasofamerica.com (accessed on 7 May 2021).
108. Tripathy, P.K.; Tripathy, A.K.; Agarwal, A.; Mohanty, S.P. MyGreen: An IoT-Enabled Smart Greenhouse for Sustainable Agriculture. *IEEE Consum. Electron. Mag.* **2021**, *10*, 57–62. [CrossRef]
109. Geng, X.; Zhang, Q.; Wei, Q.; Zhang, T.; Cai, Y.; Liang, Y.; Sun, X. A Mobile Greenhouse Environment Monitoring System Based on the Internet of Things. *IEEE Access* **2019**, *7*, 135832–135844. [CrossRef]
110. Fei, X.; Xiao, W.; Yong, X. Development of Energy Saving and Rapid Temperature Control Technology for Intelligent Greenhouses. *IEEE Access* **2021**, *9*, 29677–29685. [CrossRef]
111. Subahi, A.F.; Bouazza, K.E. An Intelligent IoT-Based System Design for Controlling and Monitoring Greenhouse Temperature. *IEEE Access* **2020**, *8*, 125488–125500. [CrossRef]
112. Misra, N.N.; Dixit, Y.; Al-Mallahi, A.; Bhullar, M.S.; Upadhyay, R.; Martynenko, A. IoT, Big Data and Artificial Intelligence in Agriculture and Food Industry. *IEEE Internet Things J.* **2020**, 1. [CrossRef]
113. Ullah, I.; Fayaz, M.; Naveed, N.; Kim, D. ANN Based Learning to Kalman Filter Algorithm for Indoor Environment Prediction in Smart Greenhouse. *IEEE Access* **2020**, *8*, 159371–159388. [CrossRef]

114. Muñoz, M.; Guzmán, J.L.; Sánchez, J.A.; Rodríguez, F.; Torres, M.; Berenguel, M. A New IoT-based Platform for Greenhouse Crop Production. *IEEE Internet Things J.* **2020**, 1. [CrossRef]
115. Bai, X.; Wang, Z.; Sheng, L.; Wang, Z. Reliable Data Fusion of Hierarchical Wireless Sensor Networks with Asynchronous Measurement for Greenhouse Monitoring. *IEEE Trans. Control Syst. Technol.* **2019**, *27*, 1036–1046. [CrossRef]
116. Turgut, D.; Boloni, L. Value of Information and Cost of Privacy in the Internet of Things. *IEEE Commun. Mag.* **2017**, *55*, 62–66. [CrossRef]
117. Rousi, M.; Sitokonstantinou, V.; Meditskos, G.; Papoutsis, I.; Gialampoukidis, I.; Koukos, A.; Karathanassi, V.; Drivas, T.; Vrochidis, S.; Kontoes, C.; et al. Semantically Enriched Crop Type Classification and Linked Earth Observation Data to Support the Common Agricultural Policy Monitoring. *IEEE J. Sel. Top. Appl. Earth Obs. Remote Sens.* **2021**, *14*, 529–552. [CrossRef]
118. Ammad Uddin, M.; Ayaz, M.; Aggoune, E.-H.M.; Mansour, A.; Le Jeune, D. Affordable Broad Agile Farming System for Rural and Remote Area. *IEEE Access* **2019**, *7*, 127098–127116. [CrossRef]
119. Joshi, A.; Dandekar, I.; Hargude, N.; Shrotri, A.P.; Dandekar, A.R. Application of Internet of the Things(IOT) for the Water Conservation and Entrepreneurship in the Rural Area. In Proceedings of the 2019 IEEE Pune Section International Conference (PuneCon), Pune, India, 18–20 December 2019; pp. 1–4. [CrossRef]
120. De Lima, C.; Belot, D.; Berkvens, R.; Bourdoux, A.; Dardari, D.; Guillaud, M.; Isomursu, M.; Lohan, E.-S.; Miao, Y.; Barreto, A.N.; et al. Convergent Communication, Sensing and Localization in 6G Systems: An Overview of Technologies, Opportunities and Challenges. *IEEE Access* **2021**, *9*, 26902–26925. [CrossRef]
121. Sandal, Y.S.; Pusane, A.E.; Kurt, G.K.; Benedetto, F. Reputation Based Attacker Identification Policy for Multi-Access Edge Computing in Internet of Things. *IEEE Trans. Veh. Technol.* **2020**, *69*, 15346–15356. [CrossRef]
122. Wang, J.; Hao, S.; Wen, R.; Zhang, B.; Zhang, L.; Hu, H.; Lu, R. IoT-Praetor: Undesired Behaviors Detection for IoT Devices. *IEEE Internet Things J.* **2021**, *8*, 927–940. [CrossRef]
123. Jia, Y.; Zhong, F.; Alrawais, A.; Gong, B.; Cheng, X. FlowGuard: An Intelligent Edge Defense Mechanism against IoT DDoS Attacks. *IEEE Internet Things J.* **2020**, *7*, 9552–9562. [CrossRef]
124. Agrawal, N.; Tapaswi, S. Defense Mechanisms against DDoS Attacks in a Cloud Computing Environment: State-of-the-Art and Research Challenges. *IEEE Commun. Surv. Tutor.* **2019**, *21*, 3769–3795. [CrossRef]
125. Somani, G.; Gaur, M.S.; Sanghi, D.; Conti, M.; Rajarajan, M.; Buyya, R. Combating DDoS Attacks in the Cloud: Requirements, Trends, and Future Directions. *IEEE Cloud Comput.* **2017**, *4*, 22–32. [CrossRef]
126. Wang, J.; Zhu, R.; Liu, S. A Differentially Private Unscented Kalman Filter for Streaming Data in IoT. *IEEE Access* **2018**, *6*, 6487–6495. [CrossRef]
127. Yu, Y.; Ding, Y.; Zhao, Y.; Li, Y.; Zhao, Y.; Du, X.; Guizani, M. LRCoin: Leakage-Resilient Cryptocurrency Based on Bitcoin for Data Trading in IoT. *IEEE Internet Things J.* **2019**, *6*, 4702–4710. [CrossRef]
128. Ali, I.; Kaur, S.; Khamparia, A.; Gupta, D.; Kumar, S.; Khanna, A.; Al-Turjman, F. Security Challenges and Cyber Forensic Ecosystem in IoT Driven BYOD Environment. *IEEE Access* **2020**, *8*, 172770–172782. [CrossRef]
129. Neshenko, N.; Bou-Harb, E.; Crichigno, J.; Kaddoum, G.; Ghani, N. Demystifying IoT Security: An Exhaustive Survey on IoT Vulnerabilities and a First Empirical Look on Internet-Scale IoT Exploitations. *IEEE Commun. Surv. Tutor.* **2019**, *21*, 2702–2733. [CrossRef]
130. Chaterji, S.; DeLay, N.; Evans, J.; Mosier, N.; Engel, B.; Buckmaster, D.; Ladisch, M.R.; Chandra, R. Lattice: A Vision for Machine Learning, Data Engineering, and Policy Considerations for Digital Agriculture at Scale. *IEEE Open J. Comput. Soc.* **2021**, *2*, 227–240. [CrossRef]
131. Jin, B.; Jiang, D.; Xiong, J.; Chen, L.; Li, Q. D2D Data Privacy Protection Mechanism Based on Reliability and Homomorphic Encryption. *IEEE Access* **2018**, *6*, 51140–51150. [CrossRef]
132. Harn, L.; Hsu, C.-F.; Xia, Z.; He, Z. He Lightweight Aggregated Data Encryption for Wireless Sensor Networks (WSNs). *IEEE Sens. Lett.* **2021**, *5*, 1–4. [CrossRef]
133. Jia, B.; Zhang, X.; Liu, J.; Zhang, Y.; Huang, K.; Liang, Y. Liang Blockchain-Enabled Federated Learning Data Protection Aggregation Scheme With Differential Privacy and Homomorphic Encryption in IioT. *IEEE Trans. Ind. Inform.* **2022**, *18*, 4049–4058. [CrossRef]
134. Chen, L.; Thombre, S.; Jarvinen, K.; Lohan, E.S.; Alén-Savikko, A.; Leppakoski, H.; Bhuiyan, M.Z.H.; Bu-Pasha, S.; Ferrara, G.N.; Honkala, S.; et al. Robustness, Security and Privacy in Location-Based Services for Future IoT: A Survey. *IEEE Access* **2017**, *5*, 8956–8977. [CrossRef]
135. Ballal, K.D.; Dittmann, L.; Ruepp, S.; Petersen, M.N. IoT Devices Reliability Study: Multi-RAT Communication. In Proceedings of the 2020 IEEE 6th World Forum on Internet of Things (WF-IoT), New Orleans, LA, USA, 2–16 June 2020; pp. 1–2. [CrossRef]

applied sciences

MDPI

Article

Application of Remote Sensing Tools to Assess the Land Use and Land Cover Change in Coatzacoalcos, Veracruz, Mexico

Josept David Revuelta-Acosta [1], Edna Suhail Guerrero-Luis [1], Jose Eduardo Terrazas-Rodriguez [1], Cristian Gomez-Rodriguez [2] and Gerardo Alcalá Perea [3,*]

1 Department of Civil Engineering, Universidad Veracruzana, Coatzacoalcos, Veracruz 96535, Mexico; jrevuelta@uv.mx (J.D.R.-A.); zS16014243@estudiantes.uv.mx (E.S.G.-L.); eterrazas@uv.mx (J.E.T.-R.)
2 Department of Mechanical Engineering, Universidad Veracruzana, Coatzacoalcos, Veracruz 96535, Mexico; crisgomez@uv.mx
3 Centro de Investigacion en Recursos Energeticos y Sustentables, Universidad Veracruzana, Coatzacoalcos, Veracruz 96535, Mexico
* Correspondence: galcala@uv.mx

Citation: Revuelta-Acosta, J.D.; Guerrero-Luis, E.S.; Terrazas-Rodriguez, J.E.; Gomez-Rodriguez, C.; Alcalá Perea, G. Application of Remote Sensing Tools to Assess the Land Use and Land Cover Change in Coatzacoalcos, Veracruz, Mexico. *Appl. Sci.* **2022**, *12*, 1882. https://doi.org/10.3390/app12041882

Academic Editors: Dimitrios S. Paraforos and Anselme Muzirafuti

Received: 24 December 2021
Accepted: 4 February 2022
Published: 11 February 2022

Abstract: Land use and land cover (LULC) change has become an important research topic for global environmental change and sustainable development. As an important part of worldwide land conservation, sustainable development and management of water resources, developing countries must ensure the use of innovative technology and tools that support their various decision making systems. This study provides the most recent LULC change analysis for the last six years (2015–2021) of Coatzacoalcos, Veracruz, Mexico, one of the most important petrochemical cities in the world and host of the ongoing Interoceanic Corridor project. The analysis was carried out using Landsat 8 Operational Land Imager (OLI) satellite images, ancillary data and ground-based surveys and the Normalized Difference Vegetation Index (NDVI) to identify and to ameliorate the discrimination between four main macro-classes and fourteen classes. The LULC classification was performed using the maximum likelihood classifier (MLC) to produce maps for each year, as it was found to be the best approach when compared to minimum distance (MDM) and spectral angle mapping (SAM) methods. The macro-classes were water, built-up, vegetation and bare soil, whereas the classes were an improved classification within those. Our study achieved both user accuracy (UA) and producer accuracy (PA) above 90% for the proposed macro-classes and classes. The average Kappa coefficient for macro-classes was 0.93, while for classes it was 0.96, both comparable to previous studies. The results from the LULC analysis show that residential, industry and commercial areas slowed down their growth throughout the study period. These changes were associated with socio-economical drivers such as insecurity and lack of economic investments. Groves and trees presented steady behaviors, with small increments during the five-year period. Swamps, on the other hand, significantly degraded, being about 2% of the study area in 2015 and 0.93% in 2021. Dunes and medium and high vegetation densities ($\sim$ 80%) transitioned mostly to low vegetation densities. This behavior is associated with rainfall below the annual reference and increments of surface runoff due to the loss of vegetation cover. Lastly, the present study seeks to highlight the importance of remote sensing for a better understanding of the dynamics between human–nature interactions and to provide information to assist planners and decision-makers for more sustainable land development.

Keywords: remote sensing; land use classification; GIS; Coatzacoalcos

1. Introduction

Land use and land cover (LULC) changes are among the research topics more recurrent in remote sensing [1–4]. Remote sensing provides comprehensive and extensive information to understand the interaction between terrestrial ecosystems and their responses to environmental factors [5–11]. This technology is considered a powerful source to obtain information from terrestrial surface characteristics at different temporal and spatial scales [12–16]. For more than half a century, scientists have used several types of remote sensing data. From those, the most commonly used source is the Landsat satellite based-data [17]. One of main uses of the Landsat-based satellite images is the identification or classification of LULC changes [18–22]. Some interesting conclusions have been driven through the use of remote sensing in LULC changes. It has been observed that changes in land uses are dynamic, non-linear interactions between humans and nature and are driven by complex stochastic processes [23]; worldwide changes in LULC for the last 300 years have shown gains in agriculture and losses in forest [24]; rapid population growth in Africa has been identified as the main driving force for the expansion of agricultural land, whereas in developing countries, urbanization dynamics are attributed to demographic factors [25] and transitions from cropland to urban and water bodies areas have been identified as an important trend in Europe [26].

Some of the more remarkable research found in literature is the work conducted by Yuan et al. [27]. Their research consists of the use of multi-temporal Landsat Thematic Mapper data to map and quantify the LULC changes in seven counties of the Twin Cities located in the metropolitan area of the state of Minnesota, United States. They used satellite images corresponding to the years 1986, 1991, 1998 and 2002. Yuan et al. demonstrated an urban zone increment from 23.7 to 32.8% in the study area. Surfaces such as rural areas, croplands, forests and wetlands decreased from 69.6 to 60.5%. Another unique example is the work carried out by Demissie et al. [28], where Landsat satellite data from 1973 to 2015 were classified to study land use change and its possible causes in Gonder, Ethiopia. The study showed that about 60.1% of the area experienced land use changes within the study period.

In Mexico, little research related to the quantification of LULC changes has been found within the literary review. One of the few works found was carried out by Colditz et al. [29]. Their research presents a methodology to develop a land use map in Mexico for the year 2005. The scheme was based on time series from a Moderate Resolution Imaging Spectroradiometer (MODIS) with a 250 m resolution and an extensive sampling of data for the different geographic zones of the Mexican surface. The results showed a map with an overall precision of 82.5% (Kappa statistic = 0.79). A further evaluation with 780 randomly generated samples within the classes with referenced field data indicated a precision equal to 83.4% (Kappa statistic = 0.80). Another study found in the literature includes the generation of a land use map of the Latin American and Caribbean zone for the year 2008. This project was developed within the framework of the project of the Latina American Network for Monitoring and Studying of Natural Resources (SERENA) [30]. Similar to Colditz et al. [29], time series and decision trees with MODIS data (250 m) were used for the classification of land use changes. The discrete SERENA model showed an overall accuracy of 84%. Other uses of remote sensing in Mexico include the detection of chinampas in the Xaltocan area in the Northern Basin of Mexico [31], the evolution of sea temperature on the west coast of Baja California [32] and/or hydropower assessment [33].

Due to the limitations observed in the use of remote sensing in Mexico, such as limited research using remote sensing and map course resolution, this study has integrated remote sensing, Geographic Information Systems (GIS) analysis, field sampling and image revision to assess the potential of the use of Landsat 8 satellite data to assess and monitor land use changes and coverage changes in the southeastern region of the state of Veracruz, Mexico, located in the southern coastal area of the Gulf of Mexico. Three widely used classification methods are explored to select the more appropriate algorithm for LULC classification under the same level of information previously extracted from field data, digital maps and ancillary data. This comparison allows for conclusions on the advantages and drawbacks of current classification algorithms. The analysis encompasses the definitions of macro-classes of land use and their derived classes. The Normalized Difference Vegetation Index, also known as NDVI, was used to identify and improve discrimination between macro-classes and classes within the study area. A supervised classification method, the Maximum

Likelihood Classification (MLC), was applied in this study due to its availability within the GIS applications in addition to it not requiring an extensive training process [34,35]. The most important advantage of the MLC as a parametric classifier is that it takes into account the variance within the selected classes and that, for normally distributed data, the MLC method performs a better qualification than other methods, namely, decision trees [34,36].

Additionally, this study proposes a methodology to obtain updated temporal information of the change in land use and land cover for the Coatzacoalcos area using Landsat 8 satellite imagery, which has a higher resolution with the available MODIS maps. The mapping of land use and its changes are critical and important factors for sustainable development as well as the monitoring of environmental impacts. The final product of this research will help government agencies to make decisions on urban development and preservation of available natural resources, since one of the emblematic projects in Mexico known as the Interoceanic Corridor is being executed in the present year. Lastly, this piece of research is going to serve as the baseline for additional contributions on LULC change prediction with more complex analyses of the drivers forcing these landscape modifications.

2. Materials and Methods

2.1. Study Area

The study area is the southeastern region of the state of Veracruz, Mexico, in which one of the most important pretrochemical ports of the country is located at the city of Coatzacoalcos. This city is surrounded by the Coatzacoalcos and Tonalá rivers and extends between latitudes 18.06° and 18.21° and longitudes −94.22° and −94.64° (Figure 1). The area of study extends over an area about 220 km^2. The port of Coatzacoalcos is dominated by the petrochemical sector [37]. Four of the petrochemical complexes located near the city make the city one of the most important oil areas in the world [38]. Coatzacoalcos has been a transportation hub for hundreds of years; it is connected by air, land, sea and railroad to the rest of the world. Currently, the Mexican government is working on a project called the Interoceanic Corridor of the Isthmus of Tehuantepec, which represents the modernization and installation of industrial parks inside the industrial areas of the ports of Coatzacoalcos and Salina Cruz, Mexico [39]. This project will represent economic and population growth within the city. As a result, land use assessments and projections have become priorities to help both state and federal government make decisions for the projected economical growth.

Physiographically, the study area has an average elevation of 10 meters above sea level. The soils found in the region are clayey lateritic Acrisol and Luvisol in high elevations and Gleysol, Cambisol, Vertisol and Nitosol in plains [40]. The vegetation consists of riparian vegetation, swamps and wetlands in floodable areas, mesophilic mountain pine oak forests in high areas and high evergreen forests in hills and acahual areas in abandoned and cultivated pastures. The climate of the region is a warm humid climate with abundant rains in summer. The average annual temperature varies between 24 and 25 °C, while the average annual precipitation varies between 1500 and 2500 mm [41].

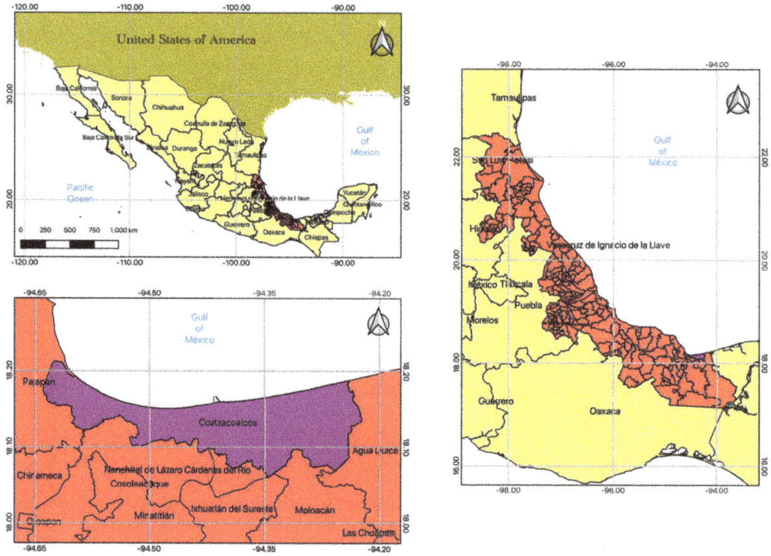

Figure 1. Location of Coatzacoalcos city in Veracruz, Mexico.

2.2. Data Acquisition and Processing

The images used in this study were obtained from the Landsat 8 satellite, which captures multispectral images of 30 m resolution sampled with the OLI (Operational Land Imager) sensor. The Landsat 8 satellite also provides a 15 m panchromatic band, allowing for a greater spatial resolution. All images were geometrically corrected and acquired in level 1 (L1). The images can be directly downloaded from the EarthExplorer gateway of the United States Geological Survey (USGS) [17]. Table 1 displays the details of the images used in this study.

Table 1. Images of Landsat 8 OLI used in the present study.

Satellite	Acquisition Date	Bands	Spectral Resolution	Path/Row	Cloudiness
Landsat 8 OLI	14/06/2015	2–7	30 m	023/047	30%
Landsat 8 OLI	26/06/2016	2–7	30 m	023/047	25%
Landsat 8 OLI	02/06/2017	2–7	30 m	023/047	9%
Landsat 8 OLI	23/06/2018	2–7	30 m	023/047	13%
Landsat 8 OLI	10/06/2019	2–7	30 m	023/047	0%
Landsat 8 OLI	12/06/2020	2–7	30 m	023/047	5%
Landsat 8 OLI	23/06/2021	2–7	30 m	023/047	0%

The Landsat 8 OLI products used consist of band 2 (0.45–0.51 µm), band 3 (0.53–0.59 µm), band 4 (0.64–0.67 µm), band 5 (0.85–0.88 µm), band 6 (1.57–1.65 µm) and band 7 (2.11–2.29 µm) [42]. All images were processed using the free software Quantum GIS (QGIS) version 3.14.16. Atmospheric correction was performed through the Semi-Automatic Classification Plugin (SCP) tool based on the Dark Object Subtraction (DOS1) algorithm [43]. The panchromatic band was used to provide higher spatial resolution. This is achieved using Gram–Schmidt Pan-Sharpening to produce a high-resolution color image that improves the training set and the classification process. Google Earth images were acquired according to acquisition dates (Table 1) and geoferenced using QGIS. The closest dates to the acquisition date were selected. Field data included collection of georeferenced points and their corresponding land use for the last two years of the analyzed periods. Figure 2 shows the methodology used for the study.

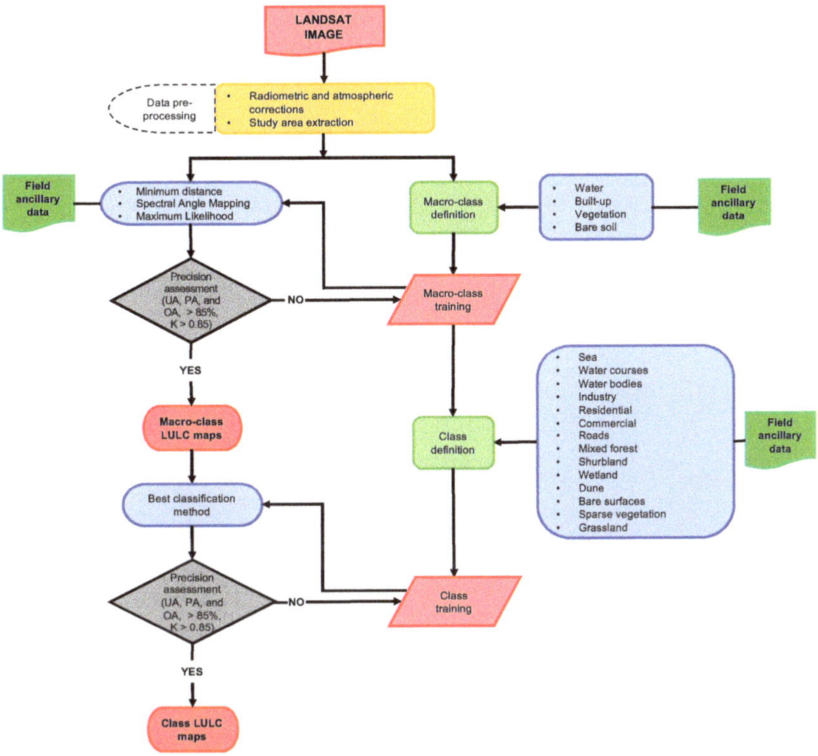

Figure 2. Flow diagram of the LULC classification methodology implemented in the SCP tool.

The DOS1 atmospheric correction method was selected, as it is the most widely used method for LULC change detection [42]. This method uses the properties of the images. For instance, pixels conforming elements such as shadows, water or forests are considered dark objects when the magnitude of the reflectance is close to zero, namely, when the reflectance is less than or equal to 1%, and all the analyzed pixels are assumed dark elements. This assumption indicates that pixels receiving low values of solar radiation (~100% shade) registered by the satellite correspond to atmospheric dispersion, which generally is caused by effects of the topography [44]. Once a dark object is found in the image, the minimum reflectance value of the total Digital Number (DN) histogram is assigned to the object. This minimum is then modified by the effects of atmospheric correction [45]. The surface reflectance is calculated using the following expression:

$$\rho_s = \frac{\pi(L_\lambda - L_p)d^2}{ESUN_\lambda \cos\theta_s} \tag{1}$$

where ρ_s is the surface reflectance, d is the distance to the sun, $ESUN_\lambda$ is the mean solar exo-atmospheric irradiance, L_λ is the solar spectral radiance to the satellite, θ_s is the zenith angle and λ_p is the path radiance given by [46]

$$L_p = L_{min} - L_{DO1\%} \tag{2}$$

where L_{min} is the radiance corresponding to a digital count value for which the sum of all the pixels with digital counts lower than or equal to this value is equal to 0.01% of all the pixels from the image considered [46] and $L_{DO1\%}$ is the radiance of Dark Object assumed to have reflectance value of 0.01.

2.3. NDVI Classification

Based on the interpretation of the Normalized Difference Vegetation Index (NDVI), the analysis of Google-Earth-extracted images, RGB (red, green, blue) images, panchromatic bands and ground-based georeferenced information, NDVI thresholds were proposed to classify the area of study into four macro-classes during the analysis period (2015–2021), as shown in Table 2.

Table 2. Suitable NDVI ranges identified for the LULC macro-classes.

Macro-Class	NDVI Range
Water	−0.010–−0.133
Built-up	0.040–0.232
Bare soil	0.145–0.259
Vegetation	0.694–0.895

The NDVI thresholds implemented in the proposed classification were selected after an extensive literature review of previous studies where the NDVI was used as the spectral index of diverse land covers [47–52]. The NDVI values were calculated by the equation proposed in [53,54]:

$$NDVI = \frac{\phi_{nir} - \phi_{red}}{\phi_{nir} + \phi_{red}} \quad (3)$$

where ϕ_{nir} is the near-infrared reflectance and ϕ_{red} is the red-band reflectance. Band 4 (0.64–0.67 µm) and band 5 (0.85–0.88 µm) represent ϕ_{nir} and ϕ_{red}, respectively, in the Landsat products. The NDVI values vary from −1 to 1. Values close to 1 represent vegetation in optimal environmental conditions, whereas low values of NDVI indicate low vegetation density or a different land use. After the macro-class definition, a second classification was performed to refine the LULC characteristics through a more extensive discrimination of pixels. These new classes were defined according to the CORINE land cover inventory proposed by the European Environmental Agency [55]. This classification process required the use of the ancillary data and field data to assign new information to the training set. This assignation was repeated until a suitable spatial distribution was obtained. This new classification consisted of 14 classes, as shown in Table 3.

2.4. Classifier Comparison

In addition to its social and environmental objectives, this study seeks to evaluate the potential and the drawbacks of the minimum distance classifier [56], the spectral angle mapping classifier [57] and the maximum likelihood classifier [34,35,58] in deriving information on LULC. Although there are more classifications, such as artificial neural networks [34] and parallelepiped classification [56], the comparison was limited to the three selected methods because they are often used to classify LULC. However, limited information exists when the same training input and ancillary data are used to perform the classification process. Here, the comparison was set under the assumption that the same amount of information was available.

Table 3. Land use and land cover (LULC) classes and descriptions.

LULC Type	Description
	Water
Sea	Area of salt water that surrounds landmasses
Water courses	Natural or artificial water courses serving as water drainage channels
Water bodies	Natural or artificial water bodies with presence of standing water surface during most of the year
	Built-up
Industry	Any factory processing of raw materials and manufacture of goods
Residence	Land use in which housing predominates
Commercial	Areas for retail sales, services, offices, etc.
Roads	Roads of any artificial paved surface material (asphalt or concrete)
	Vegetation
Mixed forest	Vegetation formation composed principally of trees, including shrub and bush understorey
Shrub	A shrub or clump of shrubs with stems of moderate length
Wetland	Areas of low-lying, uncultivated ground where water collects
	Bare soil
Dune	A mound of sand or other loose sediment formed by the wind
Bare surfaces	Soil with low vegetation density
Sparse vegetation	Soil with a mix of low and high vegetation densities
Grassland	Soil with high vegetation density, such as grasslands

2.4.1. Minimum Distance Classifier

The minimum distance method (MDM) calculates the Euclidean distance $d(x,y)$ between the spectral signatures given in the training data set and the spectral signatures of the image pixels. The spectral distance is calculated through the following expression:

$$d(x,y) = \sqrt{\sum_{i=1}^{n}(x_i - y_i)^2} \tag{4}$$

where x is the spectral signature vector of an image, y is the spectral signature vector of the training area and n is the number of bands of the image. Once the spectral distance is computed for every pixel, the class with the closest spectral signature to the training set is assigned according to the following discrimination function [56]:

$$x \epsilon C_k \Longleftrightarrow d(x,y_k) < d(x,y_j) \tag{5}$$

where C_k is the land cover macro-class or class, y_k is the spectral signature of class k and y_j is the spectral signature of class j. This equation is valid when $k \neq j$.

2.4.2. Spectral Angle Mapping Classifier

The spectral angle mapping (SAM) algorithm computes the spectral angle between the spectral signatures of the image pixels and the training spectral signatures. The spectral angle θ is given by

$$\theta(x,y) = \arccos\left(\frac{\sum_{i=1}^{n} x_i y_i}{\sqrt{\sum_{i=1}^{n} x_i^2}\sqrt{\sum_{i=1}^{n} y_i^2}}\right) \tag{6}$$

Thus, a pixel resides in the macro-class or class that has the lowest spectral angle, as provided by

$$x \epsilon C_k \Longleftrightarrow \theta(x,y_k) < \theta(x,y_j) \tag{7}$$

where $k \neq j$.

2.4.3. Maximum Likelihood Classifier

The maximum likelihood classifier (MLC) is based on the probability that a pixel belongs to or is within macro-classes or a particular class [59,60].

The MLC algorithm calculates the weighted distance or probability D that an unknown value in the vector M_p belongs to one of the macro-classes or classes M_c. This likelihood is based on the Bayesian equation [61]:

$$D = \ln(a_c) - \frac{\ln(|Cov_c|)}{2} - \frac{(M_p - M_c)T(Cov_c - 1)(M_p - M_c)}{2} \quad (8)$$

where D is the weighted distance or probability, c is a particular macro-class or class, M_p is the measurement vector of the candidate pixel, M_c is the mean vector of the sample of macro-class or class c, a_c is the percentage probability of any pixel belonging to the macro-class or class c, Cov_c is the covariance matrix of the pixels in the sampled macro-class or class c, $|Cov_c|$ is the determinant of the matrix Cov_c, Cov_{c-1} is the inverse of Cov_c and T is the transposition function.

2.5. Precision Assessment

The analysis of the accuracy is an important step for the evaluation of the resulting classification because the users of the information, once the classification is performed, need to know how accurate the result is in order to use the data in their decision making. A minimum level of overall precision for a selected macro-class or class of at least 85%, according to the recommendations found in similar works [49,60,62,63], was proposed in this study. The selected ratio between a training set and validation set was 70/30. This means that 70% of the available information was used for training purposes, whereas 30% was used for validation. Acceptable precision measurements used in this work included producer precision (PA > 85%), user precision (UA > 85%), overall precision (OA > 85%) and the Kappa coefficient (K) [49,64,65]. The reference values of the Kappa coefficient proposed by Viera and Garrett [66] are shown in Table 4.

Table 4. Possible ranges of map comparison and level of agreement of Kappa coefficient (K).

Level	Value	Range
1	<0	Poor
2	0.01–0.40	Fair
3	0.41–0.60	Acceptable
4	0.61–0.80	Good
5	0.81–1.00	Excellent

For precision evaluation, a set of verification points are required. The sample should be designed to achieve low standard errors in precision estimates, and this is generally achieved by random selection of points. The number of samples should be calculated by

$$N = \left(\frac{\sum_{i=0}^{N} (W_i - S_i)}{S_o} \right)^2 \quad (9)$$

where N is the sample size, W_i is the mapped area proportion of the class i, S_i is the standard deviation of the stratum i and So is the expected standard deviation of overall accuracy, often valued at 0.01 [67–69].

2.6. Validation Point Estimation

The number of samples calculated with Equation (9) is shown in Table 5. The assumption of this expression claims that randomly generated validation points are proportional to the size of the selected class. In other words, the larger the class is, the more sample points are needed to verify the correct LULC assignation. For instance, as vegetation represented the larger area of the city and its surroundings, more points were needed to validate the land use classification. This behavior can be seen in Table 5 for the macro-classes in question. Similarly, the number of samples required to validate the classes presents the same behavior; the classes representing mixed forest (vegetation) and bare surfaces (bare soil) needed 79 and 88 samples, respectively, as they are the largest land uses in the study area.

Table 5. Average number of samples for macro-class and class validation. Units in pixels.

Macro-class	Water			Built-Up				Vegetation				Bare soil		
Samples (pixel)	128			111				122				123		
Class	**Sea**	**W. C**	**W. B.**	**Industry**	**Residential**	**Commercial**	**Roads**	**Mixed forest**	**Shrubland**	**Wetland**	**Dune**	**B. S.**	**S. V.**	**Grassland**
Samples (pixel)	83	41	28	30	56	23	36	79	37	26	29	88	27	24

It is important to mention that pixels representing the sea area were not counted, as they did not change significantly during this period of time and we found no factor for any small increment of decrease that suggested any environmental effect on ties. However, it was taken into consideration to illustrate the applicability of the classification algorithm. On the other hand, rivers and water bodies might be subjected to changes due to water availability or variations in hydrologic processes.

3. Results

This section is divided into four main subsections: First, the analysis of the land use and its spatial distribution at macro-class level is presented. Second, an analysis at class level is performed, similar to the one conducted in the first section. Third, an assessment of the land use change from 2015 to 2021 is carried out as a study case, and, lastly, a discussion of the possible drivers forcing land use change is established.

3.1. LULC Classifier Assessment and Selection

The three classification methods were run for the seven-year period proposed in this analysis. Results show a similar behavior for extracting LULC information using the same level of ancillary data. This condition allowed a fair comparison between the classification algorithms. Figure 3 shows the macro-class classification for the year 2017, which was the year which presented higher discrepancy among LULC classification methods. Graphically, one can observe that algorithms performed comparably. Significant variations were found within the clusters sharing bare soil and and built-up zones, whereas a more accurate definition was found for those composed of built-up vegetation and bare soil vegetation. These variations were attributed mostly to the spectral signature dispersion and departure within the pixels defining each class and the nearby areas. Bare soil and vegetation areas, being those mostly distributed over the west and east parts of the city, respectively, presented similar spatial distributions.

Numerically, all the classification methods presented good behavior for both overall precision and Kappa values, all being above 85% and 0.85, respectively (Table 6). The spectral angle mapping (SAM) was the method that showed the smallest overall precision and Kappa values, 89% and 0.86, respectively. The highest accuracy was obtained using the maximum likelihood classifier (MLC), which presented overall precision above 90% and Kappa coefficients above 0.90 for all the years in the study period. It was observed that the MLC required less field data to obtain accuracy compared to the other classification methods.

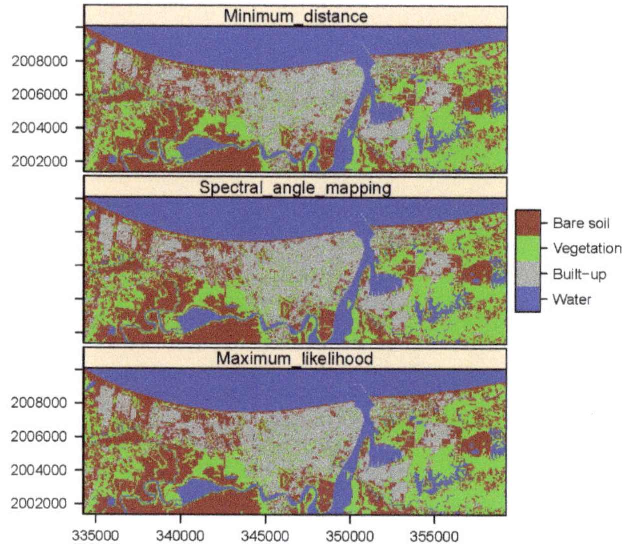

Figure 3. Maps of the macro-class-based classification for three classification methods for 2017: Top: minimum distance. Middle: spectral angle mapping. Bottom: maximum likelihood. Units: meters. Projection: WGS84 / UTM Zone 15N.

Table 6. Precision assessment of the macro-class-based classification for 2017 under three classification methods.

	MDM		SAM		MLC	
Year	**OP**	**K**	**OP**	**K**	**OP**	**K**
2015	89.12	0.85	88.89	0.85	93.28	0.91
2016	88.91	0.85	88.91	0.85	90.72	0.88
2017	89.31	0.86	89.28	0.86	95.36	0.94
2018	89.53	0.86	89.27	0.87	98.14	0.98
2019	90.13	0.87	89.90	0.86	90.02	0.87
2020	90.13	0.87	89.87	0.86	99.34	0.99
2021	89.67	0.86	89.43	0.86	96.95	0.96

The three classification methods showed the advantages of a perfect decision boundary to distinguish the macro-classes and a consistent mathematical expression based on the decision boundary for further classifications. However, these methods might overtrain the decision tree, as the training set needs several examples to cover all the possible cases within a specific class. Lastly, training tends to require significant computational time to be effective. As the MLC presented the highest level of accuracy under the same training set, it was selected as the main classification method for this study.

3.2. Macro-Class-Based LULC Classification

The classification based on macro-classes represented a course scheme to generate four of the main land uses found in Coatzacoalcos. These macro-classes are water, built-up, vegetation and bare soil. The spatial distribution of them can be observed in Figure 4. Graphically, one can see that the terrestrial surface is proportionally divided into three land uses without considering water. Most of the urbanization or built-up areas are in the north and south of the city near the river, as they were sites of the first settlers since the foundation of the city. However, the south potential flood areas and their development have been delayed, and small changes can be observed. It can also be seen that both north and south areas contain significant patches of vegetation, which, including the creation of green areas, is a good practice in urbanization development. However, urbanization in the west area of the city seemed to increment over the study period, but no vegetation buffers were observed in those zones. This indicates the absence of good practices in urbanization and government decisions in terms of territorial planning as well as the economic, social, cultural and environmental unconsciousness of new developers [70]. These zones may provoke a higher temperature sensation or heat islands, which might translate into health problems for the inhabitants [71]. Additionally, bare soil areas seemed to increase over time and were more prominent in the west area of the study. As seen in Figure 4, bare soil replaced areas of vegetation and also increased close to developed areas. These changes can be attributed to water availability during the growing season and afforestation during the development of urbanization.

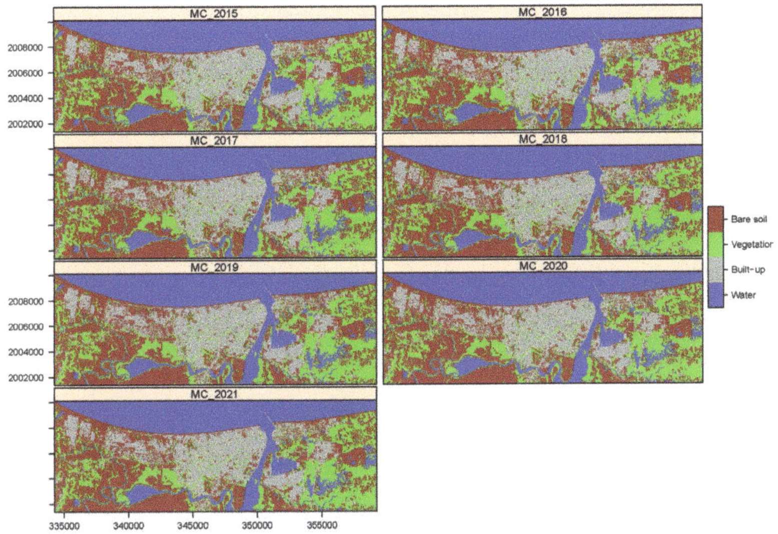

Figure 4. Maps of the macro-class-based classification for the study period (2015-2021). Units: meters. Projection: WGS84 / UTM Zone 15N.

In terms of vegetation, the east part of the city seemed to remain unaltered, except for zones near Allende Village in the northeastern and southeastern parts of the city, where the industries and petrochemical complexes are located and trigger the soil degradation. Numerically, Table 7 supports the claim observed on the above maps. Pixels representing water show small variations attributed to possible changes in the surface water balance, but a more refined classification is needed to understand how these variations occurred. The maximum percentage of water land use was 28.58% in 2017, whereas its minimum was 28.43% in 2019. Urbanization or man-made construction increased from 19.88% in 2015 to 20.21% in 2021. This increment can be explained by the trend observed in Figure 5, which shows the last six 5-year censuses. Coatzacoalcos had a mean population growth rate from 1995 to 2015 equal to 2.68%. However, we can observe that a rate equal −0.54% was shown in 2020. This means that population growth decreased in the last census. This variation can be illustrated with the reported built-up information, where man-made construction showed similar increments on a year-to-year basis. Nonetheless, the growth from 2019 to 2020 was its minimum, and no change was seen in 2021. This decrease in population and the variation within estimated in this study have socio-economic drivers. The city of Coatzalcoalcos was the second most dangerous city in 2019 [72]. This fact provoked serious economic and social issues due to extortion, racketeering and kidnapping. People moved to nearby cities to feel safe, and big companies closed their services; as a result, unemployment rate grew and investments in the city slowed down, which, in turn, decelerated urbanization and industry development.

Table 7. Percentage area estimated for the macro-class-based classification.

Macro-Class	2015	2016	2017	2018	2019	2020	2021
Water	28.45	28.44	28.50	28.47	28.43	28.44	28.44
Built-up	19.88	19.92	19.99	20.03	20.12	20.15	20.21
Vegetation	26.00	25.98	25.74	25.56	25.46	25.18	24.96
Bare soil	25.67	25.66	25.77	25.94	25.99	26.24	26.37

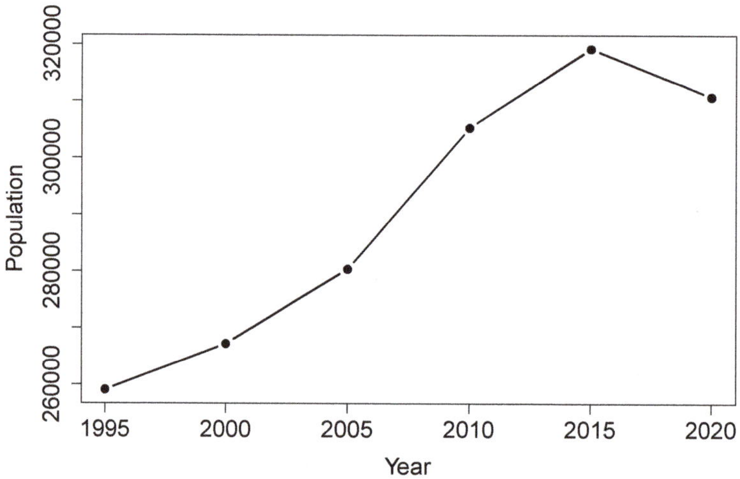

Figure 5. Last six five-year censuses in Coatzacoalcos, Veracruz. Source: National Institute of Statistics and Geography (INEGI) [73].

Bare soil also seems to increase visually in Figure 4 and is confirmed in Table 7. Bare soil variations are linked to a decrease in vegetation. These two variables are highly dependent on water availability, droughts and weather fluctuations. We can see that, in 2015, the area of study was 25.67% of bare soil, while, in 2021, it increased up to 26.37%. On the other hand, vegetation presented the inverse behavior, being 26% and 24.96% in 2015 and 2021, respectively. Analyzing the water availability, which is proportional to the precipitation, one can see in Figure 6 that rainfall tended to fluctuate from a minimum of 1107.6 mm in 2015 to a maximum of 1730.90 mm in 2017. This fluctuation, in general, tells us that, of the last three years of the period of analysis studied here, precipitation reduced drastically in 2019, which can explain the cause of the bare soil increment.

Additionally, Figure 7 shows that although minimum temperature tended to be steady from 2013 to 2021, mean temperature and maximum temperature increased during the study period. For instance, mean temperature was 26.60 °C in 2018, whereas the maximum temperature increased from 28.20 to 30.20 °C in 2018 and 2020, respectively. Increments in temperature increases soil evaporation and plant transpiration. If the latter is limited due to lack of water, water stress occurs in plants and they do not reach maturity, which might explain changes in pixel color and the move from vegetation to bare soil. It can be seen that the use of remote sensing for LULC evolution can be related and explained with inland information. This not only promotes the use of remote sensing but also validates its implementation. All the macro-class-based classification showed average user accuracy (UA) and producer accuracy (PA) higher than 90% (Table 8).

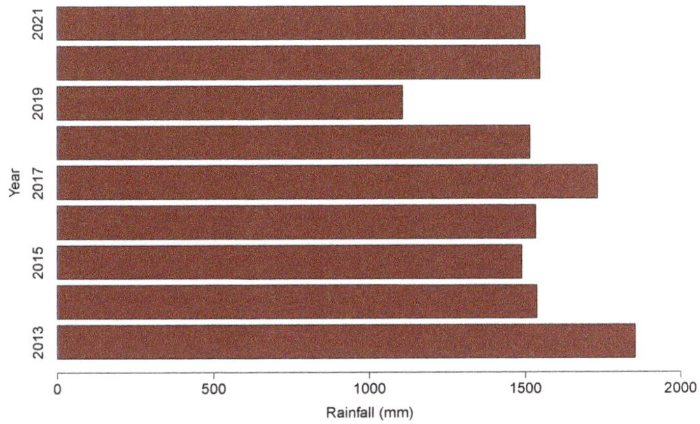

Figure 6. Annual precipitation in Coatzacoalcos, Veracruz. Source: Comision Nacional del Agua (CONAGUA) [74].

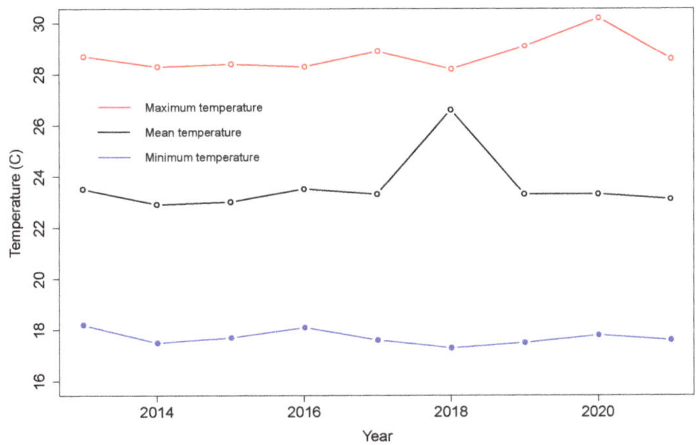

Figure 7. Annual precipitation in Coatzacoalcos, Veracruz. Source: Comision Nacional del Agua (CONAGUA) [74].

Table 8. Overall precision and Kappa coefficient (*K*) for macro-class-based classification.

Measure	2015	2016	2017	2018	2019	2020	2021
UA	93.50	91.00	95.50	98.00	90.20	99.25	96.68
PA	93.06	91.45	96.04	98.06	90.32	99.36	96.60
Overall precision	93.38	90.72	95.36	98.14	90.02	99.34	96.75
K	0.91	0.88	0.94	0.98	0.87	0.99	0.96

This means that both producers and users of these maps can rely on this classification with at least 90% confidence. These values are comparable to the ones found in [47,48]. The mean overall precision for the study period is 94.81%. The Kappa coefficients for all the studied years are located in an excellent range, as they all are higher than 0.81, according to Viera and Garret [66]. Although this classification was appropriate for the proposed macro-classes, a more refined algorithm was implemented to understand more about the evolution of the LULC of Coatzacoalcos, as it is necessary for decision making purposes.

3.3. Class-Based LULC Classification

A class-based LULC classification was conducted to identify a more discrete evolution of the land use within the study period. This identification will help prioritize the location of areas that require ecological attention, the inclusion of best management practices (BMP) for water and soil conservation as well as a more appropriate location of the upcoming urbanization due to the Interoceanic Corridor project execution. At this level, we identified that pixels representing the sea did not change significantly. This behavior can be observed in Figure 8 and Table 9, in which the percentage area representing the sea is about 20.1% for the entire period. However, water courses and water bodies seemed to change over time. Although these changes might not be significant, these changes can be confusing. We observed in Figure 6 that the annual precipitation throughout the study period decreased. As a result, bare soil area increased due to lack of water. One could assume that water bodies and water courses cannot increase in area. Nonetheless, this claim might not be entirely true, as increments of bare soil decrease infiltration and enhance the overland flow of the areas draining to the closest water courses. For that reason, water courses and water bodies in western areas increased due to increment of overland runoff in areas classified into bare soil with different vegetation densities (i.e., bare surfaces, sparse vegetation or grassland) and some over-flooded wetlands (i.e., swamps).

Man-made construction was divided into industry, residential, commercial and roads classes. Industry shows a constant percentage from 2016 to 2021. The only change occurred in 2015. The industrial area called Etileno XXI, the largest petrochemical complex in Latin America, concluded 99.2% of its construction in 2015 [75]. This, once again, confirms the accuracy of the generated map. Residence land use increased from 11.85% in 2015 to 12.24% in 2019 and maintained a steady value in 2021 (Table 9). This, once again, matches the period of violence described above that eventually provoked a decrease in population, as seen in Figure 5, which means that residence land use remained as it was in the previous years. Commercial areas showed fluctuations due to the city's economical variation. It can be observed that, in 2018, commercial areas reached their maximum of 0.55%. However, the value decreased at the end of the study period (0.43%). A small increment was observed in 2021, when commercial areas increased to 0.50%. The city experienced transitions from high commercial zones to abandoned places due to the socio-economical problems previously described. These areas transformed either to low density vegetation areas or abandoned buildings. Lastly, in this class, roads were the most difficult to identify due to the Landsat satellite image resolution equaling 30 m. Roads close to the coast line and main avenues crossing from north to south and west to east were detected easily. However, narrow streets were not characterized as roads and they were either characterized as residence, commercial or industry classes. For that reason, although roads showed increments due to increase in population and urbanization development, no claim is expressed in order to avoid confusion in this class.

Vegetation plays an important role in our natural ecosystem and also holds up the biosphere in various ways. Vegetation helps to regulate the flow of numerous biogeochemical cycles, most importantly those of nitrogen, carbon and water. It also contributes to the local and global energy balances. In this study, the classes within vegetation included mixed forest, shrubland, wetland and dune. Figure 8 shows that mixed forest is the largest extension of the study area. One can see that the eastern and southwestern areas are dominated by this class, especially for those areas where no urbanization is present. Mixed forest occupied about 18% of city and its metropolitan area for the analyzed period, reaching its minimum in 2019 when the precipitation reached its lowest value (Figure 6). As previously mentioned, at this point, one can confirm that vegetation established in the urban areas were mostly mixed forest, having more pixels where the city was initially developed than where the more recent residential areas in the west of the city have been developed. Shrublands, in general, showed an increment over time. They reached their maximum percentage area, equal to 5.31%, in 2020 with some fluctuations in 2017, a year preceded by annual precipitation below 1500 mm, which represents the driest year of the analyzed period.

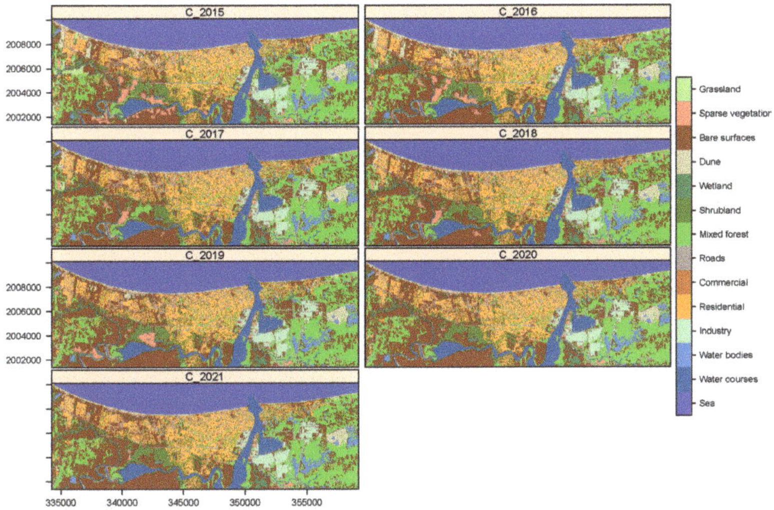

Figure 8. Maps of the class-based classification for the study period (2015–2021). Units: meters. Projection: WGS84 / UTM Zone 15N.

Table 9. Percentage area estimated for the class-based classification.

Class	2015	2016	2017	2018	2019	2020	2021
Sea	20.10	20.1	20.1	20.09	20.1	20.1	20.1
Water courses	6.10	6.31	6.31	6.31	6.32	6.33	6.34
Water bodies	1.91	1.96	2.00	1.98	1.93	2.01	2.00
Industry	2.74	2.76	2.76	2.76	2.76	2.76	2.76
Residential	11.85	11.86	12.16	12.07	12.24	12.24	12.20
Commercial	0.52	0.51	0.46	0.55	0.45	0.44	0.50
Roads	4.77	4.78	4.84	4.84	4.81	4.83	4.87
Mixed forest	18.56	18.60	19.02	18.68	18.14	18.67	18.60
Shrubland	5.15	5.21	4.82	5.18	5.16	5.31	5.28
Wetland	1.86	1.40	1.10	1.13	1.13	0.93	0.92
Dune	2.83	2.79	2.28	2.30	2.15	1.96	1.96
Bare surfaces	19.31	20.71	22.12	22.02	22.61	23.19	23.20
Sparse vegetation	2.79	2.19	1.31	1.36	1.72	0.79	0.84
Grassland	1.51	0.81	0.64	0.66	0.41	0.41	0.42

One of the most representative classes in this classification is wetland (i.e, swamps) because it represents multiple biological, economic and social values. Swamps provide services to ecological well-being, such as groundwater recharge, water purification, microclimate regulation, food resources, biodiversity and carbon storage [76–78]. In the last decades, in Coatzacoalcos, urbanization has degraded swamps indiscriminately through industrial development or residential areas. These developments followed the erroneous idea that swamps were areas with dangerous species, such as snakes, alligators, mosquitoes, etc., which represented risks to the nearby areas [79]. This study evidences how wetland or swamp degradation continues. Visually, swamps located south and southeast of the city have mostly transitioned to some degree of vegetation density, either from water stress or landfill. In 2015, swamps represented about 2% of the study area, whereas, in 2021, this fraction reached 0.92%. This fact makes this study a good indicator to prevent and preserve swamps and wetlands within Coatzacoalcos and its metropolitan area due to the important ecological services that they represent.

Dunes, in Figure 8, are located along the coastline of the study area. In addition, there are some banks utilized by the industry (sandblasting) in the east area across the

Coatzacoalcos river. One can see that dunes were replaced mainly by residential areas in the northwest part of the city. In 2015, dunes occupied almost 3% of the area in question. However, this area declined to 1.96% in 2021. The most significant evolution of urbanization occurred from 2017 to 2019. Additionally, the western coastline presented a transition from dunes to bare surfaces because that area is naturally preserved and the absence of tourism has improved the growth of native vegetation. Lastly, bare surfaces are the dominant land use along with mixed forest and residential areas (Figure 8). This land use increased from 19.31 to 23.19% through the study period in Table 9. Visually, most of the transitions were from sparse vegetation and high grassland to bare surfaces. This might have occurred because of the decrease in infiltration and precipitation previously mentioned and explored in Figures 6 and 7. Sparse vegetation decreased from 2.79% to 0.79% and increased to 0.84% in 2021, whereas grasslands reduced from 1.51% to 0.42%.

Table A1 shows the UA, PA and K values from each class throughout the analyzed period. User accuracy and producer accuracy show remarkable performance, with values within 90–100% and 80.48–100%, respectively. These individual accuracies, PA and UA, represent how well referenced pixels of the ground cover class are classified and the probability that a pixel classified into a given category actually represents that category on the ground, respectively. As expected, water and its classes showed the best performance, as it is the macro-class that presented the least variation while the other classes presented more significant variability. The Kappa coefficient shown for all classes was located in the range of excellent, according to the suggested values by Viera and Garret [66]. One can expect from these results that all the classes provide at least 90% confidence to the users of this information. The mean UA and PA values are presented in Table 10. Both UA and PA also are greater than 90%, as shown in each of the selected classes. The overall precision of the maps is also above 90%, which indicates that more than 90% of the reference pixels were correctly classified, and the K values validate the quality of this study, as they are all above 0.95.

Table 10. Overall precision and Kappa coefficient (K) for class-based classification.

Measure	2015	2016	2017	2018	2019	2020	2021
UA	96.48	96.13	96.46	97.11	97.33	96.82	95.13
PA	94.45	96.97	97.05	97.38	94.57	97.59	90.41
Overall precision	96.18	96.64	96.28	97.22	97.59	97.03	95.19
K	0.95	0.96	0.96	0.97	0.97	0.95	0.94

3.4. *Land Use and Land Cover Change through LULC Transition Matrix*

The last part of the analysis of these results includes the generation of the LULC transition matrix (Table A2), which indicates the transitions of each of the classes with respect to each other, for instance, how much area of the dunes converted to low vegetation density areas [69]. This matrix summarizes the changes already validated by the precision matrix but in terms of actual surface area. One disadvantage of this analysis is that it does not consider what happened throughout the two isolated years. However, it brings up an excellent tool for studies where LULC changes want to be predicted for future scenarios along with the possible drivers forcing those changes. The main diagonal contains the reference areas of the classes between the analyzed years. Columns named loss and gain represent the area lost or gained, respectively, for a particular class. Total gain or loss are calculated by adding up columns or rows of the reference area for each class. Particularly, this study selected the ends of the period, the years 2015 and 2021. The analysis of this matrix is straightforward. For instance, the water bodies gained 0.24 km^2 but lost 0.0297 km^2. Gains were associated with transitions of wetland (0.22 km^2), shrubland (0.0063 km^2), mixed forest (0.0009 km^2) and bare surfaces (0.0081 km^2) to water bodies. On the other hand, losses are due to a mixture of transitions, including sparse vegetation (0.009 km^2), bare surfaces (0.009 km^2) and wetlands (0.00117 km^2). All these transitions occurred between the two selected classifications.

For the sake of simplicity, only the most significant changes are discussed here. Residential land use grew only 0.88 km^2 during the last six years, which reflects a very small growth in comparison with that of industrial cities in Mexico, which has been characterized as about 5% per year [80], while Coatzacoalcos only grew 1% per year. Commercial areas reduced by 0.16 km^2, resulting in abandoned areas due to the previous mentioned socio-economic issues. One of the prominent changes was in areas of mixed forest, which gained 0.52 km^2 and lost 0.43 km^2, indicating reforestation practices and more sustainable urbanization development in some new residential, industrial and commercial areas but some activities of deforestation. Swamps, on the other hand, lost 2.09 km^2. This loss warns us of the possible future loss of ecological services provided by wetlands and their habitat, which are essential for biochemical processes and water purification because some of the non-point pollution in the city is contained by them. Lastly, vegetation densities fluctuated the most among them, namely, grassland and sparse vegetation density areas transitioned to bare surfaces. The latter gained 9.12 km^2, whereas the former two lost 2.48 and 4.66 km^2, respectively, which represents 80% of the bare surfaces gain.

4. Discussion

LULC changes measure the transitions of different land uses in complex interactions between humans and the physical environment [81,82]. Analysing LULC changes helps facilitate sustainable land use planning to protect and conserve the natural habitat and resources. The present study applied available remote sensing technology to classified course land uses, similarly to what has been presented in several studies [23,81,83–86]. The macro-classes identified were water, built-up, vegetation and bare soil. This first division allowed us to discriminate NDVI spectral indexes to improve a latter classification. Fractions of the area of study divided into macro-classes help us to observe the evolution of the city and its metropolitan area development and to explore the possible drivers forcing the changes in this initial classification. Broadly, some socio-economic factors, climate and topography were proposed as possible drivers. These drivers have also been identified as important factors for land use evolution [23,87–89]. This paper identified that security, precipitation, climate and topography were the main drivers causing LULC changes. Measurements such as producer accuracy (PA) and user accuracy (UA) showed high confidence, since they were higher than 90% for most of the macro-classes. These values are comparable and even superior to previous studies found in literature [49,84,89,90]. One can observe that the maximum likelihood classifier (MLC) is a suitable classification method to obtain acceptable results, as cited by several scientific contributions [34,35,58]. After the analysis of the macro-classes, 14 classes were exhaustively found to analyze and identify the land use in detail. Among those classes, some discussions are driven. Industry has experienced no change since the last petrochemical complex opened in 2015. Commercial areas have declined and residential zones present a slow increment due to the above-mentioned socio-economic drivers, which served to confirm the proposed hypothesis. Mixed forest and bare surfaces with low vegetation density tended to increase in surface area because reforestation programs and vegetation implemented in the coastline by the government have been a priority, as wind erosion has provoked much damage to the city's infrastructure [91]. Swamps (i.e., wetlands) also showed an important decrease in spatial contribution, losing about 2 km^2. This, in turn, might become an important environmental issue, since swamps are seen as important regulators of water pollution, carbon storage and habitat of species [92–95]. For that reason, actions need to be considered to avoid the continuous degradation of local wetlands and swamps. Additionally, it was observed that about 80% of the bare surface areas came from sparse vegetation and grassland areas. Associated changes in precipitation and high temperatures throughout the assessment period were responsible for these changes, as shown in results section. As a result of increments of bare surfaces, infiltration decreases and overland flow is exacerbated because soil properties such density, porosity and hydraulic conductivity depend on the level of established vegetation [96,97]. Lastly, a LULC transition matrix reflected the changes in surface area for each of the classes identified in this study. This is the baseline for future research that involves local problems in the city of Coatzacoalcos. A subsequent prediction of the land use using more extended spatial information of the drivers forcing the changes observed here will be proposed as an extension of this work. Additionally, the city is facing an issue with the relocation of solid residues, which naturally correlated with the actual land use. As a result, this study and methodology using higher-resolution imagery (i.e., sentinel satellite) help to study possible and suitable landfill sites.

5. Conclusions

This paper determined land use and land cover (LULC) over the period 2015–2021 for the city of Coatzacoalcos and its metropolitan area, located in the state of Veracruz, Mexico. Based on images from Landsat 8, MLC, Geographic Information Systems (GIS), NVDI spectral index and field data, the annual land use variability was produced up-to-date as reliable information for decision making and ecosystem preservation during the execution of the ongoing project called the Interoceanic Corridor. This project represents Mexico's narrowest stretch between the Pacific and Atlantic oceans and the expansion and modernization of Coatzacoalcos and Salina Cruz ports.

The objective of this study was to provide the most recent information on land use spatial distribution that Coatzacoalcos experienced in the last six years and to improve the current available course resolution maps. The satisfactory results can be summarized in four main aspects: (1) built-up areas, including industry, residential and commercial areas, have slowed down their growth due to socio-economical drivers such as security and null monetary investments; (2) vegetation such as mixed forest and low density vegetation (bare surfaces, sparse vegetation and grassland) has been sustained and increased over time due to reforestation or migration from other classes; (3) swamps experienced considerable degradation over the past five years and (4) high and medium vegetation densities have transformed mostly to low vegetation densities due to climate drives such as low precipitation and possible high soil evaporation, which might also increase the overland flow for those areas.

Lastly, this study demonstrated the use of free available Landsat data and their processing by open source tools. It provided an accurate approach to mapping and assessing LULC changes over time. This methodology can be applied similarly for longer periods of time and other satellite products and contributes to improving the number of applications of remote sensing and research in Mexico and other developing countries.

Author Contributions: Conceptualization, J.D.R.-A., E.S.G.-L., J.E.T.-R., C.G.-R. and G.A.P.; methodology, J.D.R.-A., E.S.G.-L., J.E.T.-R., C.G.-R. and G.A.P.; validation, J.D.R.-A., E.S.G.-L., J.E.T.-R., C.G.-R. and G.A.P.; formal analysis, J.D.R.-A., E.S.G.-L., J.E.T.-R., C.G.-R. and G.A.P.; investigation, J.D.R.-A., E.S.G.-L., J.E.T.-R., C.G.-R. and G.A.P.; resources, J.D.R.-A., E.S.G.-L., J.E.T.-R., C.G.-R. and G.A.P.; data curation, J.D.R.-A., E.S.G.-L., J.E.T.-R., C.G.-R. and G.A.P.; writing—original draft preparation, J.D.R.-A., E.S.G.-L., J.E.T.-R., C.G.-R. and G.A.P. All authors have read and agreed to the published version of the manuscript.

Funding: This research received no external funding.

Institutional Review Board Statement: Not applicable.

Informed Consent Statement: Not applicable.

Data Availability Statement: Some or all data, models, or code that support the findings of this study are available from the corresponding author upon reasonable request.

Acknowledgments: The authors would like to thank the anonymous reviewers for their valuable comments and feedback on this article.

Conflicts of Interest: The authors declare no conflict of interest.

Appendix A

Table A1. Precision assessment of the LULC maps of the city of Coatzacoalcos.

Class	2015			2016			2017			2018			2019			2020			2021		
	UA	PA	K	UA	PA	K	UA	PA	K	UA	PA	K	UA	PA	K	UA	PA	K	UA	PA	K
Sea	100	99.86	1	100	100	1	100	100	1	100	98.72	1	100	100	1	100	100	1	100	100	1
Water courses	98	100	0.97	99	100	0.99	99	100	0.99	95.88	99.15	0.96	97	99.93	0.97	97	99.15	0.97	100	99.53	1
Water bodies	94.95	100	0.95	91.92	100	0.92	96.97	100	0.97	96	100	0.96	98	100	0.98	97.98	100	0.98	97.96	98.15	0.98
Industry	95.88	100	0.96	97	100	0.97	100	100	1	97.96	98.23	0.98	93.94	100	0.94	97	100	0.97	97.96	98.15	0.98
Residential	90	92.74	0.89	90	93.24	0.89	91	93.21	0.89	94	93.34	0.93	97	95.27	0.96	94	91.94	0.93	93	93.58	0.98
Commercial	94.84	80.48	0.95	95.24	100	0.95	95.88	100	0.96	96.81	100	0.97	97.96	91.02	0.98	98.97	100	0.99	97.89	93.41	0.98
Roads	100	86.49	1	96	86.41	0.96	95.96	86.71	0.96	96.97	88.32	0.97	99	90.09	0.99	94.95	86.95	0.95	97	83.78	0.97
Mixed forest	97	98.93	0.96	98	95.99	0.97	98	94.90	0.98	96	97.62	0.95	97	97.82	0.96	98	96.06	0.97	94	99	0.93
Shrubland	96	88.81	0.95	95	94.54	0.95	90.91	88.34	0.90	91	97.04	0.91	95	98.18	0.95	95	99.06	0.95	95	81.10	0.95
Wetland	98.98	91.39	0.99	96.84	100	0.97	98	100	0.98	98.97	96.54	0.98	100	94.72	1	97.92	99.06	0.98	96.91	98.07	0.97
Dune	99	100	0.99	98	99.28	0.98	96.94	98.22	0.97	99	100	0.99	96.94	98.18	0.97	98	95.44	0.98	98.98	88.96	0.98
Bare surfaces	93	95.80	0.91	96	97.19	0.95	94	97.29	0.92	99	98.50	0.98	97	98.76	0.96	96	98.57	0.95	91	97.14	0.89
Sparse vegetation	99	91.24	0.98	97.95	90.89	0.98	98.98	100	0.99	98.98	100	0.99	98.99	100	0.99	96.94	100	0.96	83.33	98.73	0.83
Grassland	94	96.49	0.94	94.90	100	0.95	94.85	100	0.95	98.93	95.90	0.98	94.85	100	0.95	93.75	100	0.94	94.62	100	0.95

Table A2. LULC matrix representing the class transitions. Units in km^2 .

Class	Sea	Water Courses	Water Bodies	Industry	Residential	Commercial	Roads	Mixed Forest	Shrubland	Wetland	Dune	Bare Surfaces	Sparse Vegeta-tion	Grassland	Loss
Sea	43.85	0	0	0	0	0	0	0	0	0	0	0	0	0	0
Water courses	0	13.31	0	0	0	0	0.0009	0.0036	0	0.0009	0	0	0	0	−0.0054
Water bodies	0	0	4.13	0	0	0	0	0	0	0.0117	0	0.0009	0.0009	0	−0.0297
Industry	0	0	0	5.99	0	0	0	0	0	0	0	0	0	0	0
Residential	0	0	0	0	25.72	0.0909	0.0261	0	0	0	0	0.009	0	0	−0.126
Commercial	0	0	0	0	0.1449	0.95	0.0261	0.0009	0	0	0	0.0054	0	0	−0.1773
Roads	0	0	0	0.0027	0.0702	0.0216	10.30	0.0009	0	0	0	0.0108	0	0	−0.1062
Mixed forest	0	0	0.0009	0	0.0018	0.0009	0.0009	40.05	0.26	0	0	0.1188	0.0009	0.0432	−0.4383
Shrubland	0	0.0045	0.0063	0.0009	0.0261	0.0009	0	0.081	10.83	0	0	0.26	0.0261	0	−0.4095
Wetland	0	0.50	0.23	0	0	0	0.0063	0.31	0.29	1.97	0	0.7488	0.0117	0.0189	−2.0952
Dune	0.0189	0.0045	0	0.0135	0.31	0.0171	0.1575	0.0126	0.0324	0	4.26	1.323	0.0243	0.0063	−1.9206
Bare surfaces	0	0.0108	0.0081	0.0171	0.3087	0.0207	0.0468	0.0693	0.0135	0.0099	0.018	41.51	0.0756	0.0063	−0.6318
Sparse vegeta-tion	0	0	0	0.0126	0.0081	0.0054	0.0306	0.0207	0.0288	0.0009	0	4.55	1.4319	0.0045	−4.6593
Grassland	0	0.009	0	0	0.0045	0.0009	0.0216	0.0279	0.0765	0.0072	0.0018	2.08	0.25	0.8046	−2.4795
Gain	0.0189	0.5238	0.2439	0.0486	0.882	0.1584	0.3159	0.522	0.6948	0.0306	0.0198	9.1179	0.396	0.1062	218.1617

References

1. Giri, C.; Wang, Y. *Remote Sensing of Land Use and Land Cover*; CRC Press: Boca Raton, FL, USA, 2012.
2. Gong, P.; Wang, J.; Yu, L.; Zhao, Y.; Zhao, Y.; Liang, L.; Niu, Z.; Huang, X.; Fu, H.; Liu, S.; et al. Finer resolution observation and monitoring of global land cover: First mapping results with Landsat TM and ETM+ data. *Int. J. Remote Sens.* **2013**, *34*, 2607–2654. [CrossRef]
3. Chowdhury, S.; Chao, D.K.; Shipman, T.C.; Wulder, M.A. Utilization of Landsat data to quantify land-use and land-cover changes related to oil and gas activities in West-Central Alberta from 2005 to 2013. *GISci. Remote Sens.* **2017**, *54*, 700–720. [CrossRef]
4. Ye, J.; Hu, Y.; Zhen, L.; Wang, H.; Zhang, Y. Analysis on Land-Use Change and Its Driving Mechanism in Xilingol, China, during 2000–2020 Using the Google Earth Engine. *Remote Sens.* **2021**, *13*, 5134. [CrossRef]
5. Muttitanon, W.; Tripathi, N.K. Land use/land cover changes in the coastal zone of Ban Don Bay, Thailand using Landsat 5 TM data. *Int. J. Remote Sens.* **2005**, *26*, 2311–2323. [CrossRef]
6. Gerard, F.; Petit, S.; Smith, G.; Thomson, A.; Brown, N.; Manchester, S.; Wadsworth, R.; Bugar, G.; Halada, L.; Bezak, P. Land cover change in Europe between 1950 and 2000 determined employing aerial photography. *Prog. Phys. Geogr.* **2010** *34*, 183–205. [CrossRef]
7. Baumann, M.; Ozdogan, M; Wolter, P.T.; Krylov, A.; Vladimirova, N.; Radeloff, V.C. Landsat remote sensing of forest windfall disturbance. *Remote Sens. Environ.* **2012**, *143*, 171–179. [CrossRef]
8. Clerici, N.; Paracchini, M.L.; Maes, J. Land-cover change dynamics and insights into ecosystem services in European stream riparian zones. *Ecohydrol. Hydrobiol.* **2014**, *14*, 107–120. [CrossRef]
9. Pervez, M.S.; Henebry, G.M. Assessing the impacts of climate and land use and land cover change on the freshwater availability in the Brahmaputra River basin. *J. Hydrol. Reg. Stud.* **2015**, *3*, 285–311. [CrossRef]
10. Balthazar, V.; Vanacker, V.; Molina, A.; Lambin, E.F. Impacts of forest cover change on ecosystem services in high Andean mountains. *Ecol. Indic.* **2015**, *48*, 63–75. [CrossRef]
11. Pasha, S.V.; Reddy, C.S.; Jha, C.S.; Rao, P.P.; Dadhwal, V.K. Assessment of land cover change hotspots in Gulf of Kachchh, India using multi-temporal remote sensing data and GIS. *J. Indian Soc. Remote. Sens.* **2016**, *44*, 905–913. [CrossRef]
12. Satyanarayana, B.; Thierry, B.; Seen, D.L.; Raman, A.V.; Muthusankar, G. Remote sensing in mangrove research-relationship between vegetation indices and dendrometric parameters: A case for Coringa, east coast of India. In Proceeding of the 22nd Asian Conference on Remote Sensing, Singapore, 5–9 November 2001; Volume 5, p. 9.
13. Liang, D.; Zuo, Y.; Huang, L.; Zhao, J.; Teng, L.; Yang, F. Evaluation of the consistency of MODIS Land Cover Product (MCD12Q1) based on Chinese 30 m GlobeLand30 datasets: A case study in Anhui Province, China. *ISPRS Int. J. Geo-Inf.* **2015**, *4*, 2519–2541. [CrossRef]
14. Randazzo, G.; Italiano, F.; Micallef, A.; Tomasello, A.; Cassetti, F.P.; Zammit, A.; D'Amico, S.; Saliba, O.; Cascio, M.; Cavallaro, F.; et al. WebGIS Implementation for Dynamic Mapping and Visualization of Coastal Geospatial Data: A Case Study of BESS Project. *Appl. Sci.* **2021**, *11*, 8233. [CrossRef]
15. Randazzo, G.; Cascio, M.; Fontana, M.; Gregorio, F.; Lanza, S.; Muzirafuti, A. Mapping of Sicilian Pocket Beaches Land Use/Land Cover with Sentinel-2 Imagery: A Case Study of Messina Province. *Land* **2021**, *10*, 678. [CrossRef]
16. Ghassemi, B.; Dujakovic, A.; Żółtak, M.; Immitzer, M.; Atzberger, C.; Vuolo, F. Designing a European-Wide Crop Type Mapping Approach Based on Machine Learning Algorithms Using LUCAS Field Survey and Sentinel-2 Data. *Remote Sens.* **2022**, *14*, 541. [CrossRef]
17. USGS. Available online: https://earthexplorer.usgs.gov (accessed on 10 December 2021).
18. Townshend, J.R.G.; Gayler, J.R.; Hardy, J.R.; Jackson, M.J.; Baker, J.R. Preliminary analysis of LANDSAT-4 Thematic Mapper. *Int. J. Remote Sens.* **1983**, *4*, 817–828. [CrossRef]
19. Lunetta, R.S.; Lyon, J.G.; Guindon, B.; Elvidge, C.D. North American landscape characterization dataset development and data fusion issues. *Photogramm. Eng. Remote Sens.* **1998**, *64*, 821–828.
20. Cohen, W.B.; Goward, S.N. Landsat's role in ecological applications of remote sensing. *Bioscience* **2004**, *54*, 535–545. [CrossRef]
21. Taylor, P.; Horler, D.N.H.; Ahern, F.J. Forestry information content of Thematic Mapper data. *Int. J. Remote Sens.* **2007**, *7*, 37–41.
22. Campbell, J.B. *Introduction to Remote Sensing*, 4th ed.; The Guildford Press: New York, NY, USA, 2007.
23. Leta, M.K.; Demissie, T.A.; Tränckner, J. Modeling and Prediction of Land Use Land Cover Change Dynamics Based on Land Change Modeler (LCM) in Nashe Watershed, Upper Blue Nile Basin, Ethiopia. *Sustainability* **2021**, *13*, 3740. [CrossRef]
24. Pérez-Vega, A.; Mas, J.F.; Ligmann-Zielinska, A. Comparing two approaches to land use/cover change modeling and their implications for the assessment of biodiversity loss in a deciduous tropical forest. *Environ. Model. Softw.* **2012**, *29*, 11–23. [CrossRef]
25. Riad, P.; Graefe, S.; Hussein, H.; Buerkert, A. Landscape transformation processes in two large and two small cities in Egypt and Jordan over the last five decades using remote sensing data. *Landsc. Urban Plan.* **2020**, *197*, 103766. [CrossRef]
26. Tarasovicova, Z.; Saksa, M.; Blazik, T.; Faltan, V. Changes in agricultural land use in the context of ongoing transformational processes in Slovakia. *Agriculture* **2013**, *59*, 49–64.
27. Yuan, F.; Sawaya, K.E.; Loeffelholz, B.C.; Bauer, M.E. Land cover classification and change analysis of the Twin Cities (Minnesota) Metropolitan Area by multitemporal Landsat remote sensing. *Remote Sens. Environ.* **2005**, *98*, 317–328. [CrossRef]
28. Demissie, F.; Yeshitila, K.; Kindu, M.; Schneider, T. Land use/Land cover changes and their causes in Libokemkem District of South Gonder, Ethiopia. *Remote Sens. Appl. Soc. Environ.* **2017**, *8*, 224–230. [CrossRef]

29. Colditz, R.R.; Saldaña, G.L.; Maeda, P.; Espinoza, J.A.; Tovar, C.M.; Hernández, A.V.; Benítez, C.Z.; López, I.C.; Ressl, R. Generation and analysis of the 2005 land cover map for Mexico using 250 m MODIS data. *Remote Sens. Environ.* **2012**, *123*, 541–552. [CrossRef]
30. Blanco, P.D.; Colditz, R.R.; Saldaña, G.L.; Hardtke, L.A.; Llamas, R.M.; Mari, N.A.; Fischer, A.; Caride, C.; Aceñolaza, P.G.; del Valle, H.F.; et al. A land cover map of Latin America and the Caribbean in the framework of the SERENA project. *Remote Sens. Environ.* **2013**, *132*, 13–31. [CrossRef]
31. Morehart, C.T. Mapping ancient chinampa landscapes in the Basin of Mexico: a remote sensing and GIS approach. *J. Archaeol. Sci.* **2012** *39*, 2541–2551. [CrossRef]
32. Robinson, C.J. Evolution of the 2014–2015 sea surface temperature warming in the central west coast of Baja California, Mexico, recorded by remote sensing. *Geophys. Res. Lett.* **2016**, *43*, 7066–7071. [CrossRef]
33. Alcalá, G.; Grisales-Noreña, L.F.; Hernandez-Escobedo, Q.; Muñoz-Criollo, J.J.; Revuelta-Acosta, J.D. SHP Assessment for a Run-of-River (RoR) Scheme Using a Rectangular Mesh Sweeping Approach (MSA) Based on GIS. *Energies* **2021**, *14*, 3095. [CrossRef]
34. Pal, M.; Mather, P.M. An assessment of the effectiveness of decision tree methods for land cover classification. *Remote Sens. Environ.* **2003**, *86*, 554-–565. [CrossRef]
35. Osunmadewa, B.A.; Gebrehiwot, W.Z.; Csaplovics, E.; Adeofun, O.C. Spatio-temporal monitoring of vegetation phenology in the dry sub-humid region of Nigeria using time series of AVHRR NDVI and TAMSAT datasets. *Open Geosci.* **2018**, *10*, 1–11. [CrossRef]
36. Rawat, J.S.; Kumar, M. Monitoring land use/cover change using remote sensing and GIS techniques: A case study of Hawalbagh block, district Almora, Uttarakhand, India. *Egypt. J. Remote. Sens.* **2015**, *18*, 77–84. [CrossRef]
37. González-Mille, D.J.; Ilizaliturri-Hernández, C.A.; Espinosa-Reyes, G.; Costilla-Salazar, R.; Díaz-Barriga, F.; Ize-Lema, I.; Mejía-Saavedra, J. Exposure to persistent organic pollutants (POPs) and DNA damage as an indicator of environmental stress in fish of different feeding habits of Coatzacoalcos, Veracruz, Mexico. *Ecotoxicology* **2010**, *19*, 1238–1248. [CrossRef]
38. Morales, M.A.; Herrero V.M.; Martinez S.A.; Rodriguez M.G., Valdivieso E.; Garcia G, Elias M.A. Cleaner production and methodological proposal of eco-efficiency measurement in a Mexican petrochemical complex. *Water Sci. Tech.* **2006**, *53*, 11–16. [CrossRef]
39. BNAMERICAS. United States. 2021. Available online: https://www.bnamericas.com/en/features/mexicos-us1bn-interoceanic-corridor-to-start-operations-in-2023 (accessed on 10 December 2021).
40. INEGI. Available online: https://www.inegi.org.mx/temas/edafologia/ (accessed on 10 December 2021).
41. CONABIO. Available online: http://www.conabio.gob.mx/informacion/gis/ (accessed on 10 December 2021).
42. Prieto-Amparan J.A.; Villarreal-Guerrero F.; Martinez-Salvador M.; Manjarrez-Domínguez C.; Santellano-Estrada E.; Pinedo-Alvarez A. Atmospheric and radiometric correction algorithms for the multitemporal assessment of grasslands productivity. *Remote Sens.* **2018**, *10*, 219. [CrossRef]
43. Campbell, J.B. Evaluation of the dark-object subtraction technique for adjustment of multispectral remote-sensing data. In *Digital Image Processing and Visual Communications Technologies in the Earth and Atmospheric Sciences II*; International Society for Optics and Photonics: Bellingham, DC, USA, 1993; Volume 1819, pp. 176–188.
44. Chavez, P.S., Jr. An improved dark-object subtraction technique for atmospheric scattering correction of multispectral data. *Remote Sens. Environ.* **1988**, *24*, 459–479. [CrossRef]
45. Cui, Q.Y.; Gaillard, M.J.; Lemdahl, G.; Stenberg, L.; Sugita, S.; Zernova, G. Historical land-use and landscape change in southern Sweden and implications for present and future biodiversity. *Ecol. Evol.* **2014**, *4*, 3555–3570. [CrossRef]
46. Sobrino, J.A.; Jimenez-Munoz, J.C.; Paolini, L. Land surface temperature retrieval from LANDSAT TM 5. *Remote Sens. Environ.* **2004**, *90*, 434–440. [CrossRef]
47. El Hafyani, M.; Essahlaoui, A.; Fung-Loy, K.; Hubbart, J.A.; Van Rompaey, A. Assessment of Agricultural Water Requirements for Semi-Arid Areas: A Case Study of the Boufakrane River Watershed (Morocco). *Appl. Sci.* **2021**, *11*, 10379. [CrossRef]
48. Akbar, T.A.; Hassan, Q.K.; Ishaq, S.; Batool, M.; Butt, H.J.; Jabbar, H. Investigative spatial distribution and modelling of existing and future urban land changes and its impact on urbanization and economy. *Remote Sens.* **2019**, *11*, 105. [CrossRef]
49. Mohajane, M.; Essahlaoui, A.; Oudija, F.; Hafyani, M.E.; Hmaidi, A.E.; Ouali, A.E.; Teodoro, A.C. Land use/land cover (LULC) using landsat data series (MSS, TM, ETM+ and OLI) in Azrou Forest, in the Central Middle Atlas of Morocco. *Environments* **2018**, *5*, 131. [CrossRef]
50. Ding, Y.; Zheng, X.; Zhao, K.; Xin, X.; Liu, H. Quantifying the impact of NDVIsoil determination methods and NDVIsoil variability on the estimation of fractional vegetation cover in Northeast China. *Remote Sens.* **2016**, *8*, 29. [CrossRef]
51. Aburas, M.M.; Abdullah, S.H.; Ramli, M.F.; Ashaari, Z.H. Measuring land cover change in Seremban, Malaysia using NDVI index. *Procedia Environ. Sci.* **2015**, *30*, 238–243. [CrossRef]
52. Anim, D.O.; Kabo-bah, A.T.; Nkrumah, P.N.; Murava, R.T. Evaluation of NDVI using Spot-5 satellite data for northern Ghana. *Environ. Manag. Sustain. Dev.* **2013**, *2*, 167.
53. Song, C.; Woodcock, C.E.; Seto, K.C.; Lenney, M.P.; Macomber, S.A. Classification and change detection using Landsat TM data: when and how to correct atmospheric effects? *Remote Sens. Environ.* **2001**, *75*, 230–244. [CrossRef]
54. Richards, J.A. *Remote Sensing Digital Image Analysis*; Springer: Berlin, Germany, 1999.
55. Kosztra, B.; Büttner, G.; Hazeu, G.; Arnold, S. *Updated CLC Illustrated Nomenclature Guidelines*; European Environment Agency: Wien, Austria, 2017; pp. 1–124.

56. Richards, J.A.; Jia, X. *Remote Sensing Digital Image Analysis: An Introduction*; Springer: Berlin, Germany, 2006.
57. Kruse, F.A.; Lefkoff, A.B.; Boardman, J.W.; Heidebrecht, K.B.; Shapiro, A.T.; Barloon, P.J.; Goetz, A.F.H. The spectral image processing system (SIPS)—interactive visualization and analysis of imaging spectrometer data. *Remote Sens. Environ.* **1993**, *44*, 145–163. [CrossRef]
58. Currit, N. Development of a remotely sensed, historical land-cover change database for rural Chihuahua, Mexico. *Int. J. Appl. Earth Obs. Geoinf.* **2005**, *7*, 232–247. [CrossRef]
59. Otukei, J.R.; Blaschke, T. Land cover change assessment using decision trees, support vector machines and maximum likelihood classification algorithms. *Int. J. Appl. Earth Obs. Geoinf.* **2010**, *12*, S27–S31. [CrossRef]
60. Foody, G.M. Status of land cover classification accuracy assessment. *Remote Sens. Environ.* **2002**, *80*, 185–201. [CrossRef]
61. Smith, C.; Brown, N. *Erdas Field Guide. Revised and Expanded*; 5th ed.; eClass: Atlanta, GA, USA, 1999.
62. Thomlinson, J.R.; Bolstad, P.V.; Cohen, W. B. Coordinating methodologies for scaling landcover classifications from site-specific to global: Steps toward validating global map products. *Remote Sens. Environ.* **1999**, *70*, 16–28. [CrossRef]
63. Anderson, J.R. *A Land Use and Land Cover Classification System for Use with Remote Sensor Data*; US Government Printing Office: Washington, DC, USA, 1976; Volume 964.
64. Lu, D.; Weng, Q. A survey of image classification methods and techniques for improving classification performance. *Int. J. Remote Sens.* **2007** *28*, 823–870. [CrossRef]
65. Congalton, R.G. A review of assessing the accuracy of classifications of remotely sensed data. *Remote Sens. Environ.* **1991**, *37*, 35–46. [CrossRef]
66. Viera, A.J.; Garrett, J.M. Understanding interobserver agreement: the kappa statistic. *Fam. Med.* **2005**, *37*, 360–363.
67. Congedo, L. Semi-Automatic Classification Plugin: A Python tool for the download and processing of remote sensing images in QGIS. *J. Open Source Softw.* **2021**, *6*, 3172. [CrossRef]
68. Alshari, E.A.; Gawali, B.W. Development of classification system for LULC using remote sensing and GIS. *Global Transitions Proceedings* **2021**, *2*, 8–17. [CrossRef]
69. Olofsson, P.; Foody, G.M.; Herold, M.; Stehman, S.V.; Woodcock, C.E.; Wulder, M.A. Good practices for estimating area and assessing accuracy of land change. *Remote Sens. Environ.* **2014**, *148*, 42–57. [CrossRef]
70. Casado-Izquierdo, J.M.; Sánchez-Salazar, M.T. Coatzacoalcos: Reestructuración urbana e inversión privada en una ciudad media mexicana. *EURE (Santiago)* **2013**, *39*, 91–116. [CrossRef]
71. Tan, J., Zheng; Y., Tang, X.; Guo, C.; Li, L.; Song, G.; Chen, H. The urban heat island and its impact on heat waves and human health in Shanghai. *Int. J. Biometeorol.* **2010**, *54*, 75–84. [CrossRef]
72. Diario de Xalapa. Mexico. 2019. Available online: https://www.diariodexalapa.com.mx/local/habitantes-viven-inseguros-coatzacoalcos-ciudad-segundo-lugar-3914790.html/ (accessed on 21 December 2021).
73. INEGI. Censos y Conteos de Población y Vivienda, 1995 a 2020. Encuesta Intercensal 2015. 2020. Available online: http://ceieg.veracruz.gob.mx/wp-content/uploads/sites/21/2021/06/COATZACOALCOS_2021.pdf (accessed on 10 December 2021).
74. CONAGUA. Sistema Metereológico Nacional. México. 2021. Available online: https://smn.conagua.gob.mx/es/climatologia/temperaturas-y-lluvias/resumenes-mensuales-de-temperaturas-y-lluvias/ (accessed on 21 December 2021).
75. Braskem Idesa. Historia. Available online: https://www.braskem.com.br/idesa/historia (accessed on 21 December 2021).
76. Brasil, M.C.O.; de Magalhães-Filho, R.; do Espírito-Santo, M.M.; Leite, M.E.; Veloso, M.D.D.M.; Falcão, L.A.D. Land-cover changes and drivers of palm swamp degradation in southeastern Brazil from 1984 to 2018. *Appl. Geogr.* **2021**, *137*, 102604. [CrossRef]
77. Quintas-Soriano, C.; Castro, A.J.; Castro, H.; García-Llorente, M. Impacts of land use change on ecosystem services and implications for human well-being in Spanish drylands. *Land Use Policy* **2016**, 54, 534–548. [CrossRef]
78. Tiner, R.W.; Lang, M.W.; Klemas, V.V. *Remote sensing of Wetlands: Applications and Advances*; CRC Press: Boca Raton, FL, USA, 2015.
79. Toledo, A.; Vázquez-Botello, A.; Herzig. *El Pantano: Una Riqueza Que se Destruye*; de Ecodesarrollo, C., Toledo, A., Eds.; Centro de Ecodesarrollo: México, 1987.
80. Sobrino, J. Desempeño industrial en las principales ciudades de México, 1980–2003. *Estud. DemográFicos Urbanos* **2007**, *22*, 243–290. [CrossRef]
81. Nayak, P.M.; Byrne, M.L. Impact of land use land cover change on a sand dune ecosystem in Northwest Beach, Point Pelee National Park, Canada. *J. Great Lakes Res.* **2019**, *45*, 1047–1054. [CrossRef]
82. Pielke Sr, R.A.; Pitman, A.; Niyogi, D.; Mahmood, R.; McAlpine, C.; Hossain, F.; de Noblet, N. Land use/land cover changes and climate: modeling analysis and observational evidence. *Wiley Interdiscip. Rev. Clim. Chang.* **2011**, 2, 828–850. [CrossRef]
83. Liping, C.; Yujun, S.; Saeed, S. Monitoring and predicting land use and land cover changes using remote sensing and GIS techniques—A case study of a hilly area, Jiangle, China. *PLoS ONE* **2018**, *13*, e0200493. [CrossRef]
84. Yirsaw, E.; Wu, W.; Shi, X.; Temesgen, H.; Bekele, B. Land use/land cover change modeling and the prediction of subsequent changes in ecosystem service values in a coastal area of China, the Su-Xi-Chang Region. *Sustainability* **2017**, *9*, 1204. [CrossRef]
85. Mohan, M.; Kikegawa, Y.; Gurjar, B.R.; Bhati, S.; Kolli, N.R. Assessment of urban heat island effect for different land use–land cover from micrometeorological measurements and remote sensing data for megacity Delhi. *Theor. Appl. Climatol.* **2013**, *112*, 647–658. [CrossRef]
86. Huang, W.; Liu, H.; Luan, Q.; Jiang, Q.; Liu, J.; Liu, H. Detection and prediction of land use change in Beijing based on remote sensing and GIS. *Int. Arch. Photogramm. Remote Sens. Spat. Inf. Sci* **2008**, *37*, 75–82.

87. Sarif, M.O.; Gupta, R.D. Spatiotemporal mapping of Land Use/Land Cover dynamics using Remote Sensing and GIS approach: A case study of Prayagraj City, India (1988–2018). *Environ. Dev. Sustain.* **2021**, *24*, 888–920. [CrossRef]
88. Kolb, M.; Mas, J.F.; Galicia, L. Evaluating drivers of land-use change and transition potential models in a complex landscape in Southern Mexico. *Int. J. Geogr. Inf. Syst.* **2013**, *27*, 1804–1827. [CrossRef]
89. Serra, P.; Pons, X.; Saurí, D. Land-cover and land-use change in a Mediterranean landscape: a spatial analysis of driving forces integrating biophysical and human factors. *Appl. Geogr.* **2008**, *28*, 189–209. [CrossRef]
90. Martínez, S.; Mollicone, D. From land cover to land use: A methodology to assess land use from remote sensing data. *Remote Sens.* **2012**, *4*, 1024–1045. [CrossRef]
91. Universo. Sistema de noticias de la UV. Mexico. 2019. Available online: https://www.uv.mx/prensa/regiones/coatzacoalcos/uv-presenta-propuesta-para-minimizar-arribo-de-arena-en-malecon-de-coatzacoalcos/ (accessed on 21 December 2021).
92. Xiao, K.; Li, H.; Shananan, M.; Zhang, X.; Wang, X.; Zhang, Y.; Liu, H. Coastal water quality assessment and groundwater transport in a subtropical mangrove swamp in Daya Bay, China. *Sci. Total Environ.* **2019**, *646*, 1419–1432. [CrossRef]
93. Yule, C.M. Loss of biodiversity and ecosystem functioning in Indo-Malayan peat swamp forests. *Biodivers. Conserv.* **2010**, *19*, 393–409. [CrossRef]
94. Sabo, M.J.; Bryan, C.F.; Kelso, W.E.; Rutherford, D.A. Hydrology and aquatic habitat characteristics of a riverine swamp: I. Influence of flow on water temperature and chemistry. *Regul. River* **1999**, *15*, 505–523. [CrossRef]
95. Sabo, M.J.; Bryan, C.F.; Kelso, W.E.; Rutherford, D.A. Hydrology and aquatic habitat characteristics of a riverine swamp: II. Hydrology and the occurrence of chronic hypoxia. *Regul. River* **1999**, *15*, 525–544. [CrossRef]
96. Thompson, S.E.; Harman, C.J.; Heine, P.; Katul, G.G. Vegetation-infiltration relationships across climatic and soil type gradients. *J. Geophys. Res. Biogeosci.* **2010**, *115*. [CrossRef]
97. Perkins, K.S.; Stock, J.D.; Nimmo, J.R. Vegetation influences on infiltration in Hawaiian soils. *Ecohydrology* **2018**, *11*, e1973. [CrossRef]

applied sciences

MDPI

Article

Three-Dimensional Convolutional Neural Network on Multi-Temporal Synthetic Aperture Radar Images for Urban Flood Potential Mapping in Jakarta

Indra Riyanto [1], Mia Rizkinia [1], Rahmat Arief [2] and Dodi Sudiana [1,*]

[1] Department of Electrical Engineering, Faculty of Engineering, Universitas Indonesia, Depok 16424, Indonesia; indra.riyanto81@ui.ac.id (I.R.); mia@ui.ac.id (M.R.)
[2] Aeronautics and Space Research Organization, National Research and Innovation Agency, Jakarta 13710, Indonesia; rahmat.arief@lapan.go.id
* Correspondence: dodi.sudiana@ui.ac.id

Abstract: Flooding in urban areas is counted as a significant disaster that must be correctly mitigated due to the huge amount of affected people, material losses, hampered economic activity, and flood-related diseases. One of the technologies available for disaster mitigation and prevention is satellites providing image data on previously flooded areas. In most cases, floods occur in conjunction with heavy rain. Thus, from a satellite's optical sensor, the flood area is mostly covered with clouds which indicates ineffective observation. One solution to this problem is to use Synthetic Aperture Radar (SAR) sensors by observing backscatter differences before and after flood events. This research proposes mapping the flood-prone areas using machine learning to classify the areas using the 3D CNN method. The method was applied on a combination of co-polarized and cross-polarized SAR multi-temporal image datasets covering Jakarta City and the coastal area of Bekasi Regency. Testing with multiple combinations of training/testing data proportion split and a different number of epochs gave the optimum performance at an 80/20 split with 150 epochs achieving an overall accuracy of 0.71 after training in 283 min.

Keywords: urban flood; Sentinel-1a; Synthetic Aperture Radar (SAR); 3D Convolutional Neural Network; multi-temporal data

Citation: Riyanto, I.; Rizkinia, M.; Arief, R.; Sudiana, D. Three-Dimensional Convolutional Neural Network on Multi-Temporal Synthetic Aperture Radar Images for Urban Flood Potential Mapping in Jakarta. *Appl. Sci.* **2022**, *12*, 1679. https://doi.org/10.3390/app12031679

Academic Editors: Anselme Muzirafuti and Dimitrios S. Paraforos

Received: 20 December 2021
Accepted: 31 January 2022
Published: 6 February 2022

1. Introduction

Flooding is one of the most detrimental disasters, especially in cities such as Jakarta, because it affects a large number of residents in ways such as material losses resulting from damaged properties due to flood inundation and diseases caused by degraded sanitation in the flooded area. A major flood in Jakarta results in 8.7 trillion IDR or 625 million USD of losses and recovery efforts [1]. At present, most of the flood mapping in Indonesia has not fully utilized satellite spatial data because it still relies on data reported by the local government in the form of numerical data [2]. The visualization of the flood map is based on tabulated data in the area map that does not represent the actual conditions, resulting in a discrepancy between the reported flood area and the actual area. This difference will affect the handling of floods, such as calculating the impact of damage, the number of residents affected by the flood, and the inefficient distribution of aid. Problems that arise due to limited spatial information regarding floods can be solved by using multi-sensor remote sensing satellite data. Many technologies have been developed to predict, prevent and mitigate flood disasters more accurately, including remote sensing technology using images obtained from airborne and spaceborne platforms [3–5]. The earlier and most common form of remote sensing is optical photography, with overhead images providing information on the affected area.

When a disaster occurs, urban floods usually coincide with rain, so when observed using optical sensors on remote sensing satellites, the area is covered with clouds. With this occlusion, satellite optical sensors are not effectively used for flood observation. One solution to observing floods in cloud-covered areas is to use Synthetic Aperture Radar (SAR) sensors such as Sentinel-1, ALOS PALSAR, TerraSAR-X, and other radar sensors. The image produced by SAR is a monochromatic image containing reflectance information from the observation area by observing the difference in the backscatter before and after the incident to identify the flooded area [6].

Wide-scale Earth monitoring satellites begin with the Landsat program (Land Satellite) to monitor the Earth's surface. At present, many optical satellite systems operate at high resolution. The most widely used images are from QuickBird, SPOT-5, and Worldview Series. Despite having the ability to detect very sharp objects, optical satellite systems have the potential to be unable to detect objects on the Earth's surface due to cloud cover. Until recently, SAR data processing was mostly used for rural areas rather than in urban areas. For urban areas, SAR data has its problems, namely speckle noise, because, in urban areas, many buildings cause radar waves to experience much scattering, and the reading of the reflected waves is disturbed by multipath interference [7,8]. The double-bounce characteristic of radar signals caused by buildings is a challenge related to its contribution to SAR image speckle noise. However, it can be used to detect the presence of buildings and distinguish them from other surfaces such as soil, vegetation, and water [9–11]. The solution to detect the presence of floods through SAR images is multi-temporal filtering, which is filtering based on changes in the backscattering characteristics of SAR images taken at different times. The method used to identify and distinguish the occurrence of flood from SAR data is generally divided into two groups, namely polarization and interference. The polarization method detects the presence of water, based on changes in backscatter polarization caused by specular reflections from the water surface [10,12–14], while the interference method detects water based on changes of coherence due to changes in the spatial distribution pattern of the object that produces the backscatter [10,15–17].

Several studies related to floods in settlements used SAR and Light Detection and Ranging (LiDAR) data for small towns located on one side of a river and areas with homogeneous slopes [18–20]. Another study used the Decision Tree Classifier (DTC) to differentiate surface types based on light reflection [21,22], a condition that is difficult to meet because flooding occurs with cloudy skies. A seasonal disaster mapping system based on field observations is practical for generating initial data [23,24] but impractical for periodic events. Research on radar satellites for disaster management to date still has elevation resolution in the range of 2 m [6,7,25,26]. For Jakarta, especially in the city center to the northern coast, which often experiences floods, this is not appropriate because the flood depth is less than 2 m [27].

To map floods, previous methods were initially dominated by thresholding [6,11,13,14,28], Probability Density Function (PDF) [9,10], and more recently, Logistic Regression [29] and the Storm Water Management Model (SWMM) [30]. More specifically, in remote sensing image segmentation applications, some researchers use the Normalized Difference Vegetation Index (NDVI) to classify vegetation, classify water levels with the Normalized Difference Water Index (NDWI), and the classification of floods using Normalized Difference Flood Index (NDFI) [31,32]. Other researchers detected changes in land surface with radar images using the principle of interferometry to find coherence between images [15,16,33]. Research in the last three years has led to the use of Machine Learning (ML) in order to segment and classify increasingly complex datasets, such as the Adaptive Neuro-fuzzy Inference System (ANFIS) [34], Support Vector Machine (SVM) [35,36], Convolutional Neural Network (CNN) [36–38], and more recently with various Swarm Intelligence (SI) variants such as Particle Swarm Optimization (PSO) [39,40]. Sameen and Pradhan used a Residual Neural Network to detect the potential for landslides, where this method is intended to detect changes in soil texture as the initiation of landslides [38]. The CNN model with this residual block is used to process LiDAR data. The neurons on CNN are activated using the Rectified

Linear Unit (ReLU). This study proposes three-dimensional CNN in mapping flood areas to filter and weigh neurons and map the potential flood areas in urban areas with better accuracy and fewer number of images.

Yu Li et al. proposed an Active Self Learning method on CNN to detect floods in urban areas from the SAR image ensemble [37]. The dataset used is four TerraSAR-X images of HH polarization with the composition of one pre-event image, one co-event image, and two post-event images. Linyi Li et al. proposed a high-resolution urban flood mapping method (Super-Resolution Mapping of Urban Flood, SMUF) with the fusion of the Support Vector Machine and General Regression Neural Network (FSVM-GRNN) [35]. Because the urban flood area in the observation area is not very dense, the accuracy of this FSVM-GRNN is 80.2%.

Shen et al. proposed a machine learning process to make corrections to the mapping of flood inundation areas in near-real-time (NRT) using SAR, where the observation area is an open area without many obstacles on the surface [41]. At the time of segmentation, there are difficulties in classifying areas that are flooded with areas that have surface reflection properties similar to those of the water surface. ML is performed to correct speckle noise and another scattering, which can interfere with data reading and classification. The filtering method is used in most SAR image processing but its effect is to reduce the effective resolution and change the signal statistics and cannot completely remove noise. To overcome this, Shen et al. used the Logistic Binary Classifier (LBC) in a correction step to practice detecting the presence of water in the pixels contained in the water bodies and the surrounding buffer areas.

The objective of this work is to investigate the mapping of flood potential in Jakarta and nearby coastal areas using three-dimensional CNN on co-polarized (VV) and cross-polarized (VH) Sentinel-1a SAR images. A three-dimensional classification combines the two-dimensional image and one-dimension multi-temporal processes into a single convolution. The images are then pre-processed into grayscale images to be converted into a vector data format. The 2 January 2020 images were also sampled as flood and non-flood target sub-images, along with the corresponding locations from other images, to form the multi-temporal value changes of the flooded locations along with the consistency of the non-flooded locations. The CNN training is performed with training/test percentage values of 70/30; 80/20; and 90/10 with varying epochs between 100 and 160 iterations to obtain the best combination with the highest accuracy and the shortest processing time.

2. Materials and Methods

2.1. Location and Data

A radar image is generated from the reflection of active microwaves emitted from the radar vehicle (airplane or satellite). The transmitter in a radar system is an antenna array consisting of a waveguide and emitting a very narrow beam of microwave waves. The radar sensor moves along a trajectory, and the area highlighted by the radar moves (known as the footprint) along the surface being swept to form an image. A digital radar image consists of many pixel dots representing the backscatter or backscatter of a point on the surface. Figure 1 shows an example of a SAR image from 2 January 2020 that is free from cloud cover with the bright dots showing high backscattering while the dark dots represent low backscattering, while Figure 2 is the optical image from the same date showing cloud cover.

The radar system generally has a wavelength undisturbed by interference from water particles and water vapor in the air (clouds and rain). Because they are not dependent on illumination (irradiation) from the sun or other sources, the radar system can function day and night and in all weather. Synthetic Aperture Radar (SAR) works by detecting the phase-change of reflected signal caused by the movement of the platform to obtain the surface image with good resolution (i.e., visually discernible). The SAR system is generally divided into two wavelengths, namely short (C-band and X-band) and long (L-band and P-band) waves. Early SAR satellite systems use a single platform such as

Radarsat. Currently, the most commonly used are satellite constellation systems such as TerraSAR-X and TanDEM-X pair, four-satellites Cosmo-SkyMed in X-band, three-satellites Radarsat Constellation Mission, and Sentinel 1 satellite pair, which give shorter revisit time and higher temporal resolution [20,41–43]. Figure 3 shows the backscatter mechanism of shortwave radar (illustrated with black arrows) and long (illustrated arrows in blue) on various surfaces under normal conditions and during a flood. On the grass surface, there are surface reflections at both wavelengths due to the relative roughness of the surface. For short waves, the scattering is due to the thickness of the grass, while long waves can penetrate deeper [44].

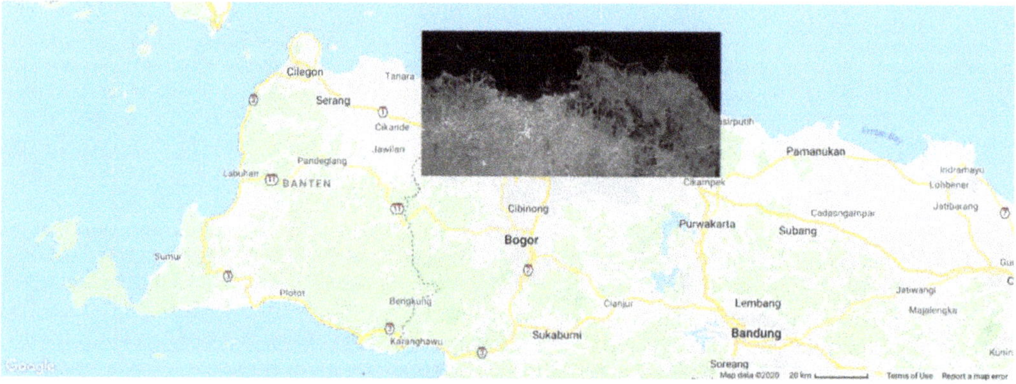

Figure 1. SAR image of Jakarta on 2 January 2020.

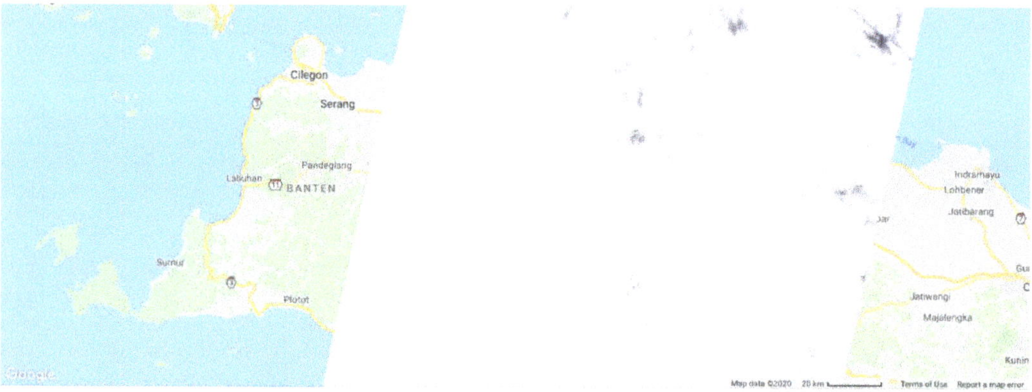

Figure 2. An optical image from Landsat-8 of Jakarta on 2 January 2020.

When a flood occurs, specular reflection occurs in both types of waves. On objects in the form of trees or forests, the reflection is dominated by the scattering volume. For short waves, the scattering comes from the canopy (leaves) of the trees, while the longwave scattering by the branches and other tree structures is added by a double-bounce, which hits the ground surface and then the tree trunk or vice versa. When there is a flood, the double reflection will get more significant due to the specular reflection on the water surface (shown as a thick line of the direction of the reflection). In urban areas, the reflection on both waves is dominated by multiple reflections, although the surface will appear coarser on short waves. When there is a flood, this double reflection will be significantly strengthened due to the specular reflection on the water surface (shown as a thick line of the direction of the reflection).

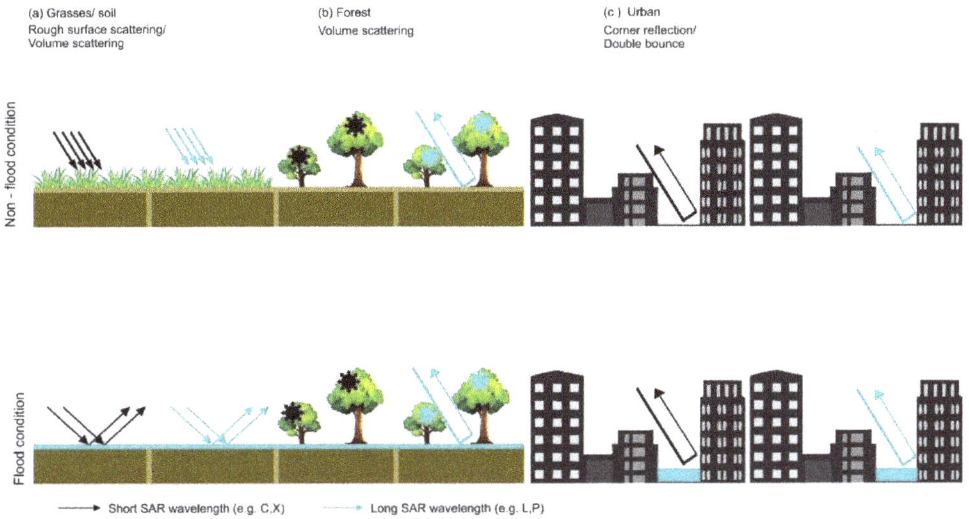

Figure 3. Radar wave backscatter mechanism on the surface of the object (**a**) Grass/Land, (**b**) Vegetation, and (**c**) Urban in normal and flood conditions, for short wavelengths (C- and X-band) and long waves (L- and P-band).

In this study, the flood data collected came from Sentinel-1a remote sensing satellite. The data are downloaded through Google Earth Engine by the Copernicus catalog and selecting available dates from the archive. The selected mode is Interferometric Wide Swath (IW) with 250 km swath and 5 m × 20 m spatial resolution [45]. For our model, the pixel resolution is preset at 10 m × 10 m. Remote sensing data combined with GIS data are integrated to create a flood hazard and potential map. Based on information obtained from remote sensing and GIS databases, the ML method can be applied for spatial modeling of flood vulnerability.

The data shown in Figure 4 are divided by the date of acquisition into three categories, namely: pre-event, consists of data from November to December 2019, which represent conditions before the major flood occurred; co-event is data taken on or near the 2 January 2020 flood, and the rest is categorized as post-event data. In Figure 5, the SAR images are set into a dataset, which contains 39 cross-polarized and 39 co-polarized images from Sentinel-1a between November 2019 and October 2020, with co-event images designated as the target image. The dataset SAR was collected using Google Earth Engine and consisted of Sentinel-1a VV and VH images between November 2019 and December 2020 as RGB composite TIF images. All images are resized into 946 × 681 pixels covering the Jakarta area and part of the Bekasi and Tangerang Regencies that flooded. The target image is further broken down as flood markers to make a 25 × 25 pixels-sized kernel. The previous individual images shown in Figure 5 are combined into three-dimensional data with a 946 × 681 × 78 pixel dataset and 25 × 25 × 20 pixel kernel.

2.2. Image Segmentation and Classification

In a digital image processing application, the primary process is segmentation to detect and identify objects and components within the image. The segmentation process divides the image into parts known as constituent objects. Automatic segmentation is generally the most challenging image processing [12]. With the development of image processing algorithms, image segmentation is also developed using region growing and merging, namely by expanding pixels so that the object becomes larger. In the end, some objects close to the same value will merge into one other, bigger object. This mathematical algorithm is the basis for developing an image segmentation algorithm that carries out the unsupervised segmentation process without human intervention.

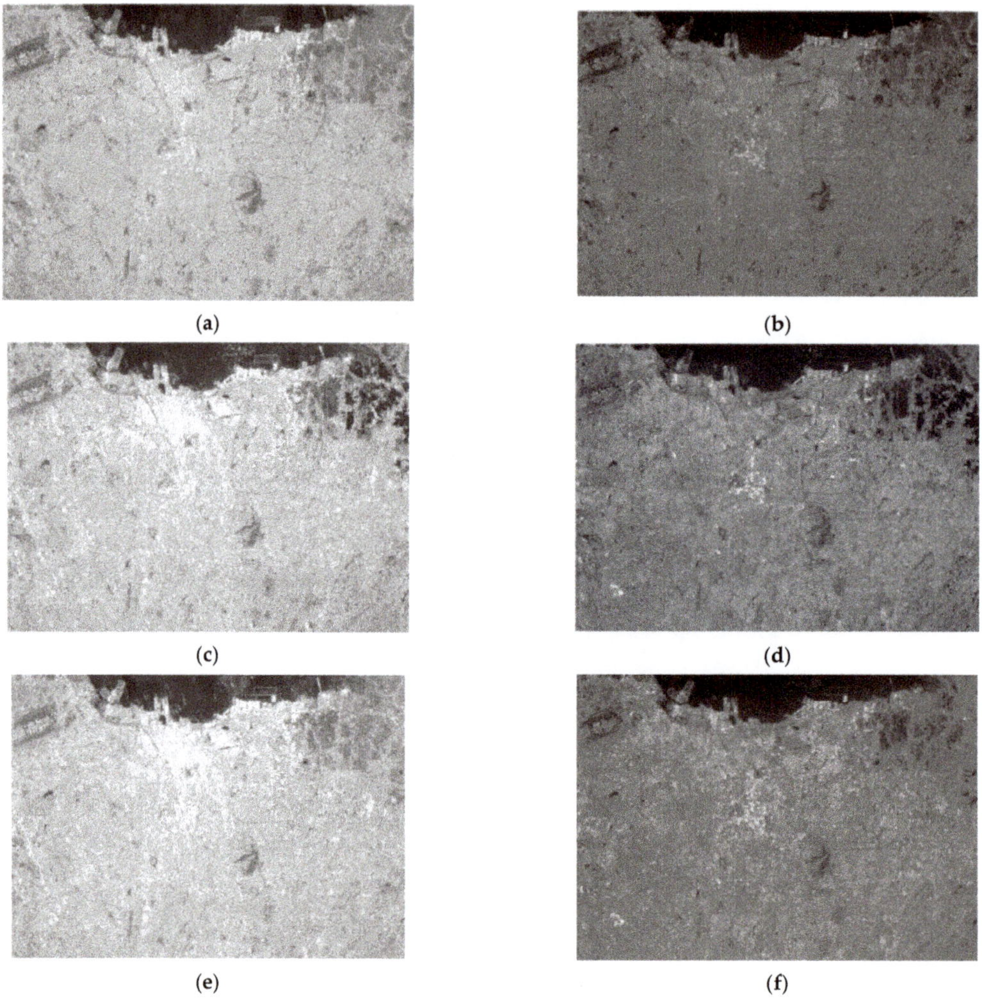

Figure 4. Sample images from co-polarization (VV) and cross-polarization (VH): (**a**,**b**) Pre-flood event; (**c**,**d**) Co-event during January 2020 flood; (**e**,**f**) Post-flood event.

Kwak et al. created a SAR satellite data processing algorithm to detect urban floods near-real-time using data before and after a flood event. Furthermore, the image is classified using a supervised classification to obtain the flood area based on building classes. The developed Probability Density Function (PDF) method can reduce the maximum backscatter intensity difference for rice fields and open areas by 35 dB; however, for urban areas, it has increased by 25 dB [9]. Further development of this method can reduce variance by 12 dB and increase urban areas by 15 dB [10]. In comparison, Liang et al. [46] used PDF to estimate the maximum similarity before thresholding by comparing the Otsu, Split-KI (Kettler and Illingworth), and Local Thresholding (LT) methods. The Overall Accuracy (OA) results obtained from the Sentinel-1a image classification in the Louisiana plain were 98.12% (Otsu), 98.55% (Split-KI), and 98.91% (LT), respectively.

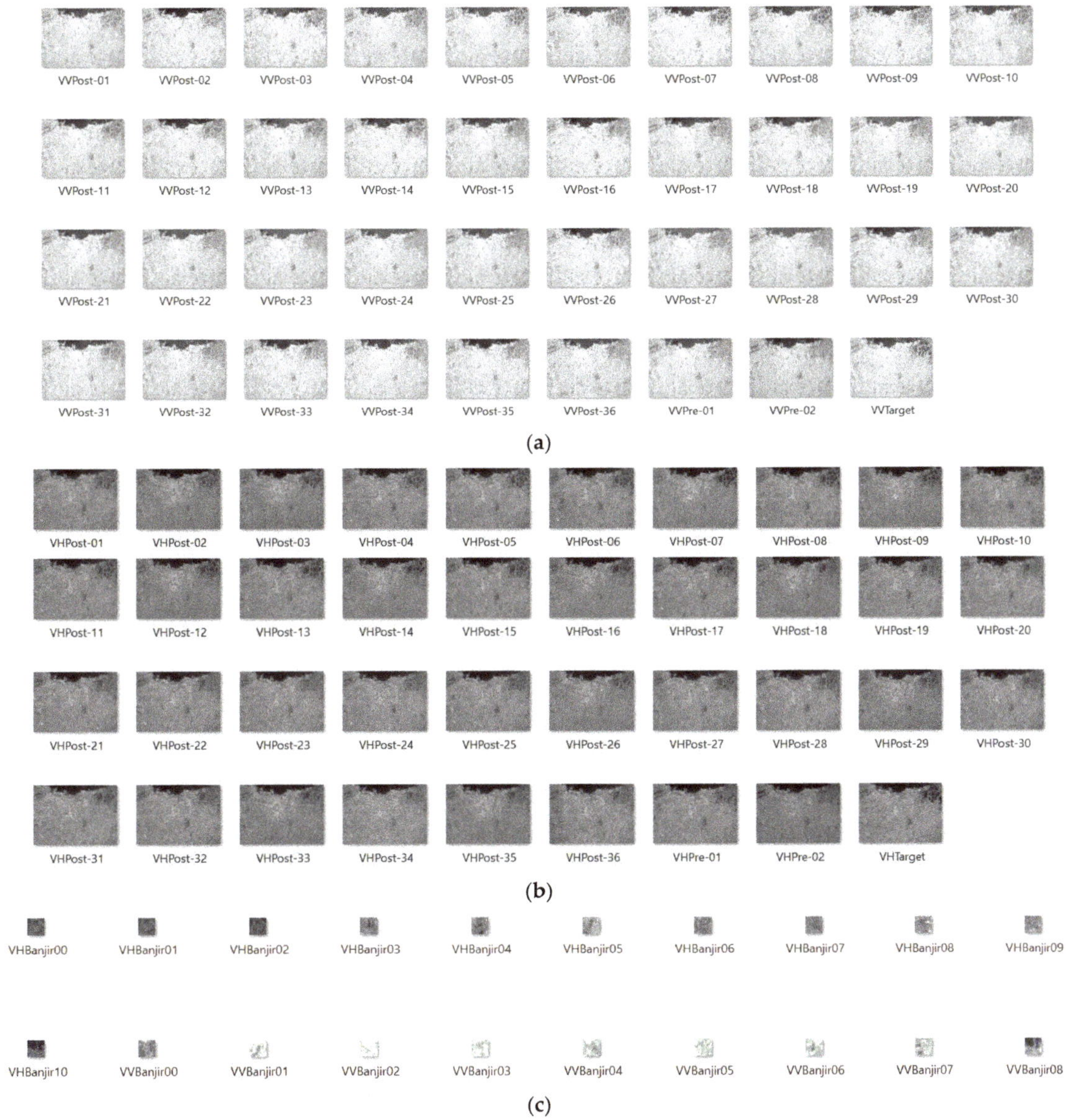

Figure 5. Sentinel-1a images: (**a**) co-polarized (VV); (**b**) cross-polarized (VH); (**c**) flood markers.

Pelich et al. proposed the creation of a large-scale global database for flood inundation maps derived from the SAR dataset [28]. The method used is histogram thresholding to delineate quickly, then the level of flood distribution is extracted from the SAR backscatter using the Probability Density Function (PDF). Thresholding is performed using the Hierarchical Split-Based Approach (HSBA) to identify pixels with a bimodal distribution on the sub-pixels, which indicates that there is an immersion limit on these pixels [47]. The accuracy of the results obtained from flood detection in rural areas is 35%.

Another technique in flood detection is to utilize the polarization characteristics of radar signals, namely the Interferometric SAR (InSAR) method. The principle of stable scatterer or persistent scatterer is used to detect areas that do not experience changes in reflection characteristics, while changes in reflection characteristics result in low coherence between image data and are assumed to be flooded. The mapping is built by creating 20 interferometric pairs from 22 consecutive Sentinel-1a images with a composition of 17 pairs of pre-event images, a pair of images during a flood, and two pairs of post-event images [48]. Chini et al. also integrated intensity data using InSAR coherence, normalized cross-correlation to detect the presence of water in urban areas and mapping of double-bounce-producing objects using histogram thresholding and region growing. Pixels are categorized as floods when there is a decrease in coherence on the RGB composite channel [16].

In line with the development of the field of artificial intelligence, image processing methods also develop by making use of artificial intelligence functions. Several artificial intelligence methods that are widely used in image processing are Artificial Neural Networks or ANNs. The method that has recently begun to be applied in studies of mapping flood potential and vulnerability is to use machine learning. Some of the methods that were implemented include Adaptive Neuro-Fuzzy [34], Support Vector Machine (SVM) [35,36], Convolutional Neural Network (CNN) [36,38], and Swarm Intelligence [39,40,49]. Dasgupta et al. used Gamma Maximum A-Posteriori (GMAP) to filter out speckle noise from SAR images, then performed surface texture analysis using the Gray Level Co-Occurrence Matrix (GLCM) [34].

Although being the most common basic method on flood mapping, NDFI/NDWI as the most straightforward method tends to amplify noise greatly. Otsu thresholding suffers from high computational requirements since it is an early optimization method. The SMUF, SVM, GRNN [34,35], and most recently CNN [50] still perform the classification process in a 2D plane and then perform the 1D multi-temporal process. Due to the complexity of the factors that influence the occurrence of floods in urban areas, the most effective and efficient classification method is needed. As a classification technology developed based on feature matching, the ML method produces a more accurate recognition than feature matching. However, it has limited extraction features that can cause errors in the computation process. This study proposes three-dimensional CNN in mapping flood areas to filter and weigh neurons and map the potential flood areas in urban areas with fewer images compared to the previous study [36,51]. CNN features unsupervised feature extraction compared to Artificial Neural Network (ANN), in which the process is achieved through the training phase to recognize flood areas. In ANN, all neurons of a layer are fully connected to every neuron from other layers, whereas in CNN, only the last layer of neurons is fully connected due to the parameter-sharing nature of the CNN, therefore the computational load of CNN is less than ANN.

2.3. Deep Learning Neural Network

Recent developments in the Deep Learning Neural Network (DNN) are increasingly opening up great opportunities in flood mapping research. Deep Learning as one of the Machine Learning models has shown promising results in image processing and pattern recognition. Therefore, this research will propose mapping the potential flood areas using the DNN algorithm. DNN is based on Artificial Neural Network and generally consists of an input layer, with more than one hidden layer and one output layer [52]. Figure 6 shows the conceptual structure of the DNN model used for flood vulnerability mapping. The input layer is the factor that affects flood ($F1$–Fn). The information is processed and analyzed in the hidden layer to determine the weight and classification of each pixel. The final result of the classification is an indication that there is a flood in the output layer with two possible labels: Flood (positive class) and Others (negative class).

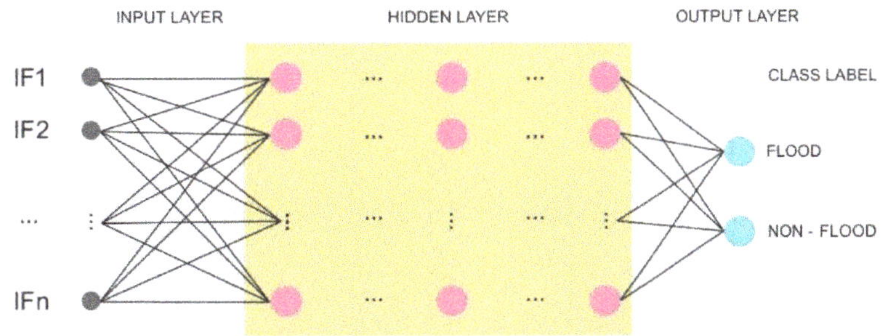

Figure 6. DNN structure concept for mapping flood potential.

DNN is a feed-forward network and is trained using the back-propagation method. However, more hidden layers will make the network challenging to train because of the different adjustment speeds in the hidden layer. DNN was implemented successfully in various applications, especially in automatic image recognition, speech recognition, language processing, and some applications in remote sensing. There is no rule of thumb about the number of hidden layers and neurons in each layer since it depends on the complexity of the problem and the conditions of the dataset.

The number of hidden layers in DNN has the advantage of representing a very complex relationship between factors. The hidden layer on DNN has neurons that are activated with the Rectified Linear Unit (ReLU) function as an alternative whose computation is more straightforward when compared to the sigmoid. Because DNN is trained on the principle of back-propagation, ReLU can minimize the decrease in learning gradient, hindering the learning process. Mathematically, the ReLU activation function can be expressed as $h'(x)$ in Equation (1).

$$h'(x) = \begin{cases} 1if_x > 0 \\ 0if_x \leq 0 \end{cases} \tag{1}$$

Hidden layers in DNN perform increasingly complex feature transformations to produce a more discriminatory feature abstraction. The classification results displayed in the output layer are based on the most abstract features obtained in the last hidden layer. During the DNN learning phase, the connection weights between layers are adjusted to reduce the difference between observed and predicted results. The back-propagation process trains DNN by providing feedback on the error results to the hidden layer. The deviation between the observed and predicted results is expressed in the loss function between entropies, as expressed in Equation (2).

$$L = -\frac{1}{N_D}\sum_{n=1}^{N_D} Tln(Y) + (1-T)\ln(1-Y) \tag{2}$$

where N_D is the number of training data points, T represents the observed output, and Y represents the predicted output. The back-propagation learning gradient used for the training sample of m is formulated in Equation (3):

$$g = \frac{1}{m}\sum_{i=1}^{C}\frac{\partial L}{\partial w} \tag{3}$$

where L is the loss function, w represents the network weight, and C = 2 represents the number of output classes used (flood and others).

2.4. Convolutional Neural Network (CNN)

CNN is one type of DNN that uses the convolutional principle in its data processing. The basic concept of CNN architecture is to utilize a convolutional layer to detect the relationship between the features of objects and a pooling layer to similar group features. The CNN architecture consists of a series of layers, namely the Convolutional Layer (CL), which functions to transform a set of activations with a differential function, a Pooling Layer, and the final result is a Fully Connected Layer (FCL). Unlike other neural networks where all neurons are fully connected with every other neuron of the next layer, CNN disregards zero-valued parameters and makes fewer connections between layers. The non-zero parameters can be shared to be used by more than one connection in the layer to reduce the number of connections. This characteristic is useful for recognizing features.

The pooling layer function is used to reduce the size of an image by downsampling it and summarizing the features. The common pooling methods to achieve grouping are average pooling, where the summary is the dominant feature, and maximum pooling by summarizing the strongest feature [53]. Average pooling produces a smooth feature that is useful to extract the most relevant value, such as the color of a surface, where a small variation in isolated points within a region does not affect the overall value. On the other hand, max-pooling extracts high contrast data, such as edges or points.

The problem with a sampling matrix (and an image) in CL is that pixels at and near the edge are sampled less than pixels farther from the edge. This sometimes results in sampling inaccuracy. To prevent this, the kernel filter is padded, with extra rows and columns to allow for more information to be collected from the edge pixels. For two-dimensional data, there are two types of padding: same padding and valid padding. Same padding maintains the sample size at the same as the original matrix; basically, it resamples the image. Valid padding considers all pixels valid, so the model considers the value. This is useful for keeping the information from corner pixels since the simple model considers it invalid due to being less sampled compared to other pixels.

The extracted features compose the feature map that the FCL will use to classify the result. This approach makes CNN a method with fewer computational requirements than the fully connected ANN structure. The CL calculation is formulated in Equation (4):

$$h_{i,j} = \sum_{k=1}^{m}\sum_{l=1}^{m} w_{(k,l)} x_{(i+k-1),(j+l-1)} \tag{4}$$

and the pooling layer (max pooling) is stated in Equation (5):

$$h_{i,j} = max\left\{ x_{(i+k-1),(j+l-1)} \forall l \leq k \; and \; 1 \leq l \leq m \right\} \tag{5}$$

with fully connected layer h formulated in Equation (6):

$$h = \sum_{i} w_i x_i \tag{6}$$

where $h_{i,j}$ is the output at point (i, j) on the layer with input x and filter w, and m denotes the width and height of the filter. Non-linear functions are used in CL and FCL to convert negative values to zero, including Sigmoid, Hyperbolic Tangent (*Tanh*), and Rectified Linear Unit (ReLU).

Three-dimensional CNN is a CNN structure whose input is a set of square matrices, $s \times (n \times n)$, so it is a suitable method for image segmentation and classification. In this study, the dataset used is multi-sensor, multi-temporal data derived from SAR and optical images, rainfall data, and ground surface contour data, as shown in Figure 7.

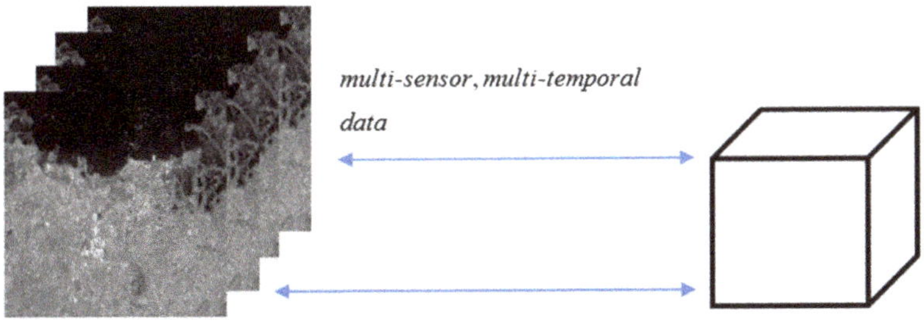

Figure 7. Representation of the multi-temporal 2D dataset into 3D data.

2.5. Proposed Method

Segmentation and classification of flooded areas using 3D CNN for the SAR image dataset and the flood factor consists of three-dimensional dataset segmentation stages using three-dimensional CNN to get initial segmentation results. These results are used to weight neuron connections to perform n-dimensional optimization so that we get the classification of pixels into flood or other categories.

In the three-dimensional CNN shown in Figure 8, several CLs with dimensions a × a × a are used to filter the input data to obtain a feature map. The input data used are shown in Table 1. The images are down sampled using a pooling layer by summarizing the features present in the images. In this model, the pooling layer uses max pooling, which summarizes the most dominant value in the sample. To prevent edge and corner pixels from being omitted by the model, valid padding is used on the input layer and the CL. The padding basically left the image unchanged but allows edge and corner pixels to be more sampled as it is now placed further from the edge. Furthermore, the pooling layer measuring b × b × b is used to reduce the map, so those neuron connections are formed to compile the information obtained, which is then formed into FCL. FCL stored the different feature values and compiled them into a feature map with two output categories, namely flood pixels and non-flood pixels.

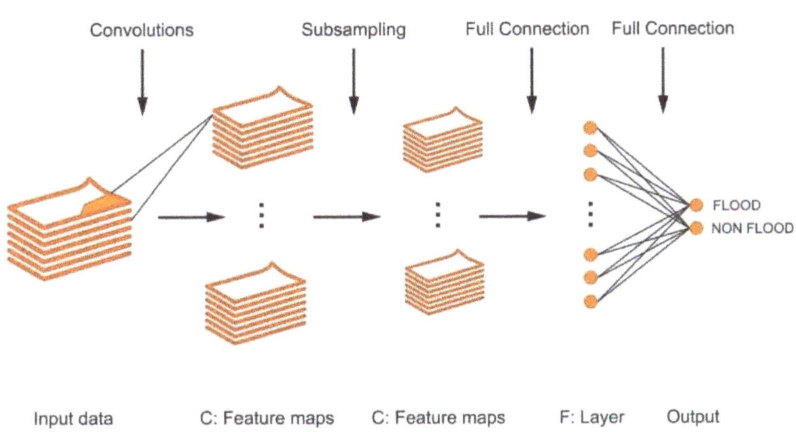

Figure 8. Representation of 3D-CNN process.

Table 1. Input data used for the 3D-CNN.

Type	Source	Resolution/Scale	Acquisition Date
Co-polarization (VV) SAR data	Sentinel-1a	10 m	21 November 2019 to 20 October 2020
Cross-polarization (VH) SAR data	Sentinel-1a	10 m	21 November 2019 to 20 October 2020

The stages carried out in this study began with an inventory of the data used for classification, namely the SAR image dataset. The pre-processing stage is comprised of registering the image data to ensure that the coordinates are consistent between different images. As the images are in RGB TIF format with r × c × 3 dimensions, they must be converted first into grayscale images, and then samples of sub-images were selected that represent flood and non-flood targets. The data are then divided into training and validation sets. The Feature Learning stage, or training, provides training data for the model to store known flood data. The commonly used proportion between the two sets is 70:30 [54], but we also include 80:20 and 90:10 for comparisons. Training data are used to train 3D CNN [36] in determining the parameters' optimized values. The next stage is to conduct training on the classification by three-dimensional CNN to detect the presence of water surfaces and differentiate them from other surfaces by the variance of the pixel values since dry land and permanent water bodies have consistent values. The ReLU plays a significant part in this phase since flood areas tend to change values, the possibility of dry land changes to a water surface and then back to dry land will result in a negative value. The ReLU rectifies this problem and prevents the neuron with a negative output from being contributed to the network. The Classification stage presents the system to other data for recognizing if there are flood features present in the images using feedback from the results of the Training stage. The overall process in the research is shown in Figure 9.

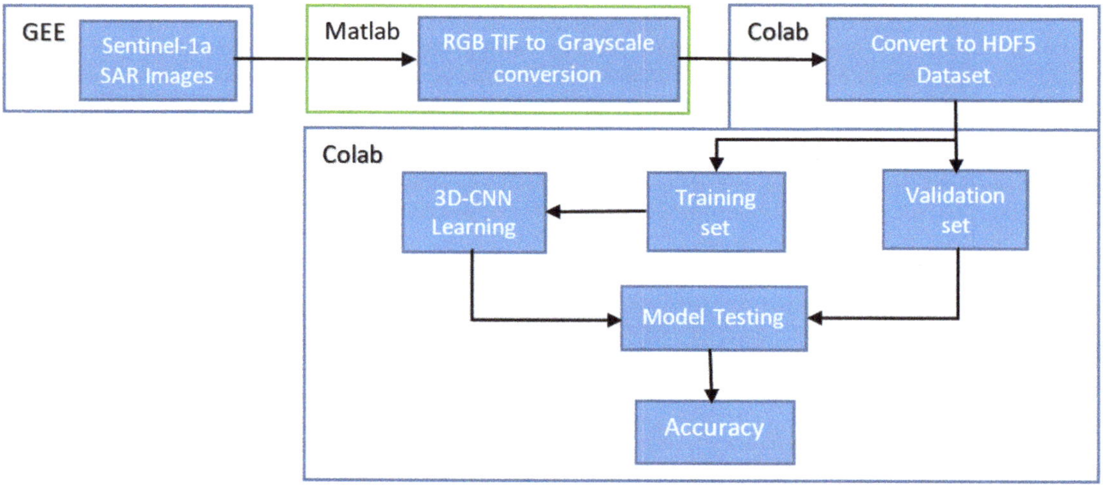

Figure 9. Workflow of the research.

3. Results and Discussion

The three-dimensional CNN model is trained with two main hyperparameters, namely: epoch, which is the complete iteration of convolution feed-forward before starting over the next iteration; and validation-split, which is the proportion of the training data used for validating the result of the training. In this research, we use the combination of training/validation split of 70/30, 80/20, and 90/10 with epochs of 100 and 150 iterations. The elapsed time and resulting accuracy for each combination are shown in Table 2 and the graphic plot in Figure 10. Accuracy is defined as the percentage of correct predictions for the test data calculated by dividing the number of correct predictions by the number of total predictions, while elapsed time counts the total time needed to perform the training with the corresponding proportion and epochs.

Table 2. The elapsed time, accuracy, and RMSE of the 3D-CNN model.

No. of Epochs	Performance Metrics	Training/Validation Split 70/30	80/20	90/10
100	Time	163 min	188 min	140 min
	Accuracy	0.667	0.659	0.685
	RMSE	0.284	0.282	0.203
150	Time	243 min	257 min	302 min
	Accuracy	0.672	0.692	0.674
	RMSE	0.288	0.314	0.296

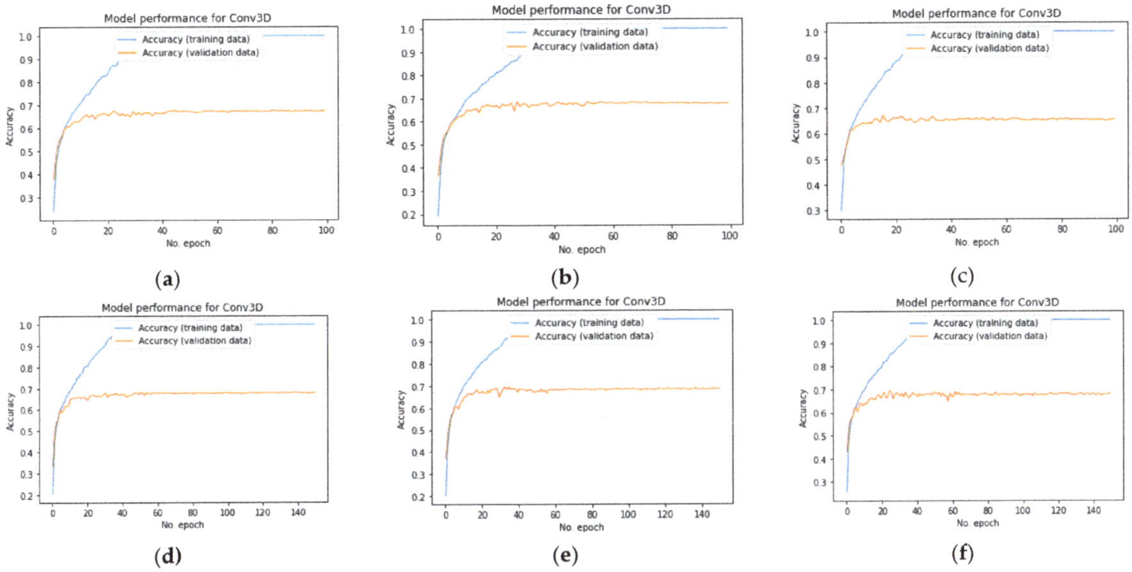

Figure 10. Graphic plot from the testing and validation of the 3D-CNN model: (**a**) 100 epochs, 70/30 split; (**b**) 100 epochs, 80/20 split; (**c**) 100 epochs, 90/10 split; (**d**) 150 epochs, 70/30 split; (**e**) 150 epochs, 80/20 split; and (**f**) 150 epochs 90/10 split.

During model testing with 100 epochs, the algorithm quickly reaches 100% training accuracy under 40 epochs, and in the first half of the epoch, the testing accuracy increases, but in the second half, it does not rise significantly, being around 0.667; 0.672; and 0.685 for 70/30; 80/20; and 90/10 data split. The overall accuracy achieved by the model is between 0.667 to 0.692 for 100 epochs completed between 140 and 183 min. Root Mean Square Error (RMSE) for 70/30 and 80/20 is around 0.28, while for 90/10 is lower at 0.2, which is consistent with higher accuracy. For 150 epochs, the accuracy of 0.672; 0.692; and 0.674 with RMSE of 0.288; 0.314; and 0.296 for the corresponding data split in 70/30; 80/20; and 90/10 ratios, respectively. The process was completed in 4 h and 3 min. Figure 10 shows that the validation accuracy quickly becomes stable after 20 to 25 epochs while the training accuracy is still increasing until it reaches 100%. This condition indicates that the model was overfitting during testing. Overfitting resulted from a vast set of neural connections, which often reduces the system fitness due to non-common cases included in learning data [55].

We readjusted the model to eliminate and reduce overfitting and then tested it with similar hyperparameters. Overfitting correction is performed by randomly deactivating some neurons on each layer, so they are not used during forward- and back-propagation training. This causes the learning process to spread out connection weights without focusing on specific neurons. In this research, the deactivation probability is set at 0.5, which means there is an equal chance of each unused neuron in the learning process. Low deactivation probability will not reduce overfitting, while high probability will cause the system to underachieve. Reducing neurons results in a smaller, simpler, and more regulated connection network, which means outlying or widely different results will be disregarded. In this manner, the overall error could be reduced by averaging errors from different connections.

The adjusted model yields the result shown in Table 3 and Figure 11. The results indicate that computation time takes 50 min longer for an 80/20 and 90/10 split, with the resulting accuracy reaching over 0.7 than the previous test. The most significant increase in accuracy is for 150 epochs with a 90/10 split of testing and validation data, which shows an increase of 0.045 for accuracy of 0.719, the lowest RMSE achieved by 70/30 split with 100 epochs with a drop of RMSE value from 0.284 to 0.024. The fastest computing time of 165 min is achieved with 100 epochs and 70/30 split data. This result is consistent that fewer training data corresponds with faster computing but lower accuracy, while a higher percentage of training data took longer but with higher accuracy.

Table 3. The elapsed time, accuracy, and RMSE of the adjusted 3D-CNN model.

No. of Epochs	Performance Metrics	Testing/Validation Split		
		70/30	80/20	90/10
100	Time	165 min	235 min	195 min
	Accuracy	0.691	0.708	0.705
	RMSE	0.024	0.051	0.078
150	Time	172 min	283 min	302 min
	Accuracy	0.699	0.709	0.719
	RMSE	0.082	0.093	0.112

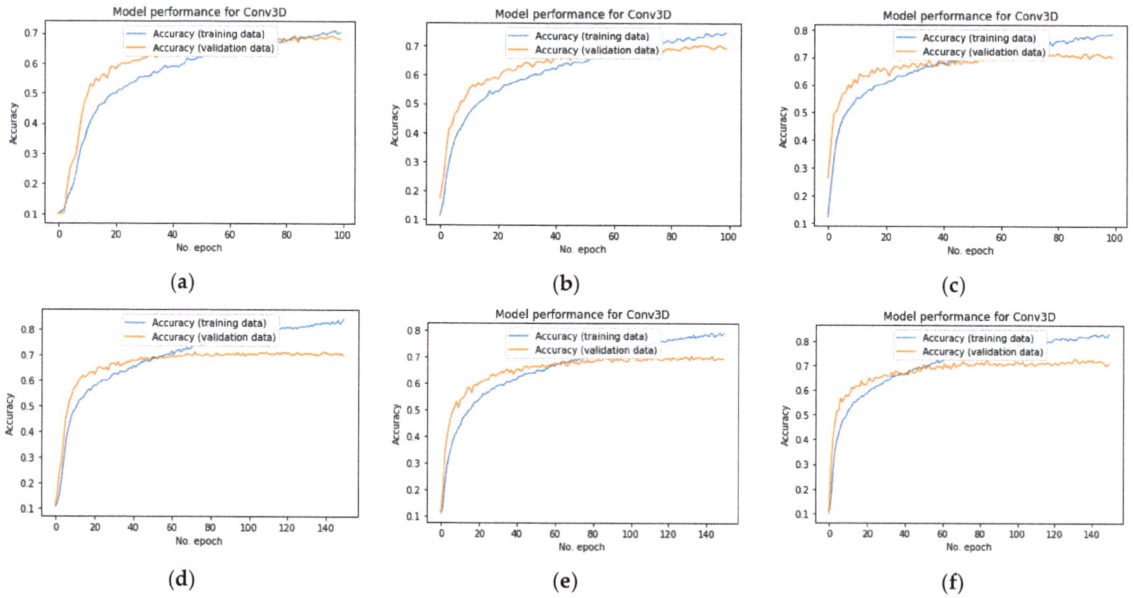

Figure 11. The testing and validation accuracy of the tuned 3D-CNN models: (**a**) 100 epochs, 70/30 split; (**b**) 100 epochs, 80/20 split; (**c**) 100 epochs, 90/10 split; (**d**) 150 epochs, 70/30 split; (**e**) 150 epochs, 80/20 split; and (**f**) 150 epochs, 90/10 split.

Further testing with 140, 145, 155, and 160 epochs to investigate the optimum combination of accuracy with shorter time yields the results shown in Figure 12. Since the testing accuracy is greater with 150 epochs than 100 epochs, we assume that accuracy will improve within 150 ± 10 epochs. Testing with a 70/30 split confirms that as epochs increased from 140 to 160, accuracy gradually improved by 13.6% from 0.567 to 0.703, while computing time increased by 12% from 240 min to 269 min. A similar trend is also observed during testing using an 80/20 data split with an accuracy increase by 13.5% from 0.577 to 0.712 but with a much longer computing time from 243 min to 304 min, representing an increase of 25.1%. The more significant increase is due to the additional time needed to perform more training for the 80/20 than the 70/30. As for the testing with a 90/10 data split, the peak accuracy performance is achieved at 150 epochs with 0.719. Testing with 140, 145, 155, and 160 epochs gives lower results.

In Table 4, the three-dimensional CNN without any combinations with other methods results in higher accuracy than what was achieved by Wang et al. at 0.685 [36]. It is comparable to Grimaldi et al. [11] on open trees flood accuracy at a range of 0.55 to 0.70, which is similar in conditions to flooded areas in Jakarta. Figure 12 shows the flood map of the proposed model compared to the SAR image, as shown in Figure 2, where the model could detect most of the dark areas of the flood while leaving out the similarly dark Jakarta Bay. Compared to the sub-district-level flood map publicly released by the government [2], it is also shown that the flood has occurred in the reported sub-districts. There are discrepancies between the detected and reported areas since the report classifies floods as a whole sub-district coverage.

Table 4. Comparison between models.

KERRYPNX	Methods		
	3D-CNN	**CNN-SVM**	**Fuzzy Logic**
Accuracy	0.719	0.685	0.669
Location	Dense Urban	Rural	Rural
Characteristic	Flat	Hill	Flat

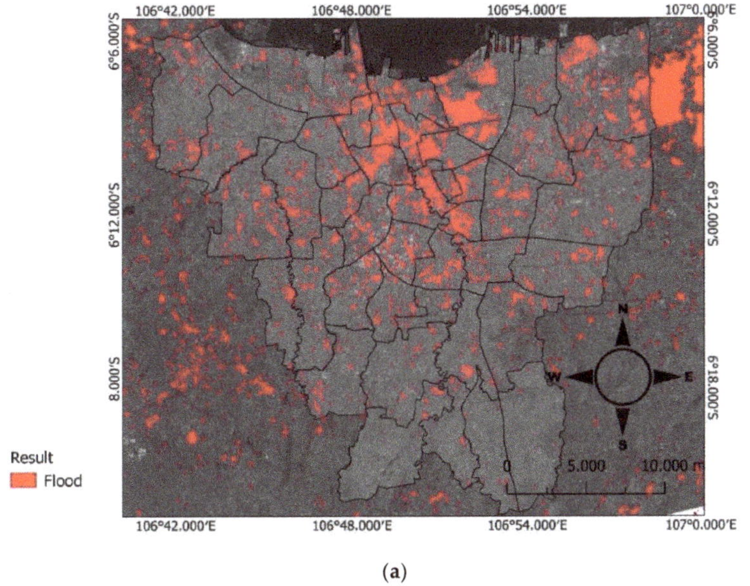

(a)

Figure 12. *Cont.*

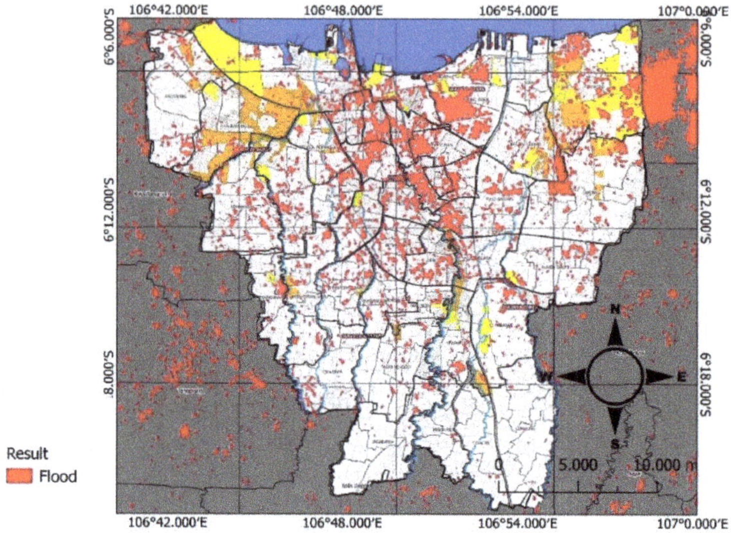

(b)

Figure 12. Result map compared to (**a**) 2 January 2020 SAR image, and (**b**) Jakarta Flood Map released by the government report.

4. Conclusions

In this study, an application of a three-dimensional Convolutional Neural Network for flood mapping is proposed. The deactivation factor minimizes the overfitting problem to reduce the number of neurons and simplify the connections. The research results are that the 3D-CNN method enables the analysis of multi-temporal images for flood detection and classification instead of using multiple image pairs with multiple classification levels. For three combinations of splitting training/test data, the highest overall accuracy of 0.72 was achieved for a split of 90/10 and 150 epochs in 302 min. Regarding computation time, the best performance is achieved with an 80/20 split and 150 epochs with an accuracy of 0.71 in 283 min. Another test with epochs other than 150 showed that accuracy gradually decreases with a 90/10 split, but with a lower training function, the accuracy improves as the number of epochs increases.

Author Contributions: The research was led by D.S. and M.R.; R.A. provided satellite data and insights on satellite data processing; I.R. was responsible for developing methods and analysis. All authors have read and agreed to the published version of the manuscript.

Funding: This research was funded by the Doctoral Program Research Indexed Publication Grant of Universitas Indonesia (PUTI Doktor UI) 2020 under Grant NKB-3321/UN2.RST/HKP.05.00/2020.

Institutional Review Board Statement: Not applicable.

Informed Consent Statement: Not applicable.

Data Availability Statement: Not applicable.

Acknowledgments: The authors would like to thank Rokhmatuloh and Ardiansyah of the Faculty of Mathematics and Natural Sciences, Universitas Indonesia, for satellite data processing support.

Conflicts of Interest: The authors declare no conflict of interest. The funders had no role in the design of the study; in the collection, analyses, or interpretation of data; in the writing of the manuscript, or in the decision to publish the results.

References

1. National Development Planning Agency. *JABODETABEK February 2007 Post-Flood Damage and Loss Estimation Report*; Ministry of National Development Planning: Jakarta, Indonesia, 2018.
2. Jakarta Disaster Mitigation Agency. *Jakarta Historical Flood Map, in Jakarta Historical Flood Map 2013–2017*; DKI Jakarta Disaster Mitigation Agency: Jakarta, Indonesia, 2017.
3. Vanama, V.S.K.; Rao, Y.S. Change Detection Based Flood Mapping of 2015 Flood Event of Chennai City Using Sentinel-1 SAR Images. In Proceedings of the IGARSS 2019—2019 IEEE International Geoscience and Remote Sensing Symposium, Yokohama, Japan, 28 July–2 August 2019; Institute of Electrical and Electronics Engineers (IEEE): Piscataway, NJ, USA, 2019; pp. 9729–9732.
4. Eini, M.; Kaboli, H.S.; Rashidian, M.; Hedayat, H. Hazard and vulnerability in urban flood risk mapping: Machine learning techniques and considering the role of urban districts. *Int. J. Disaster Risk Reduct.* **2020**, *50*, 101687. [CrossRef]
5. Wu, Z.; Shen, Y.; Wang, H.; Wu, M. Urban flood disaster risk evaluation based on ontology and Bayesian Network. *J. Hydrol.* **2020**, *583*, 124596. [CrossRef]
6. Ciecholewski, M. River channel segmentation in polarimetric SAR images: Watershed transform combined with average contrast maximisation. *Expert Syst. Appl.* **2017**, *82*, 196–215. [CrossRef]
7. Martinis, S.; Kersten, J.; Twele, A. A fully automated TerraSAR-X based flood service. *ISPRS J. Photogramm. Remote Sens.* **2015**, *104*, 203–212. [CrossRef]
8. Schumann, G.J.P.; Moller, D.K. Microwave remote sensing of flood inundation. *Phys. Chem. Earth Parts A/B/C* **2015**, *83–84*, 84–95. [CrossRef]
9. Kwak, Y.; Yun, S.; Iwami, Y. A new approach for rapid urban flood mapping using ALOS-2/PALSAR-2 in 2015 Kinu River Flood, Japan. In Proceedings of the 2017 IEEE International Geoscience and Remote Sensing Symposium (IGARSS), Ft. Worth, TX, USA, 23–28 July 2017; IEEE: Piscataway, NJ, USA, 2017; pp. 1880–1883.
10. Kwak, Y.; Natsuaki, R.; Yun, S. Effect of Building Orientation on Urban Flood Mapping Using Alos-2 Amplitude Images. In Proceedings of the IGARSS 2018—2018 IEEE International Geoscience and Remote Sensing Symposium, Valencia, Spain, 22–27 July 2018; IEEE: Piscataway, NJ, USA, 2018; pp. 2350–2353.
11. Grimaldi, S.; Xu, J.; Li, Y.; Pauwels, V.; Walker, J. Flood mapping under vegetation using single SAR acquisitions. *Remote Sens. Environ.* **2020**, *237*, 111582. [CrossRef]
12. Tanguy, M.; Chokmani, K.; Bernier, M.; Poulin, J.; Raymond, S. River flood mapping in urban areas combining Radarsat-2 data and flood return period data. *Remote Sens. Environ.* **2017**, *198*, 442–459. [CrossRef]
13. Rahman, M.R.; Thakur, P.K. Detecting, mapping and analysing of flood water propagation using synthetic aperture radar (SAR) satellite data and GIS: A case study from the Kendrapara District of Orissa State of India. *Egypt. J. Remote Sens. Space Sci.* **2018**, *21*, S37–S41. [CrossRef]
14. Jo, M.; Osmanoglu, B. Rapid Generation of Flood Maps Using Dual-Polarimetric Synthetic Aperture Radar Imagery. In Proceedings of the IGARSS 2019—2019 IEEE International Geoscience and Remote Sensing Symposium, Yokohama, Japan, 28 July–2 August 2019; IEEE: Piscataway, NJ, USA, 2019; pp. 9764–9767.
15. Pulvirenti, L.; Chini, M.; Pierdicca, N.; Boni, G. Detection of flooded urban areas using sar: An approach based on the coherence of stable scatterers. In Proceedings of the 2017 IEEE International Geoscience and Remote Sensing Symposium (IGARSS), Fort Worth, TX, USA, 23–28 July 2017; Institute of Electrical and Electronics Engineers (IEEE): Piscataway, NJ, USA, 2017; pp. 5701–5704.
16. Chini, M.; Pulvirenti, L.; Pelich, R.; Pierdicca, N.; Hostache, R.; Matgen, P. Monitoring Urban Floods Using SAR Interferometric Observations. In Proceedings of the IGARSS 2018—2018 IEEE International Geoscience and Remote Sensing Symposium, Valencia, Spain, 22–27 July 2018; Institute of Electrical and Electronics Engineers (IEEE): Piscataway, NJ, USA, 2018; pp. 8785–8788.
17. Chini, M.; Hostache, R.; Pelich, R.-M.; Matgen, P.; Pulvirenti, L.; Pierdicca, N. Probabilistic Urban Flood Mapping Using SAR Data. In Proceedings of the IGARSS 2019—2019 IEEE International Geoscience and Remote Sensing Symposium, Yokohama, Japan, 28 July–2 August 2019; Institute of Electrical and Electronics Engineers (IEEE): Piscataway, NJ, USA, 2019; pp. 4643–4645.
18. Refice, A.; D'Addabbo, A.; Pasquariello, G.; Lovergine, F.P.; Capolongo, D.; Manfreda, S. Towards high-precision flood mapping: Multi-temporal SAR/InSAR data, Bayesian inference, and hydrologic modeling. In Proceedings of the 2015 IEEE International Geoscience and Remote Sensing Symposium (IGARSS), Milan, Italy, 26–31 July 2015; Institute of Electrical and Electronics Engineers (IEEE): Piscataway, NJ, USA, 2015; pp. 1381–1384.
19. Mason, D.C.; Trigg, M.; Garcia-Pintado, J.; Cloke, H.; Neal, J.; Bates, P. Improving the TanDEM-X Digital Elevation Model for flood modelling using flood extents from Synthetic Aperture Radar images. *Remote Sens. Environ.* **2016**, *173*, 15–28. [CrossRef]
20. Guimarães, U.S.; Narvaes, I.S.; Galo, M.L.B.T.; da Silva, A.Q.; Camargo, P.O. Radargrammetric approaches to the flat relief of the amazon coast using COSMO-SkyMed and TerraSAR-X datasets. *ISPRS J. Photogramm. Remote Sens.* **2018**, *145*, 284–296. [CrossRef]
21. Yang, J.; He, Y.; Caspersen, J. A multi-band watershed segmentation method for individual tree crown delineation from high resolution multispectral aerial image. In Proceedings of the 2014 IEEE Geoscience and Remote Sensing Symposium, Quebec City, QC, Canada, 13–18 July 2014; Institute of Electrical and Electronics Engineers (IEEE): Piscataway, NJ, USA, 2014; pp. 1588–1591.
22. Boni, G.; Pulvirenti, L.; Silvestro, F.; Squicciarino, G.; Pagliara, P.; Onori, R.; Proietti, C.; Candela, L.; Pisani, A.R.; Zoffoli, S. User oriented multidisciplinary approach to flood mapping: The experience of the Italian Civil Protection System. In Proceedings of the 2015 IEEE International Geoscience and Remote Sensing Symposium (IGARSS), Milan, Italy, 26–31 July 2015; Institute of Electrical and Electronics Engineers (IEEE): Piscataway, NJ, USA, 2015; pp. 834–837.

23. Xie, J.; Yu, W.; Li, G. An inter-agency collaborative computing framework for fast flood mapping using distributed remote sensing data. In Proceedings of the 2016 Fifth International Conference on Agro-Geoinformatics (Agro-Geoinformatics) 2016, Tianjin, China, 18–20 July 2016; 2016; pp. 1–5. [CrossRef]
24. Yang, J.; He, Y.; Caspersen, J.P.; Jones, T.A. Delineating Individual Tree Crowns in an Uneven-Aged, Mixed Broadleaf Forest Using Multispectral Watershed Segmentation and Multiscale Fitting. *IEEE J. Sel. Top. Appl. Earth Obs. Remote Sens.* **2016**, *10*, 1390–1401. [CrossRef]
25. Chini, M.; Papastergios, A.; Pulvirenti, L.; Pierdicca, N.; Matgen, P.; Parcharidis, I. SAR coherence and polarimetric information for improving flood mapping. In Proceedings of the 2016 IEEE International Geoscience and Remote Sensing Symposium (IGARSS), Beijing, China, 10–15 July 2016; Institute of Electrical and Electronics Engineers (IEEE): Piscataway, NJ, USA, 2016; pp. 7577–7580.
26. Duan, Y.; Fang, L.; Licheng, J.; Peng, Z.; Lu, Z. SAR Image segmentation based on convolutional-wavelet neural network and Markov random field. *Pattern Recognit.* **2017**, *64*, 255–267. [CrossRef]
27. Burle, S. FloodMap.net. Available online: https://www.floodmap.net/Elevation/ElevationMap/?gi=1642911 (accessed on 1 August 2020).
28. Pelich, R.; Chini, M.; Hostache, R.; Matgen, P.; Delgado, J.M.; Sabatino, G. Towards a global flood frequency map from SAR data. In Proceedings of the 2017 IEEE International Geoscience and Remote Sensing Symposium (IGARSS), Fort Worth, TX, USA, 23–28 July 2017; Institute of Electrical and Electronics Engineers (IEEE): Piscataway, NJ, USA, 2017; pp. 4024–4027.
29. Lee, J.-Y.; Kim, J.-S. Detecting Areas Vulnerable to Flooding Using Hydrological-Topographic Factors and Logistic Regression. *Appl. Sci.* **2021**, *11*, 5652. [CrossRef]
30. Sidek, L.M.; Chua, L.H.C.; Azizi, A.S.M.; Basri, H.; Jaafar, A.S.; Moon, W.C. Application of PCSWMM for the 1-D and 1-D–2-D Modeling of Urban Flooding in Damansara Catchment, Malaysia. *Appl. Sci.* **2021**, *11*, 9300. [CrossRef]
31. Cian, F.; Marconcini, M.; Ceccato, P. Normalized Difference Flood Index for rapid flood mapping: Taking advantage of EO big data. *Remote Sens. Environ.* **2018**, *209*, 712–730. [CrossRef]
32. Cian, F.; Marconcini, M.; Ceccato, P.; Giupponi, C. Flood Depth Estimation by Means of High-Resolution SAR Images and LiDAR_Data_ResearchGate. Available online: https://www.researchgate.net/publication/326067701_Flood_depth_estimation_by_means_of_high-resolution_SAR_images_and_LiDAR_data (accessed on 16 September 2020).
33. Iglesias, R.; Garcia-Boadas, E.; Vicente-Guijalba, F.; Centolanza, G.; Duro, J. Towards Unsupervised Flood Mapping Generation Using Automatic Thresholding and Classification Approaches. In Proceedings of the IGARSS 2018—2018 IEEE International Geoscience and Remote Sensing Symposium, Valencia, Spain, 22–27 July 2018; IEEE: Piscataway, NJ, USA, 2018; pp. 4927–4930.
34. Dasgupta, A.; Grimaldi, S.; Ramsankaran, R.; Pauwels, V.; Walker, J. Towards operational SAR-based flood mapping using neuro-fuzzy texture-based approaches. *Remote Sens. Environ.* **2018**, *215*, 313–329. [CrossRef]
35. Li, L.; Chen, Y.; Xu, T.; Shi, K.; Huang, C.; Liu, R.; Lu, B.; Meng, L. Enhanced Super-Resolution Mapping of Urban Floods Based on the Fusion of Support Vector Machine and General Regression Neural Network. *IEEE Geosci. Remote Sens. Lett.* **2019**, *16*, 1269–1273. [CrossRef]
36. Wang, Y.; Fang, Z.; Hong, H.; Peng, L. Flood susceptibility mapping using convolutional neural network frameworks. *J. Hydrol.* **2020**, *582*, 124482. [CrossRef]
37. Li, Y.; Martinis, S.; Wieland, M. Urban flood mapping with an active self-learning convolutional neural network based on TerraSAR-X intensity and interferometric coherence. *ISPRS J. Photogramm. Remote Sens.* **2019**, *152*, 178–191. [CrossRef]
38. Sameen, M.I.; Pradhan, B. Landslide Detection Using Residual Networks and the Fusion of Spectral and Topographic Information. *IEEE Access* **2019**, *7*, 114363–114373. [CrossRef]
39. Bui, Q.-T.; Nguyen, Q.-H.; Nguyen, X.L.; Pham, V.D.; Nguyen, H.D.; Pham, V.-M. Verification of novel integrations of swarm intelligence algorithms into deep learning neural network for flood susceptibility mapping. *J. Hydrol.* **2020**, *581*, 124379. [CrossRef]
40. Bui, D.T.; Hoang, N.-D.; Martínez-Álvarez, F.; Ngo, P.-T.T.; Hoa, P.V.; Pham, T.D.; Samui, P.; Costache, R. A novel deep learning neural network approach for predicting flash flood susceptibility: A case study at a high frequency tropical storm area. *Sci. Total Environ.* **2020**, *701*, 134413. [CrossRef]
41. Shen, X.; Anagnostou, E.N.; Allen, G.H.; Brakenridge, G.R.; Kettner, A. Near-real-time non-obstructed flood inundation mapping using synthetic aperture radar. *Remote Sens. Environ.* **2019**, *221*, 302–315. [CrossRef]
42. Ali, S.; Arief, R.; Dyatmika, H.S.; Maulana, R.; Rahayu, M.I.; Sondita, A.; Setiyoko, A.; Maryanto, A.; Budiono, M.E.; Sudiana, D. Digital Elevation Model (DEM) Generation with Repeat Pass Interferometry Method Using TerraSAR-X/Tandem-X (Study Case in Bandung Area). In *IOP Conference Series: Earth and Environmental Science*; IOP Publishing: Bristol, UK, 2019; Volume 280, p. 012019.
43. Landuyt, L.; Van Wesemael, A.; Schumann, G.J.-P.; Hostache, R.; Verhoest, N.E.C.; Van Coillie, F.M.B. Flood Mapping Based on Synthetic Aperture Radar: An Assessment of Established Approaches. *IEEE Trans. Geosci. Remote Sens.* **2019**, *57*, 722–739. [CrossRef]
44. Felegari, S.; Sharifi, A.; Moravej, K.; Amin, M.; Golchin, A.; Muzirafuti, A.; Tariq, A.; Zhao, N. Integration of Sentinel 1 and Sentinel 2 Satellite Images for Crop Mapping. *Appl. Sci.* **2021**, *11*, 10104. [CrossRef]
45. Schubert, A.; Miranda, N.; Geudtner, D.; Small, D. Sentinel-1A/B Combined Product Geolocation Accuracy. *Remote Sens.* **2017**, *9*, 607. [CrossRef]
46. Liang, J.; Liu, D. A local thresholding approach to flood water delineation using Sentinel-1 SAR imagery. *ISPRS J. Photogramm. Remote Sens.* **2020**, *159*, 53–62. [CrossRef]

47. Chini, M.; Hostache, R.; Giustarini, L.; Matgen, P. A Hierarchical Split-Based Approach for Parametric Thresholding of SAR Images: Flood Inundation as a Test Case. *IEEE Trans. Geosci. Remote Sens.* **2017**, *55*, 6975–6988. [CrossRef]
48. Pulvirenti, L.; Chini, M.; Pierdicca, N.; Boni, G. Flood Detection in Urban Areas: Analysis of Time Series of Coherence Data in Stable Scatterers. In Proceedings of the IGARSS 2019—2019 IEEE International Geoscience and Remote Sensing Symposium, Yokohama, Japan, 28 July–2 August 2019; Institute of Electrical and Electronics Engineers (IEEE): Piscataway, NJ, USA, 2019; pp. 9745–9747.
49. Pham, V.D.; Nguyen, Q.-H.; Nguyen, H.-D.; Pham, V.-M.; Vu, V.M.; Bui, Q.-T. Convolutional Neural Network—Optimized Moth Flame Algorithm for Shallow Landslide Susceptible Analysis. *IEEE Access* **2020**, *8*, 32727–32736. [CrossRef]
50. Munawar, H.S.; Ullah, F.; Qayyum, S.; Heravi, A. Application of Deep Learning on UAV-Based Aerial Images for Flood Detection. *Smart Cities* **2021**, *4*, 1220–1242. [CrossRef]
51. Wang, Y.; Hong, H.; Chen, W.; Li, S.; Panahi, M.; Khosravi, K.; Shirzadi, A.; Shahabi, H.; Panahi, S.; Costache, R. Flood susceptibility mapping in Dingnan County (China) using adaptive neuro-fuzzy inference system with biogeography based optimization and imperialistic competitive algorithm. *J. Environ. Manag.* **2019**, *247*, 712–729. [CrossRef] [PubMed]
52. Muhadi, N.A.; Abdullah, A.F.; Bejo, S.K.; Mahadi, M.R.; Mijic, A. Deep Learning Semantic Segmentation for Water Level Estimation Using Surveillance Camera. *Appl. Sci.* **2021**, *11*, 9691. [CrossRef]
53. Scherer, D.; Müller, A.; Behnke, S. Evaluation of Pooling Operations in Convolutional Architectures for Object Recognition. In Proceedings of the International Conference on Artificial Neural Networks, Thessaloniki, Greece, 15–18 September 2010.
54. James, G.; Witten, D.; Hastie, T.; Tibshirani, R. *An Introduction to Statistical Learning*, 7th ed.; Springer: New York, NY, USA, 2017.
55. Srivastava, N.; Hinton, G.; Krizhevsky, A.; Sutskever, I.; Salakhutdinov, R. Dropout: A Simple Way to Prevent Neural Networks from Overfitting. *J. Mach. Learn. Res.* **2014**, *15*, 1929–1958.

applied sciences

MDPI

Article

Modeling and Spatiotemporal Mapping of Water Quality through Remote Sensing Techniques: A Case Study of the Hassan Addakhil Dam

Anas El Ouali [1,*], Mohammed El Hafyani [2], Allal Roubil [2], Abderrahim Lahrach [1], Ali Essahlaoui [2], Fatima Ezzahra Hamid [3], Anselme Muzirafuti [4], Dimitrios S. Paraforos [5], Stefania Lanza [6] and Giovanni Randazzo [7]

1 Functional Ecology and Environmental Engineering Laboratory, Faculty of Science and Technology, Sidi Mohamed Ben Abdellah University, Fez BP2202, Morocco; abderrahim.lahrach@usmba.ac.ma
2 Department of Geology, Laboratory of Geoengineering and Environment, Research Group "Water Sciences and Environment Engineering", Faculty of Sciences, Moulay Ismail University Zitoune, Meknes BP11201, Morocco; m.elhafyani@edu.umi.ac.ma (M.E.H.); allal.roubil@edu.umi.ac.ma (A.R.); a.essahlaoui@fs-umi.ac.ma (A.E.)
3 Guir-Ziz-Rheris Hydraulic Basin Agency, Errachidia BP52000, Morocco; f.ezzahrahamid@gmail.com
4 Interreg Italia-Malta-Project, Pocket Beach Management & Remote Surveillance System (BESS), University of Messina, 31-98166 Messina, Italy; a.muzirafuti@edu.umi.ac.ma
5 Institute of Agricultural Engineering, University of Hohenheim, 70599 Stuttgart, Germany; d.paraforos@uni-hohenheim.de
6 GeoloGIS s.r.l. Spin Off of Università degli Studi di Messina, 31-98166 Messina, Italy; stefania.lanza@unime.it
7 Dipartimento di Scienze Matematiche e Informatiche, Scienze Fisiche e Scienze della Terra, Università degli Studi di Messina, 31-98166 Messina, Italy; giovanni.randazzo@unime.it
* Correspondence: anas.elouali@usmba.ac.ma; Tel.: +212-061-495-3445

Citation: El Ouali, A.; El Hafyani, M.; Roubil, A.; Lahrach, A.; Essahlaoui, A.; Hamid, F.E.; Muzirafuti, A.; Paraforos, D.S.; Lanza, S.; Randazzo, G. Modeling and Spatiotemporal Mapping of Water Quality through Remote Sensing Techniques: A Case Study of the Hassan Addakhil Dam. *Appl. Sci.* **2021**, *11*, 9297. https://doi.org/10.3390/app11199297

Academic Editors: José Miguel Molina Martínez and António José Madeira Nogueira

Received: 11 July 2021
Accepted: 30 September 2021
Published: 7 October 2021

Abstract: With its high water potential, the Ziz basin is one of the most important basins in Morocco. This paper aims to develop a methodology for spatiotemporal monitoring of the water quality of the Hassan Addakhil dam using remote sensing techniques combined with a modeling approach. Firstly, several models were established for the different water quality parameters (nitrate, dissolved oxygen and chlorophyll a) by combining field and satellite data. In a second step, the calibration and validation of the selected models were performed based on the following statistical parameters: compliance index R^2, the root mean square error and *p*-value. Finally, the satellite data were used to carry out spatiotemporal monitoring of the water quality. The field results show excellent quality for most of the samples. In terms of the modeling approach, the selected models for the three parameters (nitrate, dissolved oxygen and chlorophyll a) have shown a good correlation between the measured and estimated values with compliance index values of 0.62, 0.56 and 0.58 and root mean square error values of 0.16 mg/L, 0.65 mg/L and 0.07 μg/L for nitrate, dissolved oxygen and chlorophyll a, respectively. After the calibration, the validation and the selection of the models, the spatiotemporal variation of water quality was determined thanks to the multitemporal satellite data. The results show that this approach is an effective and valid methodology for the modeling and spatiotemporal mapping of water quality in the reservoir of the Hassan Addakhil dam. It can also provide valuable support for decision-makers in water quality monitoring as it can be applied to other regions with similar conditions.

Keywords: Ziz basin; water quality; satellite image analysis; modeling approach; nitrate; dissolved oxygen; chlorophyll a; climate change; time series analysis; environmental monitoring

1. Introduction

Over the last two decades, Morocco, as a Mediterranean country affected by climate change, has pursued an economic and social policy characterized by numerous development programs such as the policy of dam construction [1]. These hydraulic infrastructures provide a variety of services to both humans and the environment by organizing agricultural practice [2,3], as well as ensuring an efficient mobilization of water resources and improving the living conditions and environment of citizens [3]. These dams also provide habitat for fauna and flora and play a very important role in the global carbon cycle and climate change [4,5]. However, they are facing the interannual variability of precipitation and the succession of droughts and floods [6–9]. In order to monitor the water quality and observe the biophysical and biochemical conditions of the Hassan Addakhil dam and to prevent serious damage from occurring to the ecological system, the Guir-Ziz-Rheris Hydraulic Basin Agency (HBAGZR), in charge of water resources management in the Errachidia region, conducts in situ measurement surveys. The implemented system of in situ measurement (Figure 1) and monitoring is not practical due to its limitations in time and space [10]. It is expensive and has deficiencies that prevent accurate and complete results. Therefore, it is essential to have a complete, accurate, fast and inexpensive monitoring system to follow the water quality of the dam in order to avoid any degradation by applying prompt treatments.

(**a**) (**b**)

Figure 1. In situ measurement and monitoring of the Hassan Addakhil dam with (**a**) fieldwork data collection equipment and (**b**) laboratory data analysis.

Recently, geospatial tools have been widely used for the spatiotemporal monitoring of environmental phenomena [6,11,12], especially the monitoring of lake water quality parameters [3,13–30]. Such application is mainly enabled by the high spatial resolution data [21,24,26] as well as the temporal resolution. However, this aspect has always encountered problems due to the lack of appropriate sensors [31,32]. Moreover, moderate resolution sensors that are characterized by frequent revisit time and high radiometric resolution have been used [31], but the spatial resolution of these sensors does not allow for small lakes [31]. Several works have been carried out using Landsat TM and ETM+ data, but these satellites are limited in terms of revisit time [21,31] for very frequent monitoring. However, with the availability of new satellites with higher spatial, spectral and temporal resolution, such as Landsat OLI and Sentinel-2, retrieval and mapping of water quality from the satellite orbit has become more accessible. In 2008, Kallio et al. [31] conducted a study with the main purpose of monitoring turbidity and colored dissolved organic matter (CDOM) through ETM+ images in lakes in two river basins in southern Finland. The results showed that despite limitations in spectral and radiometric resolution, these images can be an effective and useful tool for water quality monitoring of small lakes (<1 km^2). Toming and his collaborators [18] conducted a study in Estonia in which they evaluated Sentinel-2 Multispectral Imager (MSI) data in the mapping of different lake water quality parameters such as chlorophyll a (Chl-a), water color, CDOM and dissolved organic carbon (DOC). Therefore, field data of different parameters were compared to the Sentinel-2 derived band ratio algorithms. The obtained results showed a strong correlation between the Sentinel-2 MSI ratio bands and the different lake water quality parameters such as Chl-a (R^2 = 0.83). In the Czech Republic, a study was carried out by Saberioon and his collaborators [33]. It aimed at developing a semiempirical model for predicting water quality parameters such as Chl-a and total suspended solids (TSS) by combining Sentinel-2A data and machine learning methods. The results showed an adequate prediction accuracy for both Chl-a (R^2 = 0.85, $RMSE_p$ = 48.57) and TSS (R^2= 0.80, $RMSE_p$ = 19.55).

Jerry C. Ritchie et al. [29] conducted a study aimed at providing the capability of remote sensing technology in mapping water quality parameters (suspended sediments (turbidity), chlorophyll and temperature). As a result, in situ measurements have been used to assess water quality, and empirical relationships between spectral properties and water quality parameters have been established. Another study was carried out by Carly Hyatt Hansen et al. [30] in the USA at three lakes in the Great Salt Lake surface water system (namely the Great Salt Lake, Farmington Bay and Utah Lake), the objective of which was to improve techniques for the development of algal mapping models through the use of field sampling methods. This study has shown that Landsat, SENTINEL-2 and MODIS sensors are suitable for monitoring water quality in the lake system. In some cases, temporal variability may be an obstacle to detecting short-term events, but it may be sufficient in other areas where short-term variability is lower.

In Morocco and in another context, El Hafyani et al. [34] conducted a study in the Tafilalet plain aiming at modeling and mapping soil salinity through Landsat Oli images. The results showed a strong fitting of this technique with R^2 of models ranging from 0.53 to 0.75 and root mean square error of 0.62 to 0.82 dS/m. Karaoui et al. [3] carried out a study aiming at estimating and mapping the water quality parameters in the Bin El Ouidane reservoir through better understanding the relationship between the latter and digital data. The correlation results showed that all the studied parameters have an R^2 greater than 0.52 and that they can be transformed into predictive models by stepwise regression. This work carried out at the Bin El Ouidane reservoir is of considerable importance for the water resource managers of the Oum Er-Rbia Hydraulic Agency. Thus, the present study was carried out at the level of the Hassan Addakhil dam, in collaboration with the Guir-Ziz-Rheris Hydraulic Basin Agency (HBAGZR). It aims at the validation of this method and the strengthening of its results by comparing them with other studies in the same context.

The objective of this study is to conduct modeling and spatiotemporal mapping of water quality of the Hassan Addakhil reservoir by combining the high spatial resolution data (Sentinel-2) and field measurements. In fact, 20 samples were collected on 14 March 2021, at the same Sentinel-2 satellite transit time. Measurements of nitrate, dissolved oxygen and Chl-a were carried out. Next, a statistical study was performed to select the bands correlated with the quality measurements, and a stepwise regression analysis was elaborated to model each parameter. Finally, a spatiotemporal mapping was made for water quality.

2. Materials and Methods

2.1. Study Area

The Hassan Addakhil reservoir is located in the southeast of Morocco at a longitude of 4°28′50.98″ W, latitude of 31′01′00.44″ N and altitude of 1125 m (Figure 2). It accurately lays at Foum Rhiour on the Ziz River, to the north of Errachidia city. It was built in 1970, five years after the devastating flood of October 1965, which ravaged the Ziz valley, leaving 25,000 people homeless. Its retention capacity is 312.8 million m^3. The objective of its construction was to ensure protection against floods and to achieve agricultural development of the Ziz valley and the Tafilalet plain by regulating its floods. This dam receives the water of the Ziz River and its tributaries, which drain the Upper Ziz watershed.

The latter is rich in fertile valleys but with low development of perennial courses. The geology is of Jurassic type [35] with limestone and dolomitic limestone formations that constitute good water reservoirs [36–39]. The climate is semiarid with short and brutal precipitation. The rainwater that escapes infiltration and evapotranspiration flows into the Hassan Addakhil dam (Figure 3). Downstream of the dam, aridity increases and evaporation phenomena increase. The dry period often lasts up to eight months, with maximum temperatures obtained during the months of June, July and August. The winter is relatively wet and very cold with minimum temperatures in January [40].

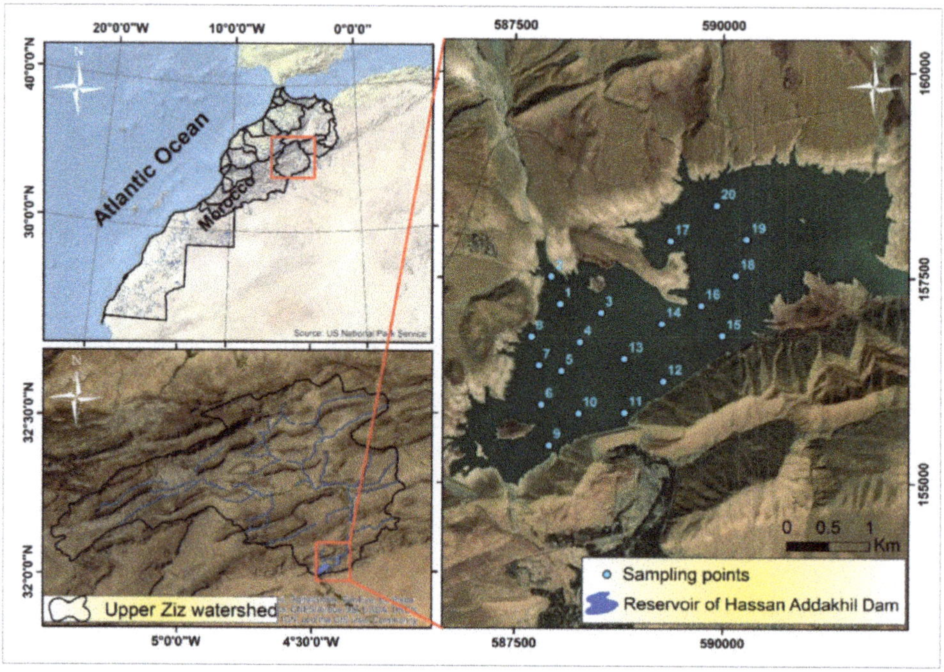

Figure 2. Location of study area and sampling points.

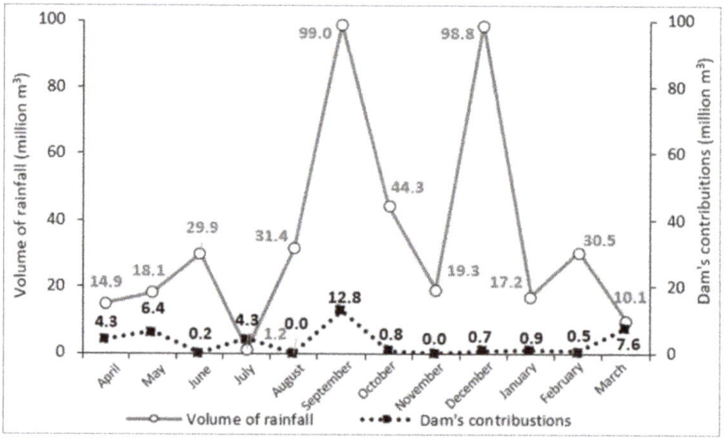

Figure 3. Correlation between volume of rainfall and dam's contributions.

2.2. Data

2.2.1. Ground Data

Ground truth samples were taken at 20 points distributed over the reservoir of Hassan Addakhil dam (Figure 2), where nitrate, dissolved oxygen and Chl-a were measured by chemical process at the laboratory of the Guir-Ziz-Rheris Hydraulic Basin Agency in Errachidia and Gaya Laboratory in Rabat, Morocco, according to the Moroccan law adopted for aquatic waters [41] (Table 1). The Chl-a indicates the stage of eutrophication in the reservoir, while the nitrates' concentration is directly related to the agricultural practices upstream of the reservoir, as well as to wastewater discharge.

Table 1. On-ground data sample characterizations.

Sampling Data ID	Dissolved Oxygen (mg/L)	Nitrates (mg/L)	Chl-a (µg/L)	Sampling Data ID	Dissolved Oxygen (mg/L)	Nitrates (mg/L)	Chl-a (µg/L)
01	7.4	1.82	0.55	11	7.5	1.57	0.47
02	9	1.57	0.77	12	9	1.56	0.55
03	7.6	1.56	0.51	13	9.1	1.6	0.69
04	8.2	1.65	0.61	14	9	1.82	0.57
05	7.9	1.31	0.48	15	8.6	1.54	0.64
06	5.8	1.45	0.77	16	9.7	1.78	0.52
07	8.1	1.5	0.54	17	7.4	1.74	0.72
08	7.4	1.42	0.60	18	7.5	1.96	0.68
09	7.4	1.6	0.49	19	7.2	1.71	0.77
10	7.2	0.8	0.53	20	8.9	1.85	0.55

Dissolved oxygen was measured in situ using a portable dissolved oxygen meter (BANTE Instruments 821). For the determination of Chl-a, a volume of samples between 0.1 and 2 L was first filtered under vacuum through a glass fiber filter without organic binder with a diameter greater than 1 µm, depending on the algal content, after shaking. Then we proceed to the extraction step by pouring a small volume of acetone (20 mL to 30 mL) into the tube containing the filtered pieces. This step was followed by shaking the extract contained in the extraction tubes for at least 3 min. Finally, we proceeded to the reading of a part of the clear extract by UV-Vis spectrophotometry (Lovibond), which provides double-beam operation with a scattered light rate of 0.01%, wavelength accuracy of +/−0.1 nm and stability of 0.00015. The measurements were made at two wavelengths, A1 = 665 nm and A2 = 750 nm, by comparison with a reference cell filled with acetone.

The determination of nitrates was done by UV-Vis spectrophotometric calibration (Lovibond). In fact, after the preparation of the solution noted, it was smothered by dissolving 129 mg of ammonium nitrate (of raw formula NH_4NO_3) in 1.0 L of distilled water. A solution of mass concentration (or content) equal to 100 mg L^{-1} was then obtained. Then, we subtracted the absorbance of the blank from the absorbance of each standard solution and plotted the calibration curve showing absorbance versus mass of nitrate, in milligrams per liter. Finally, the nitrate concentration C was determined from the UV–visible calibration curve, established following the Beer–Lambert law.

The analyses of the samples were measured in three replicates, and the average was calculated. The Table 1 shows the average of the three measured values.

2.2.2. Satellite Data

Twelve images obtained from the Sentinel-2 sensor of the European Space Agency (https://sentinel.esa.int/web/sentinel/sentinel-data-access (accessed on 10 August 2021)) were used in this study. These images are characterized by a high spatial resolution of 10 to 60 m from the visible to mid-infrared range and a revisit time of 10 days (Table 2). The March image was used for calibration with field data and model validation, while the other images were used for spatiotemporal monitoring of different parameters. These images have been uploaded for the period April 2020–March 2021 (Table 3). The QGIS software was used to process the Sentinel-2 satellite images through the interface (Semi-Automatic Classification Plugin) developed by Luca Congedo [42]. There were several preprocessing steps, including the conversion of digital number (DN) to top of atmosphere reflectance (TOA) and the subsequent atmospheric correction by the dark object subtraction (DOS) algorithm [43].

Table 2. Sentinel-2 satellite image characteristics.

Sentinel-2 Bands	Wavelength (nm)	Spatial Resolution (m)
Coastal Aerosol	442.7	60
Blue	492.4	10
Green	559.8	10
Red	664.6	10
Vegetation red edge	704.1	20
Vegetation red edge	740.5	20
Vegetation red edge	782.8	20
NIR	832.8	10
Narrow NIR	864.7	20
Water vapor	945.1	60
SWIR-Cirrus	1373.5	60
SWIR	1613.7	20
SWIR	2202.4	20

Table 3. Sentinel-2 satellite image acquisition dates.

Image	Acquisition Dates
1	28 April 2020
2	3 May 2020
3	6 June 2020
4	7 July 2020
5	6 August 2020
6	20 September 2020
7	28 October 2020
8	19 November 2020
9	19 December 2020
10	13 January 2021
11	17 February 2021
12	14 March 2021

2.3. Methodology

Figure 4 shows the different phases of this work. A field mission was carried out in the Hassan Addakhil dam on the same day of the satellite visit in order to calibrate the extracted models for the different water quality parameters, for which the image of 14 March 2021 has been used. This mission was done in collaboration with the staff of the Guir-Ziz-Rheris Hydraulic Basin Agency, and the analyses were realized in its laboratory. Later on, a statistical study was carried out to extract the different correlated bands with the different parameters, and a multiple stepwise analysis modeling approach was used in order to set the models. Several models were extracted for the different parameters, and the selection of a suitable one was made on the basis of the compliance index R^2, the root mean square error (RMSE) and p-value. Finally, after the models' validation and the extraction of their equations, spatiotemporal monitoring of the reservoir water quality was performed through multitemporal images.

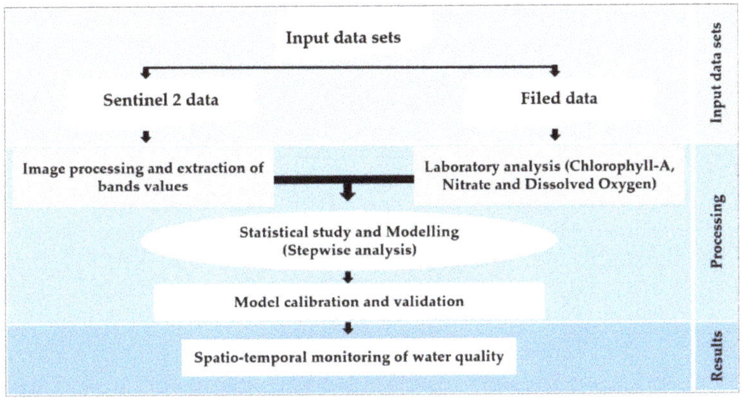

Figure 4. Flowchart of the methodology used.

3. Results

3.1. Model Assessment and Validation

Modeling is a representation of reality in order to demonstrate some of its properties. Therefore, there are several types of models, such as stochastic models, optimization models, dynamic simulation models and empirical statistical models, that allow predicting the outcome of a categorical variable using a set of quantitative and/or qualitative predictors.

In this case, a statistical model based on a stepwise multiple regression analysis was developed for the models' creation in order to estimate different water quality parameters using the RStudio open source software, using the following equation:

$$Y = b_0 + b_1 \times X_1 + b_2 \times X_2 + \ldots + b_k \times X_k$$

Y is the predicted variable with regression coefficients b_1 to $_k$ and Y-intercept b_0 when the values for the predictor variables are X_1 to $_k$.

Firstly, a correlation study was conducted between the different water quality parameters and the satellite image bands in order to select the appropriate bands for the elaboration of the models (Table 4). For Chl-a, bands B5, B6 and B7 showed a strong correlation with this band with correlation coefficients of 0.81, 0.71 and 0.73, respectively. The bands B1, B3 and B4 showed a strong correlation with nitrates with correlation coefficients of 0.73, 0.69, and 0.73, respectively. Dissolved oxygen measurements showed a positive correlation with bands B2 and B3 with correlation coefficients of 0.71 and 0.75, respectively. The bands that were chosen in the first step were later integrated into the equations of the different models (Table 5). The choice of the suitable model was based on the three statistical parameters, namely the compliance index R^2, the root mean square error (RMSE) and *p*-value. The priority of choice was given to the models that have the strongest conformity index and the lowest root mean square error, while the threshold of *p*-value was fixed at a value of 0.05. Table 5 represents the different developed models along with their equations and the different statistical parameters for each. For dissolved oxygen, the chosen model is the one that combines band 2 and band 3, with a compliance index R^2 of 0.56, a root mean square error of about 0.65 mg/L and a *p*-value of about 0.0009. For nitrates, the chosen model is the one that combines band 1, band 3 and band 4, with a compliance index R^2 of 0.62, a root mean square error of about 0.16 mg/L and a *p*-value of about 0.0011. Lastly, for Chl-a, the chosen model is the one that groups band 5, band 6 and band 8, with a compliance index R^2 of 0.58, a root mean square error of the order of 0.07 µg/L and a *p*-value of the order of 0.0024.

In order to verify the accuracy of the proposed models, the measured values in the field and the observed values of the different parameters were presented with their equations (Figure 5).

Table 4. Correlation between satellite bands and water quality parameters.

	B1	B2	B3	B4	B5	B6	B8
Chl-a	-	-	-	-	0.81	0.71	0.73
Nitrates	0.73	-	0.69	0.73	-	-	-
Dissolved Oxygen	-	0.71	0.75	-	-	-	-

Table 5. Statistical parameters of the best performance models.

	Model Equations	RMSE	R^2	*p*-Value
Nitrate	$0.00372 \times \mathbf{B01} - 3.05$	0.19 (mg/L)	0.39	0.0030
	$0.00895 \times \mathbf{B03} + 0.31$	0.24 (mg/L)	0.06	0.2945
	$0.00114 \times \mathbf{B04} + 0.630$	0.19 (mg/L)	0.39	0.0031
	$0.00372 \times \mathbf{B01} - (3.056 \times 10^{-7}) \times \mathbf{B03} - 3.0593$	0.20 (mg/L)	0.39	0.0140
	$0.00273 \times \mathbf{B01} + 0.00084 \times \mathbf{B04} - 2.53264$	0.17 (mg/L)	0.58	0.0067
	$0.000412 \times \mathbf{B03} + 0.00125 \times \mathbf{B04} + 1.1273$	0.20 (mg/L)	0.40	0.0127
	$\mathbf{0.003099 \times B01 - 0.000944 \times B03 + 0.0010509 \times B04 - 1.81}$	**0.16(mg/L)**	**0.62**	**0.0011**
Dissolved Oxygen	$0.0128 \times \mathbf{B02} - 8.324$	0.75 (mg/L)	0.30	0.0040
	$0.0102 \times \mathbf{B03} - 6.704$	0.63 (mg/L)	0.55	0.0001
	$\mathbf{0.00075 \times B02 + 0.00989 \times B03 - 7.09}$	**0.65(mg/L)**	**0.56**	**0.0009**
Chl-a	$0.0007071 \times \mathbf{B05} + 0.184$	0.07 (μg/L)	0.41	0.0022
	$0.0010213 \times \mathbf{B06} + 0.3066$	0.08 (μg/L)	0.37	0.0042
	$0.002332 \times \mathbf{B08} - 0.15655$	0.07 (μg/L)	0.45	0.0011
	$0.0004766 \times \mathbf{B05} + 0.000591 \times \mathbf{B06} + 0.144903$	0.07 (μg/L)	0.49	0.0029
	$0.000423 \times \mathbf{B05} + 0.001572 \times \mathbf{B08} - 0.1589$	0.07 (μg/L)	0.55	0.0010
	$0.000574 \times \mathbf{B06} + 0.00167 \times \mathbf{B08} - 0.1076$	0.07 (μg/L)	0.53	0.0015
	$\mathbf{0.0003214 \times B05 + 0.000378 \times B06 + 0.0013207 \times B08 - 0.126}$	**0.07(μg/L)**	**0.58**	**0.0024**

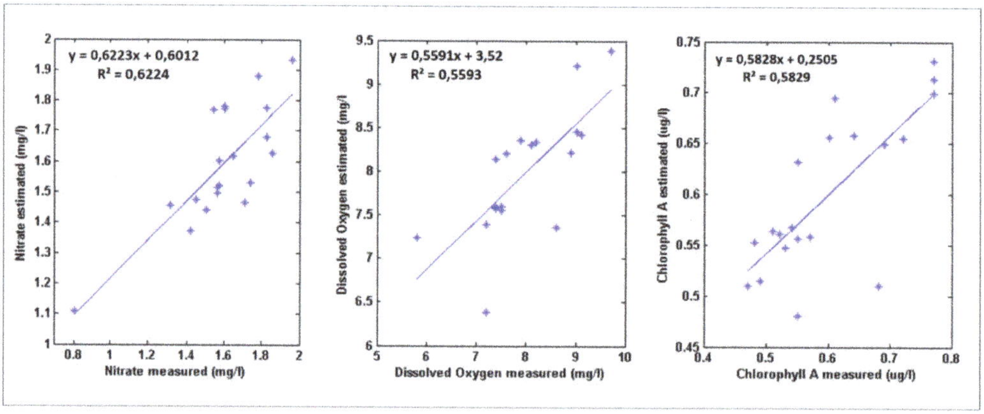

Figure 5. Water quality parameters measured versus estimated through the models (best models' performance).

3.2. Spatial Variation of Water Quality

The dissolved oxygen levels measured during the field campaign vary between 5.8 and 9.7 mg/L, while the values estimated by the model show a minimum value of 6.39 mg/L and a maximum one of 9.39 mg/L (Figure 6). The spatial variation of this parameter shows a well-oxygenated zone in the northeast of the reservoir. This area represents the water inlet to the reservoir. For the spatiotemporal variation, the maps of different months show high values in the northeastern part of the reservoir with a decrease moving away from this area (Figure 6). Except for few months such as January, October and December, this variation can be explained by the coincidence of these periods with that of water supply for agriculture downstream of the dam which allows the movement of water in the reservoir and consequently an agitation of the water leading to an increase in the values of this parameter.

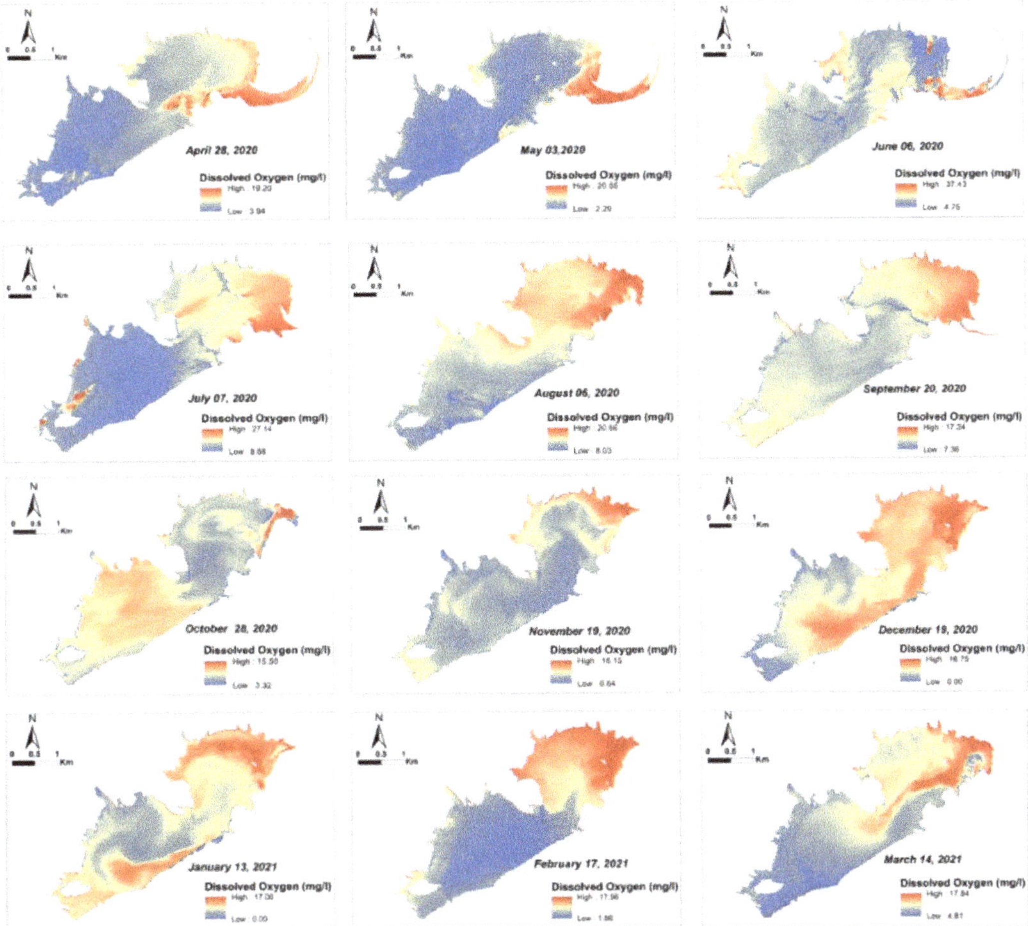

Figure 6. Obtained maps of dissolved oxygen using best model.

Generally, the samples showed an excellent quality from the point of view of the nitrate parameter with values ranging between 0.8 and 1.96 mg/L. The estimated values for the model are between 1.11 and 1.96 mg/L (Figure 7). The spatial variation of nitrate shows a decrease in values from the northeastern part of the reservoir, representing the outlet, to the southeastern part (Figure 7).

The temporal variation shows that nitrate values do not exceed 10 mg/L throughout the year. This proves the excellent water quality of this reservoir. This variation in the reservoir can be explained by the leaching from agricultural soils and also by domestic discharges of the agglomerations upstream.

Generally, the samples show an excellent quality compared to the quality standards for surface water in Morocco with Chl-a concentrations varying between 0.47 and 0.77 µg/L. The estimated values for the model range between 0.48 and 0.73 µg/L (Figure 8).

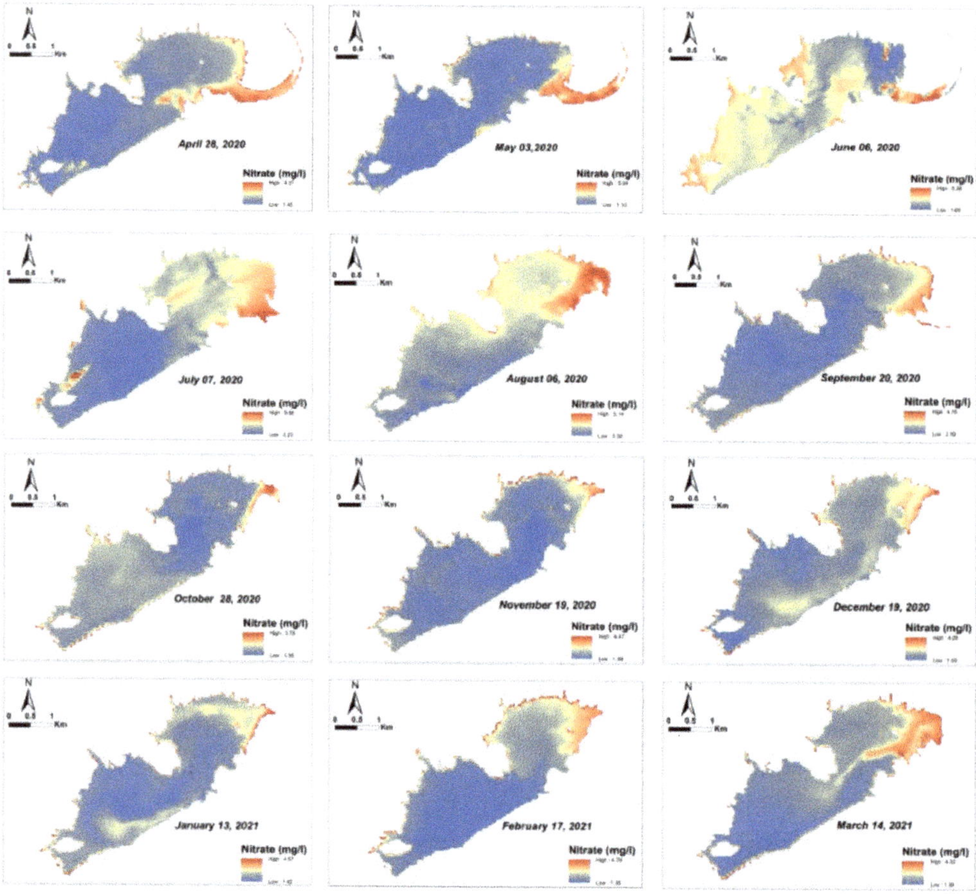

Figure 7. Obtained maps of nitrate using best model.

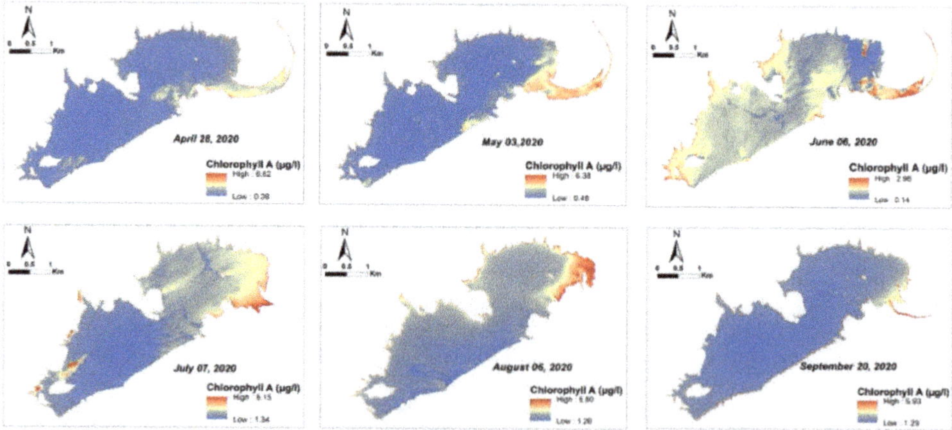

Figure 8. *Cont.*

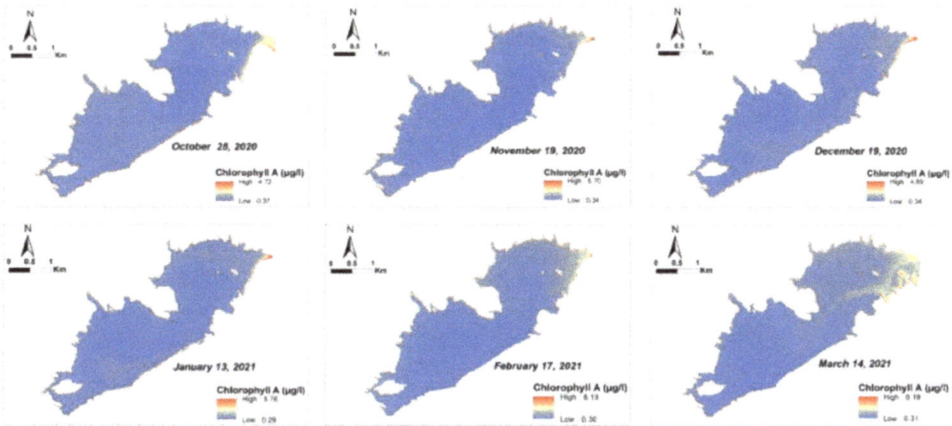

Figure 8. Obtained maps of Chl-a using best model.

4. Discussion

The combination of field data, high spatial resolution images and modeling shows a strong efficiency in the spatiotemporal monitoring of water quality at the reservoir scale. On the other hand, the measured values of the different proposed models showed in most cases a strong correlation with those measured in the field. Three water quality parameters were selected in this study: dissolved oxygen, nitrates and Chl-a. The choice of these parameters is based on their importance in the eutrophication of fresh or coastal water [44,45]. The most visible phenomena are the appearance in spring and summer of the green tides in coastal marine water and the water in lakes and rivers. These manifestations correspond to an ecological imbalance linked to excessive inputs of phosphorus (including in the form of phosphate $PO_4{}^{3}$-) and nitrogen (nitrate NO^{3}-). In fact, these inputs lead to an explosion in the development of aquatic plants, which leads to an excessive local accumulation of biomass and is the cause of various undesirable effects such as impoverishment of biodiversity, visual and olfactory nuisance, inconvenience for bathing, difficulties in water treatment (drinking water), gas emissions and colonization by algae producing toxins such as certain Cyanophyceae.

Several estimation models have been developed based on the multiple stepwise regression analysis, while the choice of the suitable model was based on the largest compliance index (R^2) and the smallest root mean square error (RMSE). The estimation of nitrate was done by applying the model that groups bands B1, B3 and B4 with the largest R^2 and the smallest RMSE among all the extracted models (R^2 = 0.62, RMSE = 0.16 mg/L). For the estimation of dissolved oxygen, the model chosen is the one that combines the two bands B2 and B3 (with R^2 = 0.56 and RMSE = 0.65 mg/L). For Chl-a, the model chosen is the one that includes the bands B5, B6 and B8 (with R2 = 0.58 and RMSE= 0.07 µg/L).

The choice of the bands integrated into the model was made on the basis of a statistical study that was carried out between in situ measurements and satellite data. For chlorophyll a, a correlation was obtained between the in situ measurements of this parameter and band 5 of the Sentinel-2 sensor, which is located in the 704.1 nm spectral range. This result is in agreement with the results obtained by Toming [18], who used the peak reflectance between 700 and 720 nm for the estimation of this parameter. For nitrate, this study showed that the estimation of this parameter is very efficient when using the spectral interval from 442.7 to 664.6 nm (Table 4). For the estimation of the dissolved oxygen, the results show that the spectral interval of 492.4 to 559.8 nm is more appropriate (Table 4). Another work that was carried out by Vanhellemont and Ruddick [23] has shown that one of the main advantages of Sentinel-2 over Landsat-8 is the presence of the band B5 (704.1 nm) with a spatial resolution of 20 m to determine chlorophyll absorption. Eventually, these images will be useful for many aquatic water quality monitoring applications, and they can also be combined into a virtual constellation to improve temporal coverage.

In Morocco, satellite images have been widely used for modeling and spatiotemporal monitoring of several environmental phenomena, but the application of satellite images in spatiotemporal monitoring of lake water quality is still not well developed. The only study is the one carried out by Karaoui and his collaborators in 2019 [3], which aimed to map water quality parameters using Sentinel images. Indeed, the spectral interval from 559.8 to 740.5 nm for the mapping of Chl-a, the spectral interval from 832.8 to 1373.5 nm for dissolved oxygen and the spectral interval from 442.7 to 864.7 nm for nitrate had R^2 indices of 0.78, 0.74 and 0.67, respectively. Therefore, our work involves a complementarity and a recognition of what has been done by these authors, obviously applying it with a different approach and in a different context, adding the aspect of spatiotemporal monitoring which allows for continuous survey throughout the year.

In addition, the results can contribute indirectly to the quantification of the impact of both the agriculture and the discharges in the upstream part of the reservoir. This system represents a very effective and economical solution for monitoring water quality and could be applied by hydraulic basin agencies under several restrictions to travel or other activities or in periods that require remote work (e.g., at the time of COVID-19). Ultimately, this approach is more efficient and not only can be used under similar conditions but also provides vital information on water quality parameters in a faster, more accurate and less computationally expensive way. As a perspective, a seasonal analysis is required to evaluate, calibrate and validate the models obtained in a temporal way.

5. Conclusions

In this study, a new method of combining high-resolution and field data was applied for the spatiotemporal mapping of certain surface water quality parameters, namely nitrate, dissolved oxygen and Chl-a in the Hassan Addakhil dam in southeastern Morocco. The field results show an excellent quality for most of the samples. In terms of the modeling approach, the models selected for the three parameters have shown a good correlation between the measured and estimated values with compliance index values of 0.62, 0.56 and 0.58 and root mean square error values of 0.16 mg/L, 0.65 mg/L and 0.07 µg/L for nitrate, dissolved oxygen and Chl-a, respectively. After the calibration, the validation and the selection of the models, the spatiotemporal variation of water quality was determined thanks to the multitemporal satellite data.

In summary, this research represents an efficient and useful solution for the hydraulic basin agency in charge of water resources management in the region. Indeed, it will help to minimize the costs of quality surveys carried out throughout the year. It can also contribute to decision-making regarding agricultural profitability and its relation with water quality, as well as to the development of strategies for efficient water resources management.

Author Contributions: Conceptualization, A.E.O., A.L., M.E.H. and A.R.; data curation, A.E.O. and F.E.H.; formal analysis, A.E.O., M.E.H. and A.R.; methodology, A.E.O., M.E.H. and A.R.; project administration, A.L. and A.E.; resources, A.E.O.; software, A.E.O., M.E.H. and A.R.; supervision, A.L. and A.E.; validation, A.E.O., M.E.H. and A.R.; visualization, A.E.O., M.E.H., S.L., G.R., D.S.P., A.M. and A.R.; writing—original draft, A.E.O. and A.R.; writing—review and editing, A.E.O., M.E.H., A.R., A.E., A.L., S.L., G.R., D.S.P. and A.M. All authors have read and agreed to the published version of the manuscript.

Funding: This research received no external funding.

Institutional Review Board Statement: Not Applicable.

Informed Consent Statement: Not Applicable.

Acknowledgments: The authors gratefully acknowledge the support of the staff of the Guir-Ziz-Rheris Hydraulic Basin Agency and Gaya Laboratory in Rabat for equipment and field mission assistance.

Conflicts of Interest: The authors declare that they have no conflict of interest.

References

1. Haut-Commissariat au Plan. *Prospective Maroc 2030—Eveil aux problématiques du Maroc 2030—Rapport;* Haut-Commissariat au Plan: Casablanca, Morocco, 2007.
2. Brönmark, C.; Hansson, L.A. Environmental issues in lakes and ponds: Current state and perspectives. *Environ. Conserv.* **2002**, *29*, 290–307. [CrossRef]
3. Karaoui, I.; Abdelghani, B.; Arioua, A.; Hssaisoune, M.; Sabri, E.M.; Ait Ouhamchich, K.; Elhamdouni, D.; El Amrani, I.; Nouaim, W. Evaluating the potential of Sentinel-2 satellite images for water quality characterization of artificial reservoirs: The Bin El Ouidane Reservoir case study (Morocco). *Meteorol. Hydrol. Water Manag.* **2019**, *7*, 31–39. [CrossRef]
4. Tranvik, L.J.; Downing, J.A.; Cotner, J.B.; Loiselle, S.A.; Striegl, R.G.; Ballatore, T.J.; Dillon, P.; Knoll, L.B.; Kutser, T.; Larsen, S.; et al. Lakes and reservoirs as regulators of carbon cycling and climate. *Limn. Oceanogr.* **2009**, *56*, 2298–2314. [CrossRef]
5. Moss, T. Spatial fit, from panacea to practice: Implementing the EU Water Framework Directive. *Ecol. Soc.* **2012**, *17*, 1–12. [CrossRef]
6. El Hafyani, M.; Essahlaoui, A.; Van Rompaey, A.; Mohajane, M.; El Hmaidi, A.; El Ouali, A.; Moudden, F.; Serrhini, N.-E. Assessing Regional Scale Water Balances through Remote Sensing Techniques: A Case Study of Boufakrane River Watershed, Meknes Region, Morocco. *Water* **2020**, *12*, 320. [CrossRef]
7. Driouech, F.; Déqué, M.; Sánchez-Gómez, E. Weather Regimes-Moroccan Precipitation Link in a Regional Climate Change Simulation. *Glob. Planet. Chang.* **2010**, *72*, 1–10. [CrossRef]
8. El Moçayd, N.; Kang, S.; Eltahir, E.A.B. Climate Change impacts on the Water Highway project in Morocco. *Hydrol. Earth Syst. Sci. Discuss.* **2019**, *24*, 1467–1483. [CrossRef]
9. Ouatiki, H.; Boudhar, A.; Tramblay, Y.; Jarlan, L.; Benabdelouhab, T.; Hanich, L.; El Meslouhi, M.R.; Chehbouni, A. Evaluation of TRMM 3B42 V7 Rainfall Product over the Oum Er Rbia Watershed in Morocco. *Climate* **2017**, *5*, 1. [CrossRef]
10. Philipson, P.; Kratzer, S.; Ben Mustapha, S.; Strombeck, N.; Stelzer, K. Satellite-based water quality monitoring in Lake Vanern, Sweden. *Int. J. Remote Sens.* **2016**, *37*, 3938–3960. [CrossRef]
11. Krtolica, I.; Cvijanović, D.; Obradović, Đ.; Novković, M.; Milošević, D.; Savić, D.; Vojinović-Miloradov, M.; Radulović, S. Water quality and macrophytes in the Danube River: Artificial neural network modelling. *Ecol. Indic.* **2021**, *121*, 107076. [CrossRef]
12. Liarico, A.J.M.; Dávila, V.O.R.; Tapia, Á.M.F.; Valdivia, J.Q.; Medina, S.A.Z. Bioenergetic and water quality modeling for eutrophication assessment of El Pañe Reservoir, Peru. *Ecohyd. Hydrobiol.* **2021**, *21*, 114–128. [CrossRef]
13. Basnyat, P.; Teeter, L.D.; Lockaby, B.G.; Flynn, K.M. The use of remote sensing and GIS in watershed level analyses of non-point source pollution problems. *For. Ecol. Manag.* **2000**, *128*, 65–73. [CrossRef]
14. Oseke, F.I.; Anornu, G.K.; Adjei, K.A.; Eduvie, M.O. Assessment of water quality using GIS techniques and water quality index in reservoirs affected by water diversion. *Water-Energy Nexus* **2021**, *4*, 25–34. [CrossRef]
15. Pierson, D.C.; Strömbeck, N. A modelling approach to evaluate preliminary remote sensing algorithms: Use of water quality data from Swedish Great Lakes. *Geophysica* **2000**, *36*, 177–202.
16. Wang, X.; Ma, T. Application of remote sensing techniques in monitoring and assessing the water quality of Taihu Lake. *Bull. Environ. Contam. Toxicol.* **2001**, *67*, 863–870. [CrossRef]
17. Yang, M.; Merry, C.J.; Sykes, R.M. Integration of water quality modeling, remote sensing, and GIS. *J. Am. Water Resour. Assoc.* **1999**, *35*, 253–263. [CrossRef]
18. Toming, K.; Kutser, T.; Laas, A.; Sepp, M.; Paavel, B.; Nõges, T. First Experiences in Mapping Lake Water Quality Parameters with Sentinel-2 MSI Imagery. *Remote Sens.* **2016**, *8*, 640. [CrossRef]
19. Olmanson, L.G.; Bauer, M.E.; Brezonik, P.L. A 20-year Landsat water clarity census of Minnesota's 10,000 lakes. *Remote Sens. Environ.* **2008**, *112*, 4086–4097. [CrossRef]
20. Palmer, S.C.J.; Kutser, T.; Hunter, P.D. Remote sensing of inland waters: Challenges, progress and future directions. *Remote Sens. Environ.* **2015**, *157*, 1–8. [CrossRef]
21. Brezonik, P.; Menken, K.D.; Bauer, M. Landsat-based remote sensing of lake water quality characteristics, including chlorophyll and colored dissolved organic matter (CDOM). *Lake Reserv. Manag.* **2005**, *21*, 373–382. [CrossRef]
22. Kallio, K.; Kutser, T.; Hannonen, T.; Koponen, S.; Pulliainen, J.; Vepsälainen, J.; Pyhälahti, T. Retrieval of water quality from airborne imaging spectrometry of various lake types in different seasons. *Sci. Total Environ.* **2001**, *268*, 59–77. [CrossRef]
23. Vanhellemont, Q.; Ruddick, K. Acolite for Sentinel-2: Aquatic applications of MSI imagery. In Proceedings of the ESA Living Planet Symposium, Prague, Czech Republic, 9–13 May 2016; pp. 9–13.
24. Ansper, A.; Alikas, K. Retrieval of Chlorophyll *a* from Sentinel-2 MSI Data for the European Union Water Framework Directive Reporting Purposes. *Remote Sens.* **2019**, *11*, 64. [CrossRef]
25. Brezonik, P.; Olmanson, L.G.; Finlay, J.C.; Bauer, M. Factors affecting the measurement of CDOM by remote sensing of optically complex inland waters. *Remote Sens. Environ.* **2015**, *157*, 199–215. [CrossRef]
26. Toming, K.; Kutser, T.; Uiboupin, R.; Arikas, A.; Vahter, K.; Paavel, B. Mapping Water Quality Parameters with Sentinel-3 Ocean and Land Colour Instrument imagery in the Baltic Sea. *Remote Sens.* **2017**, *9*, 1070. [CrossRef]
27. Soomets, T.; Uudeberg, K.; Jakovels, D.; Brauns, A.; Zagars, M.; Kutser, T. Validation and Comparison of Water Quality Products in Baltic Lakes Using Sentinel-2 MSI and Sentinel-3 OLCI Data. *Sensors* **2020**, *20*, 742. [CrossRef]
28. Hansen, C.H.; Burian, S.J.; Dennison, P.E.; Williams, G.P. Spatiotemporal Variability of Lake Water Quality in the Context of Remote Sensing Models. *Remote Sens.* **2017**, *9*, 409. [CrossRef]

29. Ritchie, J.C.; Zimba, P.V.; Everitt, J.H. Remote sensing techniques to assess water quality. Photogrammetric engineering & remote sensing. *Photogramm. Eng. Remote Sens.* **2003**, *69*, 695–704.
30. Gholizadeh, M.H.; Melesse, A.M.; Reddi, L. A Comprehensive Review on Water Quality Parameters Estimation Using Remote Sensing Techniques. *Sensors* **2016**, *16*, 1298. [CrossRef]
31. Kallio, K.; Attila, J.; Härmä, P.; Koponen, S.; Pulliainen, J.; Hyytiäinen, U.M.; Pyhälahti, T. Landsat ETM+ images in the estimation of seasonal lake water quality in boreal river basins. *Environ. Manag.* **2008**, *42*, 511–522. [CrossRef]
32. Verpoorter, C.; Kutser, T.; Seekel, D.; Tranvik, L.J. A global inventory of lakes based on high-resolution satellite imagery. *Geophys. Res. Lett.* **2014**, *41*, 639–642. [CrossRef]
33. Saberioon, M.; Brom, J.; Nedbal, V.; Soucek, P.; Cisar, P. Chlorophyll-a and total suspended solids retrieval and mapping using Sentinel-2A and machine learning for inland waters. *Ecol. Indic.* **2020**, *113*, 106236. [CrossRef]
34. El Hafyani, M.; Essahlaoui, A.; Teodoro, A.C.; Mohajane, M. Modeling and mapping of soil salinity in Tafilalet plain (Morocco). *Arab. J. Geosci.* **2019**, *12*, 35. [CrossRef]
35. El Ouali, A.; Mudry, J.; Mania, J.; Chauve, P.; Elyamine, N.; Marzouk, M. Present recharge of an aquifer in a semi-arid region: An example from the Turonian limestones of the Errachidia basin, Morocco. *Environ. Geol.* **1999**, *38*, 171–176. [CrossRef]
36. Nouayti, N.; Khattach, D.; Hilali, M. Cartographie des zones potentielles pour le stockage des eaux souterraines dans le haut bassin du Ziz (Maroc): Apport de la télédétection et du système d'information géographique. *Bull. Inst. Sci. Rabat Sect. Sci. Terre* **2017**, *39*, 45–57.
37. Muzirafuti, A.; Boualoul, M.; Barreca, G.; Allaoui, A.; Bouikbane, H.; Lanza, S.; Crupi, A.; Randazzo, G. Fusion of Remote Sensing and Applied Geophysics for Sinkholes Identification in Tabular Middle Atlas of Morocco (the Causse of El Hajeb): Impact on the Protection of Water Resource. *Resources* **2020**, *9*, 51. [CrossRef]
38. El-Ouardi, H.; Boualoul, M.; Ouhaddach, H.; Habib, M.; Muzarafuti, A.; Allaoui, A.; Amine, A. Fault analysis and its relationship with karst structures affecting lower Jurassic limestone in the Agourai plateau (Middle Atlas, Morocco). *Geogaceta* **2018**, *63*, 119–122.
39. Muzirafuti, A.; Boualoul, M.; Randazzo, G.; Lanza, S.; Allaoui, A.; El Ouardi, H.; Habibi, M.; Ouhaddach, H. The use of remote sensing for water protection in the karst environment of the Tabular Middle Atlas/the cause of El Hajeb/Morocco. In *Earth Observation Advancements in a Changing World*; AIT Series, Trends in Earth Observation; Chirici, G., Gianinetto, M., Eds.; Italian Society of Remote Sensing: Firenze, Italy, 2019.
40. Lehner, B.; Liermann, C.R.; Revenga, C.; Vörösmarty, C.; Fekete, B.; Crouzet, P.; Döll, P.; Endejan, M.; Frenken, K.; Magome, J.; et al. High-resolution mapping of the world's reservoirs and dams for sustainable river-flow management. *Front. Ecol. Environ.* **2011**, *9*, 494–502. [CrossRef]
41. Dahir n°1-16-113 du 6 Kaada 1437 Portant Promulgation de la loi n°36-15 Relative à l'eau. Available online: http://www.sgg.gov.ma/Portals/0/lois/Loi_36-15_Fr.pdf?ver=2018-11-28-153115-853 (accessed on 24 September 2021).
42. Congedo, L. Semi-Automatic Classification Plugin Documentation. Available online: http://rgdoi.net/10.13140/RG.2.2.25480.65286/1 (accessed on 1 October 2021).
43. Chavez, P.S., Jr. An Improved Dark-Object Subtraction Technique for Atmospheric Scattering Correction of Multispectral Data. *Remote Sens. Environ.* **1988**, *24*, 459–479. [CrossRef]
44. Schalles, J.F.; Gitelson, A.A.; Yacobi, Y.Z.; Kroenke, A.E. Estimation of Chlorophyll *a* from time series measurements of high spectral resolution reflectance in an eutrophic lake. *J. Phycol.* **1998**, *34*, 383–390. [CrossRef]
45. Honeywill, C.; Paterson, D.N.; Hagerthey, S.E. Determination of microphytobenthic biomass using pulse-amplitude modulated minimum fluorescence. *Eur. J. Phycol.* **2002**, *37*, 485–492. [CrossRef]

applied sciences

MDPI

Article

Assessment of Agricultural Water Requirements for Semi-Arid Areas: A Case Study of the Boufakrane River Watershed (Morocco)

Mohammed El Hafyani [1,*], Ali Essahlaoui [1], Kimberley Fung-Loy [2,3], Jason A. Hubbart [4,5,6] and Anton Van Rompaey [2]

1 Department of Geology, Laboratory of Geoengineering and Environment, Research Group "Water Sciences and Environment Engineering", Faculty of Sciences, Moulay Ismail University, Zitoune, Meknes BP11201, Morocco; a.essahlaoui@fs-umi.ac.ma
2 Department of Earth and Environmental Sciences, Geography and Tourism Research Group, KU Leuven, Celestijnenlaan 200E, 3001 Heverlee, Belgium; kimberley.fung-loy@uvs.edu (K.F.-L.); anton.vanrompaey@kuleuven.be (A.V.R.)
3 Department of Sustainable Management of Natural Resources, Anton de Kom University of Suriname, Paramaribo, Suriname
4 Institute of Water Security and Science, West Virginia University, 3109 Agricultural Sciences Building, Morgantown, WV 26506, USA; jason.hubbart@mail.wvu.edu
5 Division of Plant and Soil Sciences, Davis College of Agriculture, Natural Resources and Design, West Virginia University, Agricultural Sciences Building, Morgantown, WV 26506, USA
6 Division of Forestry and Natural Resources, Davis College of Agriculture, Natural Resources and Design, West Virginia University, Agricultural Sciences Building, Morgantown, WV 26506, USA
* Correspondence: m.elhafyani@edu.umi.ac.ma

Citation: El Hafyani, M.; Essahlaoui, A.; Fung-Loy, K.; Hubbart, J.A.; Van Rompaey, A. Assessment of Agricultural Water Requirements for Semi-Arid Areas: A Case Study of the Boufakrane River Watershed (Morocco). *Appl. Sci.* **2021**, *11*, 10379. https://doi.org/10.3390/app112110379

Academic Editors: Dimitrios S. Paraforos and Manuel Armada

Received: 27 August 2021
Accepted: 26 October 2021
Published: 5 November 2021

Abstract: This work was undertaken to develop a low-cost but reliable assessment method for agricultural water requirements in semi-arid locations based on remote sensing data/techniques. In semi-arid locations, water resources are often limited, and long-term water consumption may exceed the natural replenishment rates of groundwater reservoirs. Sustainable land management in these locations must include tools that facilitate assessment of the impact of potential future land use changes. Agricultural practices in the Boufakrane River watershed (Morocco) were used as a case study application. Land use practices were mapped at the thematic resolution of individual crops, using a total of 13 images generated from the Sentinel-2 satellites. Using a supervised classification scheme, crop types were identified as cereals, other crops followed by cereals, vegetables, olive trees, and fruit trees. Two classifiers were used, namely Support vector machine (SVM) and Random forest (RF). A validation of the classified parcels showed a high overall accuracy of 89.76% for SVM and 84.03% for RF. Results showed that cereal is the most represented species, covering 8870.43 ha and representing 52.42% of the total area, followed by olive trees with 4323.18 ha and a coverage rate of 25%. Vegetables and other crops followed by cereals cover 1530.06 ha and 1661.45 ha, respectively, representing 9.4% and 9.8% of the total area. In the last rank, fruit trees occupy only 3.67% of the total area, with 621.06 ha. The Food and Agriculture Organization (FAO) free software was used to overlay satellite data images with those of climate for agricultural water resources management in the region. This process facilitated estimations of irrigation water requirements for all crop types, taking into account total potential evapotranspiration, effective rainfall, and irrigation water requirements. Results showed that olive trees, fruit trees, and other crops followed by cereals are the most water demanding, with irrigation requirements exceeding 500 mm. The irrigation requirements of cereals and vegetables are lower than those of other classes, with amounts of 300 mm and 150 mm, respectively.

Keywords: Sentinel-2; SVM; RF; Boufakrane River watershed; irrigation requirements; water resources; sustainable land use; agriculture

1. Introduction

According to World Bank data, the global rural population declined from 66.39% of the total population in 1960 to 44.72% in 2016. Meanwhile, agricultural added value, as a percentage of GDP (gross domestic product), decreased from 7.59 to 3.43% during the period 1994–2017. Despite these changes, the continued growth of the global human population has resulted in phases of deforestation together with innovations, increasing the efficiency of agricultural systems, and subsequent crises of unprecedented demographic, economic and urban expansion [1,2].

In semi-arid regions like Morocco, agricultural practices are facing a series of challenges not limited to climate change, which is reflected in an increasingly warmer and drier climate [3], coupled with the increasingly random spatiotemporal variability of rainfall, and associated droughts and floods [4–8]. These issues are exacerbated by increasingly complex land use and land cover practices that in turn adversely affect socio-economic development [9]. Agriculture, as an important economic sector, is therefore deeply impacted, given crop production dependence on the annual rainfall distribution [4–8]. It is clear that humans have created and are now witnessing a great agricultural ecosystem disturbance [8].

Recently, geospatial technologies have been used extensively for spatiotemporal monitoring of environmental phenomena, including land use/land cover changes [6,9–15], understanding the ecosystem functions [16,17], identifying agricultural systems and crop mapping [3,8,18,19], estimating fractional crop cover and crop residue [20], estimating the impacts of urbanization on agricultural dynamics [3], identifying the karst cavities in agricultural areas [21–23], and water balance assessments at regional and local scales [6,24]. Many investigations have shown a great deal of potential in terms of different machine learning approaches in imagery classification, such as vector machine support [3,25–29] and random forest [30–32]. These innovative data interrogation and modeling approaches are critically important for estimating agricultural crop water use. This is important in order to implement effective strategies for advanced water resource management for agriculture in response to contemporary water budget challenges. Several works have been published supporting these needs in recent years [33–41], particularly in semi-arid zones [42]. Ofentse Moseki et al. [42] used the CROPWAT model to determine the irrigation needs of the Jatropha crop in Botswana. They used the CROPWAT model to estimate baseline evapotranspiration (ETo), evapotranspiration (ETc), irrigation water requirements (IWR) and yield response to irrigation scheduling in Botswana. The results showed that the annual ETo from 2014 to 2016 at the station was 1456 mm. The lowest monthly ETo (50.10 mm) was observed in June and the highest (182.59 mm) in January.

This model is widely used, especially in understanding the changes in crop water requirements [38], which are defined as the depth of water required to meet the evapotranspiration water loss (ETc) of a disease-free crop growing in large fields; this parameter is important for promoting sustainable development. In particular, the model is used in the determination of crop water and of the effects of irrigation programming on the crop [40,41]. This model allows calculation of the water requirements of the different crops using soil data, climatic data, and data on the crops themselves. Therefore, to determine the crop's water requirement, several parameters were calculated.

Calculation of Potential Evaporation of Crop ETc: Before calculating the ETc, specific studies on the water requirements of crops in the area should be examined; the meteorological and research stations and the environment should also be visited. The calculation of this parameter is done by the following two main steps:

(a) *Reference evapotranspiration (ET_0):* Collect climate data and choose the method for calculating ET_0 for each 30 or 10 day period using the average climate data.
(b) *Crop coefficient (kc):* Determine the timing of planting or sowing, the rate of crop development, the duration of crop development stages and the growing season. Choose the kc for a given crop plan and crop development stage under prevailing climatic conditions, and prepare a crop coefficient curve for each one.
(c) *Crop evapotranspiration (ETcrop):* Calculate ETcrop for each 30- or 10-day period:

$$ETc = Kc * ET_0$$

Irrigation requirements: Part of the crop water requirements is met by rainfall (*Pe*), groundwater (*Ge*) and stored soil water (*Wb*); or

$$Irr.Req = ETc - Pe - Ge - Wb$$

and is determined on a monthly basis.

The specific objectives of this work were to (i) use high spatial resolution Sentinel-2 images to map crop types in the Boufakrane watershed; (ii) evaluate machine learning (ML) methods such as support vector machine (SVM) classifier and random forest (RF) in crop species' mapping; (iii) use the CROPWAT 8.0 model to estimate water demand for agriculture in the study area through the calculation of potential evaporation and effective rainfall. Finally, irrigation water requirements were estimated.

2. Materials and Methods

2.1. Study Area

The Boufakrane River watershed is located in the headwaters region of the Great Sebou Basin between longitudes 5°25′46.13″ and 5°37′49.71″ W and between latitudes 33°28′54.40″ and 33°58′32.93″ N (Figure 1). Locally, the area is part of the Fez-Meknes region, which is one of the most important and productive areas for agriculture in the region, given its relatively high water availability and good quality soils. The regional Useful Agricultural Area (UAA) is estimated to be approximately 1,340,826 hectares, representing 15% of the national total area. The UAA is dominated by cereals (816,000 ha) and olive trees (350,000 ha), and 14% of the area is irrigated (184,162 ha). Climatically, the region is characterized by a semi-arid climate, with a mean annual rainfall of 500 mm, a mean annual reference evapotranspiration of 907 mm, and a dry season extending from June to October (Figure 2).

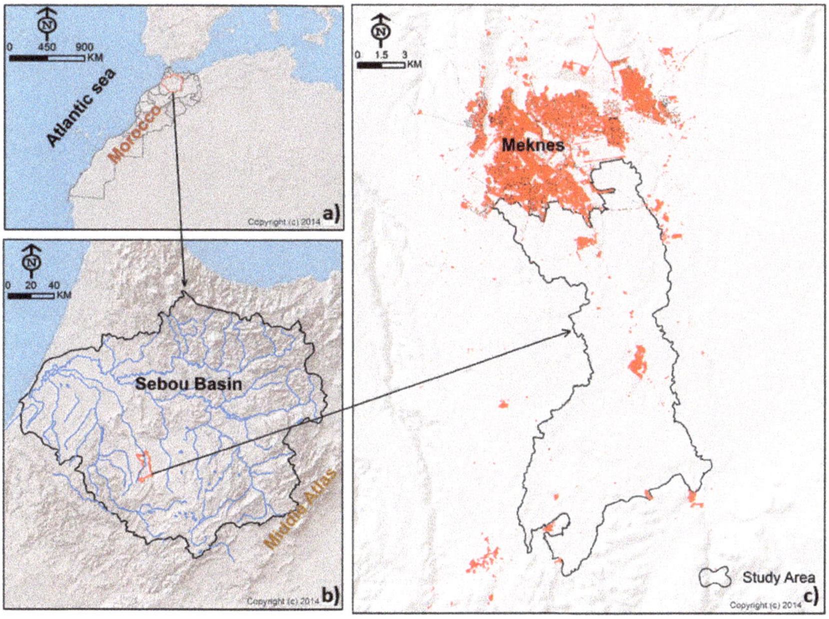

Figure 1. Study area: (**a**) Kingdom of Morocco, (**b**) Sebou Basin, (**c**) Boufakrane watershed.

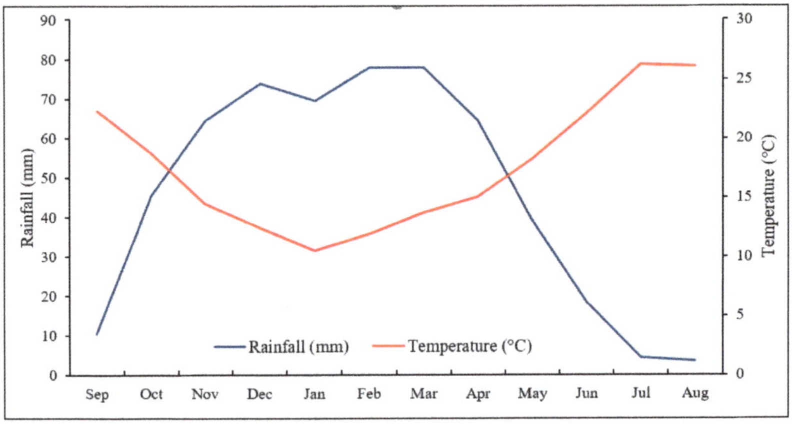

Figure 2. Ombrothermic diagram of the meteorological station at Meknes. Rainfall is mean monthly rain depth and temperature is mean monthly air temperature (1998–2018).

2.2. Data

The regions of interest used for land use classification were developed based on a series of field missions throughout the study area along with high-resolution Google Earth visualizations. A total of 88,546 pixels with a resolution of 10 m^2 was used for classification; 65% were used as reference data and 35% for the validation. The sampling was done in a random way; it was chosen to be representative and well distributed in the area. Figure 3 shows the field data used for this work. Weather data were collected for the study period of September 2018 to August 2019. These data included monthly mean rainfall, monthly average minimum temperature, monthly average maximum temperature, humidity (%), wind speed (m/s), and sunshine intensity (hours). Figure 4 shows climate data collected by the climate station at the Faculty of Science of Meknes, Moulay Ismail University, coordinates: Latitude: 33°52′11.12″ N, Longitude: 5°32′35.11″ W, Z = 554 m.

For the satellite data, a total of 13 satellite images covering a whole crop year were used to carry out this work. All these images were obtained from the Sentinel-2 sensor of the European Space Agency (https://sentinel.esa.int/web/sentinel/sentinel-data-access (accessed on 15 August 2021)). This mission was launched in June 2015, with a revisit time (i.e., image interval) of 10 days and image spatial resolution of 10 m to 60 m in thirteen spectral bands from visible to mid-infrared. Images were downloaded from https://scihub.copernicus.eu/dhus/#/home (accessed on 15 August 2021) for the period August 2018–August 2019 (Table 1).

Table 1. Sentinel-2 satellite image acquisition dates.

Image	Acquisition Dates
1	22 August 2018
2	1 September 2018
3	21 October 2018
4	15 November 2018
5	15 December 2018
6	14 January 2019
7	13 February 2019
8	15 March 2019
9	29 April 2019
10	14 May 2019
11	8 June 2019
12	18 July 2019
13	22 August 2019

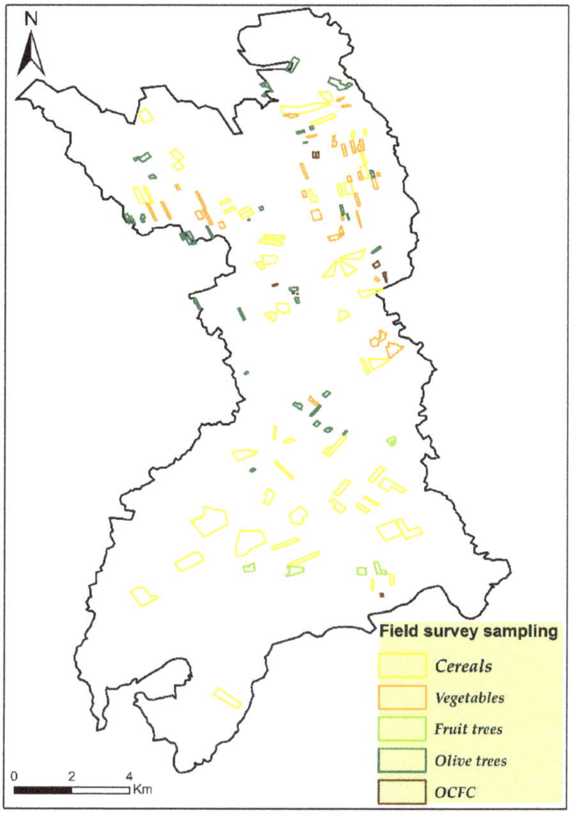

Figure 3. Field survey sampling. OCFC: Other crops follow by cereals.

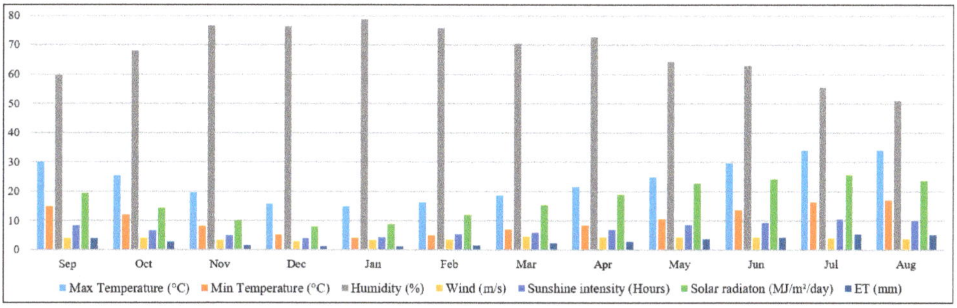

Figure 4. Climatic data.

2.3. Methodolgy

Normalized difference vegetation index (*NDVI*) was calculated to construct *NDVI* time-series images. In parallel, several surveys and field missions were conducted in the region to collect reference (validation) points for the different agricultural crops. These data were combined with the high spatial resolution Google Earth images. Spectral profiles were then constructed and used as input data for a machine learning approach to map different crop species in the region. As a final step, and to associate satellite data with observed water management practices, CROPWAT 8.0 software was used to estimate crop water requirements (Figure 5). The simulation was carried out on three parameters: potential evapotranspiration for the different crop types (ETc), effective rainfall (ER), and irrigation water requirements (IWR). These three parameters are dependent on each other. The ETc is the amount of water that should be transpired in a given time by the crop, while the ER is defined as the rainfall fraction that responds to the crops' water needs [43]. During the rainy months, rainfall covers the water requirements of the crops, while during the dry months, rainfall must be supplemented by irrigation water to cover water requirements.

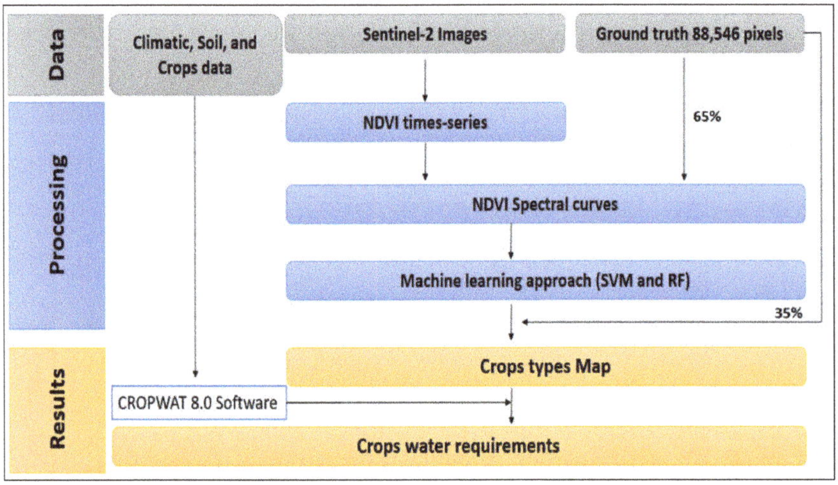

Figure 5. Flowchart of the methodology.

2.4. Support Vector Machine (SVM)

SVMs belong to a family of algorithms that use supervised learning and are specialized in solving mathematical discrimination and regression problems. They were developed in 1998 by Vladimir Vapnik [44]. Support vector machines (SVMs) represent a group of theoretically superior machine learning algorithms. The development of this method was initially triggered by the exploration and formalization of machine learning capacity control and over-fitting problems [44] and represents an efficient technique, with reduced data and processing demands. The method avoids the problems of over-adjustment and does not require any hypothesis on the type of data. Although non-parametric, the method is capable of developing efficient decision limits and can therefore minimize classification errors. This is done by searching for the optimal separation between classes [45]. Their work was quickly adopted because of their ability to work with large data, their theoretical guarantees and the good results achieved in practice. Requiring a small number of parameters, SVMs are appreciated for their simplicity of use.

2.5. Random Forest (RF) Classifier

Random forest (RF) was developed by [46]. It is a supervised non-parametric method applicable for both classification and prediction [47,48]. Model subroutines are composed of a combination of decision trees used independently to assign the most frequent class to the input data, and the majority vote of the trees determines the class prediction. The part of the data not used in tree training is used for performance evaluation.

For the current investigation, after extracting the crop type characteristic based on the data collected in the field, twelve decision trees were constructed and were the basis for RF classifier (Figure 6). These decision trees make it possible to predict the different classes.

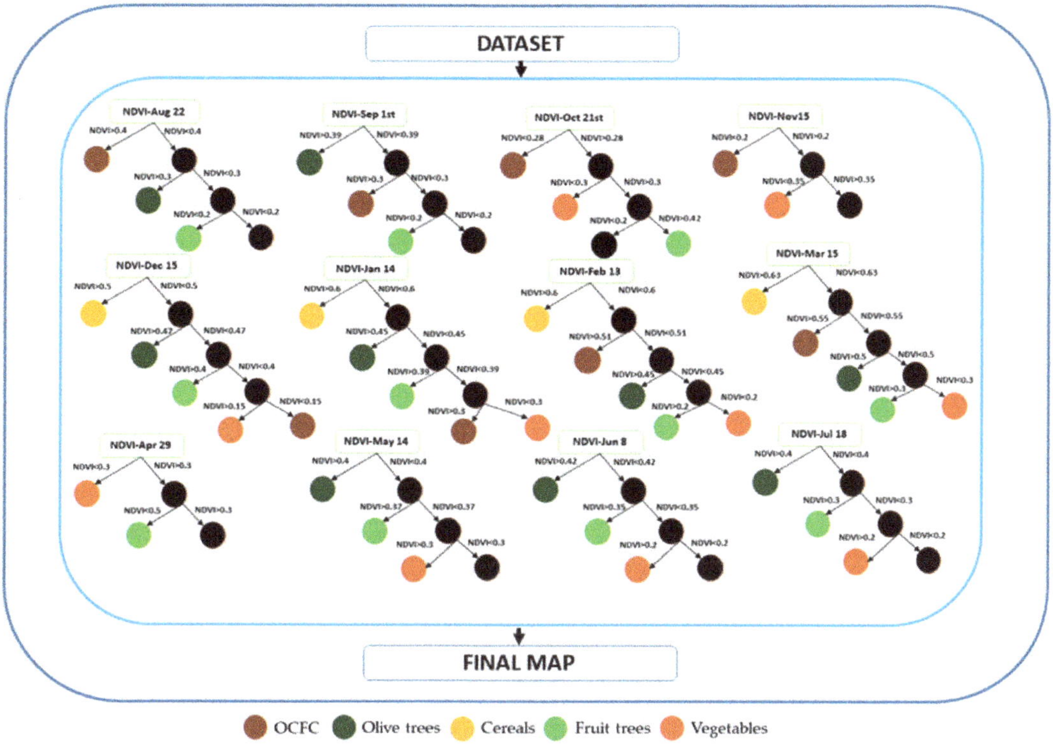

Figure 6. Decision trees used for random forest classification. OCFC = Other Crops Followed by Cereals.

2.6. NDVI Time-Series Spectral Profile Curves

Field data and data collected from the Regional Directorate of Agriculture showed the presence of five main cropping systems in the region, including cereals, other crops followed by cereals, vegetables (onion, potatoes, tomatoes), olive trees and fruit trees. More than two hundred profiles were developed for the different crop types. These profiles were associated with the field data and the visualization of high spatial resolution Google Earth images in order to collect input data for classification (Figures 7 and 8).

From Figure 8, it is possible to discriminate the spectral characteristics of the different crops in relation to *NDVI* values during the year. This index, proposed for the first time by Rouse et al., 1973 [49], is widely used and provides information on the quantity and vigor of vegetation, taking into account the near infrared (*NIR*) and visible red bands of the electromagnetic spectrum [49,50] calculated by the following equation:

$$NDVI = \rho_{NIR} - \rho_{RED} / \rho_{NIR} + \rho_{RED}$$

where ρ_{NIR} : the reflectance in the near − infrared reflectance, ρ_{RED} : the reflectance in the red band.

For example, for an olive pixel, the *NDVI* value did not change significantly throughout the year, with an increase around February. While for a pixel of cereals, the *NDVI* values did increase with the crop growth cycle.

Figure 7. Different types of crops in the study area. (**a**) Olive trees, (**b**) Cereals, (**c**) Fruit trees, (**d**) Vegetables (Source: Esri, Maxar, GeoEye, Earthstar Geographics, CNES/Airbus DS, USDA, USGS, AeroGRID, IGN, and the GIS User Community).

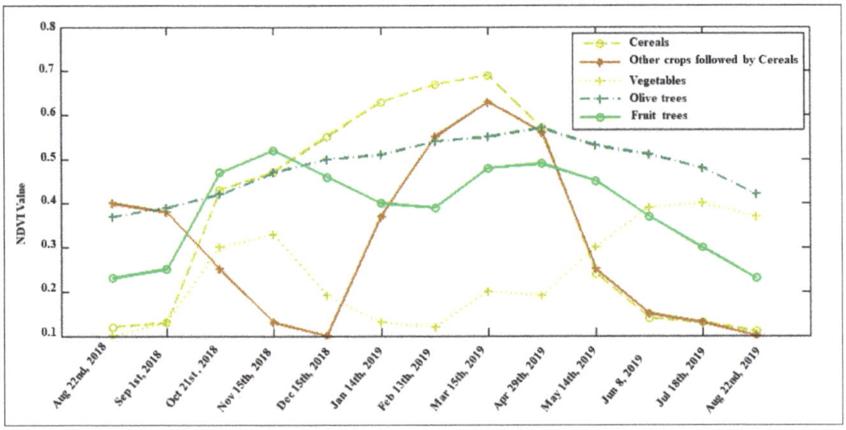

Figure 8. *NDVI* time-series spectral profile curves at different crop developmental stages.

3. Results

3.1. Overall Accuracy

The classification approaches used in this work were selected based on the confusion matrix and the Kappa index [51–53], whose the overall accuracy is the proportion of the area mapped correctly. It provides the user of the map with the probability that a randomly selected location on the map is correctly classified. The Kappa coefficient measures the agreement between the resulting classes of the classifier and the true values [52,54], with values ranging from 0 to 1, where 0 represents no agreement and 1 represents perfect agreement. The Table 2 shows the confusion matrix calculated from the reference data and the map classes; the reference data is presented in the row, and the map classes in the columns. The results showed that the two approaches showed high classification accuracies. The overall accuracy for the SVM classifier exceeded 89.76%, and a significant agreement by the Kappa index of 0.79 was obtained. The overall accuracy for RF was 84.03%, with a Kappa index of 0.68. For the validation, 57,554 pixels were used. In most cases, it showed a high accuracy of this classification for most crop species. Few confusions between classes were recorded using the RF approach (e.g., crops followed by cereals, cereals, vegetables, and olive trees).

Table 2. Confusion Matrix. OCFC = Other Crops Followed by Cereals.

	Reference Data (%)											
Map Classes	**Cereals**		**OCFC**		**Vegetables**		**Olive Trees**		**Fruit Trees**		**Total (%)**	
Classifier	**SVM**	**RF**	**SVM**	**RF**	**SVM**	**RF**	**SVM**	**RF**	**SVM**	**RF**	**SVM**	**RF**
Cereals	88.23	87.40	6.05	24.43	2.05	1.52	1.21	5.17	0.65	6.17	64.17	64.32
OCFC	5.27	4.48	88.10	41.86	4.80	10.76	1.66	7.39	0.48	0.62	5.54	5.81
Vegetables	3.97	3.30	1.88	10.96	92.61	84.30	0.97	0.39	0.39	0.06	14.63	13.13
Olive trees	2.20	3.41	3.76	13.47	0.43	0.88	94.79	70.05	2.36	23.69	11.42	10.79
Fruit trees	0.33	1.40	0.21	9.29	0.11	2.53	1.37	17.01	96.13	69.46	4.23	5.94
											100	100

In addition to the overall accuracy and the Kappa index, other types of accuracy and errors were calculated for both classifiers, including the producer's accuracy, the user's accuracy, the commission error, and the omission error (Table 3) [53].

(i) Producer's accuracy is defined as the probability that a value in a reference dataset was correctly classified. Producer's accuracy is the complement to the probability of omission error.
(ii) User's accuracy represents the probability that a resulting value in a certain class is really that class. User's accuracy is the complement to the probability of commission error.
(iii) Commission errors represent the fraction of the resulting values in a class that does not belong to that class.
(iv) Omission errors represent the fraction of values that belongs to one class but was predicted in a different class.

Table 3. Commission, omission, producer's and user's accuracy for the SVM and RF classifiers.

Classes	**Commission (%)**		**Omission (%)**		**Producer's Accuracy (%)**		**User's Accuracy (%)**	
Classifier	**SVM**	**RF**	**SVM**	**RF**	**SVM**	**RF**	**SVM**	**RF**
Cereals	0.74	1.91	11.77	12.60	88.23	87.40	99.26	98.09
OCFC	82.93	92.26	11.90	58.14	88.10	41.86	17.07	7.74
Vegetables	20.49	19.38	7.39	15.70	92.61	84.30	79.51	80.62
Olive trees	15.56	33.95	5.21	29.95	94.79	70.05	84.44	66.05
Fruit trees	9.27	53.29	3.87	30.54	96.13	69.46	90.73	46.71

In terms of accuracy, the producer's accuracy and user's accuracy confirm the results found for the overall accuracy. The producer's accuracy showed very high values, with more than 80% for all classes in the both classifiers SVM and RF, except the class other crops followed by cereals, which presented a value of 41.86% for the RF classifier. The user's accuracy showed very high values, with more than 79% for all classes in the SVM classifier, except for other crops followed by cereals, which presented a value 17.07%. For the RF classifier, this accuracy shows high values for the three classes of cereals, vegetables, and olive trees: 98.09%, 80.62%, and 66.05%, respectively. The other two classes, fruit trees and other crops followed by cereals, represent low precision, with values of 46.71% and 7.74%, respectively.

In terms of commission error, the results showed that the other crops followed by cereals represents the highest values for the two classifiers SVM and RF, with values of 82.93% and 92.26%, respectively, followed by vegetables and fruit trees. The other two classes represented a low value for this error. For the omission error, cereals and other crops followed by cereals represented the highest value for the SVM, whereas for RF, cereals and other crops followed by cereals represent a high value of 58.14%, followed by fruit trees with 30.54% and olive trees with 29.95%.

3.2. Crop Mapping

Crop mapping was performed based on field data combined with a detailed study of the chlorophyll response (*NDVI*) for each of crop type. The crops determined in the region were cereals, other crops followed by cereals, vegetables (onions, potatoes, tomatoes), olive trees, and fruit trees. In order to calculate the areas of each class, pixel size was used. Therefore, after obtaining the classes, the area of each class was obtained by multiplying the number of pixels and the pixel size (10 m × 10 m). For the crop year 2018–2019, the areas determined for the cereals represented the largest class, with an area of 8870.43 ha, followed by olive trees with an area of 4323.18 ha. Classes of other crops followed by cereals and vegetables represented 1661.45 ha and 1530.06 ha, respectively. The least represented class was that of fruit trees, with only 661.05 ha (Figure 9).

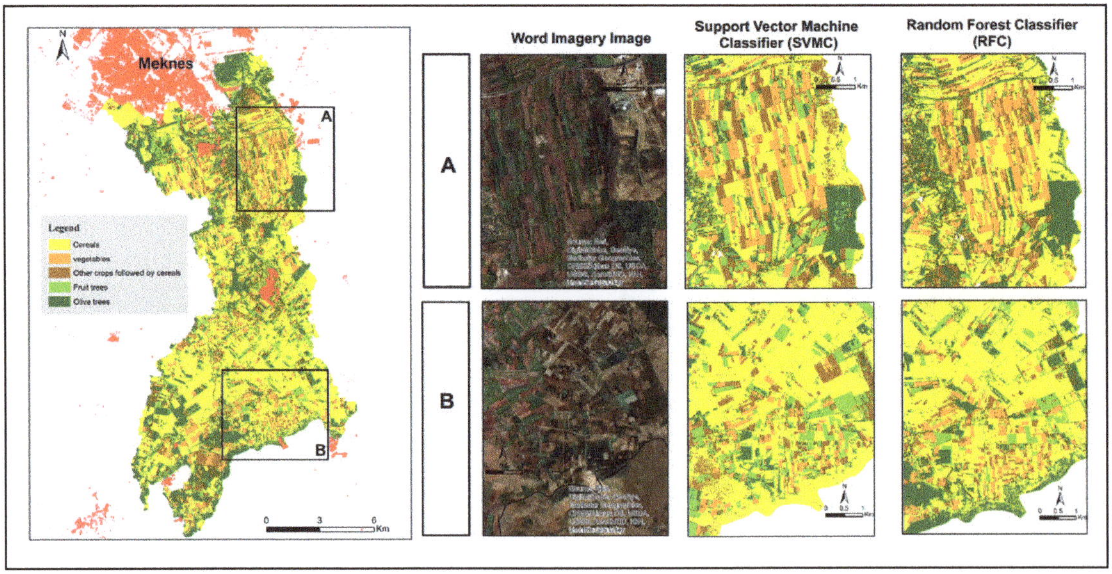

Figure 9. Crop mapping using SVM and RF classification. (**A**) Northern part (**B**) Southern part.

3.3. CROPWAT for Water Crop Requirements

This section includes calculation of the crop water requirements using the FAO free software CROPWAT 8.0, based on climate, soil, and crop data. Thus, three main variables were estimated by the units of water depth (mm): ETc, ER, and IWR. For preliminary planning, monthly data are frequently used, and the total of the data of the different crops over the area constitutes the basis for determining the supply.

The climatic data were used to calculate the reference evapotranspiration ET_0, and by determining the timing of planting or sowing, the rate of crop development, the duration of crop development stages, and the growing season kc for a given crop were chosen. Then, the ETc for each crop type was calculated for each 10 day period. Figure 10 shows the different crop types' evapotranspiration from October 2018 to August 2019. The curves show that this parameter increases in the driest months (July and August) for all crop types, while it decreases in the rainiest months (December to February). For the class of other crops followed by cereals, for example, the potential evapotranspiration reached up to 70 mm.

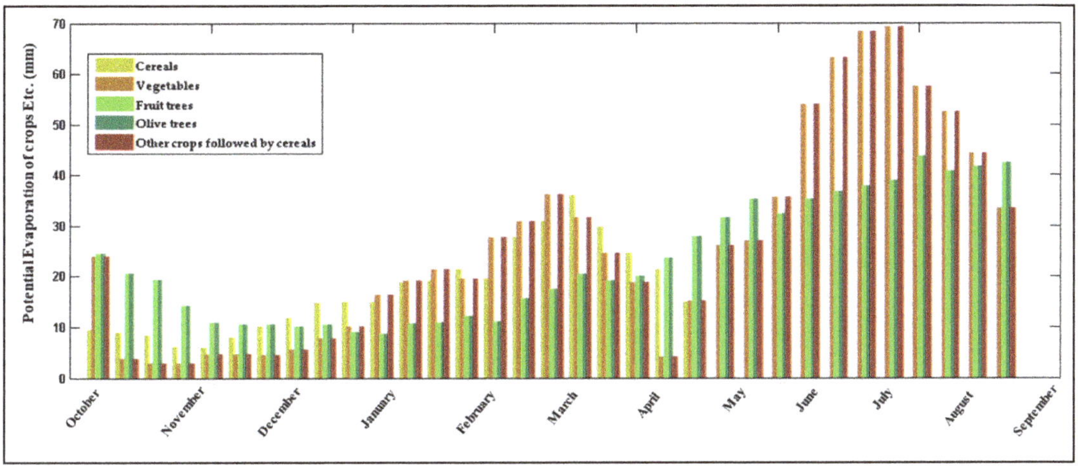

Figure 10. Potential evaporation of crops ETc.

Not all precipitation is effective, and in the most cases some of this precipitation can be lost through surface runoff, deep percolation, or evaporation. Only part of the high-intensity rain can penetrate and be stored in the root zone. These rains can be 100% effective when the vegetation cover is complete, while they can be only 60% effective with a low percentage of vegetation cover. The relationship between the average monthly ER and the average monthly rainfall is shown for different values of the average monthly ETc [55]. Figure 11 shows the evolution of ER for the different types of crops. The evolution curves of this parameter show that the crops' needs were met in the rainy months. However, in the dry months, these crops suffered from water stress.

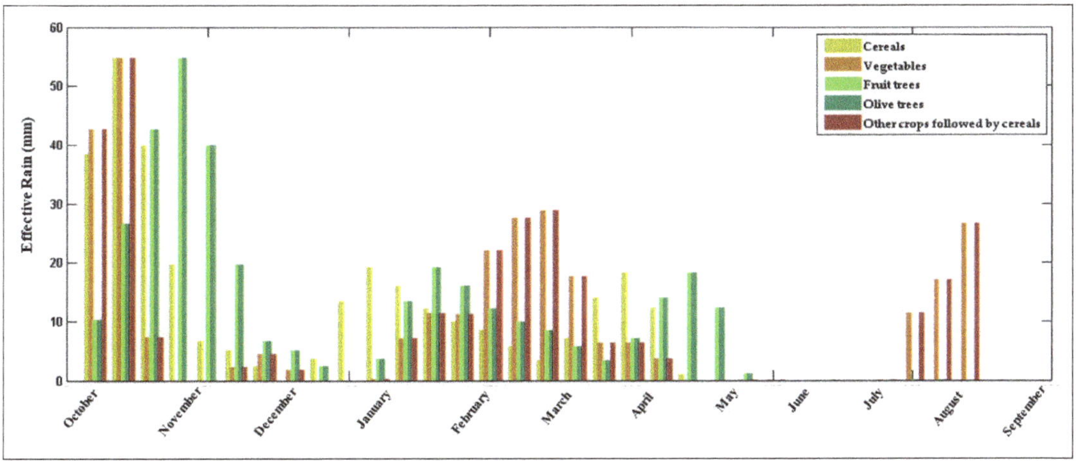

Figure 11. Effective rain.

Irrigation water requirements are calculated using the field water balance, based on ETc and ER. They allow for optimal production in a given growing environment.

Figure 12 shows the evolution of water requirements for agriculture for the different types of crops. In the rainy months, rainfall covers the water requirements of the crops; during this time, the water requirements for irrigation are expected to be very low. This parameter is strongly related to the climatic conditions and is directly influenced by variations in conditions. It is inversely correlated with the daily rainfall. Thus, it is high in the dry months and low in the rainy months.

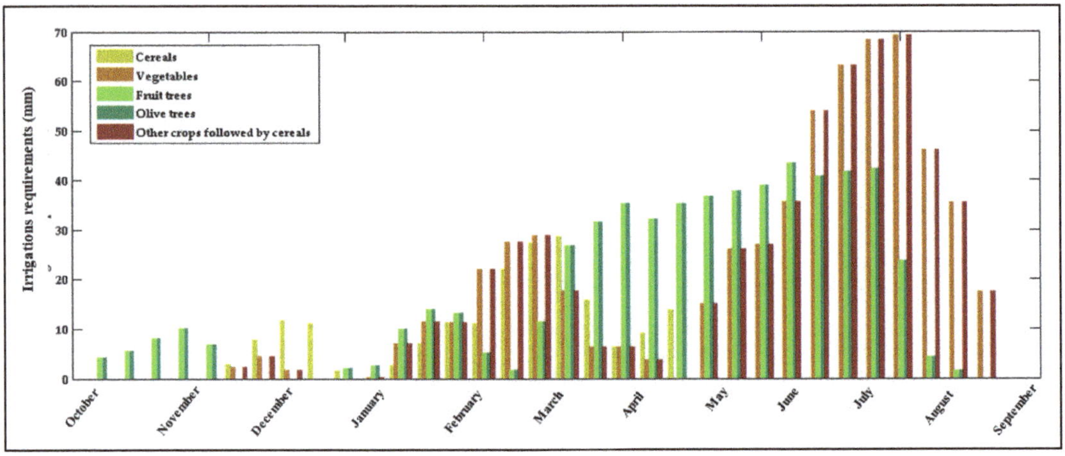

Figure 12. Irrigation requirements.

Figure 13 shows the total for the year of the three estimated parameters (ETc, ER, and IWR) for the different types of crops. For water irrigation requirements, other crops followed by cereals, olive trees, and fruit trees were the three types of crops that required a very large quantity of water, exceeding 500 mm. Vegetables required about 450 mm; the demand of cereals did not exceed 200 mm. The potential evapotranspiration was strongly correlated with the water demand; the crop types that demanded a lot of water were those that recorded high values of evapotranspiration. For the ER, the class that recorded the lowest values was vegetables, while the other classes recorded values higher than 300 mm.

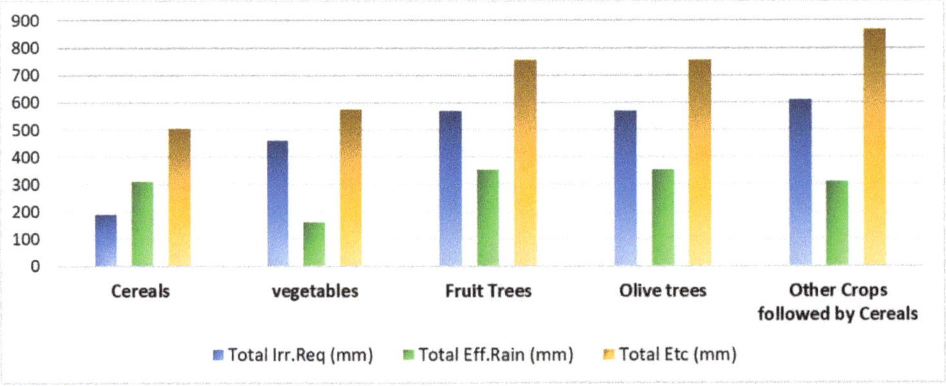

Figure 13. Total Etc, Eff.Rain, and Irr.Req.

4. Discussion

Using the classification approach described earlier, and based on spectral analysis results, a map of the different agricultural species was produced for the study area (Figure 9). Results generally showed that the classification using both classifiers was satisfactory, with the exception of some confusion between a few classes, which is likely due to the spectral similarity of the crops.

Previous investigations using geospatial techniques have shown a strong efficiency in land use/land cover change monitoring [6,11], crop mapping [3,18], and identification of agricultural systems [8]. Ouzemou et al. [3] carried out an insightful study in the plains of Tadla, Morocco. The objectives included mapping different agricultural species using high-resolution satellite data and machine learning approaches and comparing the different used approaches. Their study showed an overall accuracy of 89.26%, 85.27%, and 57.17%, respectively, for random forest, support vector machine, and spectral angle mapper, with a Kappa index of 0.85, 0.80, and 0.4, respectively. Comparing our result with this one, our study also showed a high overall accuracy of 89.76% for SVM and 84.03% for RF, and a Kappa index of 0.79 and 0.68, respectively.

The CROPWAT model was used to determine water demand for the different agricultural species in the region. Three parameters were calculated, namely, the crops' potential evaporation, the effective rain, and the irrigation requirements. The results obtained by [40] showed that the irrigation requirements varied according to the location, whereas the required water quantity per palm varied between 115 and 200 liters per day. Comparing our result with this one, our study showed that olive trees, fruit trees, and other crops followed by cereals are the most water demanding, with needs exceeding 500 mm. The water demands of cereals and vegetables are lower than that of other classes, with amounts of 300 mm and 150 mm, respectively.

As explained, water consumption is increasing in the Saiss plain. This is mainly due to excessive exploitation. According to the 1939–2002 groundwater data record, there is a constant deficit of approximately 100 Mm^3/year, with an estimated inflow of 242 Mm^3/year and an estimated outflow of 342 Mm^3/year. While the output includes abstraction with 260 Mm^3/year, and rivers and springs with 82 Mm^3/year, 22% of the water balance is dedicated to human drinking water supplies and 78% to private irrigation [56].

5. Conclusions

The assessment and estimation of water demand for agriculture is crucial to improve water resource management in a given region. The final objective was to determine the water demand for agriculture in the Boufakrane River watershed through several steps. First, a map of the different crop types was produced using the SVM and RF machine learning algorithms, based on field data combined with the high spatial resolution Google Earth images. Five crop types were mapped, including cereals, other crops followed by cereals, vegetables, olive trees, and fruit trees. Then, the evaluation of the classification map was made based on the Kappa index and the overall accuracy. Finally, the satellite data were combined with climate, soil, and crop data before being used as inputs for CROPWAT software to estimate the water requirements for agriculture.

The mapping results showed a strong potential of high-resolution satellite data in agricultural species mapping. The evaluation of the two classifiers used (RF and SVM) showed a Kappa index higher than 0.67 and an overall accuracy exceeding 83%. The irrigation requirements showed that the other crops followed by cereals, olive trees, and fruit trees were the three types of crops that required a very large quantity of water, exceeding 500 mm. Vegetables required an amount of about 450 mm; the demand of cereals did not exceed 200 mm.

The method developed in this work facilitates estimations of the agriculture water demand in the study area, thereby promoting sustainable water resource management. Through this study, we recommend a combination of these methods with existing real data for the implementation of a system for the quantification of water resources for crops throughout Morocco, which would allow validation using global crop yield data. It was also an opportunity to see the link between water demand and known groundwater reserves and existing data on actual evapotranspiration in this area.

Author Contributions: Conceptualization, M.E.H., A.E. and A.V.R.; Data curation, M.E.H., A.E. and A.V.R.; Formal analysis, M.E.H., A.E. and A.V.R.; Funding acquisition, A.E. and A.V.R.; Investigation, A.E. and A.V.R.; Methodology, M.E.H., A.E. and A.V.R.; Project administration, A.E. and A.V.R.; Resources, A.E. and A.V.R.; Software, M.E.H.; Supervision, A.E. and A.V.R.; Validation, M.E.H., A.E. and A.V.R.; Visualization, M.E.H.; Writing—original draft, M.E.H., A.E. and A.V.R.; Writing—review and editing, M.E.H., K.F.-L. and J.A.H. All authors have read and agreed to the published version of the manuscript.

Funding: This research received no external funding.

Institutional Review Board Statement: Not Applicable.

Informed Consent Statement: Not Applicable.

Acknowledgments: The authors would like to thank the Thematic Project 4, Integrated Water Resources Management of the Institutional University Cooperation, and VLIR-UOS for the financial support, equipment and mission at KU Leuven, Belgium.

Conflicts of Interest: The authors declare no conflict of interest.

References

1. Beckers, V.; Poelmans, L.; Van Rompaey, A.; Dendoncker, N. The Impact of Urbanization on Agricultural Dynamics: A Case Study in Belgium. *J. Land Use Sci.* **2020**, *15*, 626–643. [CrossRef]
2. Mazoyer, M.; Roudart, L. *A History of World Agriculture: From the Neolithic Age to the Current Crisis*; NYU Press: Manhattan, NY, USA, 2006.
3. Ouzemou, J.-E.; El Harti, A.; Lhissou, R.; El Moujahid, A.; Bouch, N.; El Ouazzani, R.; Bachaoui, E.M.; El Ghmari, A. Crop Type Mapping from Pansharpened Landsat 8 NDVI Data: A Case of a Highly Fragmented and Intensive Agricultural System. *Remote Sens. Appl. Soc. Environ.* **2018**, *11*, 94–103. [CrossRef]
4. Almazroui, M.; Nazrul Islam, M.; Saeed, S.; Alkhalaf, A.K.; Dambul, R. Assessment of Uncertainties in Projected Temperature and Precipitation over the Arabian Peninsula Using Three Categories of Cmip5 Multimodel Ensembles. *Earth Syst. Environ.* **2017**, *1*, 1–20. [CrossRef]
5. Driouech, F.; Déqué, M.; Sánchez-Gómez, E. Weather Regimes—Moroccan Precipitation Link in a Regional Climate Change Simulation. *Glob. Planet. Chang.* **2010**, *72*, 1–10. [CrossRef]
6. El Hafyani, M.; Essahlaoui, A.; Van Rompaey, A.; Mohajane, M.; El Hmaidi, A.; El Ouali, A.; Moudden, F.; Serrhini, N.-E. Assessing Regional Scale Water Balances through Remote Sensing Techniques: A Case Study of Boufakrane River Watershed, Meknes Region, Morocco. *Water* **2020**, *12*, 320. [CrossRef]
7. El Moçayd, N.; Kang, S.; Eltahir, E.A.B. Climate Change Impacts on the Water Highway Project in Morocco. *Hydrol. Earth Syst. Sci.* **2020**, *24*, 1467–1483. [CrossRef]
8. Lebrini, Y.; Boudhar, A.; Hadria, R.; Lionboui, H.; Elmansouri, L.; Arrach, R.; Ceccato, P.; Benabdelouahab, T. Identifying Agricultural Systems Using SVM Classification Approach Based on Phenological Metrics in a Semi-Arid Region of Morocco. *Earth Syst. Environ.* **2019**, *3*, 277–288. [CrossRef]
9. Barakat, A.; Ouargaf, Z.; Khellouk, R.; El Jazouli, A.; Touhami, F. Land Use/Land Cover Change and Environmental Impact Assessment in Béni-Mellal District (Morocco) Using Remote Sensing and GIS. *Earth Syst. Environ.* **2019**, *3*, 113–125. [CrossRef]
10. Mohajane, M.; Essahlaoui, A.; Oudija, F.; El Hafyani, M.; Hmaidi, A.E.; El Ouali, A.; Randazzo, G.; Teodoro, A.C. Land Use/Land Cover (LULC) Using Landsat Data Series (MSS, TM, ETM+ and OLI) in Azrou Forest, in the Central Middle Atlas of Morocco. *Environments* **2018**, *5*, 131. [CrossRef]
11. Rogan, J.; Chen, D. Remote Sensing Technology for Mapping and Monitoring Land-Cover and Land-Use Change. *Prog. Plan.* **2004**, *61*, 301–325. [CrossRef]
12. Sciortino, M.; De Felice, M.; De Cecco, L.; Borfecchia, F. Remote Sensing for Monitoring and Mapping Land Productivity in Italy: A Rapid Assessment Methodology. *CATENA* **2020**, *188*, 104375. [CrossRef]
13. Wu, Q.; Li, H.; Wang, R.; Paulussen, J.; He, Y.; Wang, M.; Wang, B.; Wang, Z. Monitoring and Predicting Land Use Change in Beijing Using Remote Sensing and GIS. *Landsc. Urban Plan.* **2006**, *78*, 322–333. [CrossRef]
14. Zhang, Y.; Ye, A. Spatial and Temporal Variations in Vegetation Coverage Observed Using AVHRR GIMMS and Terra MODIS Data in the Mainland of China. *Int. J. Remote Sens.* **2020**, *41*, 4238–4268. [CrossRef]
15. Randazzo, G.; Cascio, M.; Fontana, M.; Gregorio, F.; Lanza, S.; Muzirafuti, A. Mapping of Sicilian Pocket Beaches Land Use/Land Cover with Sentinel-2 Imagery: A Case Study of Messina Province. *Land* **2021**, *10*, 678. [CrossRef]

16. Gerard, F.; Petit, S.; Smith, G.; Thomson, A.; Brown, N.; Manchester, S.; Wadsworth, R.; Bugar, G.; Halada, L.; Bezák, P.; et al. Land Cover Change in Europe between 1950 and 2000 Determined Employing Aerial Photography. *Prog. Phys. Geogr.* **2010**, *34*, 183–205. [CrossRef]
17. Tavares, P.A.; Beltrão, N.; Guimarães, U.S.; Teodoro, A.; Gonçalves, P. Urban Ecosystem Services Quantification through Remote Sensing Approach: A Systematic Review. *Environments* **2019**, *6*, 51. [CrossRef]
18. Waldhoff, G.; Lussem, U.; Bareth, G. Multi-Data Approach for Remote Sensing-Based Regional Crop Rotation Mapping: A Case Study for the Rur Catchment, Germany. *Int. J. Appl. Earth Obs. Geoinf.* **2017**, *61*, 55–69. [CrossRef]
19. Wu, M.; Huang, W.; Niu, Z.; Wang, Y.; Wang, C.; Li, W.; Hao, P.; Yu, B. Fine Crop Mapping by Combining High Spectral and High Spatial Resolution Remote Sensing Data in Complex Heterogeneous Areas. *Comput. Electron. Agric.* **2017**, *139*, 1–9. [CrossRef]
20. Yue, J.; Tian, Q. Estimating Fractional Cover of Crop, Crop Residue, and Soil in Cropland Using Broadband Remote Sensing Data and Machine Learning. *Int. J. Appl. Earth Obs. Geoinf.* **2020**, *89*, 102089. [CrossRef]
21. Boualoul, M.; Randazzo, G.; Lanza, S.; Allaoui, A.; Ouardi, H.E.; Habibi, H.; Ouhaddach, H. The use of remote sensing for water protection in the karst environment of the Tabular Middle Atlas/the causse of El Hajeb/Morocco. In Proceedings of the IX Conference of the Italian Society of Remote Sensing, Firenze, Italy, 30 November–1 December 2018; Volume 2000, p. 16.
22. Muzirafuti, A.; Boualoul, M.; Barreca, G.; Allaoui, A.; Bouikbane, H.; Lanza, S.; Crupi, A.; Randazzo, G. Fusion of Remote Sensing and Applied Geophysics for Sinkholes Identification in Tabular Middle Atlas of Morocco (the Causse of El Hajeb): Impact on the Protection of Water Resource. *Resources* **2020**, *9*, 51. [CrossRef]
23. Ouardi, H.E.; Boualoul, M.; Ouhaddach, H.; Habibi, M.; Muzarafuti, A.; Allaoui, A.; Amine, A. Fault Analysis and Its Relationship with Karst Structures Affecting Lower Jurassic Limestones in the Agourai Plateau (Middle Atlas, Morocco). *Geogaceta* **2018**, *63*, 119–122.
24. Wieland, M.; Martinis, S. Large-Scale Surface Water Change Observed by Sentinel-2 during the 2018 Drought in Germany. *Int. J. Remote Sens.* **2020**, *41*, 4742–4756. [CrossRef]
25. Bachri, I.; Hakdaoui, M.; Raji, M.; Teodoro, A.C.; Benbouziane, A. Machine Learning Algorithms for Automatic Lithological Mapping Using Remote Sensing Data: A Case Study from Souk Arbaa Sahel, Sidi Ifni Inlier, Western Anti-Atlas, Morocco. *ISPRS Int. J. Geo-Inf.* **2019**, *8*, 248. [CrossRef]
26. Du, P.; Tan, K.; Xing, X. Wavelet SVM in Reproducing Kernel Hilbert Space for Hyperspectral Remote Sensing Image Classification. *Opt. Commun.* **2010**, *283*, 4978–4984. [CrossRef]
27. Liu, Y.; Zhang, B.; Wang, L.; Wang, N. A Self-Trained Semisupervised SVM Approach to the Remote Sensing Land Cover Classification. *Comput. Geosci.* **2013**, *59*, 98–107. [CrossRef]
28. Maulik, U.; Chakraborty, D. A Self-Trained Ensemble with Semisupervised SVM: An Application to Pixel Classification of Remote Sensing Imagery. *Pattern Recognit.* **2011**, *44*, 615–623. [CrossRef]
29. Mohajane, M.; Costache, R.; Karimi, F.; Bao Pham, Q.; Essahlaoui, A.; Nguyen, H.; Laneve, G.; Oudija, F. Application of Remote Sensing and Machine Learning Algorithms for Forest Fire Mapping in a Mediterranean Area. *Ecol. Indic.* **2021**, *129*, 107869. [CrossRef]
30. Fathizad, H.; Ali Hakimzadeh Ardakani, M.; Sodaiezadeh, H.; Kerry, R.; Taghizadeh-Mehrjardi, R. Investigation of the Spatial and Temporal Variation of Soil Salinity Using Random Forests in the Central Desert of Iran. *Geoderma* **2020**, *365*, 114233. [CrossRef]
31. Iqbal, F.; Lucieer, A.; Barry, K. Poppy Crop Capsule Volume Estimation Using UAS Remote Sensing and Random Forest Regression. *Int. J. Appl. Earth Obs. Geoinf.* **2018**, *73*, 362–373. [CrossRef]
32. Izquierdo-Verdiguier, E.; Zurita-Milla, R. An Evaluation of Guided Regularized Random Forest for Classification and Regression Tasks in Remote Sensing. *Int. J. Appl. Earth Obs. Geoinf.* **2020**, *88*, 102051. [CrossRef]
33. Bekele, A.A.; Pingale, S.M.; Hatiye, S.D.; Tilahun, A.K. Impact of Climate Change on Surface Water Availability and Crop Water Demand for the Sub-Watershed of Abbay Basin, Ethiopia. *Sustain. Water Resour. Manag.* **2019**, *5*, 1859–1875. [CrossRef]
34. Birhanu, A.; Murlidhar Pingale, S.; Soundharajan, B.; Singh, P. GIS-Based Surface Irrigation Potential Assessment for Ethiopian River Basin. *Irrig. Drain.* **2019**, *68*, 607–616. [CrossRef]
35. Grammatikopoulou, I.; Sylla, M.; Zoumides, C. Economic Evaluation of Green Water in Cereal Crop Production: A Production Function Approach. *Water Resour. Econ.* **2020**, *29*, 100148. [CrossRef]
36. Lee, S.K.; Dang, T.A. Predicting the Water Use-Demand as a Climate Change Adaptation Strategy for Rice Planting Crops in the Long Xuyen Quadrangle Delta. *Paddy Water Environ.* **2019**, *17*, 561–570. [CrossRef]
37. Paymard, P.; Yaghoubi, F.; Nouri, M.; Bannayan, M. Projecting Climate Change Impacts on Rainfed Wheat Yield, Water Demand, and Water Use Efficiency in Northeast Iran. *Theor. Appl. Climatol.* **2019**, *138*, 1361–1373. [CrossRef]
38. Ruan, H.; Yu, J.; Wang, P.; Wang, T. Increased Crop Water Requirements Have Exacerbated Water Stress in the Arid Transboundary Rivers of Central Asia. *Sci. Total Environ.* **2020**, *713*, 136585. [CrossRef]
39. Severo Santos, J.F.; Naval, L.P. Spatial and Temporal Dynamics of Water Footprint for Soybean Production in Areas of Recent Agricultural Expansion of the Brazilian Savannah (Cerrado). *J. Clean. Prod.* **2020**, *251*, 119482. [CrossRef]
40. Surendran, U.; Sushanth, C.M.; Joseph, E.J.; Al-Ansari, N.; Yaseen, Z.M. FAO CROPWAT Model-Based Irrigation Requirements for Coconut to Improve Crop and Water Productivity in Kerala, India. *Sustainability* **2019**, *11*, 5132. [CrossRef]
41. Surendran, U.; Sushanth, C.M.; Mammen, G.; Joseph, E.J. Modelling the Crop Water Requirement Using FAO-CROPWAT and Assessment of Water Resources for Sustainable Water Resource Management: A Case Study in Palakkad District of Humid Tropical Kerala, India. *Aquat. Procedia* **2015**, *4*, 1211–1219. [CrossRef]

42. Moseki, O.; Murray-Hudson, M.; Kashe, K. Crop Water and Irrigation Requirements of *Jatropha curcas* L. in Semi-Arid Conditions of Botswana: Applying the CROPWAT Model. *Agric. Water Manag.* **2019**, *225*, 105754. [CrossRef]
43. Brouwer, C.; Heibloem, M. Irrigation Water Management: Irrigation Water Needs. *Train. Man.* **1986**, *3*. Available online: https://www.fao.org/3/s2022e/s2022e00.htm (accessed on 20 August 2021).
44. Vapnik, V. The Support Vector Method of Function Estimation. In *Nonlinear Modeling*; Springer: Boston, MA, USA, 1998; pp. 55–85.
45. Bahari, N.I.S.; Ahmad, A.; Aboobaider, B.M. Application of Support Vector Machine for Classification of Multispectral Data. *IOP Conf. Ser. Earth Environ. Sci.* **2014**, *20*, 012038. [CrossRef]
46. Breiman, L. Random Forests. *Mach. Learn.* **2001**, *45*, 5–32. [CrossRef]
47. Costache, R.; Hong, H.; Pham, Q.B. Comparative Assessment of the Flash-Flood Potential within Small Mountain Catchments Using Bivariate Statistics and Their Novel Hybrid Integration with Machine Learning Models. *Sci. Total Environ.* **2020**, *711*, 134514. [CrossRef]
48. Nefeslioglu, H.A.; Sezer, E.; Gokceoglu, C.; Bozkir, A.S.; Duman, T.Y. Assessment of Landslide Susceptibility by Decision Trees in the Metropolitan Area of Istanbul, Turkey. *Math. Probl. Eng.* **2010**, *2010*, 1–15. [CrossRef]
49. Rouse, J.W., Jr. Monitoring the Vernal Advancement and Retrogradation of Natural Vegetation. NASA/GSFCT Type II Report, Greenbelt, MD, USA. 1973. Available online: https://ntrs.nasa.gov/citations/19730020508 (accessed on 22 August 2021).
50. Rouse, J.W., Jr. Monitoring the Vernal Advancement and Retrogradation (Greenwave Effect) of Natural Vegetatioa NASA/GSFCT Type III Final Report, Greenbelt, MD, USA. 1974. Available online: https://ntrs.nasa.gov/citations/19740022555 (accessed on 22 August 2021).
51. Congalton, R.G. A Review of Assessing the Accuracy of Classifications of Remotely Sensed Data. *Remote Sens. Environ.* **1991**, *37*, 35–46. [CrossRef]
52. Stehman, S.V. Estimating the Kappa Coefficient and Its Variance under Stratified Random Sampling. *Photogramm. Eng. Remote Sens.* **1996**, *62*, 401–407.
53. Olofsson, P.; Foody, G.M.; Stehman, S.V.; Woodcock, C.E. Making Better Use of Accuracy Data in Land Change Studies: Estimating Accuracy and Area and Quantifying Uncertainty Using Stratified Estimation. *Remote Sens. Environ.* **2013**, *129*, 122–131. [CrossRef]
54. Foody, G.M. Explaining the Unsuitability of the Kappa Coefficient in the Assessment and Comparison of the Accuracy of Thematic Maps Obtained by Image Classification. *Remote Sens. Environ.* **2020**, *239*, 111630. [CrossRef]
55. Doorenbos, J.; Pruitt, W.O. *Guidelines for Predicting Crop Water Requirements*; FAO irrigation and drainage paper; Food and Agriculture Organization of the United Nations: Rome, Italy, 1977; ISBN 978-92-5-100279-7.
56. ABHS. *Etude D'actualisation du Plan Directeurd'Aménagements Intégrés des Ressources en Eau (PDAIRE) du Bassin Hydraulique de Sebou*; ABHS: Fez, Morocco, 2011.

applied sciences

MDPI

Article

Comparative Analysis of Machine Learning Algorithms in Automatic Identification and Extraction of Water Boundaries

Aimin Li [1,*], Meng Fan [2], Guangduo Qin [2], Youcheng Xu [2] and Hailong Wang [2]

[1] School of Geo-Science and Technology, Zhengzhou University, Zhengzhou 450001, China
[2] School of Water Conservancy Engineering, Zhengzhou University, Zhengzhou 450001, China; fanm15755409160@163.com (M.F.); gd13223022138@163.com (G.Q.); chengyouxu1998@163.com (Y.X.); whl18307065356@163.com (H.W.)
* Correspondence: aiminli@zzu.edu.cn

Abstract: Monitoring open water bodies accurately is important for assessing the role of ecosystem services in the context of human survival and climate change. There are many methods available for water body extraction based on remote sensing images, such as the normalized difference water index (NDWI), modified NDWI (MNDWI), and machine learning algorithms. Based on Landsat-8 remote sensing images, this study focuses on the effects of six machine learning algorithms and three threshold methods used to extract water bodies, evaluates the transfer performance of models applied to remote sensing images in different periods, and compares the differences among these models. The results are as follows. (1) Various algorithms require different numbers of samples to reach their optimal consequence. The logistic regression algorithm requires a minimum of 110 samples. As the number of samples increases, the order of the optimal model is support vector machine, neural network, random forest, decision tree, and XGBoost. (2) The accuracy evaluation performance of each machine learning on the test set cannot represent the local area performance. (3) When these models are directly applied to remote sensing images in different periods, the AUC indicators of each machine learning algorithm for three regions all show a significant decline, with a decrease range of 0.33–66.52%, and the differences among the different algorithm performances in the three areas are obvious. Generally, the decision tree algorithm has good transfer performance among the machine learning algorithms with area under curve (AUC) indexes of 0.790, 0.518, and 0.697 in the three areas, respectively, and the average value is 0.668. The Otsu threshold algorithm is the optimal among threshold methods, with AUC indexes of 0.970, 0.617, and 0.908 in the three regions respectively and an average AUC of 0.832.

Keywords: water extraction; modified normalized difference water index (MNDWI); remote sensing; machine learning algorithm

Citation: Li, A.; Fan, M.; Qin, G.; Xu, Y.; Wang, H. Comparative Analysis of Machine Learning Algorithms in Automatic Identification and Extraction of Water Boundaries. *Appl. Sci.* **2021**, *11*, 10062. https://doi.org/10.3390/app112110062

Academic Editors: Dimitrios S. Paraforos and Anselme Muzirafuti

Received: 31 August 2021
Accepted: 25 October 2021
Published: 27 October 2021

1. Introduction

Water is the source of life: the earth's surface open water body accounts for about 74% of the total earth area, it is an important resource for all life survival, and it is also the most important component of living organisms [1,2]. In China, the distribution of water resources is quite uneven, and the pollution situation is serious. So, how to identify water bodies efficiently and accurately has become a severe issue [3,4].

With the rapid development of aviation and aerospace technology, remote sensing technology has provided advanced support for many fields, including resource survey, environmental monitoring, mapping, and geography [5,6].

The development of remote sensing technology makes it possible to extract water information quickly and accurately, which is substantially different from conventional field survey methods employed in the past [7–10].

Monitoring open water bodies accurately is an important and basic application in remote sensing. Various water body mapping approaches have been developed to extract water bodies from multispectral images [11–13]. Using remote sensing images to monitor a water body is mainly based on spectral bands and each image's spatial feature, so the identification methods can be categorized into three types from different perspectives.

(1) Water body index method: This method is based on the spectral curves of water bodies, and thresholds are utilized to effectively distinguish water bodies from the background [14]. Different water indexes have already been proposed in the past few decades. Specifically, in 1996, McFeeters [15] introduced the normalized difference water index (NDWI) model to extract water bodies. However, this model is unable to distinguish between dark shadow and water bodies. To overcome the shortcomings of NDWI, in 2006, Xu [16] proposed the modification of normalized difference water index (MNDWI) to enhance open water features in remotely sensed imagery, and this model has better results for urban water bodies extraction. The water body index method has the characteristics of high precision and low computational cost, which has been widely used in practical applications. In the last few decades, the MNDWI of Xu is one of the most widely used water indices for various fields, including surface water mapping, land use/cover change analyses, and ecological research [17–20].

(2) Machine learning methods: These methods feature pixel-based pattern recognition analysis, mainly including supervised and unsupervised classification techniques. The supervised methods mainly include neural network [21–25], support vector machine (SVM) [26–28], logistic regression [29,30], and random forest [31–33], and the unsupervised classification methods mainly include K-means clustering [34] and ISODATA clustering [35,36] methods. The machine learning algorithm has been widely used in remote sensing water extraction due to its high accuracy.

(3) Object-based image analysis methods (OBIA): Due to the limitations of pixel-based classification methods, such as the salt and pepper phenomenon in classification results, object-based classification techniques have been increasingly applied in remote sensing classification in recent years [37,38]. Many successful cases of water body extraction using OBIA methods have been reported [39–43]. Given that urban functional zones (UFZs) are composed of diverse geographic objects, Du et al. [44] presented a novel object-based UFZ mapping method using very-high-resolution (VHR) remote sensing images. Based on object-oriented analysis technology and multi-source data, Guo et al. [45] proposed a multi-level classification scheme based on goals and rules to study the changes of glacier environments.

In addition, some studies also have used synthetic aperture radar (SAR) data to monitor the surface dynamics, because these data are insensitive to clouds [14,46,47]; the area of surface water can be extracted from SAR data based on textural analysis [48], change detection [49], automatic segmentation [50], and classification [51].

At present, machine learning algorithms to extract water bodies mainly include neural networks, support vector machines, and random forest algorithms. The studies carried out in the past have identified the best performing classification algorithm by comparing different classification algorithms. However, none of them provides a comprehensive comparative analysis of some popular classification algorithms [37,52].

There are few studies on the evaluation of the transfer performance of each machine learning algorithm applied to remote sensing images in different periods. Based on Landsat-8 images, this study uses machine learning algorithms such as decision tree, logistic regression, random forest, and neural network to extract water bodies. First of all, the effect of each machine learning algorithm on the test set is discussed. After that, each machine learning algorithm is applied to three different local areas, and its effect on each local area is evaluated. At last, each machine learning algorithm is applied to remote sensing images in different periods to evaluate the model transfer performance of each machine learning algorithm, and three threshold methods are compared. The results could shed light on the future work of water body extraction based on remote sensing.

2. Data and Pre-Processing

2.1. Data

Landsat-8 data from the website (http://glovis.usgs.gov/ (accessed on 20 October 2021)) of the United States Geological Survey are used. Landsat-8, launched as a collaboration between the United States Geologic Survey (USGS) and National Aeronautics and Space Administration (NASA) on 11 February 2013, carries onboard the OLI push broom multispectral radiometer [53]. As shown in Table 1, the Landsat-8 OLI/TIRS imagery has 11 spectral bands in total, including eight spectral bands (i.e., three visible bands, two bands for describing aerosol, water vapor, and cirrus clouds, two short-wave infrared bands (SWIR) and near infrared (NIR)) with spatial resolution of 30 m, one panchromatic spectral band with a spatial resolution of 15 m, and two thermal spectral bands with a spatial resolution of 100 m [54]. Landsat-8 remote sensing images (path 123; raw 039) of the same area acquired on 4 October 2019 and 20 October 2019 are used in our experiment. Specifically, the data on 20 October 2019 are used to establish the model and compare the effect of each algorithm, and the data on 4 October 2019 are used to examine the performance of model transfer. Three different areas with different surface features are selected from remote sensing images. As shown in Figure 1, Area1 has a large area of water with relatively simple surface object types, while Area2 has a small water area and complex surface environment, and its water extraction is affected by numerous vegetation and mountain shadow. Area3 is located in the urban built-up area and has multiple contiguous water bodies; thus, the water extraction is affected by nearby buildings and roads.

To avoid the effects of too many clouds and aerosol, images with fewer clouds are selected here. All original data are processed by converting the original digital number (DN) value into spectral radiance, through Equation (1) [55]. The formula is given as follows:

$$L_\lambda = M_L \cdot Q_{cal} + A_L \tag{1}$$

where:

L_λ = spectral radiance($W/m^2 \cdot sr \cdot um$);

M_L = radiance multiplicative scaling factor for the spectral band(radiance_mult_band_n from the metadata);

A_L = radiance additive scaling factor for the spectral band(radiance_add_band_n from the metadata);

Q_{cal} = raw digital numbers (DN).

Table 1. Spectral band spatial resolution and wavelength of the Landsat-8 image.

Landsat-8 OLI and TIRS Bands	Wavelength (um)	Spatial Resolution (m)
Coastal/Aerosol	0.435–0.451	30
Blue	0.452–0.512	30
Green	0.533–0.590	30
Red	0.636–0.673	30
NIR	0.851–0.879	30
SWIR-1	1.566–1.651	30
TIR-1	10.60–11.19	100
TIR-2	11.50–12.51	100
SWIR-2	2.107–2.294	30
Pan	0.503–0.676	15
Cirrus	1.363–1.384	30

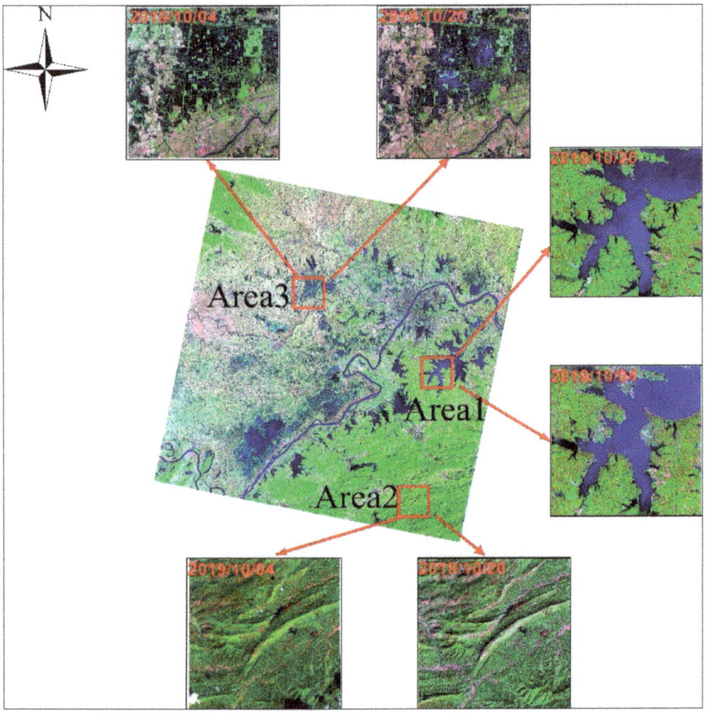

Figure 1. Landsat-8 remote sensing images are displayed in false color in bands 7, 5, and 3. Three local areas are extracted from this image. Area1 has a large area of water distribution with a simple ground environment and is only affected by vegetation; Area2 is affected by mountain shadow and vegetation; Area3 is located in the urban built-up area with scattered water distribution and is affected by roads and buildings.

2.2. Pre-Processing

By adopting spectral band combinations 7/5/4, 7/4/3, 6/5/4, and 4/3/2 combined with visual interpretation, a sample dataset is selected from Landsat images for classification; the sample set contains 340 water samples and 454 non-water samples. To avoid the influences of heterogeneous categories in the subsequent classification, the ratio of other ground object samples to the water body samples remains at 1.3:1.

The characteristics of the data, such as a large correlation between multiple spectral bands in the original images and similar information and structures between different spectral bands, generally bring significant amounts of redundancy. For this reason, principal component analysis (PCA) for dimensionality reduction is applied to remove repetitive and redundant information between various spectral bands [56]. The first and second principal components in the PCA with a cumulative variance contribution of 99% are selected as classification characteristics.

Based on the PCA, four generally used texture features, i.e., contrast, autocorrelation, dissimilarity, and entropy are extracted. The distance is set to be 1 pixel (distance of 30 m), 2 pixels (distance of 60 m), and 3 pixel (distance of 90 m), and 3×3, 5×5, 7×7, and 9×9 are selected as windows with orientations of 0°, 45°, 90°, and 135°. Optimal combined features are selected as the characteristic spectral bands for water body extraction. When the two parameters—i.e., window size and distance—increase, the edges of the images get fuzzy, and the window size shows more effects than distance. Considering the factors of ground objects correlation and image resolution, we set the distance to 1 pixel and select a 3×3 window with four orientations of 0°, 45°, 90°, and 135°.

After the size and window parameters are determined, J-M distance [57,58] and transformed divergence [59] (T-D) in many extracted texture features are used for studying the separability of ground objects; thus, the characteristics ultimately used for classification are determined as well. As shown in Table 2, the separability of the first component (PCA1) and the second component (PCA2) is compared in detail, and the separability of J-M dissimilarity in PCA2 is the optimal. Therefore, in later classifications, a total of six characteristics are selected.

Table 2. Separability of the samples.

	PCA1		PCA2	
	J-M	T-D	J-M	T-D
Contrast	1.723	1.9991	1.825	1.0000
Autocorrelation	1.404	1.6832	1.787	1.989
Dissimilarity	1.562	1.9232	1.84	1.0000
Entropy	1.634	1.9746	1.816	1.999

3. Research Methods

First of all, the performance of machine learning algorithms with a different sample number is discussed. During this process, the optimal parameters of the models are determined and the indices, such as precision and AUC, are used to evaluate the performances of algorithms in the test set. Then, according to spectral characteristics, the water indices are constructed, and on this basis, thresholds are selected; thus, water bodies and other ground objects are classified and identified. Moreover, machine learning methods, such as SVM, decision tree, and random forest, are used to extract water bodies. At last, the accuracy of the test results is verified for the same area at different times.

3.1. MNDWI

In 2006, Xu [16] presented a modification of normalized difference water index (MNDWI) (Equation (2)) by replacing the NIR spectral band used in NDWI with the SWIR spectral band to reduce the influence of building information on water bodies. By using the MNDWI water index method, the MNDWI image is binarized by selecting an appropriate threshold to achieve water bodies extraction. The determination of thresholds affected the accuracy of water body extraction, and different thresholds might be made by subjective judgments of different people. To reduce such influences, three methods for determining thresholds are used for comparison and discussion. The three threshold methods used in this article are as follows: (1) the user-defined threshold method, which is determined according to visual effect through multiple experiments; (2) the Otsu threshold method [60,61]; and (3) the adaptive threshold method, which is used to scan the image through a 3*3 window.

The MNDWI is expressed as follows:

$$\text{MNDWI} = \frac{\text{GREEN} - \text{SWIR}}{\text{GREEN} + \text{SWIR}} \tag{2}$$

where Green is the radiance of the green band, which corresponds to the 3rd Landsat-8 image band; SWIR represents the short-wave infrared band radiance, namely band 6 of the Landsat-8 image.

3.2. Machine Learning Algorithms

In this research, six machine learning algorithms are selected, all of them used the same group of sample set, and the whole samples are divided into a training set and a test set by the ratio of 7:3. Furthermore, in the process of model training, the relevant parameters of the models are further trained by using 10-fold cross-validation with hierarchical sampling of the training set. Finally, some indices, such as accuracy, recall rate [62], and AUC [63], are utilized to assess the results.

3.2.1. SVM

SVM has a simple structure but a strong generalization ability to solve problems with high-dimensionality, small sample numbers [64,65]. In this study, the Gaussian radial basis function is selected as the kernel function. By using the grid search method in combination with 10-fold cross-validation, the optimal parameters are determined as $C = 3$ and $\gamma = 0.003$.

3.2.2. Decision Tree

The decision tree determines the categories of the samples in the dataset by assigning the sample data to a certain leaf node. There are many methods for constructing the decision tree, but all of them are based on the different purity indices selected and sample attributes for classification [66]. The algorithms ID3, C4.5, C5.0, etc. are generally used. A classification and regression tree (CART) algorithm is used in this study, and pre-pruning is utilized to avoid the overfitting problem. The parameters mainly include the limited depth of the decision tree, the minimum sample number of leaf nodes, and the least sample number of separable leaf nodes. By using the grid search method and 10-fold cross-validation, the final parameters are determined as follows: the entropy is selected as the purity index and the maximum depth is 7. The lowest sample number of separable leaf nodes is 8, and the minimum sample number of leaf nodes is 1.

3.2.3. Multi-Hidden-Layer Neural Network

The neural network uses specific learning algorithms to learn from data through many learning algorithms; however, the network is generally trained by iteratively modifying connection weights and deviations until the error between the output generated by the network and the expected output is smaller than some specified threshold [21]. The input characteristics are passed to the next layer of nerve cells through a non-linear activation function and then continue to be passed down after activation of the nerve cells in this layer. That process is repeated and cycled to the output layer. The repeated superposition of these non-linear functions ensures that the neural network has sufficient non-linear fitting ability, while different activation functions can affect the output of different neural networks. By selecting a sigmoid activation function, it is determined that the neural network structure should have four layers based on multiple tests through cross-validation. Except for input and output layers, the numbers of nerve cells in the two hidden layers are eight and six, respectively.

3.2.4. Random Forest

The random forest is an ensemble method specially designed for a decision tree classifier, and the selection of random attributes is further added to its training process. Using similar parameters to those used for the decision tree, the random forest model is easy to implement and shows good effects [32,33]. In this research, parameters are determined by using cross-validation and grid search methods. The main parameters of random forest are as follows, there are 10 weak estimators in the decision tree, and the maximum depth is 4. Moreover, a Gini function is selected as the purity index.

3.2.5. XGBoost

The core of XGBoost is an ensemble algorithm based on the gradient boosting decision tree (GBDT), and it can be used for classification or regression problems. Its modelling process is as follows: a decision tree is built, and one more tree is added upon each iteration to form a strong evaluator integrating many numerical models [67,68]. The accuracy is superior to that of a weak estimator, and its calculation speed and performance are good [69]. The main parameters are set as follows: the maximum depth of each tree is 3, and a weak classification estimator with 300 decision trees is established. The learning rate is set to be 0.01.

3.2.6. Logistic Regression Algorithm

The logistic regression is a type of classification model. It establishes a regression formula for samples and a sigmoid function is used for classification. For more information, please refer to references [70,71].

4. Experiment and Analysis

4.1. *Effects of the Sample Number on Learning Algorithms*

For each classification algorithm in machine learning, the basic requirement is that the training and test set are reliable and there are enough samples for training. In this way, a good classifier can be trained. It is assumed that the samples selected by visual interpretation are reliable: namely, the various classes of the sample points are assigned to correct labels. Based on this, a small sample is randomly selected from the training set and divided into a training set and a validation set in the proportion of 7:3. By using the accuracy of the validation set of the small sample as an evaluation index, the effects of the sample number on the classification effects of each algorithm are discussed, so as to judge whether the sample number selected is sufficient to achieve the purpose of the training model.

As demonstrated in Figure 2, the accuracies of the classification algorithms in the validation set of the experiment all tend to increase with the sample number, and they show a smaller error relative to the accuracy in the training set. Moreover, the accuracies gradually tend to be equal. This indicates that there is almost no underfitting of the samples, and the parameters of each algorithm are well adjusted. The accuracy of the logistic regression algorithm is improved rapidly, approximating to the accuracy in the training set when the sample number is small, suggesting that there is almost no overfitting. As the sample number increases, the accuracy stabilizes; however, other classification algorithms need larger samples to achieve this stability, and the accuracy fluctuates (albeit within a small range), therefore, the number of training samples selected in the experiment can meet the needs of model training.

4.2. *Analysis of Performance Indices of Machine Learning Algorithms*

After testing the performance of the models when using each algorithm on sets of different sample numbers, the effect of each model in the same test set is further evaluated, so as to reflect the predictive abilities of the models to some extent and judge the generalization abilities of the algorithms. As shown in Table 3, the value of the accuracy index and recall index of each model in classifying water bodies and other ground objects are high, the accuracy index is in the range of 0.945–1, and the recall index is in the range of 0.911–1. However, the AUC index can better represent the comprehensive performances of the models and the higher the value, the better the performance [63]. There is little difference in the effect of each machine learning algorithm on the test set, and the AUC index ranges from 0.956 to 0.987; by analyzing AUC data, the logistic regression and XGBoost algorithm are found to perform best on the test set, followed by the SVM, the neural network, then the random forest, while the decision tree has (in general) the worst performance. Whether the evaluation of these algorithms in the test set can accurately represent the generalization abilities of the algorithms for classifying water bodies in the remote sensing images needs to be discussed and studied using remote sensing images acquired under different conditions.

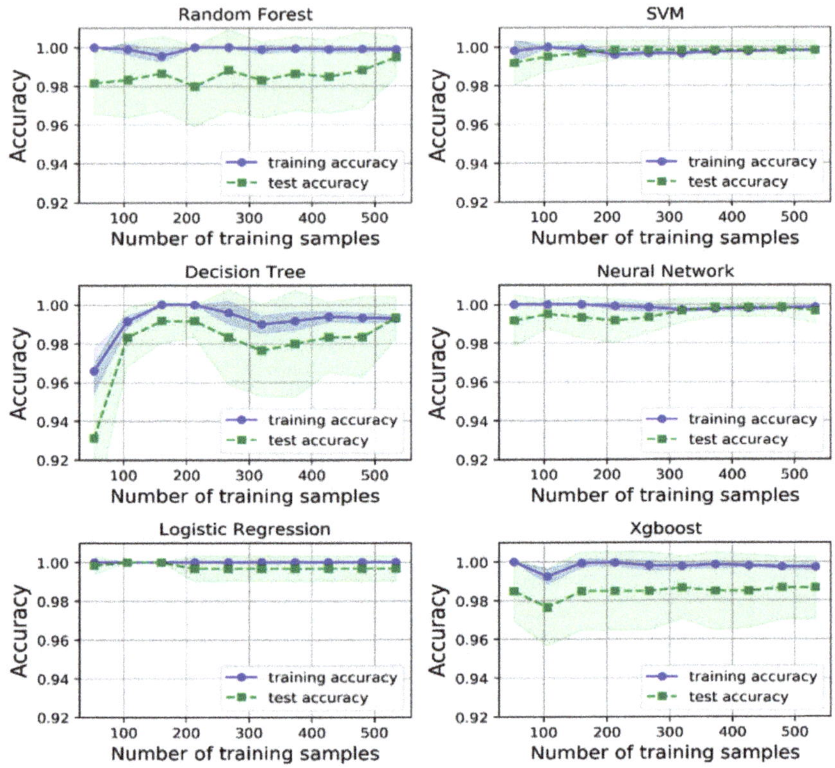

Figure 2. Effects of the sample number on performance of each algorithm.

Table 3. Analysis of performance indices of each algorithm.

		Accuracy	Recall	AUC
SVM	Water	1.000	0.949	0.983
	Other	0.968	1.000	
Random Forest	Water	0.975	0.975	0.979
	Other	0.983	0.983	
Decision Tree	Water	1.000	0.911	0.956
	Other	0.945	1.000	
Neural Network	Water	1.000	0.962	0.981
	Other	0.976	1.000	
Logistic Regression	Water	1.000	0.975	0.987
	Other	0.984	1.000	
XGBoost	Water	1.000	0.975	0.987
	Other	0.984	1.000	

4.3. Comparative Analysis of NDWI and Machine Learning Algorithms

The model established by 2019/10/20 training data is used for water extraction in three areas of 2019/10/20. Statistical results of AUC indicators of each algorithm are shown in Figure 3 (For more details, see Tables A1–A4 in the Appendix A). In general, the XGBoost algorithm has the best accuracy, with an average AUC of 0.966, and the AUC indicators in the three regions are 0.985, 0.972, and 0.941 respectively, which is followed by the random forest algorithm with an average AUC of 0.964, and the AUC indicators in the three regions are 0.985, 0.973, and 0.935; the SVM algorithm has the worst accuracy, the average AUC is 0.898 and the AUC indicators in the three regions are 0.982, 0.789, and

0.923, respectively. When each machine learning algorithm is applied to three different local regions, the average range of AUC index is 0.898–0.966 (for more details, see Table A1 in Appendix A), and the descending order of each machine learning algorithm is XGBoost, random forest, decision tree, logistic regression, neural network, and SVM according to the value of the AUC index. However, this is inconsistent with the conclusion of Section 4.2. In Section 4.2, there is little difference in the accuracy of each machine learning algorithm on the test set, and the AUC index ranges from 0.956 to 0.987. The machine learning algorithms are XGBoost, LR, SVM, NN, RF, and DT in descending order according to the value of the AUC index. It further explains that the evaluation on the test set cannot represent the effect of each algorithm applied in a local area. Among the threshold classification methods, the Otsu threshold algorithm is the best, with an average AUC of 0.957, and the AUC indicators in the three regions are 0.985, 0.922, and 0.964, respectively, followed by the custom threshold algorithm, and the worst performance among all algorithms is adaptive threshold algorithm: the average AUC is only 0.764.

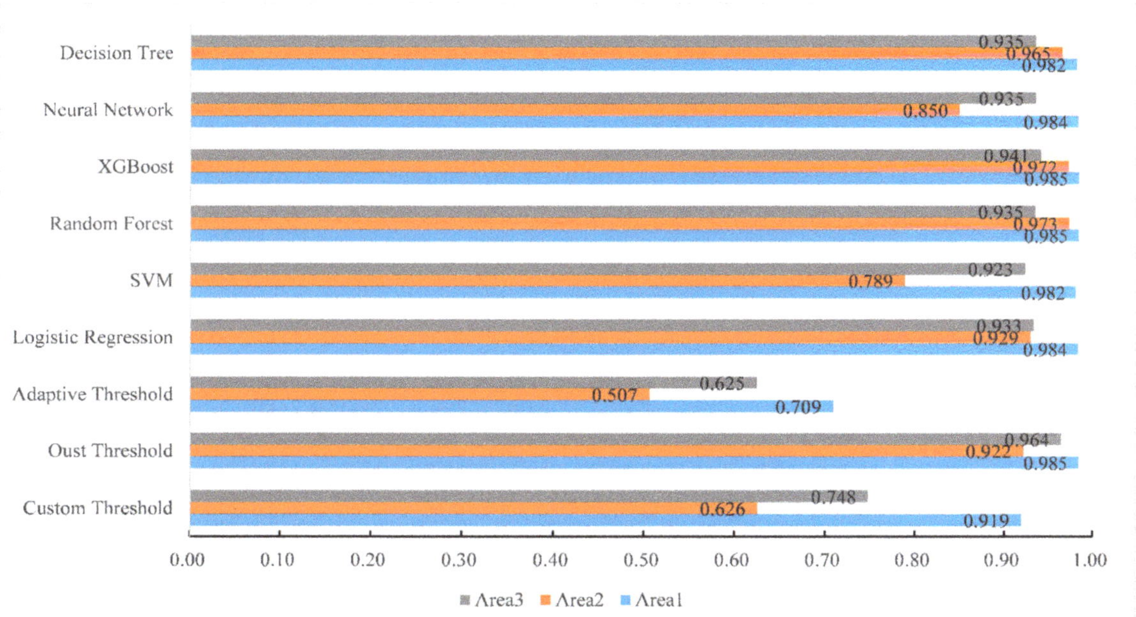

Figure 3. Statistics of the AUC index of each algorithm applied in the three regions.

The image water extraction results of each algorithm were placed in the supplementary materials, as shown in Figure S1: Classification results of each algorithm in Area1 on October 20; Figure S2: Classification results of each algorithm in Area2 on October 20; Figure S3: Classification results of each algorithm in Area3 on October 20. As can be seen from the results graph, compared with other algorithms, the salt and pepper phenomenon for the adaptive threshold and custom threshold is very serious, there is a large number of non-water body "noise", other algorithms basically have the same visual interpretation effect, and there is no obvious difference, but the edge part is slightly different due to the influence of adjacent features.

4.4. Reliability Test

To discuss the effects of the aforementioned algorithms in water body extraction from remote sensing images in different periods, a remote sensing image captured on 4 October 2019 in the same region is selected. Based on this, the water bodies are classified using the same algorithms and parameters. The aim is to verify whether the experimental results of each algorithm under different image conditions are reliable and decide whether the models are universal.

The model established by the data of 2019/10/20 is used in the data of 2019/10/04 for water body extraction. The statistical results of the AUC indicators of each algorithm are shown in Figure 4 (for more details, see Tables A5–A8 in Appendix A). As shown in Table 4, the AUC indicators of each machine learning algorithm for three regions all show a significant decline, with a decreased range of 0.33–66.52% As shown in Figure 4, the differences among the different algorithm performances in the three areas are obvious. In the surface complex Area2, the AUC index of the machine learning algorithms is near 0.5, which means it is difficult to extract water bodies accurately. In Area1 with a simple surface environment, although the accuracy of all machine learning algorithms decreases, the errors are still within an acceptable range. In general, the decision tree algorithm has better transfer performance, with an average AUC of 0.668, and the AUC indexes of the three regions are 0.790, 0.518, and 0.697 respectively. The XGBoost algorithm has an average AUC of 0.631, and its AUC index in the three regions is 0.718, 0.512, and 0.665, respectively. The logistic regression algorithm has the worst accuracy, with an average AUC of 0.392, the AUC index in the three regions is 0.329, 0.489, and 0.357, respectively, which is inconsistent with the conclusion in Sections 4.2 and 4.3. When the model is directly transferred to remote sensing images of different periods for water extraction, the generalization ability of each machine learning algorithm is different. Among the threshold classification methods, the Otsu threshold algorithm is optimal, and its average AUC is 0.832. The AUC indexes in the three regions are 0.970, 0.617, and 0.908, respectively, which exceed the accuracy of the other machine learning algorithms. For the other two threshold algorithm, custom threshold, whose average AUC is 0.700, and the AUC indexes in the three regions are 0.842, 0.549, and 0.708 respectively. The adaptive threshold algorithm has an average AUC of 0.611, and its AUC indicators in the three regions are 0.703, 0.506, and 0.623 respectively. All in all, for different periods of remote sensing images, the threshold method is better than most of the machine learning algorithms, because the sensor imaging is affected by clouds, sun angles, and sensors. Due to the influence of the angle and other factors, the characteristics of remote sensing images will be very different during the adjacent imaging time. Even if there is no major change in the surface features, the pixel value of the remote sensing image could also change significantly. Therefore, the machine learning models trained on the data of 2019/10/20 may not be suitable for different periods.

Table 4. AUC index changes statistics of each machine learning algorithm.

The Method Name		Area1	Area2	Area3	Average
Threshold Method	Custom Threshold	−8.41%	−12.27%	−5.37%	−8.68%
	Otsu Threshold	−1.49%	−33.10%	−5.78%	−13.46%
	Adaptive Threshold	−0.92%	−0.16%	−0.33%	−0.47%
Machine Learning Method	Logistic Regression	−66.52%	−47.37%	−61.74%	−58.54%
	SVM	−66.25%	−32.20%	−60.49%	−52.98%
	Random Forest	−30.12%	−49.73%	−62.68%	−47.51%
	XGBoost	−27.16%	−47.32%	−29.35%	−34.61%
	Neural Network	−30.13%	−42.96%	−62.13%	−45.07%
	Decision Tree	−19.60%	−46.28%	−25.40%	−30.43%

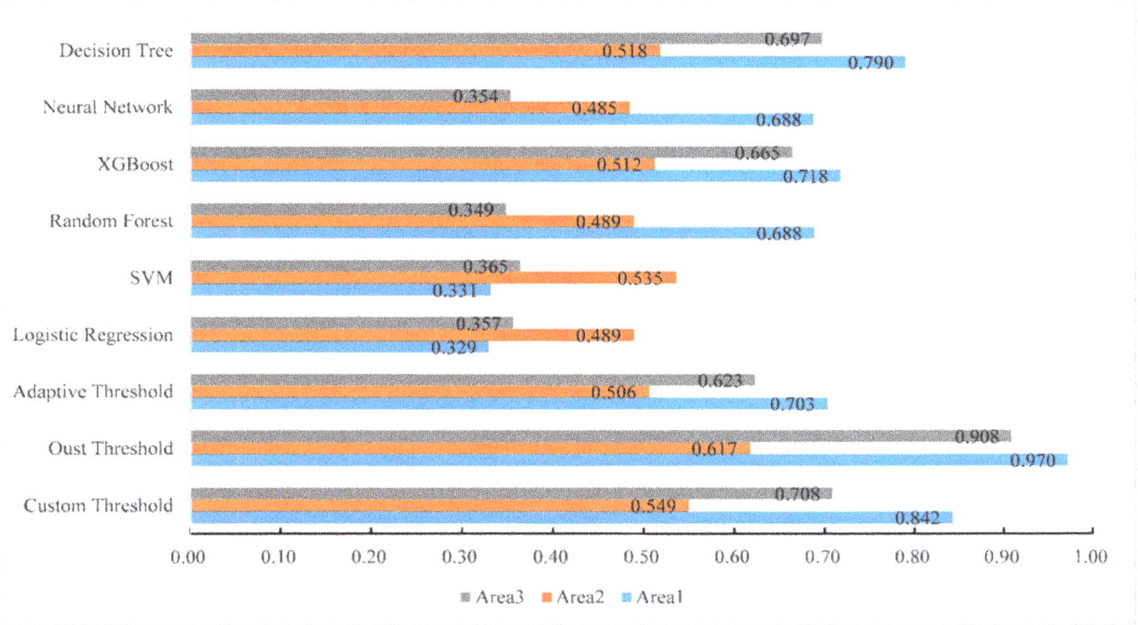

Figure 4. The AUC indexes of the three regions in different periods for each algorithm.

However, the water extraction effect of the threshold method is related to the remote sensing image data, and the water extraction effects of remote sensing images from different periods do not affect each other.

The water extraction results of each algorithm were placed in the supplementary materials, as shown in Figure S4: Classification results of each algorithm in Area1 on October 4; Figure S5: Classification results of each algorithm in Area2 on October 4; Figure S6: Classification results of each algorithm in Area3 on October 4. It can be seen from the classification result diagrams that most of the machine learning pepper and salt phenomenon is very serious, and there is a large number of non-water "noise". The visual effects of various algorithms are also significantly different.

5. Discussion

This study mainly selects neural network, support vector machine (SVM), logistic regression, random forest, decision tree, and XGBoost from machine learning algorithms, and it selects the MNDWI water index combined with three threshold methods to extract the water bodies. Michael Schmitt [72] pointed out that for a simple surface environment, only the threshold method can achieve satisfactory results, and when the surface environment is slightly more complicated, a supervised classification method, such as SVM, needs to be introduced. However, for the supervised classification method, how to choose the appropriate number of samples is a problem worthy of research. For example, Deepakrishna Somasundaram et al. [73] selected 3765 water samples and 2685 non-water samples from the four-view Landsat-8 OLI image; Wei Jiang et al. [74] selected more than 10,000 water samples and non-water samples in each study area. The choice of these large numbers of training samples brings additional costs. In order to study the influence of sample size on various algorithms, an experiment was designed in this paper, as outlined in Section 4.1. As shown in Figure 2, there are great differences in the number of samples required for various algorithms to reach their optimal. The logistic regression algorithm requires the lowest number of samples, which is close to 110. The SVM algorithm has the best performance when the number of samples reaches 150. As the number of samples increases, the order of the optimal model is neural network, random forest, decision tree, and XGBoost. The primary task of water body extraction is to select a certain number of samples for the training model. The conclusion of the sample number requirements of each machine learning algorithm in this paper can be used as a reference for other similar applications to reduce the cost of sample selection.

Most studies only use test set samples to evaluate the optimal model and use the selected model for the final classification of images. However, Liu Yang et al. [75] pointed out that in different surface environments, various types of shadows or background noises need to be considered. For example, compared with arid areas, the influence of vegetation on water extraction should be considered in humid areas. In mountainous areas, the extracted water is often mixed with mountain shadow. These types of background information have different influences on different water extraction algorithms [61,76]. For the above reasons, it is worth discussing whether the evaluation effect on the test set can explain the actual generalization performance of the model, that is, whether the evaluation effect on the test set is consistent with the evaluation effect on the local area. For this reason, three local areas with different ground conditions are selected. As shown in Figure 3, in general, the simpler the ground scene, the better the classification accuracy. If the ground scene is complex, the accuracy of various algorithms has a great difference. Generally, three algorithms (decision tree, XGBoost, and Otsu) can perform well in various scenarios. In the case of mountain shadow in the ground background, it is suggested to give priority to the XGBoost algorithm. In the case of roads and buildings in the ground background, besides the XGBoost or decision tree algorithms, a logistic regression algorithm with a relatively simple model can also be tried.

However, when multi-stage extraction research on water bodies is needed, the original model will naturally be directly used to extract water bodies from remote sensing images in other different periods. As shown in Table 4, when various machine learning algorithms are directly used to extract water bodies from remote sensing images in different periods, the AUC indicators of each machine learning algorithm for the three regions all show a significant decline, with a decrease range of 0.33–66.52%. Generally, simple ground scenes have higher accuracy, while complex ground scenes have some effects for different machine learning algorithms. As shown in Table 4, among all the machine learning algorithms, the accuracy of decision tree decreased the least in the three regions on average, and the AUC index decreased 30.43% on average, followed by XGBoost. In the threshold method, although the change of adaptive threshold is small, its accuracy is always very low, while the Otsu algorithm not only has a good accuracy, but also the average decline of the AUC index is small, which is 13.46%. The decision tree algorithm can still achieve better classification results, and the Otsu algorithm also performs well. Experiments show that it is not recommended to directly use the machine learning model to extract water from remote sensing images in different periods. The Otsu classification result can be used as a reference, so that training samples can be selected in other periods quickly and conveniently to extract water bodies using machine learning algorithms.

In summary, for water extraction from remote sensing images, although various algorithms can achieve satisfactory results under certain conditions, none of them can be applied to all remote sensing image and scenes. The factors affecting the classification accuracy of remote sensing images mainly include the complexity of the field landscape, the availability of data, the effectiveness of the processing method, and the experience judgment of the processing personnel [5,76]. Therefore, on the basis of this study, when extracting water from remote sensing images, the water index (MNDWI preferred) can be used first and combined with the Otsu algorithm to classify water bodies. This result is in agreement with the results obtained by Ya'nan Zhou et al. [38], who used the NDWI image to select water samples from the input image. However, if the accuracy does not meet the requirements of the application, on the basis of its classification, researchers can further select the number of samples that meet the requirements of various machine learning algorithms (Figure 2) and select the corresponding machine learning training model. Among the various machine learning algorithms, XGBoost, decision tree, and logistic regression algorithms are preferentially recommended.

6. Conclusions

Based on Landsat-8 images, decision tree, logistic regression, random forest, neural network, support vector machine, and XGBoost algorithms are used to extract water bodies. Firstly, the effect of each machine learning algorithm on the test set is discussed. Secondly, each machine learning algorithm is applied to three different local areas, and the consistency between the accuracy of each machine learning algorithm on the test set and the accuracy of the local area is evaluated. Finally, each machine learning algorithm is applied to remote sensing images in different periods, the model transfer performance of each machine learning algorithm is examined, and three threshold methods are compared. The following conclusions are drawn:

(1) There are great differences in the numbers of samples required for various algorithms to reach their optimal. The logistic regression algorithm requires a minimum number of samples, about 110. The SVM algorithm has the best performance when the number of samples reaches 150. As the number of samples increases, the optimal order of the model is neural network, random forest, decision tree, and XGBoost.

(2) The accuracy evaluation effect of each machine learning on the test set cannot represent the effect on the local area, because the surface complexity is not same in the three local areas. In Area1 with a single surface type, its AUC range is 0.982–0.985; in Area2 with complex surface environment (numerous vegetation and mountain shadow), its AUC range is 0.789–0.973; in Area3 with wide water distribution, its AUC range is 0.923–0.941 in an urban built-up area.

(3) When the models are directly applied to remote sensing images in different periods, the model accuracy is greatly reduced, the AUC indicators of each machine learning algorithm for three regions all show a significant decline, with a decreasing range of 0.33–66.52%. In general, among the machine learning algorithms, the decision tree algorithm has good transfer performance, with an average AUC of 0.668, and the AUC indexes in the three regions are 0.790, 0.518, and 0.697 respectively. Among the threshold methods, the Otsu threshold algorithm is the optimal, with an average AUC of 0.832 and AUC indexes in the three regions are 0.970, 0.617, and 0.908, respectively.

(4) Owing to the complex distribution of ground objects and many influential factors in the remote sensing image classification, it is difficult to collect small and dispersed water bodies in this research. This limits the performances of these models in the environment with many hill shadows and complex ground objects. The accuracy of these models needs to be further improved; more samples should be collected from images over different areas and periods to train the models in the future.

Supplementary Materials: The following are available online at https://www.mdpi.com/article/10.3390/app112110062/s1, Detailed descriptions of Figure S1: Classification results of each algorithm in Area1 on October 20; Figure S2: Classification results of each algorithm in Area2 on October 20; Figure S3: Classification results of each algorithm in Area3 on October 20; Figure S4: Classification results of each algorithm in Area1 on October 4; Figure S5: Classification results of each algorithm in Area2 on October 4; Figure S6: Classification results of each algorithm in Area3 on October 4.

Author Contributions: Supervision, A.L.; Writing—original draft, M.F.; Writing—review and editing, M.F., G.Q., Y.X. and H.W. All authors have read and agreed to the published version of the manuscript.

Funding: The work is supported by the Joint Funds of National Natural Science Foundation of China (Grant no. U1704125).

Institutional Review Board Statement: Not applicable.

Informed Consent Statement: Not applicable.

Data Availability Statement: Not applicable.

Conflicts of Interest: The authors declare no conflict of interest.

Appendix A

Table A1. Statistics of the AUC index of each algorithm applied in the three regions.

The Method Name		Area1	Area2	Area3	Average
Threshold Method	Custom Threshold	0.919	0.626	0.748	0.764
	Otsu Threshold	0.985	0.922	0.964	0.957
	Adaptive Threshold	0.709	0.507	0.625	0.614
Machine Learning Method	Logistic Regression	0.984	0.929	0.933	0.949
	SVM	0.982	0.789	0.923	0.898
	Random Forest	0.985	0.973	0.935	0.964
	XGBoost	0.985	0.972	0.941	0.966
	Neural Network	0.984	0.850	0.935	0.923
	Decision Tree	0.982	0.965	0.935	0.961

Table A2. Statistics of various indexes of each algorithm in Area1 on October 20.

Method	Category	Precision	Recall	F1-Score	AUC
Neural Network	water	0.998	0.945	0.971	0.984
	other	0.971	0.999	0.985	
Random Forest	water	0.997	0.950	0.973	0.985
	other	0.973	0.998	0.986	
SVM	water	0.999	0.933	0.965	0.982
	other	0.964	1.000	0.982	
XGBoost	water	0.996	0.952	0.974	0.985
	other	0.974	0.998	0.986	
Logistic Regression	water	0.999	0.942	0.970	0.984
	other	0.969	1.000	0.984	
Decision Tree	water	0.998	0.937	0.966	0.982
	other	0.966	0.999	0.982	
Adaptive Threshold	water	0.508	0.909	0.652	0.709
	other	0.911	0.515	0.658	
Custom Threshold	water	0.838	0.999	0.912	0.919
	other	0.999	0.894	0.944	
Otsu Threshold	water	0.979	0.983	0.981	0.985
	other	0.991	0.988	0.990	

Table A3. Statistics of various indexes of each algorithm in Area2 on October 20.

Method	Category	Precision	Recall	F1-Score	AUC
Neural Network	water	0.708	0.595	0..647	0.850
	other	0.992	0.995	0.993	
Random Forest	water	0.962	0.241	0.385	0.973
	other	0.984	1.000	0.992	
SVM	water	0.585	0.712	0.642	0.789
	other	0.994	0.989	0.992	
XGBoost	water	0.961	0.221	0.360	0.972
	other	0.984	1.000	0.992	
Logistic Regression	water	0.869	0.498	0.633	0.929
	other	0.990	0.998	0.994	
Decision Tree	water	0.948	0.144	0.250	0.965
	other	0.982	1.000	0.991	
Adaptive Threshold	water	0.027	0.670	0.052	0.507
	other	0.986	0.496	0.660	
Custom Threshold	water	0.256	0.825	0.391	0.626
	other	0.996	0.950	0.972	
Otsu Threshold	water	0.856	0.361	0.508	0.922
	other	0.987	0.999	0.993	

Table A4. Statistics of various indexes of each algorithm in Area3 on October 20.

Method	Category	Precision	Recall	F1-Score	AUC
Neural Network	water	0.982	0.690	0.810	0.935
	other	0.889	0.995	0.939	
Random Forest	water	0.999	0.628	0.771	0.935
	other	0.870	1.000	0.930	
SVM	water	0.995	0.567	0.722	0.923
	other	0.852	0.999	0.919	
XGBoost	water	0.998	0.671	0.802	0.941
	other	0.883	0.999	0.938	
Logistic Regression	water	0.997	0.624	0.768	0.933
	other	0.869	0.999	0.929	
Decision Tree	water	0.984	0.682	0.805	0.935
	other	0.886	0.995	0.938	
Adaptive Threshold	water	0.405	0.747	0.525	0.625
	other	0.846	0.558	0.673	
Custom Threshold	water	0.497	1.000	0.664	0.748
	other	1.000	0.593	0.745	
Otsu Threshold	water	0.970	0.893	0.930	0.964
	other	0.958	0.989	0.973	

Table A5. AUC index statistics of each algorithm in three regions on October 4.

The Method Name		Area1	Area2	Area3	Average
Threshold Method	Custom Threshold	0.842	0.549	0.708	0.700
	Otsu Threshold	0.970	0.617	0.908	0.832
	Adaptive Threshold	0.703	0.506	0.623	0.611
Machine Learning Method	Logistic Regression	0.329	0.489	0.357	0.392
	SVM	0.331	0.535	0.365	0.410
	Random Forest	0.688	0.489	0.349	0.509
	XGBoost	0.718	0.512	0.665	0.631
	Neural Network	0.688	0.485	0.354	0.509
	Decision Tree	0.790	0.518	0.697	0.668

Table A6. Statistics of various indexes of each algorithm in Area1 on October 4.

Method	Category	Precision	Recall	F1-Score	AUC
Neural Network	water	0.379	0.999	0.550	0.688
	other	0.996	0.097	0.177	
Random Forest	water	0.377	1.000	0.547	0.688
	other	1.000	0.087	0.160	
SVM	water	0.020	0.001	0.001	0.331
	other	0.643	0.994	0.781	
XGBoost	water	0.435	1.000	0.607	0.718
	other	1.000	0.284	0.442	
Logistic Regression	water	0.017	0.001	0.001	0.329
	other	0.642	0.990	0.779	
Decision Tree	water	0.579	1.000	0.734	0.790
	other	1.000	0.599	0.749	
Adaptive Threshold	water	0.493	0.918	0.641	0.703
	other	0.913	0.478	0.628	
Custom Threshold	water	0.685	0.998	0.812	0.842
	other	0.998	0.746	0.854	
Otsu Threshold	water	0.949	0.985	0.967	0.970
	other	0.992	0.971	0.981	

Table A7. Statistics of various indexes of each algorithm in Area2 on October 4.

Method	Category	Precision	Recall	F1-Score	AUC
Neural Network	water	0.028	0.999	0.055	0.485
	other	1.000	0.283	0.441	
Random Forest	water	0.022	1.000	0.042	0.489
	other	1.000	0.050	0.095	
SVM	water	0.090	0.039	0.054	0.535
	other	0.980	0.992	0.986	
XGBoost	water	0.024	1.000	0.048	0.512
	other	1.000	0.165	0.283	
Logistic Regression	water	0.001	0.001	0.001	0.489
	other	0.978	0.937	0.957	
Decision Tree	water	0.037	0.992	0.072	0.518
	other	1.000	0.463	0.633	
Adaptive Threshold	water	0.026	0.692	0.050	0.506
	other	0.986	0.451	0.619	
Custom Threshold	water	0.100	0.925	0.181	0.549
	other	0.998	0.826	0.904	
Otsu Threshold	water	0.248	0.314	0.277	0.617
	other	0.986	0.980	0.983	

Table A8. Statistics of various indexes of each algorithm in Area3 on October 4.

Method	Category	Precision	Recall	F1-Score	AUC
Neural Network	water	0.007	0.001	0.001	0.354
	other	0.702	0.944	0.805	
Random Forest	water	0.001	0.002	0.001	0.349
	other	0.697	0.924	0.794	
SVM	water	0.020	0.001	0.002	0.365
	other	0.709	0.978	0.822	
XGBoost	water	0.329	1.000	0.496	0.665
	other	1.000	0.183	0.309	
Logistic Regression	water	0.002	0.001	0.001	0.357
	other	0.712	0.994	0.830	
Decision Tree	water	0.395	1.000	0.567	0.697
	other	1.000	0.386	0.557	
Adaptive Threshold	water	0.399	0.756	0.523	0.623
	other	0.847	0.543	0.662	
Custom Threshold	water	0.419	0.997	0.590	0.708
	other	0.997	0.444	0.615	
Otsu Threshold	water	0.828	0.973	0.895	0.908
	other	0.988	0.919	0.952	

References

1. Fletcher, T.D.; Andrieu, H.; Hamel, P. Understanding, management and modelling of urban hydrology and its consequences for receiving waters: A state of the art. *Adv. Water Resour.* **2013**, *51*, 261–279. [CrossRef]
2. Papa, F.; Prigent, C.; Rossow, W.B. Monitoring Flood and Discharge Variations in the Large Siberian Rivers From a Multi-Satellite Technique. *Surv. Geophys.* **2008**, *29*, 297–317. [CrossRef]
3. Chen, C.; He, X.Y.; Lu, Y.; Chu, Y.L. Application of Landsat Time-Series Data in Island Ecological Environment Monitoring: A Case Study of Zhoushan Islands, China. *J. Coast. Res.* **2020**, *108*, 193–199. [CrossRef]
4. Yang, Y.; Yang, Y.; Liu, D.; Nordblom, T.; Wu, B.; Yan, N. Regional Water Balance Based on Remotely Sensed Evapotranspiration and Irrigation: An Assessment of the Haihe Plain, China. *Remote Sens.* **2014**, *6*, 2514–2533. [CrossRef]
5. Wang, Y.N.; Huang, F.; Wei, Y.C. Water Body Extraction from LANDSAT ETM plus Image Using MNDWI and K-T Transformation. In Proceedings of the 2013 21st International Conference Geoinformatics, Kaifeng, China, 20–22 June 2013.
6. Yang, X.; Zhao, S.; Qin, X.; Zhao, N.; Liang, L. Mapping of Urban Surface Water Bodies from Sentinel-2 MSI Imagery at 10 m Resolution via NDWI-Based Image Sharpening. *Remote Sens.* **2017**, *9*, 596. [CrossRef]
7. Durand, M.; Gleason, C.J.; Garambois, P.A.; Bjerklie, D.; Smith, L.C.; Roux, H.; Rodriguez, E.; Bates, P.D.; Pavelsky, T.M.; Monnier, J.; et al. An intercomparison of remote sensing river discharge estimation algorithms from measurements of river height, width, and slope. *Water Resour. Res.* **2016**, *52*, 4527–4549. [CrossRef]
8. Yang, X.; Qin, Q.; Grussenmeyer, P.; Koehl, M. Urban surface water body detection with suppressed built-up noise based on water indices from Sentinel-2 MSI imagery. *Remote Sens. Environ.* **2018**, *219*, 259–270. [CrossRef]
9. Cui, X.; Guo, X.; Wang, Y.; Wang, X.; Zhu, W.; Shi, J.; Lin, C.; Gao, X. Application of remote sensing to water environmental processes under a changing climate. *J. Hydrol.* **2019**, *574*, 892–902. [CrossRef]
10. Li, D.; Wang, G.; Qin, C.; Wu, B. River Extraction under Bankfull Discharge Conditions Based on Sentinel-2 Imagery and DEM Data. *Remote Sens.* **2021**, *13*, 2650. [CrossRef]
11. Du, Y.; Zhang, Y.H.; Ling, F.; Wang, Q.M.; Li, W.B.; Li, X.D. Water Bodies' Mapping from Sentinel-2 Imagery with Modified Normalized Difference Water Index at 10-m Spatial Resolution Produced by Sharpening the SWIR Band. *Remote Sens.* **2016**, *8*, 354. [CrossRef]
12. Donchyts, G.; Baart, F.; Winsemius, H.; Gorelick, N.; Kwadijk, J.; van de Giesen, N. Earth's surface water change over the past 30 years. *Nat. Clim. Chang.* **2016**, *6*, 810–813. [CrossRef]
13. Yang, K.; Yao, F.; Wang, J.; Luo, J.; Shen, Z.; Wang, C.; Song, C. Recent dynamics of alpine lakes on the endorheic Changtang Plateau from multi-mission satellite data. *J. Hydrol.* **2017**, *552*, 633–645. [CrossRef]
14. Cao, M.; Mao, K.; Shen, X.; Xu, T.; Yan, Y.; Yuan, Z. Monitoring the Spatial and Temporal Variations in The Water Surface and Floating Algal Bloom Areas in Dongting Lake Using a Long-Term MODIS Image Time Series. *Remote Sens.* **2020**, *12*, 3622. [CrossRef]
15. McFeeters, S.K. The use of the Normalized Difference Water Index (NDWI) in the delineation of open water features. *Int. J. Remote Sens.* **1996**, *17*, 1425–1432. [CrossRef]
16. Xu, H. Modification of normalised difference water index (NDWI) to enhance open water features in remotely sensed imagery. *Int. J. Remote Sens.* **2006**, *27*, 3025–3033. [CrossRef]

17. Feyisa, G.L.; Meilby, H.; Fensholt, R.; Proud, S.R. Automated Water Extraction Index: A new technique for surface water mapping using Landsat imagery. *Remote Sens. Environ.* **2014**, *140*, 23–35. [CrossRef]
18. Du, Z.; Li, W.; Zhou, D.; Tian, L.; Ling, F.; Wang, H.; Gui, Y.; Sun, B. Analysis of Landsat-8 OLI imagery for land surface water mapping. *Remote Sens. Lett.* **2014**, *5*, 672–681. [CrossRef]
19. Singh, K.V.; Setia, R.; Sahoo, S.; Prasad, A.; Pateriya, B. Evaluation of NDWI and MNDWI for assessment of waterlogging by integrating digital elevation model and groundwater level. *Geocarto Int.* **2015**, *30*, 650–661. [CrossRef]
20. Duan, Z.; Bastiaanssen, W.G.M. Estimating water volume variations in lakes and reservoirs from four operational satellite altimetry databases and satellite imagery data. *Remote Sens. Environ.* **2013**, *134*, 403–416. [CrossRef]
21. Zhang, L.; Wang, J.S.; An, Z.Y. Classification method of CO2 hyperspectral remote sensing data based on neural network. *Comput. Commun.* **2020**, *156*, 124–130. [CrossRef]
22. Li, Y.F.; Liu, C.C.; Zhao, W.P.; Huang, Y.F. Multi-spectral remote sensing images feature coverage classification based on improved convolutional neural network. *Math. Biosci. Eng.* **2020**, *17*, 4443–4456. [CrossRef]
23. Cui, W.; Zhou, Q.; Zheng, Z.D. Application of a Hybrid Model Based on a Convolutional Auto-Encoder and Convolutional Neural Network in Object-Oriented Remote Sensing Classification. *Algorithms* **2018**, *11*, 9. [CrossRef]
24. Tian, Y.; Jia, R.S.; Xu, S.H.; Hua, R.; Deng, M.D. Super-resolution reconstruction of remote sensing images based on convolutional neural network. *J. Appl. Remote Sens.* **2019**, *13*, 13. [CrossRef]
25. Chen, Y.; Fan, R.; Yang, X.; Wang, J.; Latif, A. Extraction of urban water bodies from high-resolution remote-sensing imagery using deep learning. *Water* **2018**, *10*, 585. [CrossRef]
26. Qi, H.N.; Huang, M.L. Research on SVM ensemble and its application to remote sensing classification. In Proceedings of the International Conference on Intelligent Systems and Knowledge Engineering (ISKE 2007), Chengdu, China, 15–16 October 2007.
27. Alimjan, G.; Sun, T.L.; Jumahun, H.; Guan, Y.; Zhou, W.T.; Sun, H.G. A Hybrid Classification Approach Based on Support Vector Machine and K-Nearest Neighbor for Remote Sensing Data. *Int. J. Pattern Recognit. Artif. Intell.* **2017**, *31*, 1750034. [CrossRef]
28. Razaque, A.; Frej, M.B.; Almi'ani, M.; Alotaibi, M.; Alotaibi, B. Improved Support Vector Machine Enabled Radial Basis Function and Linear Variants for Remote Sensing Image Classification. *Sensors* **2021**, *21*, 4431. [CrossRef] [PubMed]
29. Cheng, Q.; Varshney, P.K.; Arora, M.K. Logistic regression for feature selection and soft classification of remote sensing data. *IEEE Geosci. Remote Sens. Lett.* **2006**, *3*, 491–494. [CrossRef]
30. Hogland, J.; Billor, N.; Anderson, N. Comparison of standard maximum likelihood classification and polytomous logistic regression used in remote sensing. *Eur. J. Remote Sens.* **2013**, *46*, 623–640. [CrossRef]
31. Veerabhadraswamy, N.; Devagiri, G.M.; Khaple, A.K. Fusion of complementary information of SAR and optical data for forest cover mapping using random forest algorithm. *Curr. Sci.* **2021**, *120*, 193–199. [CrossRef]
32. Li, L.H.; Jing, W.P.; Wang, H.H. Extracting the Forest Type From Remote Sensing Images by Random Forest. *IEEE Sens. J.* **2021**, *21*, 17447–17454. [CrossRef]
33. Shetty, S.; Gupta, P.K.; Belgiu, M.; Srivastav, S.K. Assessing the Effect of Training Sampling Design on the Performance of Machine Learning Classifiers for Land Cover Mapping Using Multi-Temporal Remote Sensing Data and Google Earth Engine. *Remote Sens.* **2021**, *13*, 1433. [CrossRef]
34. Liang, G.; Zhao, X.L.; Zhao, J.H.; Zhou, F.N. Feature Selection and Mislabeled Waveform Correction for Water-Land Discrimination Using Airborne Infrared Laser. *Remote Sens.* **2021**, *13*, 3628. [CrossRef]
35. Memarsadeghi, N.; Mount, D.M.; Netanyahu, N.S.; Le Moigne, J. A fast implementation of the ISODATA clustering algorithm. *Int. J. Comput. Geom. Appl.* **2007**, *17*, 71–103. [CrossRef]
36. Mahboob, M.; Genc, B. Evaluation of ISODATA Clustering Algorithm for Surface Gold Mining Using Satellite Data. In Proceedings of the 2019 International Conference on Electrical, Communication, and Computer Engineering (ICECCE), Swat, Pakistan, 24–25 July 2019.
37. Balha, A.; Mallick, J.; Pandey, S.; Gupta, S.; Singh, C.K. A comparative analysis of different pixel and object-based classification algorithms using multi-source high spatial resolution satellite data for LULC mapping. *Earth Sci. Inform.* **2021**, 1–17. [CrossRef]
38. Zhou, Y.; Luo, J.; Shen, Z.; Hu, X.; Yang, H. Multiscale Water Body Extraction in Urban Environments From Satellite Images. *IEEE J. Sel. Top. Appl. Earth Obs. Remote Sens.* **2014**, *7*, 4301–4312. [CrossRef]
39. Janowski, L.; Wroblewski, R.; Dworniczak, J.; Kolakowski, M.; Rogowska, K.; Wojcik, M.; Gajewski, J. Offshore benthic habitat mapping based on object-based image analysis and geomorphometric approach. A case study from the Slupsk Bank, Southern Baltic Sea. *Sci. Total Environ.* **2021**, *801*, 149712. [CrossRef] [PubMed]
40. Granger, J.E.; Mahdianpari, M.; Puestow, T.; Warren, S.; Mohammadimanesh, F.; Salehi, B.; Brisco, B. Object-based random forest wetland mapping in Conne River, Newfoundland, Canada. *J. Appl. Remote Sens.* **2021**, *15*, 038506. [CrossRef]
41. Zhao, B.; Gou, P.; Yang, F.; Tang, P.P. Improving object-oriented land use/cover classification from high resolution imagery by spectral similarity-based post-classification. *Geocarto Int.* **2021**, 1–24. [CrossRef]
42. Aahlaad, M.; Mozumder, C.; Tripathi, N.; Pal, I. An Object-Based Image Analysis of WorldView-3 Image for Urban Flood Vulnerability Assessment and Dissemination Through ESRI Story Maps. *J. Indian Soc. Remote Sens.* **2021**, 1–16. [CrossRef]
43. Ma, L.; Zhu, X.X.; Qiu, C.P.; Blaschke, T.; Li, M.C. Advances of Local Climate Zone Mapping and Its Practice Using Object-Based Image Analysis. *Atmosphere* **2021**, *12*, 1146. [CrossRef]
44. Du, S.H.; Du, S.H.; Liu, B.; Zhang, X.Y. Mapping large-scale and fine-grained urban functional zones from VHR images using a multi-scale semantic segmentation network and object based approach. *Remote Sens. Environ.* **2021**, *261*, 112480. [CrossRef]

45. Guo, S.C.; Du, P.J.; Xia, J.S.; Tang, P.F.; Wang, X.; Meng, Y.P.; Wang, H. Spatiotemporal changes of glacier and seasonal snow fluctuations over the Namcha Barwa-Gyala Peri massif using object-based classification from Landsat time series. *ISPRS J. Photogramm. Remote Sens.* **2021**, *177*, 21–37. [CrossRef]
46. Liu, X.; Zhang, Y.; Ling, X.; Huang, X. Automatic and Unsupervised Water Body Extraction Based on Spectral-Spatial Features Using GF-1 Satellite Imagery. *IEEE Geosci. Remote Sens. Lett.* **2018**, *16*, 927–931. [CrossRef]
47. Bao, L.; Lv, X.; Yao, J. Water Extraction in SAR Images Using Features Analysis and Dual-Threshold Graph Cut Model. *Remote Sens.* **2021**, *13*, 3465. [CrossRef]
48. Di Baldassarre, G.; Schumann, G.; Brandimarte, L.; Bates, P. Timely Low Resolution SAR Imagery To Support Floodplain Modelling: A Case Study Review. *Surv. Geophys.* **2011**, *32*, 255–269. [CrossRef]
49. Long, S.; Fatoyinbo, T.E.; Policelli, F. Flood extent mapping for Namibia using change detection and thresholding with SAR. *Environ. Res. Lett.* **2014**, *9*, 35002. [CrossRef]
50. Pulvirenti, L.; Chini, M.; Pierdicca, N.; Guerriero, L.; Ferrazzoli, P. Flood monitoring using multi-temporal COSMO-SkyMed data: Image segmentation and signature interpretation. *Remote Sens. Environ.* **2011**, *115*, 990–1002. [CrossRef]
51. Zhao, L.; Yang, J.; Li, P.; Zhang, L. Seasonal inundation monitoring and vegetation pattern mapping of the Erguna floodplain by means of a RADARSAT-2 fully polarimetric time series. *Remote Sens. Environ.* **2014**, *152*, 426–440. [CrossRef]
52. Srivastava, P.K.; Han, D.; Rico-Ramirez, M.A.; Bray, M.; Islam, T. Selection of classification techniques for land use/land cover change investigation. *Adv. Sp. Res.* **2012**, *50*, 1250–1265. [CrossRef]
53. Kuhn, C.; de Matos Valerio, A.; Ward, N.; Loken, L.; Sawakuchi, H.O.; Kampel, M.; Richey, J.; Stadler, P.; Crawford, J.; Striegl, R.; et al. Performance of Landsat-8 and Sentinel-2 surface reflectance products for river remote sensing retrievals of chlorophyll-a and turbidity. *Remote Sens. Environ.* **2019**, *224*, 104–118. [CrossRef]
54. García-Llamas, P.; Suárez-Seoane, S.; Fernández-Guisuraga, J.M.; Fernández-García, V.; Fernández-Manso, A.; Quintano, C.; Taboada, A.; Marcos, E.; Calvo, L. Evaluation and comparison of Landsat 8, Sentinel-2 and Deimos-1 remote sensing indices for assessing burn severity in Mediterranean fire-prone ecosystems. *Int. J. Appl. Earth Obs. Geoinf.* **2019**, *80*, 137–144. [CrossRef]
55. U.S. Geological Survey. Landsat 8 Data Users Handbook. *Nasa* **2016**, *8*, 97.
56. Zhao, G.; Maclean, A.L. A comparison of canonical discriminant analysis and principal component analysis for spectral transformation. *Photogramm. Eng. Remote Sens.* **2000**, *66*, 841–847.
57. Mohajane, M.; Essahlaoui, A.; Oudija, F.; Hafyani, M.E.; Hmaidi, A.E.; Ouali, A.E.; Randazzo, G.; Teodoro, A.C. Land Use/Land Cover (LULC) Using Landsat Data Series (MSS, TM, ETM+ and OLI) in Azrou Forest, in the Central Middle Atlas of Morocco. *Environ.* **2018**, *5*, 131. [CrossRef]
58. Randazzo, G.; Cascio, M.; Fontana, M.; Gregorio, F.; Lanza, S.; Muzirafuti, A. Mapping of Sicilian Pocket Beaches Land Use/Land Cover with Sentinel-2 Imagery: A Case Study of Messina Province. *Land* **2021**, *10*, 678. [CrossRef]
59. Forestier, G.; Inglada, J.; Wemmert, C.; Gancarski, P. Comparison of optical sensors discrimination ability using spectral libraries. *Int. J. Remote Sens.* **2013**, *34*, 2327–2349. [CrossRef]
60. Otsu, N. A Threshold Selection Method from Gray-Level Histograms. *IEEE Trans. Syst. Man. Cybern.* **1979**, *9*, 62–66. [CrossRef]
61. Ji, L.; Zhang, L.; Wylie, B. Analysis of Dynamic Thresholds for the Normalized Difference Water Index. *Photogramm. Eng. Remote Sens.* **2009**, *75*, 1307–1317. [CrossRef]
62. Xie, H.; Zhang, Y.; He, Y.; You, K.; Fan, B.; Yu, D.; Li, M. Automatic and Fast Recognition of On-Road High-Emitting Vehicles Using an Optical Remote Sensing System. *Sens.* **2019**, *19*, 3540. [CrossRef] [PubMed]
63. Huo, J.; Gao, Y.; Shi, Y.; Yin, H. Cross-Modal Metric Learning for AUC Optimization. *IEEE Trans. Neural Netw. Learn. Syst.* **2018**, *29*, 4844–4856. [CrossRef]
64. Pal, M.; Mather, P.M. Support vector machines for classification in remote sensing. *Int. J. Remote Sens.* **2005**, *26*, 1007–1011. [CrossRef]
65. Min, J.H.; Lee, Y.-C. Bankruptcy prediction using support vector machine with optimal choice of kernel function parameters. *Expert Syst. Appl.* **2005**, *28*, 603–614. [CrossRef]
66. Zhang, L. Management of offshore oil pollution and logistics transportation based on decision tree. *Arab. J. Geosci.* **2021**, *14*, 1655. [CrossRef]
67. Ma, M.H.; Zhao, G.; He, B.S.; Li, Q.; Dong, H.Y.; Wang, S.G.; Wang, Z.L. XGBoost-based method for flash flood risk assessment. *J. Hydrol.* **2021**, *598*, 126382. [CrossRef]
68. Buthelezi, M.N.M.; Lottering, R.T.; Hlatshwayo, S.T.; Peerbhay, K. Comparing rotation forests and extreme gradient boosting for monitoring drought damage on KwaZulu-Natal commercial forests. *Geocarto Int.* **2020**, 1–24. [CrossRef]
69. Samat, A.; Li, E.Z.; Wang, W.; Liu, S.C.; Lin, C.; Abuduwaili, J. Meta-XGBoost for Hyperspectral Image Classification Using Extended MSER-Guided Morphological Profiles. *Remote Sens.* **2020**, *12*, 1973. [CrossRef]
70. Peng, J.; Lee, K.; Ingersoll, G. An Introduction to Logistic Regression Analysis and Reporting. *J. Educ. Res.* **2002**, *96*, 3–14. [CrossRef]
71. Mishra, V.N.; Kumar, V.; Prasad, R.; Punia, M. Geographically Weighted Method Integrated with Logistic Regression for Analyzing Spatially Varying Accuracy Measures of Remote Sensing Image Classification. *J. Indian Soc. Remote Sens.* **2021**, *49*, 1189–1199. [CrossRef]
72. Schmitt, M. Potential of Large-Scale Inland Water Body Mapping from Sentinel-1/2 Data on the Example of Bavaria's Lakes and Rivers. *PFG–J. Photogramm. Remote Sens. Geoinf. Sci.* **2020**, *88*, 271–289. [CrossRef]

73. Somasundaram, D.; Zhang, F.; Wang, S.; Ye, H.; Zhang, Z. Learning vector quantization neural network for surface water extraction from Landsat OLI images. *J. Appl. Remote Sens.* **2020**, *14*, 032605. [CrossRef]
74. Jiang, W.; He, G.; Long, T.; Ni, Y.; Liu, H.; Peng, Y.; Wang guizhou, W. guizhou Multilayer Perceptron Neural Network for Surface Water Extraction in Landsat 8 OLI Satellite Images. *Remote Sens.* **2018**, *10*, 755. [CrossRef]
75. Yang, L.; Tian, S.W.; Yu, L.; Ye, F.Y.; Qian, J.; Qian, Y.R. Deep learning for extracting water body from landsat imagery. *Int. J. Innov. Comput. Inf. Control* **2015**, *11*, 1913–1929.
76. Roli, F.; Giacinto, G.; Vernazza, G. Comparison and Combination of Statistical and Neural Network Algorithms for Remote-Sensing Image Classification. In *Neurocomputation in Remote Sensing Data Analysis*; Kanellopoulos, I., Wilkinson, G.G., Roli, F., Austin, J., Eds.; Springer: Berlin/Heidelberg, Germany, 1997; pp. 117–124.

applied sciences

MDPI

Article

A Transfer Learning Technique for Inland Chlorophyll-a Concentration Estimation Using Sentinel-3 Imagery

Muhammad Aldila Syariz [1,2], Chao-Hung Lin [1], Dewinta Heriza [1], Umboro Lasminto [2], Bangun Muljo Sukojo [3] and Lalu Muhamad Jaelani [3,*]

1 Department of Geomatics, National Cheng Kung University, Tainan 70101, Taiwan; p68077030@mail.ncku.edu.tw (M.A.S.); linhung@mail.ncku.edu.tw (C.-H.L.); p68097080@gs.ncku.edu.tw (D.H.)
2 Civil Engineering, Institut Teknologi Sepuluh Nopember, Surabaya 60111, Indonesia; umboro@ce.its.ac.id
3 Geomatics Engineering, Institut Teknologi Sepuluh Nopember, Surabaya 60111, Indonesia; bangun_ms@geodesy.its.ac.id
* Correspondence: lmjaelani@geodesy.its.ac.id; Tel.: +62-(031)-5929486

Citation: Syariz, M.A.; Lin, C.-H.; Heriza, D.; Lasminto, U.; Sukojo, B.M.; Jaelani, L.M. A Transfer Learning Technique for Inland Chlorophyll-a Concentration Estimation Using Sentinel-3 Imagery. *Appl. Sci.* **2022**, *12*, 203. https://doi.org/10.3390/app12010203

Academic Editors: Dimitrios S. Paraforos and Anselme Muzirafuti

Received: 21 November 2021
Accepted: 22 December 2021
Published: 25 December 2021

Publisher's Note: MDPI stays neutral with regard to jurisdictional claims in published maps and institutional affiliations.

Abstract: Chlorophyll-a (Chla) concentration, which serves as a phytoplankton substitute in inland waters, is one of the leading indicators for water quality. Generally, water samples are analyzed in professional laboratories, and Chla concentrations are measured regularly for the purpose of water quality monitoring. However, limited spatial water sampling and the labor-intensive nature of data collection make global and long-term monitoring difficult. The developments of remote-sensing optical sensors and technologies make the long-term monitoring of Chla concentrations for an entire water body more achievable. Many studies based on machine learning techniques, such as regression and artificial neural network (ANN) methods, have recently been proposed for Chla concentration estimation using optical satellite images. The methods based on machine learning can achieve accurate estimation. However, overfitting problems may arise because the in situ Chla dataset is generally insufficient to train a complicated machine learning model, which makes trained models inapplicable. In this study, an ANN model containing three convolutional and two fully connected layers with 4953 unknown parameters is designed. A transfer learning method, consisting of model pretraining, main-training, and fine-tuning stages, is proposed to ease the problem of insufficient in situ samples. In the model pretraining stage, the ANN model is pretrained and initialized using samples derived from an existing Chla concentration model. The pretrained ANN model is then fine-tuned using the proposed transfer learning technique with in situ samples collected in five different campaigns carried out during early 2019 from Laguna Lake, the Philippines. Before the transfer learning, data augmentation and rebalancing methods are conducted to enrich the variability and to near-uniformly distribute the in situ samples in Chla concentration space, respectively. To estimate the alleviation of model overfitting, the trained ANN model, using an in situ dataset from Laguna Lake, was tested using an in situ dataset from Lake Victoria, Uganda, obtained in 2019, which has a similar trophic state as Laguna Lake. The experimental results from Sentinel-3 imagery indicated that the overfitting problem was significantly alleviated and the trained ANN model outperformed related models in terms of the root-mean-squared error of the estimated Chla concentrations.

Keywords: chlorophyll-a concentration; artificial neural network; transfer learning; overfitting

1. Introduction

Lakes are land-surrounded water bodies that generally provide freshwater for human daily needs. For instance, water from Lake Biwa, Japan, is used as a water drinking resource for people in Osaka and Kyoto and has been maintained as a conservation ecosystem with good water quality [1]. In Indonesia, a freshwater treatment plant, namely, PDAM Kabupaten Kerinci, was built around Lake Kerinci in Jambi to take, store, filter, and distribute the water to people living nearby [2]. Meanwhile, the worldwide demand for fish products has steadily increased due to the growing need for protein and the shift in behavior towards the consumption of healthier food [3,4]. The aquaculture industry often adds nutrient fertilizers, which are useful for commercial fish, to the water somewhere around the lake body. This procedure can fulfil the consumption demands; however, algal growth may be enhanced when nutrients are oversupplied. Consequently, the penetration of sunlight, which is required for respiration in fish, is limited and may lead to the extensive deterioration of water quality and the declining availability of freshwater, harming not only the fish, but also society. Therefore, the long-term monitoring of the water quality in lakes is necessary for the authorities to develop sustainable management initiatives to prevent water quality degradation and to maintain freshwater supplies in the future.

Chlorophyll-a (Chla), a pigment found in every phytoplankton species, is considered a critical water quality parameter for many environmental issues [5–7]. The water quality and Chla concentration can be categorized into four classes based on the trophic state index: oligotrophic (less than 2.6 μg/L), mesotrophic (2.6–20 μg/L), eutrophic (20–56 μg/L), and hypertrophic (more than 56 μg/L) [8]. The water quality condition for each class is described in Table 1. Chla concentrations measured using field surveys are accurate and precise; however, the concentration data are only available at the sampling locations. Taking more measurements from the lake water body is hindered by the high labor and financial costs. Remote sensing technology enables researchers to empirically estimate the Chla concentration at the full spatial coverage of the lake water body by regressing the remote-sensing reflectance (R_{rs}) or the features with the in situ data obtained from field survey. Dall'Olmo and Gitelson [9] utilized the features of band ratios, combining R_{rs} at wavelengths 443, 490, and 560 nm (denoted as λ_{443}, λ_{490}, and λ_{560}) in a three-band model, in which the in situ samples used in training ranged from 4.4 μg/L to 217.3 μg/L. Al-Shehhi et al. [10] exchanged the R_{rs} at wavelengths λ_{560} to λ_{645}, which has been found to represent both water turbidity and algal absorption in a narrower range of in situ data (0.1–27.8 μg/L). Chen et al. [11] performed local calibration in Chinese waters resulting in an R_{rs} feature at λ_{580}, λ_{600}, and λ_{692}. Gitelson et al. [12] and Moses et al. [13] simplified the three-band model to a two-band model by removing the R_{rs} at λ_{443} due to the similar sensitivity to absorption as R_{rs} at λ_{490}. Hence, Mishra and Mishra [14] proposed a differentiate index, called the normalized differentiate Chla index (NDCI), and demonstrated that the method outperforms the three-band and two-band models in cross validation. Many researchers [15–20] searched for important features that are sensitive to Chla concentrations; however, the procedure is somewhat statistically exhaustive.

Table 1. Description of trophic state index [8].

Trophic Class	Chla Concentration Range (in μg/L)	Water Condition
Oligotrophic	0~2.6	A lake with very clear waters and high drinking water quality due to low nutrient content and algal production.
Mesotrophic	2.6~20	Commonly clear water lakes with beds of submerged aquatic plants and medium levels of nutrients.
Eutrophic	20~56	The water body will be dominated either by aquatic plants or algae.
Hypertrophic	More than 56	Highly nutrient-rich lakes characterized by frequent and severe nuisance algal blooms and low transparency.

Another promising procedure to estimate Chla concentrations is by means of an artificial neural network (ANN). Buckton et al. [21] proposed a fully connected neural network containing one hidden layer that revealed the capability of ANN for Chla concentration estimation. Similar work was also conducted by other researchers [22–24]. Hafeez et al. [25]

designed several fully connected neural networks and searched for the optimal hyperparameters, including the number of hidden layers and the number of neurons in a layer. The study also revealed that the optimal ANN model outclassed the other machine learning methods, including random forest, cubist regression, and support vector regression, in terms of Chla concentration estimation. Furthermore, several researchers utilized convolutional neural networks (CNNs), which consider neighborhood spectral information in Chla concentration modelling using convolutional layers with 3D kernels [26–29].

An ANN model requires a high number of labelled data—that is, it uses in situ Chla concentrations as outputs and their corresponding R_{rs} in satellite images as inputs, and the initial values for unknown parameters for model training. Pyo et al. [28] constructed a CNN model with more than 2000 unknown parameters. This model was trained using only 238 labelled data. Meanwhile, Aptoula and Ariman [26] utilized 320 labelled data to train a CNN model containing 2432 unknown parameters. However, overfitting problems may arise because insufficient labelled data are used to search for the optimal values of thousands of unknown parameters during model training. Nguyen et al. [30] applied data augmentation to enrich the labelled data; however, they did not consider the data imbalance problem that may affect the estimation accuracy. Furthermore, some researchers utilized simulated datasets instead of in situ Chla concentration data to deal with the labelled data insufficiency [31–33]. A simulated dataset means that the Chla concentration information is obtained from an existing known model. With this procedure, the labelled data insufficiency can be solved; however, training a neural network model with a simulated dataset may not reach the global optimum of the defined loss function. Syariz et al. [34] proposed a two-stage training method, in which the model is firstly pretrained using a simulated dataset, and the pretrained model is then retrained using an in situ dataset. The advantage of this method is that the pretraining process is able to provide good initial values for the unknown parameters before the main training process using the in situ dataset. The training process can train an ANN model rather well for Chla concentration estimation. However, the overfitting problem is not fully alleviated because of the lack of training sample variability and the problem of training sample imbalance.

In this study, the main objectives were (1) to propose a transfer learning technique using the two-stage transfer training approach for better Chla concentration estimation accuracy; (2) to enrich and balance the Chla-labelled data by performing data augmentation and rebalancing techniques; and (3) to test the ANN model trained using the improved proposed two-stage training transfer learning approach with an in situ dataset from Laguna Lake, using the in situ dataset acquired from Lake Victoria, Uganda. To evaluate the effectiveness of the proposed model learning methods, an ANN model, namely WaterNet, first proposed by Syariz et al. [34], was adopted. The input to WaterNet was a water-body image patch of the size 7 (width) $\times$ 7 (height) $\times$ 16 (bands) and the output was an estimated Chla concentration at the center pixel of the input patch. Lastly, the proposed transfer learning method can increase the accuracy of Chla concentration retrieval in the lake water body, which can later be utilized by governments to better understand the lake water state and develop a clinical management plan to prevent water quality degradation and to maintain freshwater supplies in the future. The remainder of the paper is organized as follows. Section 2 describes the study area, data material, acquisition, and preprocessing. Section 3 elaborates the proposed transfer learning technique, data augmentation, and data rebalancing. Section 4 presents the experimental results, performance, and the comparisons of the trained ANN model and related models, and Section 5 provides the conclusions and future work.

2. Data Materials and Preprocessing

The in situ dataset acquired from Laguna Lake, the Philippines, was used to train the proposed ANN model while the in situ dataset acquired from Lake Victoria, Uganda, which has a similar trophic state (i.e., mesotrophic) with Laguna Lake, was utilized to test the trained model. The acquisitions of these two datasets are described in Sections 2.1 and 2.2.

The Sentinel-3 imagery used for Chla estimations and data preprocessing are described in Section 2.3.

2.1. Laguna Lake of the Philippines

Laguna Lake, with an area of 900 km^2 and an average depth of 2.5 m, is the largest lake in the Philippines. There are more than 20 million people living in the surrounding areas of Laguna Lake, indicating the importance of the lake in providing freshwater for local daily needs [35]. However, around 17% of the lake water body (~150 km^2) is occupied by aquaculture cages, where the nutrients and hazardous substances from industrial activity may pollute Laguna Lake, in addition to the issues of rapid population growth, industrialization, and urbanization [36,37]. In this study, field measurements of Chla concentrations were conducted during five different campaigns in 2019, as shown in Figure 1. Infinity-CLW ACLW2-USB, an optical-based data logger used to measure Chla concentrations, was installed on a boat at a depth of 0.5 m below the water's surface. The data logger recorded the Chla concentrations once per second during a 5-hour field survey, collecting more than 15,000 records at each campaign. Outlier removal and data down-sampling were conducted to remove noise and to match the Chla concentration sampling resolution with the spatial resolution of the Sentinel-3 images, respectively. After the data pre-processing, 257 in situ Chla samples were obtained from the five field campaigns, as shown in Table 2, and the resulting samples were utilized to train the ANN model and related models for comparison and evaluation.

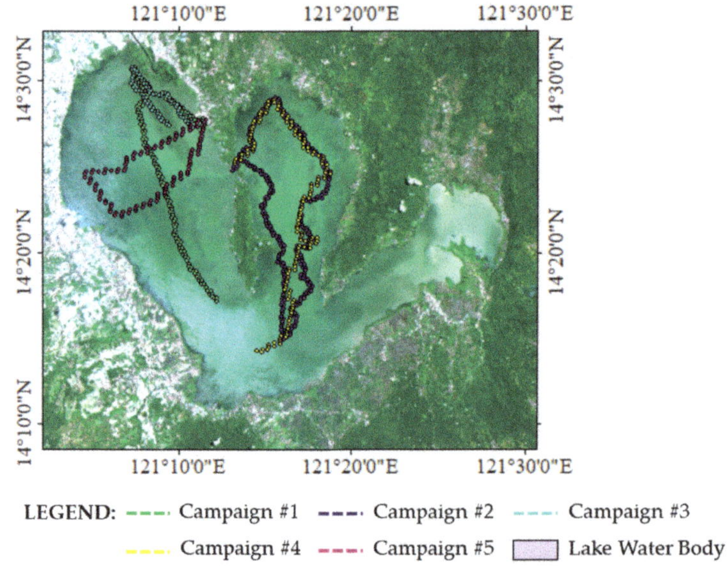

Figure 1. Laguna Lake and field campaigns for Chla concentration collection. The routes of the field campaigns and the locations of the collected samples are visualized by colors.

Table 2. Statistical summary of the in situ samples from Laguna Lake. "Min", "Max", Mean" and "Std." represent the minimum, maximum, mean, and standard deviation of the Chla concentrations, respectively.

Campaign #	Date (in 2019)	# of Samples	Chla Concentration Statistics (μg/L)			
			Min.	Max.	Mean	Std.
1	11 Jan	35	9.072	13.235	11.391	0.655
2	29 Mar	74	7.378	8.076	7.906	0.163
3	6 Apr	98	6.980	10.970	8.459	1.218
4	26 Apr	22	6.731	7.692	7.254	0.315
5	30 Apr	48	7.856	11.295	9.613	0.801

2.2. Lake Victoria of Uganda

The in situ Chla concentrations from Lake Victoria were obtained from the Mendeley Online Database (https://data.mendeley.com/ (accessed on 3 August 2021)), as provided by Deirmendjan et al. [38]. The study in [38] estimated the dissolved organic matter (DOM) under the support of the project Lake Victoria Greenhouse Gas Dynamics (LAVIGAS). In this project, there were three campaign periods: 29 March to 8 April 2018, 25 October to 4 November 2018, and 7 June to 17 June 2019. At each period, the water samples for Chla concentrations were measured daily in water depths ranging from 1 to 40 m. Considering that (1) the samples have the same trophic state as Laguna Lake (2.6–20 μg/L), the measurement depth should be similar to that for Laguna Lake (0.5 m), (2) the Chla sampling time should match with the Sentinel-3 image acquisition time, and (3) the Sentinel-3 image pixels corresponding the collected Chla samples should be cloud-free, only two in situ samples, shown in Figure 2, could be utilized. These two samples were used to evaluate the inference performance of the trained models to compare the trained model with related models.

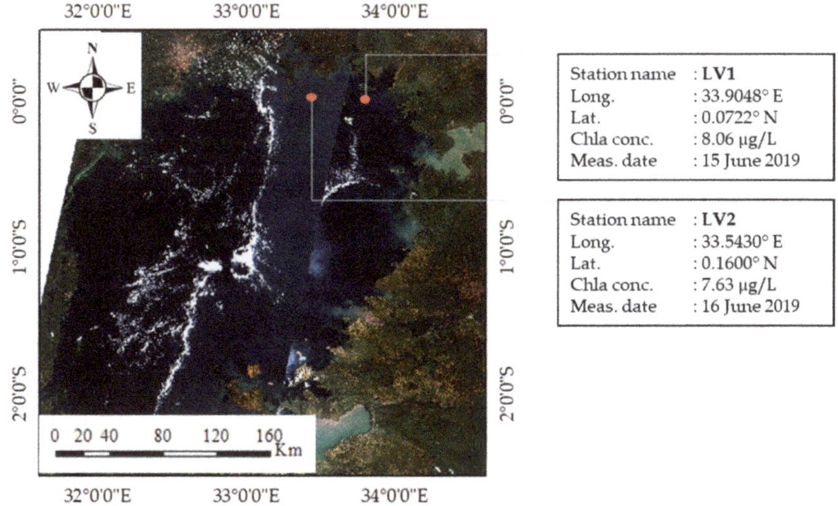

Figure 2. Lake Victoria and in situ samples. The locations of samples are marked by red dots, and the sample information is provided.

2.3. Sentinel-3 Image Dataset

Fifteen level 2 water full resolution (WFR) images of Laguna Lake, acquired by the ocean and land color instrument (OLCI) sensor of Sentinel-3, were utilized. A Sentinel-3

WFR image contains 16 atmospherically-corrected bands, excluding bands 13–15 (λ_{761}, λ_{764}, λ_{767}) and bands 18–19 (λ_{885}, λ_{900},) which are mainly designed for atmospheric correction [39]. The water-leaving reflectance in Sentinel-3 WFR images is further divided by π to derive the remote-sensing reflectance R_{rs}. In addition, the Sentinel-3 WFR product also contains several water quality parameters, including the Chla concentrations estimated by using an inverse radiative transfer model–neural network (IRTM-NN) [40]. The Chla concentrations from the IRTM-NN were regarded as a simulated dataset in this study and were used for model pretraining.

In the image data preprocessing, cloud-free water pixels in the R_{rs} images and their neighboring local patches of the spatial size 7 × 7 were extracted. Image patches containing non-water pixels, such as cloud, and pixels with negative R_{rs} values due to imprecise atmospheric correction or cloud shadow, were excluded from the dataset, forming full-water R_{rs} image patches. The summary of the R_{rs} image patches is presented in Table 3. Similarly, the cloud-free Sentinel-3 image patches corresponding to the locations with the simulated Chla concentrations generated by IRTM-NN were extracted. These image patches with simulated Chla concentrations were used in the model pretraining. The R_{rs} water patches and their corresponding simulated Chla data were used as a training set. The training set is denoted as $\{(\mathbf{P}_i, s\_chla_i)\}_{i=1}^{n}$, where n denotes the number of simulated labelled data, and $\mathbf{P}_i$ and s_chla_i represent the i-th R_{rs} water patch and its corresponding simulated Chla concentration, respectively. There were a total of 47,231 simulated labelled data. In addition, 275 in situ Chla data over Laguna Lake and their corresponding R_{rs} water patches were used as the retraining dataset. The retraining dataset is denoted as $\{(\mathbf{K}_i, t\_chla_i)\}_{i=1}^{m}$, where n represents the number of in situ Chla samples, and $\mathbf{k}_i$ and t_chla_i represent the i-th R_{rs} water patch and its corresponding in situ Chla concentration, respectively. In addition, one Sentinel-3 WFR image R_{rs} located in Lake Victoria was also obtained, and the acquisition date of the image was 15 June 2019. A R_{rs} water patch of the size 7 × 7 located at the field measurement point LV_1 was extracted. As for the field measurement LV_2, which was taken on 16 June 2019, the water patch was extracted from the Sentinel-3 image acquired on 15 June 2019. This means that the estimation was conducted using the image acquired one day before the field measurement in LV_2.

Table 3. Summary of Sentinel-3 R_{rs} image patches from Laguna Lake.

Image Acquisition Day	# of Image Patches	
	Pretraining Stage	Transfer-Learning Stage
11 Jan	1008	35
29 Mar	4715	74
6 Apr	5681	98
26 Apr	1908	22
30 Apr	2582	48
15 Jan	3471	
22 Jan	2017	
7 Feb	3722	
8 Feb	3681	
19 Feb	2877	
2 Mar	1654	
10 Mar	3393	
26 Mar	2766	
10 Apr	3984	
21 Apr	3772	
Total	47,231	275

Considering the stability of the model training, the water patches from Laguna Lake and Lake Victoria containing R_{rs} at 16 spectral bands were normalized to the range 0, 1 using the minimal and maximal R_{rs} values at each spectral wavelength. The data normalization process was also performed for the in situ and simulated Chla concentration data.

3. Methodology

3.1. Artificial Neural Network Model

An ANN model, namely WaterNet, proposed by Syariz et al. [34] was adopted. As shown in Figure 3, the input and output to the model was an image patch of the size $7 \times 7 \times 16$ and an estimated Chla concentration in the center pixel of the patch, respectively. The model is an end-to-end network structure consisting of three phases: that is, band expansion, feature extraction, and Chla concentration estimation. In the band expansion phase, there were three convolutional layers with $1 \times 1 \times 3$ kernel filters. The $1 \times 1 \times 3$ kernel filters performing convolution on the spectral domain attempt to augment spectral features from the spectral bands of the input image patch, which is also known as spectral feature extraction via band combination [41–43]. Meanwhile, two convolutional layers containing ten filters of the size $3 \times 3 \times 42$, and five filters of the size $3 \times 3 \times 10$, were utilized in the feature extraction phase. With those filters, the spatial feature information was extracted. The output to this phase was a feature map of the size $3 \times 3 \times 5$, and this output was further flattened and linked to the Chla concentration estimation phase which contained two fully connected layers. A rectified linear unit (ReLU) and sigmoid functions were used as the activation function in convolution and fully connected layers, respectively. In total, this ANN model contained 4753 unknown parameters.

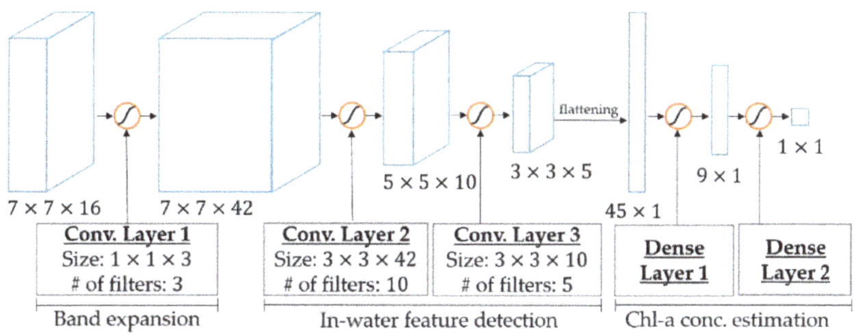

Figure 3. Network structure of WaterNet.

3.2. ANN Model Training

Utilizing insufficient in situ Chla concentration data and unsuitable initialization for the unknown parameters in ANN model training may lead to model overfitting and make the loss function difficult to converge. Syariz et al. [34] proposed a two-stage training approach consisting of pretraining and main-training, which is shown to be able to deal with the aforementioned problems. The first stage provides a better initialization for the unknown parameters before the main stage by pretraining the model with the simulated labelled data $\{(\mathbf{K}_i, s\_chla_i)\}_{i=1}^{n}$. Here, the estimation error is large and backpropagating the error could make the extraction of the spatial feature not optimum. Moreover, the convergence of the loss function may not reach its global minimum due to the utilization of the simulated data. However, this allows the model to have suitable initial values of the unknown parameters before the main training stage. Then, the pretrained model is refined with the in situ labelled data $\{(\mathbf{P}_i, t\_chla_i)\}_{i=1}^{m}$. This procedure is also known as transfer learning.

In this study, the two-stage training was adopted and the main stage part was improved by the implementation of fine-tuning, another kind of transfer learning technique. Moreover, data augmentation and rebalancing were also proposed and performed before the training in the improved main stage. The aim was to have more in situ labelled data with balanced amounts of samples in Chla concentration distribution space. Details regarding the data augmentation and rebalancing and the proposed transfer learning approach are explained below.

3.2.1. Data Augmentation and Rebalancing

To enrich the variability of the Chla in situ dataset, the data augmentation technique was implemented, as the convolutional processing is insensitive to rotation and scale [44,45]; however, the balance of data may not be considered. In this study, the data augmentation was performed on the in situ labelled data $\{(\mathbf{P}_i, t\_chla_i)\}_{i=1}^{m}$ by applying rotation to the image patches (with angles of 90°, 180°, and 270°) and flipping the rotated images from the left to right. Then, the rotated and flipped image patches were linked to their corresponding Chla concentration as a new dataset, namely, an augmented dataset: that is, $\{(\mathbf{Q}_i, n\_chla_i)\}_{i=1}^{q}$ where q is the number of rotated and flipped images (2216 data in total). The augmented dataset was further reclassified into 12 classes, with the first class starting from 6 µg/L, the last class ending at 12 µg/L, and each class covering 0.5 µg/L, as shown in Figure 4. Figure 4a implies the frequency of the in situ Chla in the augmented dataset. As seen, the difference between the Chla concentration data inter-range is huge, and indicates the imbalanced distribution of the data. Training the model with a data imbalance may reduce the optimum accuracy, and therefore data rebalancing is necessary. For that, a sample rebalancing technique was conducted by randomly removing several rotated and flipped in situ labelled data if the frequency of Chla concentration of the corresponding class was more than 100 sets (see Figure 4b). This kept the Chla concentration data at each range equal to or less than 100 sets, thus the balance of the data was achieved. In total, the data augmentation and rebalancing generated 900 rebalanced data $\{(\mathbf{R}_i, n\_chla_i)\}_{i=1}^{r}$ where r denotes the number of rebalanced in situ labelled data, $\mathbf{q}_i$ and n_chla_i represent the i-th R_{rs} water patch and its corresponding in situ Chla concentration, respectively. This also includes its original data $\{(\mathbf{P}_i, t\_chla_i)\}_{i=1}^{m}$. For simplification, the summary of dataset variations is described below.

- The simulated labelled data $\{(\mathbf{K}_i, s\_chla_i)\}_{i=1}^{n}$,
- the in situ labelled data which also refer to the original dataset $\{(\mathbf{P}_i, t\_chla_i)\}_{i=1}^{m}$
- the augmented dataset $\{(\mathbf{Q}_i, n\_chla_i)\}_{i=1}^{q}$, and
- the rebalanced dataset $\{(\mathbf{R}_i, n\_chla_i)\}_{i=1}^{r}$.

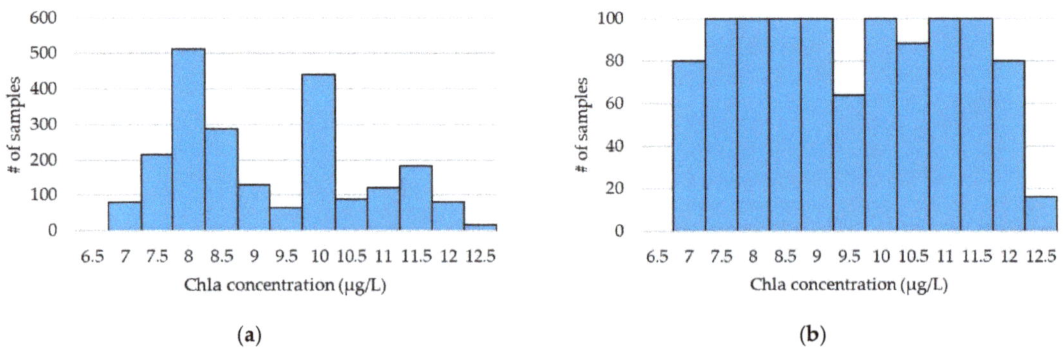

Figure 4. Amount of in situ Chla concentration after performing data augmentation (**a**) without and (**b**) with the consideration of the balance of data.

3.2.2. Transfer Learning

In this study, two-stage training was ado by the implementation of fine-tuning. The prc transfer learning is as follows. There are two s

1. Main-training stage. With the help of the suitable values of unknown parameters increases the possibility that the search can be reached. This also means that the estimation error is smaller. The error is t parameters and the spatial feature is mc
2. Fine-tuning stage. In the previous stag ready powerful, and continuing training rebalanced dataset may only endanger technique is performed in this stage by is split into two parts: the body part, c ANN model; and the head part, consisti is the Chla concentration estimation pha body part only. Inputting an image pat feature image. In machine learning, the known as a feature extractor. Moreover as the last phase with a random initial to the body part. Here, if the gradient i values all the way through the networ To prevent this problem, the layers in t of the model, are frozen or set as untr training be performed on the new head from the powerful spatial feature an optimized. Lastly, all of the layers are to the previous stage or sub-stage in rate of 0.001, the learning rate is now setting such very small rate is to obta parts. For simplification, Figure 5 sh

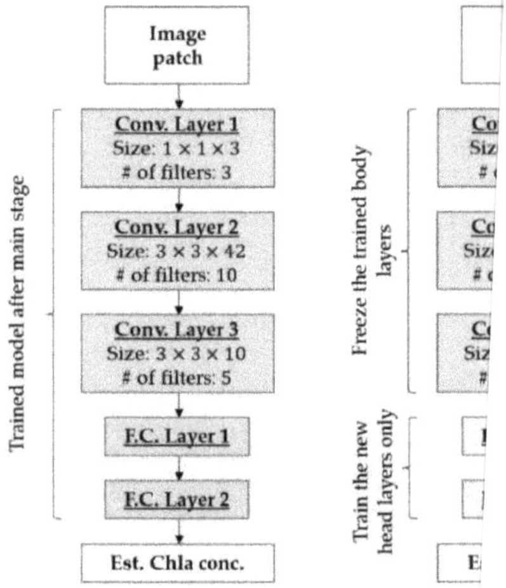

Figure 5. Training flow in the main-trainin represent trained unknown parameters in a la the green box denotes the random reinitializ

For hyperparameters, the Adam optimizer is employed due to its capability in tively tuning the learning rate and moment [46], and the mean squared error (MSE) is as the loss function L and is defined as follows:

$$L = \frac{1}{m}\sum_{i=1}^{m}(prChla_i - isChla_i)^2,$$

where $prChla_i$ is the prediction or estimation of Chla concentration from the input in age patch of the i-th labelled data. Moreover, overfitting is alleviated by adopting two regularization techniques: dropout and L_2 regularization. The dropout rate is set to 0.5, meaning that only 50% of the total unknown parameters are temporarily deactivated when computing the loss function for model convergence monitoring, whereas the L_2 regularization adds the Frobenius norm to the loss function to penalize large weights during error backpropagation for the tuning of unknown parameters. The maximum epoch is set to 30 and the trained network from an epoch with the smallest value of the loss function will be stored and used for the Chla concentration estimation.

4. Experimental Results and Discussion

This study proposed a transfer learning technique consisting of model pretraining, main-training, and fine-tuning stages for Chla ANN model training with an insufficient in situ dataset. In addition, the data augmentation and rebalancing were integrated with the transfer learning for Chla in situ data enrichment and imbalance. To evaluate the proposed method, a k-fold cross validation was performed with the Chla in situ dataset from Laguna Lake, the Philippines, where k was empirically set to 10. In this section, the results of the proposed transfer learning are presented in Section 4.1, and the effect of data imbalance to the trained ANN model is presented in Section 4.2. In addition, Section 4.3 demonstrates the comparisons between the CNN model trained by the proposed transfer learning with the related models using the dataset from Lake Victoria, Uganda. For accuracy assessment, the root mean squared error (RMSE) is employed by rooting the MSE in Equation (1).

4.1. Evaluation of the Transfer Learning

To evaluate the proposed transfer learning technique with the processes of data augmentation and rebalance, the ANN named WaterNet was used for Chla concentration estimation. For details about WaterNet, please refer to Section 3.2. To evaluate the performance of the three training stages in the transfer learning, the hyperparameters containing the batch size, the optimizer, and the number of epochs was the same and 10-fold cross validation was performed on the rebalanced dataset from Laguna Lake. The evaluation results are presented in Table 4. After the model pretraining, the accuracy of the estimated Chla concentrations was not satisfied. The range and average of RMSEs of the folds were 2.070~2.228 µg/L and 2.144 µg/L, respectively. This implies that a poor performance with high estimation errors was obtained when training the ANN model using the simulated Chla data. Although the ANN model at this stage cannot effectively retrieve the Chla concentrations, this training stage can provide suitable initial values for the unknown parameters for the coming stage. As a result, a better estimation result was obtained in the main-training stage. The range and average of the RMSEs at folds decreased to 0.4866~0.6887 µg/L and 0.5819 µg/L, respectively. Moreover, the trained ANN model was further fine-tuned in the next stage. The average RMSE improved from 0.5819 µg/L in the second stage to 0.3724 µg/L in the third stage. This was caused by setting the layers in the band extension and feature extraction phases to untrainable and only permitting the backpropagation to work on the layers in the Chla concentration phase.

Table 4. Performance of training stages in the proposed transfer learning.

Fold	Transfer Learning Performance (RMSE in µg/L)		
	Pretraining Stage	Transfer-Learning	
		Main-Training Stage	Fine-Tuning Stage
1	2.070	0.689	0.478
2	2.144	0.562	0.229
3	2.113	0.487	0.219
4	2.089	0.606	0.430
5	2.190	0.576	0.414
6	2.120	0.517	0.284
7	2.194	0.609	0.441
8	2.228	0.560	0.387
9	2.216	0.633	0.508
10	2.079	0.581	0.336
Avg.	2.144	0.582	0.372

The ANN model trained by the proposed transfer learning was applied to five Sentinel-3 images, which were acquired at similar dates with the field campaigns in Laguna Lake, Philippines. The Chla concentration maps for the water body, shown in Figure 6, are visualized by colors ranging from yellow (6 µg/L) to red (12 µg/L). In addition, the outputs from the feature extraction phase in the ANN shown in Figure 3 are convolutional feature maps of the size $3 \times 3 \times 5$. The feature maps imply the importance of spatial features for the Chla estimation. To visualize the feature maps for the whole lake body, the center pixels of the feature maps were extracted and combined to form spatial feature maps. The Chla concentrations of Laguna Lake on 6 April 2019, estimated by the trained ANN and the spatial feature maps extracted from the trained ANN, are shown in Figure 7. The spatial feature maps #1 and #3 are flashier than the others. To address this on the two spatial feature maps, the two dashed boxes are set on the maps to represent the area of interest for highlight and discussion. As shown in the brown dashed box, most of the features within this area have smaller values in feature map #1 and higher values in feature map #3. The significant differences between these two feature maps result in high Chla concentrations during the model prediction. As for those in the yellow dashed box, the opposite results are obtained, because the area is homogeneous and the pixels within this area have similar values. This observation revealed that the proposed transfer learning is able to preserve spatial features that are important in Chla concentration estimation.

4.2. *Performance of Data Augmentation and Rebalancing*

Three datasets are used and tested in this subsection, namely, original, augmented, and balanced datasets. The original dataset refers to the Chla in situ data acquired from Laguna Lake, the Philippines. The augmented and balanced datasets are the augmented in situ datasets without and with, respectively, the consideration of in situ Chla concentration unbalancing. The comparisons of the proposed transfer learning using these three datasets are shown in Figure 8. The results indicated that the RMSEs of the training using the original dataset ranged from 0.5 µg/L to 1.0 µg/L. By using the augmented dataset, the RMSEs of estimated Chla concentrations ranged from 7.5 µg/L to 9.5 µg/L. This is caused by the fact that more Chla samples in the augmented dataset are in the Chla concentration ranges 7.5~8 µg/L and 9.5~10 µg/L. Consequently, the sample imbalance on the Chla concentrations makes the performance of the trained model worse than that trained using the original dataset. When the data rebalancing that considers the distribution of samples' Chla concentrations in the augmented dataset is performed, the RMSEs of the estimated Chla concentrations are improved to 0.5~0.7 μg/L. Similar statistical results are shown in Figure 9, where the correlation coefficient between the estimated and in situ Chla concentrations was improved when the data rebalancing was performed with data augmentation.

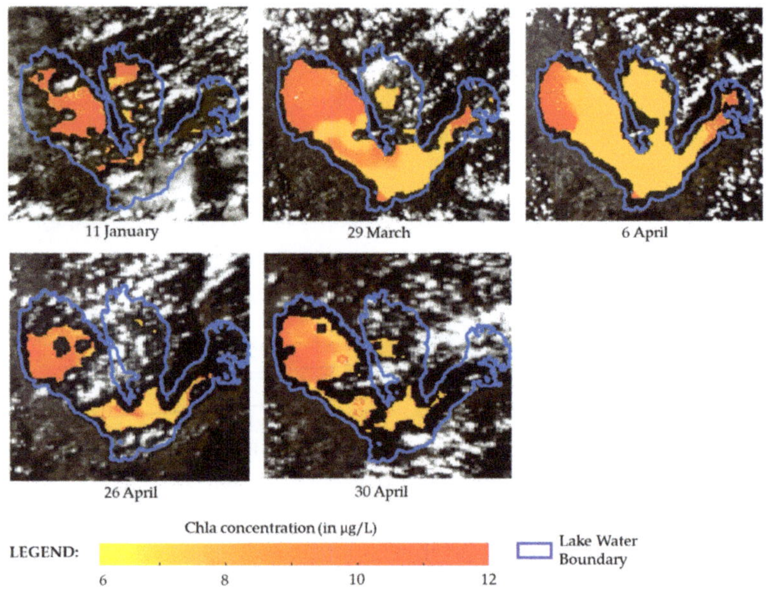

Figure 6. Maps of estimated Chla concentrations using the trained ANN model. The Sentinel-3 images are shown in false color combination (R: Band 17; G: Band 5; B: Band 3).

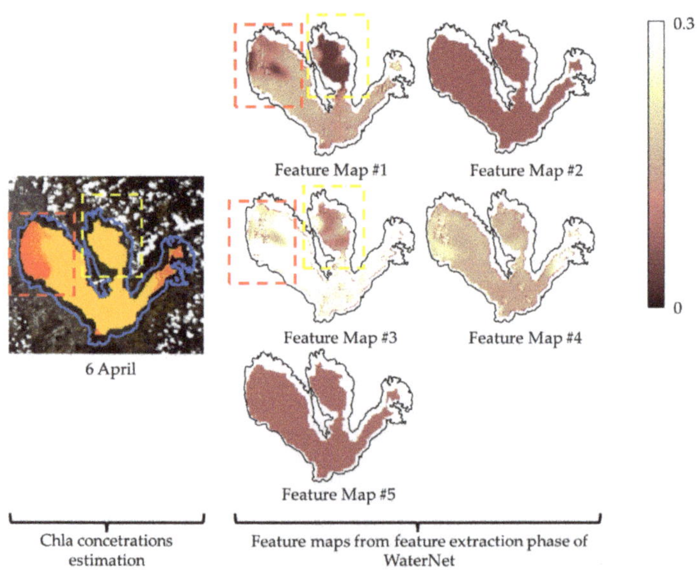

Figure 7. Feature maps in the trained ANN model. Chla concentration estimation map at 6 April 2019 (**left**) and the corresponding spatial features as the output of the feature extraction phase in WaterNet (**right**).

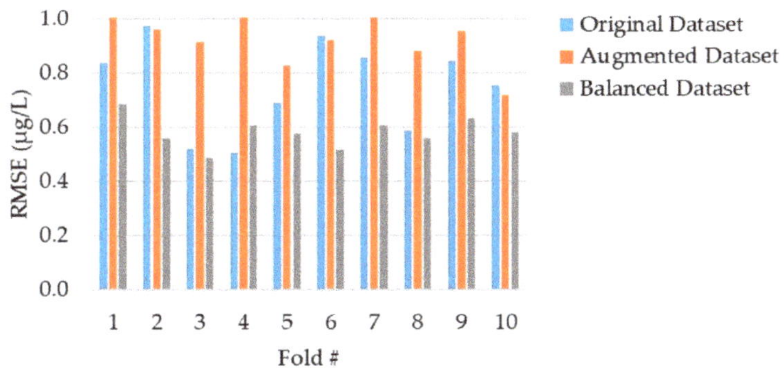

Figure 8. Comparisons of ANN model training using original, augmented, and balanced dataset.

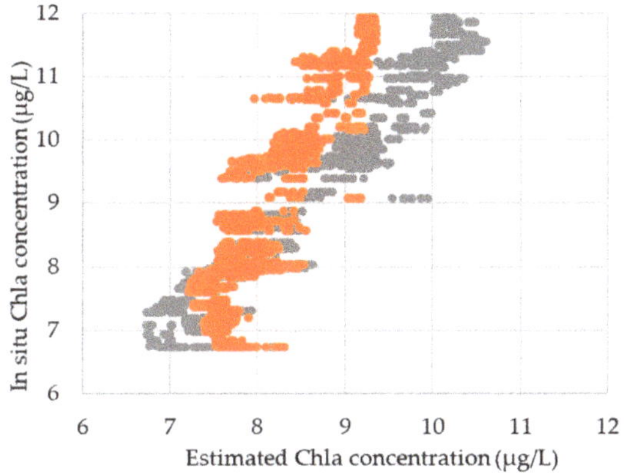

Figure 9. Performance of ANN model training with and without data rebalancing. Yellow and silver dots represent the estimated Chla concentrations using the augmented dataset with and without the process of data rebalancing.

4.3. Comparisons of Chla Estimation Models

A real model test, in which the machine-learning model is trained and tested using two geographically different and dependently corrected sample datasets, is rarely conducted due to the limited in situ Chla samples and overfitting problems. In this study, a ANN model was trained using the proposed transfer learning with the processes of data augmentation and rebalancing. The training Chla sample dataset was collected from Laguna Lake, Philippines. The trained ANN model was then applied to the Chla samples acquired from Lake Victoria, Uganda, for testing and evaluation. In addition, the trained ANN model was compared with the related models, including the three-band model [9], two-band model [13], NDCI [14], and WaterNet [34]. WaterNet is described in Section 3.1 and the other models are presented in Table 5. For fair comparisons, the three-band and two-band models were calibrated using in situ Chla-labelled data from Laguna Lake with a linear regression model. Linear regression was selected to ease overfitting problems. In addition, the hyperparameters containing the batch size, the optimizer, and the learning rate in the WaterNet training with original two-stage training are the same as that in the proposed training. Different to the other compared models, it is not necessary to calibrate the NDCI model, as the model directly outputs the estimated Chla concentrations. All of the compared models were trained using the dataset from Laguna Lake and then tested using the dataset from Lake Victoria for fair comparisons.

Table 5. Information of the compared Chla estimation models.

Model Name	Formula	Calibration Model
Three-band model	$\{[R_{rs}^{-1}(665) - R_{rs}^{-1}(709)] \times R_{rs}(754)\}$	Linear regression
Two-band model	$\{[R_{rs}(709) \div R_{rs}(665)]\}$	Linear regression
NDCI	$\left\{ \begin{array}{c} [R_{rs}(665) - R_{rs}(709)] \\ \div[R_{rs}(665) + R_{rs}(709)] \end{array} \right\}$	
WaterNet		

Table 6 shows the comparison results of WaterNet, trained using original two-stage training with original data, and the proposed method, including the improved transfer learning with data augmentation and rebalancing. The table also contains the related models using the Chla dataset from Lake Victoria. The results indicate that the three-band model with the performance RMSE = 0.588 μg/L and the two-band models with the performance RMSE = 0.509 μg/L have similar Chla concentration prediction accuracy. This may be due to the fact that these two models utilize similar R_{rs} features, that are R_{rs} at λ_{443} and at λ_{490}, which share similar sensitivity to the absorption [13]. WaterNet trained with original two-stage training and data also performed similarly, with RMSE = 0.496 μg/L. Better performances were obtained when the estimation of Chla concentrations was conducted using WaterNet with the proposed training method and NDCI. The RMSEs of the two models were 0.228 μg/L and 0.244 μg/L, and WaterNet with the proposed training was slightly better than NDCI. This means that the proposed transfer learning with the processes of data augmentation and rebalancing is able to resist the overfitting problem, and the performance of the trained model outperforms the related models.

Table 6. Comparisons of the ANN model trained by the proposed transfer learning with the related models using Chla samples acquired from Lake Victoria, Uganda.

Station Name	Estimation Error (in μg/L)				
	Three-Band Model	Two-Band Model	NDCI	WaterNet	Proposed Method
LV_1	0.746	0.653	0.304	0.645	0.302
LV_2	0.367	0.303	−0.164	0.277	0.117
RMSE	0.588	0.509	0.244	0.496	0.229

5. Conclusions and Future Work

A transfer learning method containing the stages of model pretraining, main training, and fine tuning, was proposed to train ANN models for Chla concentration estimation using Sentinel-3 images. In addition, data augmentation and rebalancing were performed not only to increase the variability of the training dataset, but also to balance the samples in terms of Chla concentrations. To evaluate the ease of overfitting and to compare with related models, the models were trained using the Chla dataset from Laguna Lake and then tested using the Chla dataset from Lake Victoria, which has the same trophic state with Laguna Lake. The quantitative assessments on the Setinel-3 WFR images demonstrate that the proposed transfer learning method is better than that of WaterNet, and the trained CNN outperforms the related models in terms of Chla estimation accuracy. Considering that the data rebalancing can provide massive effects to the performance of the model, in the near future, WaterNet will be redesigned such that the neural network can be applied to other optical satellite imagery with better spatial resolution, including Sentinel-2 and Landsat 8 images, in order to improve the extraction of important spatial features in lake water bodies. In addition, other water quality parameters, such as turbidity and total suspended matter, will be included in the modelling.

Author Contributions: Conceptualization, M.A.S., C.-H.L. and L.M.J.; data curation, M.A.S.; formal analysis, M.A.S., C.-H.L. and L.M.J.; funding acquisition, C.-H.L.; investigation, M.A.S., D.H., U.L. and B.M.S.; methodology, M.A.S.; project administration, C.-H.L.; software, M.A.S. and D.H.; supervision, C.-H.L., U.L., B.M.S. and L.M.J.; validation, M.A.S. and D.H.; visualization, M.A.S.; writing—original draft, M.A.S., C.-H.L. and L.M.J. All authors have read and agreed to the published version of the manuscript.

Funding: This research was partially funded by the Ministry of Science and Technology, Taiwan (grant numbers MOST 106-2923-M-006-003-MY3 and 109-2923-M-006-001-MY3); and the Indonesian Ministry of Research and Technology/National Agency for Research and Innovation (grant number 1377/PKS/ITS/2020).

Institutional Review Board Statement: Not applicable.

Informed Consent Statement: Not applicable.

Data Availability Statement: Not applicable.

Acknowledgments: We would like to thank Ariel C. Blanco from the University of the Philippines and Loris Deirmendjian from Paul Sabatier University and colleagues for the collection and sharing of water quality data samples from Laguna Lake and Lake Victoria, respectively. Sentinel-3 imagery courtesy of the European Space Agency.

Conflicts of Interest: The authors declare no conflict of interest.

References

1. Kira, T.; Ide, S.; Fukada, F.; Nakamura, M. Lake Biwa Experience and Lessons Learned Brief. In *Managing Lakes and Their Basins for Sustainable Use: A Report for Lake Basin Manegers and Stakeholders*; International Lake Environment Committee: Otsu, Japan, 2006; Volume 5.
2. Kementerian Lingkungan Hidup. *Profil 15 Danau Prioritas Indonesia*; Kementerian Lingkungan Hidup: Jakarta, Indonesia, 2011.
3. Ipsos Business Consultant. *Indonesia's Aquaculture-Key Sectors for Future Growth*; Ipsos Business Consultant: Jakarta, Indonesia, 2010.
4. World Bank. *Fish to 2030 Prospects for Fisheries and Aquaculture*; World Bank: Washington, DC, USA, 2013.
5. Cristina, S.; Fragoso, B.; Icely, J.; Grant, J. *Aquaspace Project Document*; Aquaspace Project: Sagaremisco, Portugal, 2018.
6. Gurlin, D.; Gitelson, A.A.; Moses, W.J. Remote estimation of chl-a concentration in turbid productive waters-Return to a simple two-band NIR-red model? *Remote Sens. Environ.* **2011**, *115*, 3479–3490. [CrossRef]
7. Moutzouris-Sidiris, I.; Topouzelis, K. Assessment of Chlorophyll-a concentration from Sentinel-3 satellite images at the Mediterranean Sea using CMEMS open source in situ data. *Open Geosci.* **2021**, *13*, 85–97. [CrossRef]
8. Carlson, R.E. A trophic state index for lakes. *Limnol. Oceanogr.* **1977**, *22*, 361–369. [CrossRef]
9. Dall'Olmo, G.; Gitelson, A.A. Effect of bio-optical parameter variability and uncertainties in reflectance measurements on the remote estimation of chlorophyll-a concentration in turbid productive waters: Modeling results. *Appl. Opt.* **2006**, *45*, 3577. [CrossRef] [PubMed]
10. Al Shehhi, M.R.; Gherboudj, I.; Zhao, J.; Ghedira, H. Improved atmospheric correction and chlorophyll-a remote sensing models for turbid waters in a dusty environment. *ISPRS J. Photogramm. Remote Sens.* **2017**, *133*, 46–60. [CrossRef]
11. Chen, J.; Zhang, X.; Quan, W. Retrieval chlorophyll-a concentration from coastal waters: Three-band semi-analytical algorithms comparison and development. *Opt. Express* **2013**, *21*, 9024. [CrossRef]
12. Gitelson, A.A.; Dall'Olmo, G.; Moses, W.; Rundquist, D.C.; Barrow, T.; Fisher, T.R.; Gurlin, D.; Holz, J. A simple semi-analytical model for remote estimation of chlorophyll-a in turbid waters: Validation. *Remote Sens. Environ.* **2008**, *112*, 3582–3593. [CrossRef]
13. Moses, W.J.; Gitelson, A.A.; Berdnikov, S.; Povazhnyy, V. Satellite estimation of chlorophyll-a concentration using the red and NIR bands of MERIS-The azov sea case study. *IEEE Geosci. Remote Sens. Lett.* **2009**, *6*, 845–849. [CrossRef]
14. Mishra, S.; Mishra, D.R. Normalized difference chlorophyll index: A novel model for remote estimation of chlorophyll-a concentration in turbid productive waters. *Remote Sens. Environ.* **2012**, *117*, 394–406. [CrossRef]
15. Andrzej Urbanski, J.; Wochna, A.; Bubak, I.; Grzybowski, W.; Lukawska-Matuszewska, K.; Łącka, M.; Śliwińska, S.; Wojtasiewicz, B.; Zajączkowski, M. Application of Landsat 8 imagery to regional-scale assessment of lake water quality. *Int. J. Appl. Earth Obs. Geoinf.* **2016**, *51*, 28–36. [CrossRef]
16. Niroumand-jadidi, M.; Bovolo, F.; Bruzzone, L. Novel Spectra-Derived Features for Empirical Retrieval of Water Quality Parameters: Demonstrations for OLI, MSI. *IEEE Trans. Geosci. Remote Sens.* **2019**, *57*, 10285–10300. [CrossRef]
17. Van Nguyen, M.; Lin, C.H.; Chu, H.J.; Jaelani, L.M.; Syariz, M.A. Spectral feature selection optimization for water quality estimation. *Int. J. Environ. Res. Public Health* **2020**, *17*, 272. [CrossRef] [PubMed]
18. Wang, X.; Zhang, F.; Ding, J. Evaluation of water quality based on a machine learning algorithm and water quality index for the Ebinur Lake Watershed. *Sci. Rep.* **2017**, *7*, 12858. [CrossRef] [PubMed]

19. Zhang, Y.; Feng, X.; Cheng, X.; Wang, C. Remote estimation of chlorophyll-a concentrations in Taihu Lake during cyanobacterial algae bloom outbreak. In Proceedings of the 2011 19th International Conference on Geoinformatics, Shanghai, China, 24–26 June 2011; pp. 1–6.
20. Guo, Y.; Liu, C.; Ye, R.; Duan, Q. Advances on water quality detection by uv-vis spectroscopy. *Appl. Sci.* **2020**, *10*, 6874. [CrossRef]
21. Buckton, D.; O'Mongain, E.; Danaher, S. The use of Neural Networks for the estimation of oceanic constituents based on the MERIS instrument. *Int. J. Remote Sens.* **1999**, *20*, 1841–1851. [CrossRef]
22. Kown, Y.S.; Baek, S.H.; Lim, Y.K.; Pyo, J.C.; Ligaray, M.; Park, Y.; Cho, K.H. Monitoring coastal chlorophyll-a concentrations in coastal areas using machine learning models. *Water* **2018**, *10*, 1020. [CrossRef]
23. Samli, R.; Sivri, N.; Sevgen, S.; Kiremitci, V.Z. Applying artificial neural networks for the estimation of chlorophyll-a concentrations along the Istanbul coast. *Pol. J. Environ. Stud.* **2014**, *23*, 1281–1287.
24. Wang, Q.; Wang, S. A predictive model of chlorophyll a in western lake erie based on artificial neural network. *Appl. Sci.* **2021**, *11*, 6529. [CrossRef]
25. Hafeez, S.; Wong, M.; Ho, H.; Nazeer, M.; Nichol, J.; Abbas, S.; Tang, D.; Lee, K.; Pun, L. Comparison of Machine Learning Algorithms for Retrieval of Water Quality Indicators in Case-II Waters: A Case Study of Hong Kong. *Remote Sens.* **2019**, *11*, 617. [CrossRef]
26. Aptoula, E.; Ariman, S. Chlorophyll-a Retrieval From Sentinel-2 Images Using Convolutional Neural Network Regression. *IEEE Geosci. Remote Sens. Lett.* **2021**, *20*, 1–5. [CrossRef]
27. Choi, J.H.; Kim, J.; Won, J.; Min, O. Modelling Chlorophyll-a Concentration using Deep Neural Networks considering Extreme Data Imbalance and Skewness. In Proceedings of the 2019 21st International Conference on Advanced Communication Technology (ICACT), Pyeongchang, Korea, 17–20 February 2019; pp. 631–634.
28. Pyo, J.C.; Duan, H.; Baek, S.; Kim, M.S.; Jeon, T.; Kwon, Y.S.; Lee, H.; Cho, K.H. A convolutional neural network regression for quantifying cyanobacteria using hyperspectral imagery. *Remote Sens. Environ.* **2019**, *233*, 111350. [CrossRef]
29. Syariz, M.A.; Lin, C.; Blanco, A.C. Chlorophyll-a Concentration Retrieval using Convolutional Neural Networks in Laugna Lake, Philippines. *Int. Arch. Photogramm. Remote Sens. Spat. Inf. Sci.* **2019**, *42*, 14–15.
30. Van Nguyen, M.; Lin, C.; Syariz, M.A.; Thu, T.; Le, H.; Blanco, A.C. Multi-task Convolution Neural Network for Season-insensitive Chlorophyll-a Estimation in Inland Water. *IEEE J. Sel. Top. Appl. Earth Obs. Remote Sens.* **2021**, *14*, 10439–10449. [CrossRef]
31. Ioannou, I.; Gilerson, A.; Gross, B.; Moshary, F.; Ahmed, S. Deriving ocean color products using neural networks. *Remote Sens. Environ.* **2013**, *134*, 78–91. [CrossRef]
32. Ioannou, I.; Gilerson, A.; Gross, B.; Moshary, F.; Ahmed, S. Neural network approach to retrieve the inherent optical properties of the ocean from observations of MODIS. *Appl. Opt.* **2011**, *50*, 3168. [CrossRef]
33. Yu, B.; Xu, L.; Peng, J.; Hu, Z. Global chlorophyll-a concentration estimation from moderate resolution imaging spectroradiometer using convolutional neural networks. *J. Appl. Remote Sens.* **2021**, *14*, 034520. [CrossRef]
34. Syariz, M.A.; Lin, C.H.; Van Nguyen, M.; Jaelani, L.M.; Blanco, A.C. WaterNet: A convolutional neural network for chlorophyll-a concentration retrieval. *Remote Sens.* **2020**, *12*, 1966. [CrossRef]
35. Saguin, K. Biographies of fish for the city: Urban metabolism of Laguna Lake aquaculture. *Geoforum* **2014**, *54*, 28–38. [CrossRef]
36. Herrera, E.; Nadaoka, K.; Blanco, A.C.; Hernandez, E.C. Hydrodynamic investigation of a shallow lake environment (Laguna Lake, Philippines) and associated implications for eutrophic vulnerability. *ASEAN Eng. J. Part C* **2015**, *4*, 48–62.
37. Tamayo-Zafaralla, M.; Santos, R.A.V.; Orozco, R.P.; Elegado, G.C.P. The ecological status of Lake Laguna de Bay, Philippines. *Aquat. Ecosyst. Health Manag.* **2002**, *5*, 127–138. [CrossRef]
38. Deirmendjian, L.; Lambert, T.; Morana, C. Dissolved organic matter composition and reactivity in Lake Victoria, the World's largest tropical lake. *Biogeochemistry* **2020**, *150*, 61–83. [CrossRef]
39. European Space Agency. *Copernicus Sentinel-3 OLCI Land User Handbook*; European Space Agency: Paris, France, 2021.
40. Bricaud, A.; Morel, A.; Babin, M.; Allali, K.; Claustre, H. Variations of light absorption by suspended particles with chlorophyll a concentration in oceanic (case 1) waters: Analysis and implications for bio-optical models. *J. Geophys. Res.* **1998**, *103*, 31033–31044. [CrossRef]
41. Ha, N.T.T.; Koike, K.; Nhuan, M.T.; Canh, B.D.; Thao, N.T.P.; Parsons, M. Landsat 8/OLI Two bands ratio algorithm for chlorophyll-a concentration mapping in hypertrophic waters: An application to west lake in Hanoi (Vietnam). *IEEE J. Sel. Top. Appl. Earth Obs. Remote Sens.* **2017**, *10*, 4919–4929. [CrossRef]
42. Ha, N.T.T.; Koike, K.; Nhuan, M.T. Improved accuracy of chlorophyll-a concentration estimates from MODIS Imagery using a two-band ratio algorithm and geostatistics: As applied to the monitoring of eutrophication processes over Tien Yen Bay (Northern Vietnam). *Remote Sens.* **2013**, *6*, 421–442. [CrossRef]
43. Menon, H.B.; Adhikari, A. Remote Sensing of Chlorophyll-A in Case II Waters: A Novel Approach With Improved Accuracy Over Widely Implemented Turbid Water Indices. *J. Geophys. Res. Ocean.* **2018**, *123*, 8138–8158. [CrossRef]
44. Kohl, S.A.A.; Romera-Paredes, B.; Meyer, C.; De Fauw, J.; Ledsam, J.R.; Maier-Hein, K.H.; Ali Eslami, S.M.; Rezende, D.J.; Ronneberger, O. A probabilistic U-net for segmentation of ambiguous images. *arxiv* **2018**, arXiv:1806.05034.
45. Patterson, J.; Gibson, A. *Deep Learning: A Practitioner's Approach*; O'Reilly Media: Sebastopol, CA, USA, 2017; ISBN 9781491914250.
46. Kingma, D.P.; Ba, J. Adam: A Method for Stochastic Optimization. In Proceedings of the ICLR, San Diego, CA, USA, 7–9 May 2015; pp. 1–15.

applied sciences

MDPI

Article

Density Estimates as Representations of Agricultural Fields for Remote Sensing-Based Monitoring of Tillage and Vegetation Cover

Markku Luotamo [1], Maria Yli-Heikkilä [2,3] and Arto Klami [1,*]

[1] Department of Computer Science, University of Helsinki, 00014 Helsinki, Finland; markku.luotamo@helsinki.fi
[2] Natural Resources Institute Finland, 31600 Jokioinen, Finland; maria.yli-heikkila@luke.fi
[3] Department of Geosciences and Geography, University of Helsinki, 00014 Helsinki, Finland
* Correspondence: arto.klami@helsinki.fi

Abstract: We consider the use of remote sensing for large-scale monitoring of agricultural land use, focusing on classification of tillage and vegetation cover for individual field parcels across large spatial areas. From the perspective of remote sensing and modelling, field parcels are challenging as objects of interest due to highly varying shape and size but relatively uniform pixel content and texture. To model such areas we need representations that can be reliably estimated already for small parcels and that are invariant to the size of the parcel. We propose representing the parcels using density estimates of remote imaging pixels and provide a computational pipeline that combines the representation with arbitrary supervised learning algorithms, while allowing easy integration of multiple imaging sources. We demonstrate the method in the task of the automatic monitoring of autumn tillage method and vegetation cover of Finnish crop fields, based on the integrated analysis of intensity of Synthetic Aperture Radar (SAR) polarity bands of the Sentinel-1 satellite and spectral indices calculated from Sentinel-2 multispectral image data. We use a collection of 127,757 field parcels monitored in April 2018 and annotated to six tillage method and vegetation cover classes, reaching 70% classification accuracy for test parcels when using both SAR and multispectral data. Besides this task, the method could also directly be applied for other agricultural monitoring tasks, such as crop yield prediction.

Keywords: machine learning; object-based classification; density estimation; histogram; land use; crop fields; soil tillage; data fusion; multispectral; SAR

Citation: Luotamo, M.; Yli-Heikkilä, M.; Klami, A. Density Estimates as Representations of Agricultural Fields for Remote Sensing-Based Monitoring of Tillage and Vegetation Cover. *Appl. Sci.* **2022**, *12*, 679. https://doi.org/10.3390/app12020679

Academic Editors: Dimitrios S. Paraforos and Anselme Muzirafuti

Received: 8 December 2021
Accepted: 6 January 2022
Published: 11 January 2022

1. Introduction

Remote sensing offers a cost-efficient approach for large-scale agricultural land use monitoring for administrative and research purposes, especially when combined with machine learning (ML) methods for estimating land use characteristics for individual crop field parcels [1–3] or other small spatial regions. These methods require a representation for each parcel derived from its pixels, either an explicitly engineered collection of features or an internal representation learnt in a data-driven fashion as in popular deep learning methods such as Convolutional Neural Networks (CNN) [4–7]. Our work is about learning good representations for crop field parcels that are often small and vary in shape. We also provide a practical computational pipeline for large-scale agricultural monitoring that can efficiently integrate information provided by multiple raster images captured at different resolutions, demonstrating it for the case of off-season soil tillage monitoring in Finland.

Previous studies have indicated object-level classification to be preferable over pixel-level information in agricultural tasks [8,9], but typically very high-level aggregate information such as the mean of individual pixel values has been used for representing parcels, making discrimination between similar classes difficult. Even though spatial features as extracted by CNNs are nowadays routinely used in remote sensing, crop field parcels have several characteristics that motivate representations focusing on the spectral distribution of sensor values instead. First of all, image content within each parcel is nearly homogeneous since the parcels are managed in a uniform way within the parcel boundaries. Hence, the prior value of spatial information at the pixel level is low. Spatial statistics are also difficult to estimate for parcels of irregular shape and with large differences in size, especially for noisy imaging sources like Synthetic Aperture Radar (SAR) as well as for cloud-occluded multispectral images (MSI). Distributional information about sensor values can, however, be reliably estimated for objects of any size and shape also in the presence of occlusion and noise. Consequently, we propose using probability density estimates (DE) of pixel values as a general purpose representation for such objects and formalize a practical computational pipeline around these representations, described in Section 2.2.

We assume the pixels of a given raster image area to be drawn from a probability distribution of pixel band values. Rough aggregate summaries, such as mean and median of the pixel values are still in active use in remote sensing due to simplicity and robustness [10–13], but our interest lies in the advantages of characterising subtle differences in the whole distribution. Histograms of pixel values have a long history as a natural representation of spectral distribution in computer vision [14–16], and they have also been considered in remote sensing [17–19]. Normalized object histograms are easy to compute but suffer from poor sample efficiency and objects with different pixel counts are not directly comparable: a small object's histogram is likely to have gaps whereas larger ones appear to be more continuous. Hence, histograms work best at a coarse bin resolution. We prefer direct modelling of the joint density using multivariate DEs [20,21], so that for each object we learn a continuous probability density over multi-band pixel values. To use the estimate as representation for subsequent processing, we collapse the density to bins resemblant of a histogram, with the advantage of inter- and extrapolation over observation gaps and reduced noise in pixel values, especially for parcels of varying size. We also consider Bayesian estimators [22] to account for uncertainty stemming from small pixel counts; in our data the field parcel size varies from tens to hundreds of pixels.

We apply the proposed method to a case of off-season soil tillage monitoring in Finland. From the standpoint of environmentally and economically sustainable agriculture, soil erosion and nutrient runoff from crop fields to surface waters is a long-standing challenge, to which soil tillage operations are a contributing factor [23,24]. Large-scale information on annual off-season tillage status of arable land is of interest for agro-environmental monitoring administration, policy makers as well as wide range of academic domains from terrestrial carbon studies to hydrological research. The problem is made challenging by the irregular shapes and small sizes of the parcels in Finland, and the limited amount of labeled training data within a single country. Furthermore, the annual off-season observation time window is limited and must take place during a relatively cloudy time of the year in Finland. The proposed computational pipeline can address all of these challenges.

Previous remote sensing studies of soil tillage detection have focused on spectral reflectance characteristics [25–27] or radar response [28–30] of soils, green vegetation, and crop residues. SAR signal penetrates cloud cover and is inherently sensitive to target 3D structure affecting backscatter mechanisms and angle. On the other hand, optical, for example. multispectral reflectance from satellite images can be fully exploited only at cloudless moments, but it can characterize a wide range of chemical and physical properties of matter as well as reveal dynamics of organic phenomena. The differences in the physical processes involved in radar backscatter and optical reflectance signals allow them to respond complementarily to the phenomena being interpreted, which results in enhanced accuracy of classification and regression for a range of remote sensing applications, for example, in agricultural or land cover contexts [31–35], or recently in soil tillage detection [36,37]. Our contribution to the topic is a practical, effort-reducing SAR-MSI data fusion technique. We use Copernicus Sentinel-1 (S1) SAR data with dimensions of two polarity bands, overlaid with polygon data of crop fields as provided by the Finnish Food Authority. Similarly to

SAR bands, we construct density estimates over two spectral indices calculated from a Sentinel-2 (S2) MSI for the same fields during a suitably narrow time window. Both sources are then merged to estimate the soil tillage category. Besides the Sentinel data used here, we expect the approach to be applicable to other SAR and MSI data sources, such as Radarsat-2, Landsat, or Modis.

To summarize, as the main contributions of this work we:

- Present a method for off-season tillage and vegetation cover detection on crop field parcels, which is a challenging task due to the heterogeneous size of the objects and a limited amount of training observations;
- Propose a representation of a free-form raster image object as a non-parametric probability density estimate, to be used for increasing robustness to variability in object size, count and missing pixel data;
- Introduce an easy-to-use framework for multi-sensor raster data fusion for sources of varying spatial resolution, applicable also outside the specific task considered here.

2. Materials and Methods

2.1. Materials

We use polygon-delineated boundaries of Finnish crop field parcels illustrated in Figure 1, collocated with mosaics of SAR and MSI satellite images over a time period from 11–23 April 2018 from Copernicus Sentinel-1 and Sentinel-2 missions, where the parcels are classified to one of multiple crop field tillage operations that affect, for example, soil properties and nutrient runoff to surface waters. The illustration reveals that the parcels have complex shapes and varying sizes, and that many of the parcels are small.

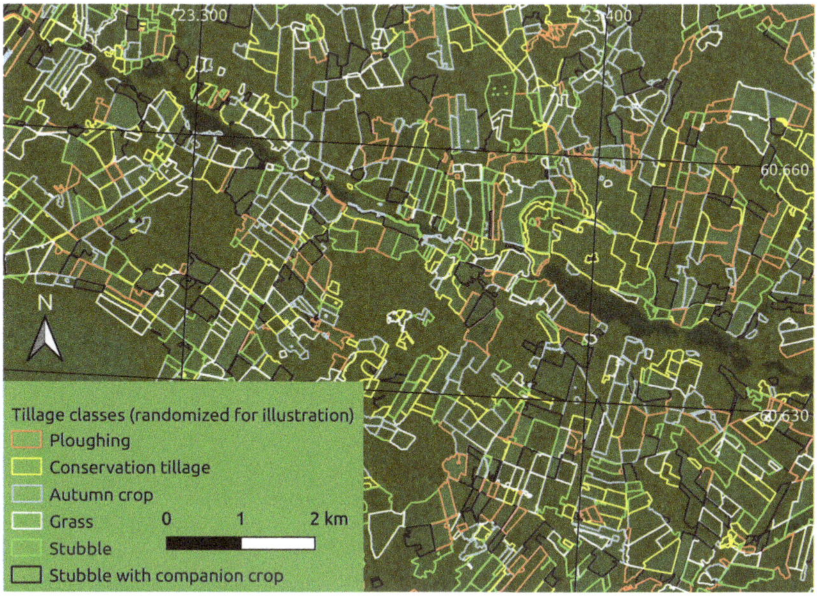

Figure 1. Foreground polygons: Autumn tillage operations annotated to six classes (colors). Background raster: Red-green rendering of a VH+VV dual polarization Sentinel-1 SAR image. Note: Due to data protection regulations, the polygons are from publicly open similar data from 2016 instead of our actual data, and the classes are randomized.

2.1.1. Crop Field Parcels and Annotations

We are interested in six categories to gauge the variety of land use and management over winter. The first class of *conventional ploughing* means mould-board ploughing in autumn to a depth of 20–25 cm. The second class of *conservation tillage* comprises tilling methods that mechanically disturb the soil to a depth less than 15 cm while retaining most of the crop residues on the surface. The last four classes include cases where the soil is either covered with crop residues (*stubble*), or with vegetation (*autumn crop*, *grass*). Soil with autumn crop has typically sparse plant cover before the growing season, and the soil surface is rough after seedbed preparation, whereas grass vegetation is typically rather thick, and the soil is covered. Stubble fields are covered with stalks and crop residues. In autumn spontaneous regrowth and weeds typically start to re-vegetate the soil. A special category of *stubble field growing catch crops* means crop fields where a companion crop (catch crop) re-vegetates the soil after the harvest of the main crop.

The region of interest (ROI) is illustrated in Figure 2 and was chosen on agrometeorological grounds: autumn tillage operations can span over many autumn months depending on the soil moisture conditions up until the soil is frozen and covered with snow. Therefore, the optimal time window to acquire images to monitor winter-time tillage status is shortly after snowmelt and before seedbed preparation in the spring. This time window is typically quite short; from two to four weeks. During this time, the soil dries out fast, but also there may occur sudden snow showers. To select the ROI, we used the regional starting dates of the thermal growing season in 2018. In this region, by mid April, the mean daily temperature permanently exceeded 5 °C, and snow had melted from open areas. The ROI was used to mask the underlying field parcels for reference data.

Reference data were annotated as follows. Information on agricultural land use in agricultural registers from two preceding growing seasons—2017 and 2018—were compared. The soil cover class was decided based on the variables of the winter-time vegetation cover related parcel-wise agri-environmental measures declared by farmers. Conservation tillage and vegetation cover are subsidised and subscribed to parcels. The different types of vegetation and crop residue cover were inferred from comparing the preceding years crop types with expertise in crop management. If a parcel was not subscribed to any measure, it was considered ploughed.

The intersection of the area of the satellite images shown in Figure 2 and of the parcel polygons yields a total of 127,757 annotated parcels. Annotations across the six classes are distributed as follows:

Conventional tillage, that is, ploughing 46,765; Conservation tillage 15,211; Autumn crop 2681; Grass 24,503; Stubble with no tillage 37,750; and Stubble with companion crop 847. We assigned each parcel exclusively to training and test sets by random sampling in a proportion of 80% for training and 20% for testing, resulting in total of 102,206 samples available for training and 25,551 for testing. However, in the computational experiments we mostly used considerably smaller subsets for studying the accuracy of models trained on less data.

2.1.2. Satellite Imagery

As the first raster component, we use polarimetric SAR intensity data (Ground Range Detected, GRD) from the Copernicus Sentinel-1 mission. Due to highly dynamic soil moisture and even plausible short-lived snow cover conditions during the time window, it is advantageous to use a mean-valued mosaic image composed of several images over the time period. The Finnish Meteorological Institute (FMI) publishes a preprocessed 11-day Sentinel-1 mean gamma-nought mosaic product [38]. See Supplement S1 for additional information and availability of the mosaic. We use an instance of a VV and VH polarity dataset from April 2018 (11th to 21st). Although the underlying spatial resolution of a Sentinel-1 Interferometric Wide Swath image is 5 $\times$ 20 m for an image of a 250 km swath [39], FMI mosaic preprocessing [38] resamples the data to 20 m spatial resolution. As a restriction, measurements prior to 2019 were quantized to 1 dB intensity intervals by the FMI data pipeline, posing a hard limit to discretization, that is, the binning of the measurements for the density estimators. Individual Sentinel-1 images that coincide with the time period and location of the mosaic and our ROI are listed in Supplement S1.

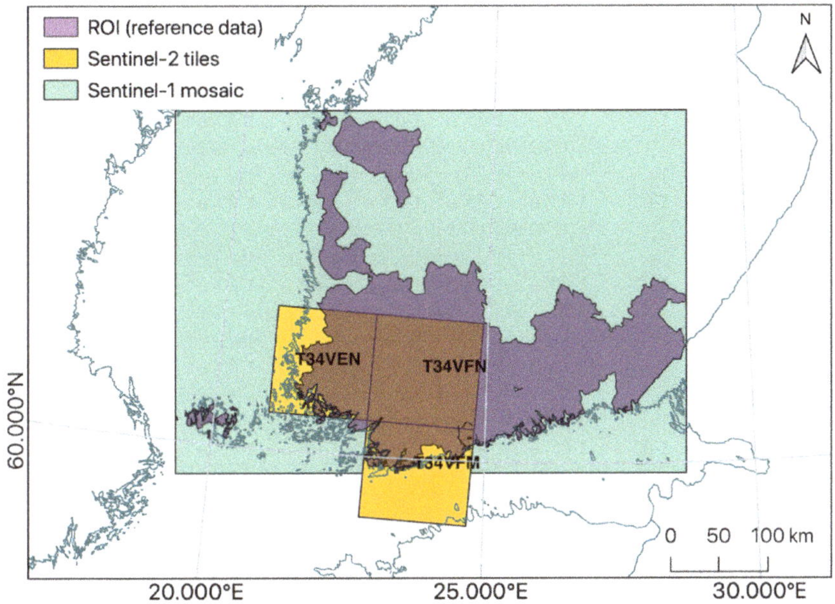

Figure 2. The region of interest (ROI) and raster data extent over southern Finland.

As a second raster component, we spatially mosaic three Sentinel-2 multispectral images selected from as close to the time period and ROI of the Sentinel-1 mosaic as possible (see Supplement S2 for the image identifiers), resampled to 10 m spatial resolution. An additional criterion for this image selection was a relatively low overall (<10%) percentage of pixels containing cloud or snow in the quality indicator (QI) metadata of the images. We also filter out individual pixels with a cloud or snow confidence value of ≥10%. This reduces the amount of pixel observations per parcel and makes the pixel sets discontinuous, but these properties do not cause problems for the proposed approach.

From the Sentinel-2 spectral channels, we calculate relevant basic spectral indices—the Normalized Differential Vegetation Index ($NDVI$) [40] and Normalized Differential Tillage Index ($NDTI$) [41,42]—as features for our density estimates. Consequently, we have $D = 2$ for MSI images. The formulas for the indices in the context of Sentinel-2 bands are:

$$NDVI = (B8 - B4)/(B8 + B4) \tag{1}$$

and

$$NDTI = (B11 - B12)/(B11 + B12), \tag{2}$$

where the Sentinel-2 band center wavelengths are: $B4$ (Red): 670 nm, $B8$ (NIR): 830 nm, $B11$ (SWIR): 1610 nm, and $B12$ (SWIR): 2200 nm.

The SAR VH/VV bands and the two optical spectral indices per pixel represent two disparate data sources at different resolutions and image extents (as seen in Figure 2). In Section 2.2 below we combine these to a common representation per parcel by extracting the pixels that coincide with the parcel delineation polygons.

2.2. Methods

2.2.1. Problem Formulation

The agricultural land monitoring task can be formulated as a machine learning problem, where we learn to predict a label $\hat{y} \in \mathcal{L}$ for a previously unseen object (field parcel) X'

given a collection of training observations $\{(X, y)\}$. For notational simplicity, we present the details for classification problems (y are discrete and mutually exclusive categories) although the representation could also be used for regression (continuous y, such as crop yield) or structured output problems.

Our focus here is on learning a suitable representation for objects that are pixel subsets of raster images. We denote individual pixels by column vectors $\mathbf{x} \in \mathbb{R}^D$ where the individual elements correspond to different channels (e.g., spectral bands of MSI or polarization channels of SAR). Each object o is defined by some subset of the pixels of an image A_i captured within a geospatial region of interest A, and hence can be represented by a matrix $X_{oi} \in \mathbb{R}^{D \times n_o}$ storing the n_o pixels for this object as its columns. Note that this formulation can be generalized in various ways; see Section 4.2.

Even though the focus of this work is on SAR and MSI data for soil tillage applications, we note that the approach is applicable to any task that satisfies the requirements of: (1) multi-banded raster data on a region of interest; (2) objects defined in terms of pixel segments of the images with a (3) class annotation on each object, using a shared coordinate reference system between the segment annotations and the rasters.

2.2.2. Data Flow: From Objects to Representations and Classification

Figure 3 shows a full data flow from raw images to predictions for the case of two remote sensing image sources. After sensor- and application-specific preprocessing and pixel-wise feature engineering of the images A_i, we extract for each object these resulting pixels from each type of image. We associate with each object an unordered pixel set per image type from within the geometric boundaries of the object shape. In the following, we represent these data from two data sources of different resolutions and extents using a shared representation of a multidimensional probability distribution per parcel.

From the object-wise pixel sets we form density estimates $p(\mathbf{x})$ for each object separately using a selected density estimation method, and then evaluate the density along a regular grid G to form the representation $\mathbf{f}$. For practical computation, this representation is formatted as a vector, which we normalize for additional robustness so that the ℓ_2 norm is one, but this normalization is not a critical part of the pipeline. This vector then becomes the representation for the supervised learning algorithm. For Bayesian density estimators, we can also consider an alternative representation that also captures the uncertainty of the estimate, explained later after describing the Logistic Gaussian Process Density Estimation (LGPDE) method.

Since our main focus is on the representation itself, we use standard classifiers readily available as a program library and in frequent use in the research field: the `scikit-learn` library's implementation of the Random Forest (RF), Support Vector Machine classifier (SVC) and a shallow feed-forward neural network (Multi-Layer Perceptron; MLP).

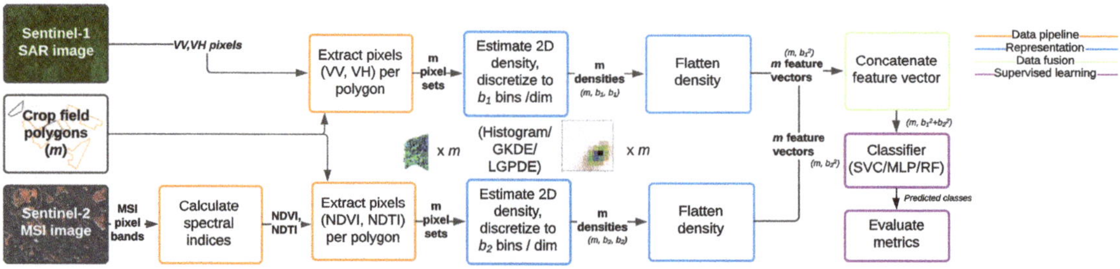

Figure 3. Process diagram with data flow from left to right.

2.2.3. Density Estimate as a Representation

A desirable object representation should condense relevant information into a similar form whether the object is spatially small or large. Put more generally, objects should be commensurately represented for an arbitrary count of observations in the measurement space. All density estimates and normalized histograms formally fulfill this requirement and we can use them to represent the object, but as discussed next, practical methods differ in terms of comparability given different amounts of pixels.

We consider a fixed-dimensional representation $\mathbf{f} = [p(\mathbf{x}_1), \ldots, p(\mathbf{x}_{B^D})]$ suitable as an input for any classifier, where the $\mathbf{x}_g$ are center points of elements (bins) in an equally spaced grid G overlaid on the density's support dimensions (pixel bands), so that G has B discretization intervals h_d in each of the D dimensions, with a total of B^D elements. $p(\boldsymbol{x})$ is a probability density that we learn based on the object's pixel collection X and then evaluate the density at the points $\mathbf{x}_g$ of G to form the representation. We consider only cases with $D = 2$, where the channels are two SAR polarisations or the two vegetation indices for MSI, so that we can directly model the joint density. For higher-dimensional cases, an alternative approach is to estimate a marginal density for each channel separately and evaluate it along a grid of B elements, resulting in a representation $\mathbf{f}_d \in \mathbb{R}^B$ for each band separately. A combined representation can then be obtained by concatenating these as $\mathbf{f} = [\mathbf{f}_1, \ldots, \mathbf{f}_D]$.

The representation can be computed for all density estimators, and next we discuss three practical alternatives and their properties.

Multivariate Histogram

As the elementary density estimate, we consider the *multivariate histogram*. For common notation with the other estimators, we formulate the normalized multivariate density histogram in the style of the univariate definition in [21] as a discretized function over G and multivariate observations $\mathbf{x} \in X$ with a total count of n:

$$p(\mathbf{x}_g) = \frac{\nu_g}{n}, \tag{3}$$

where ν_g is the number of observations $\mathbf{x}$ falling into the multivariate interval whose index is denoted by g. These intervals are defined as symmetric hybercubes around the center points of the grid.

Histograms are broadly used as representations, but are problematic for small objects with few pixels. We either need to use very small B, losing most of the resolution, or accept that the bin estimates are increasingly noisy. For large B we will typically have a significant proportion of bins with no observations at all and the non-zero bins will include only one pixel observation, and this effect becomes more severe with large D. If the pixel observations have noise comparable to or larger than the bin width h_d, a pixel often falls into one of the neighboring bins (or even further), and direct comparison of two histograms computed for two noisy realizations of the same object would indicate no similarity. Histograms also ignore uncertainty completely, which makes them poorly suited for the comparison of objects of varying size; histograms estimated from fewer pixels are noisier but this information is not captured by the estimate, and subsequent learning algorithms would falsely attribute the same amount of confidence for both.

Kernel Density Estimation

Parzen [43] formulated univariate kernel density estimation (KDE) in its modern form including the smoothing parameter, that is, bandwidth h, as:

$$p_h(x) = \frac{1}{nh} \sum_{j=1}^{n} K\left(\frac{x - x_j}{h}\right), \tag{4}$$

where the kernel K is a non-negative function and x_j are the n data points. We use an analogously defined multivariate version of KDE [20,44] with a bandwidth matrix S as:

$$p_h(\mathbf{x}) = \frac{1}{n}\sum_{j=1}^{n} K_h(\mathbf{x} - \mathbf{x}_j), \tag{5}$$

with the standard Gaussian kernel $K_h(\mathbf{x}) = (2\pi)^{-D/2}|\mathbf{S}|^{-1/2}e^{-\frac{1}{2}\mathbf{x}^{\mathbf{T}}yellow\mathbf{S}^{-1}\mathbf{x}}$ and a diagonal bandwidth matrix as the covariance matrix $\sqrt{\mathbf{S}_{dd}} = n^{\frac{-1}{D+4}}h_d$ determined by Scott's rule [21]. Note that for small objects the estimator is smoothed more, due to an inverse relationship between n and $\mathbf{S}_{dd}$. We refer to this estimate as Gaussian KDE (GKDE).

GKDE is an effective, lightweight method of providing smoothed probability density estimates for point samples independently of discretization interval or data point count. However, GKDE provides no measure of uncertainty relative to its suggested point estimate, and hence, similarly to histograms, loses information about the relative reliability of different objects.

Logistic Gaussian Process Density Estimation

For objects with only a few pixels, it becomes important to explicitly quantify the uncertainty of the density estimate itself, which neither of the above methods can achieve. For instance, for the extreme case of just one pixel, the histogram becomes a delta distribution, and while GKDE provides a smoother estimate it still suggests this single noisy pixel alone to be highly informative of the content. Bayesian estimators, instead, have the ability to explicitly model uncertainty, and in the following we describe one practical alternative building on Gaussian Processes (GP).

LGPDE, originally proposed by Leonard et al. [45], assigns a GP prior for the unnormalized logarithmic density $f(x)$ so that $\log p(x) = f(x) + C$ for any x, where C is a constant required for normalizing the density. The GP assigns a prior over the functions directly, so that for any finite collection of inputs their joint distribution is a multivariate normal, and conditioning on some pixel observations X we can then obtain the posterior distribution $p(f|X)$ that captures the uncertainty of the estimator. Due to the logistic transform, there is no closed-form analytic expression for the posterior, but both Markov Chain Monte Carlo (MCMC) sampling [46] and Laplace approximation [22] can be used for inference. We will later evaluate both the Laplace approximation as well as an MCMC implementation using the No-U-Turn Hamiltonian Monte Carlo algorithm as provided in the Stan probabilistic programming environment [47].

We use the formulation of Riihimäki et al. [22] with explicit enumeration over discretized support axes for computing the normalization term C. A prior term results from the logistic transform:

$$\log p(\boldsymbol{f}|G,\theta) = \mathcal{N}(\boldsymbol{f}|H\boldsymbol{m}, K + HMH^T), \tag{6}$$

where $\boldsymbol{f}$ is a latent function representing the density estimate surface being evaluated at points $\mathbf{x}_g$ of the discretization grid G, θ denotes the hyperparameters of the prior and the GP kernel, and $H(G)$ is a basis function that modulates the density to achieve finite support. For 2D densities we use the basis function $H(\boldsymbol{x}) = [x_1, x_1^2, x_2, x_2^2, x_1x_2]^T$. For a weakly informative prior, we parametrize a covariance adjustment of $M = 10^2 I$ and a zero mean of $\boldsymbol{m} = \mathbf{0}$. The kernel $K = K(G)$ determines a covariance matrix based on a given covariance function and a chosen multivariate bin discretization expressed by G. The posterior is formed using the likelihood

$$\log p(\boldsymbol{v}|\boldsymbol{f}) = \boldsymbol{v}^T\boldsymbol{f} - n\log\left(\sum^{B^D}\exp(f_b)\right), \tag{7}$$

where $\mathbf{v}$ is a histogram-like vector of observation counts. In the multidimensional case, f and z are vectorized to a single vector with B^D elements. The model induces a density over arbitrary $\mathbf{x}$, but the construction is already in a form explicitly represented over the grid G. Hence the representation is formed simply as the exponent of the log density.

Rather than a fixed representation $\mathbf{f}$, we now have a set of S posterior samples $\mathbf{f}^{(s)}$, either as produced by the MCMC algorithm or obtained by sampling from the Laplace approximation. They can be used within the proposed pipeline in two ways. The simplest alternative is to collapse the posterior to a point estimate as $\mathbb{E}[\exp(\{\mathbf{f}^{(s)})\}]; s \in \mathcal{S}$, to be used similarly as the results of other estimators. We call this *Point Estimate Classification* (PE-C). The other alternative, here called *Posterior Predictive Classification* (PP-C), is to pass all posterior samples of $\mathbf{f}^{(s)}$ separately to the classifier, for each object being classified. For testing, we evaluate the classifier similarly for all posterior samples and compute the posterior predictive class distribution $p(c = \hat{c} \mid \mathbf{x})$ using standard Monte Carlo approximation. This allows an end-to-end probabilistic approach for classification even if the classifier itself is designed to only produce point predictions $\hat{c}$.

3. Results

We report results for two types of experiments: Technical experiments validating the computational pipeline (Section 3.1), and evaluation of the method for the soil tillage task (Section 3.2).

3.1. Technical Validation

The core assumptions of our method are that a probability density of pixel values represents useful information about the classes of interest, and that we can learn reliable estimates of those based on individual parcels. We first validate these visually in Figure 4 for the SAR data. The top row shows that estimates computed from all pixels of a given class are visually distinct, whereas the bottom row shows that estimates computed based on pixels of individual parcels resemble the class-level information. The figure also illustrates the difficulty of the problem; the densities are distinct but highly similar in the sense that simpler representations like mean pixel value are unlikely to be sufficient for separating the classes.

For accuracy evaluation between the method variants, we use balanced subsets of parcel data described in Section 2.1.1 to make the results easier to interpret. We consider only balanced classification problems with equally many observations for each class so that classification accuracy can directly be interpreted as quality of the method, and we only consider the classes *ploughed*, *grass* and *stubble* to avoid issues with classification of the three minor classes that are difficult to separate from each other. For all of the technical experiments we use a fixed randomly chosen subset of 300 parcels per class (900 samples in total) for testing, whereas the size of the used subset of training data is a parameter for many of the experiments, seen on the horizontal axis as "Number of training samples/class". This is to investigate model performance with respect to data size.

3.1.1. Comparison of Representations

To demonstrate the effect of object representation on classifier accuracy, we compare three computationally efficient representations for three classifiers in Figure 5. The experiment was done on MSI data with NDVI and NDTI indices as image bands on data consisting of relatively small parcels (20. . . 50 px) with $B = 50$ bins per band. The parameter B controls both the amount of information we can capture and the reliability of the estimate; with small B the estimation task is easy but a majority of discriminative information is lost, whereas with large B we retain all information but can no longer reliably estimate the density from small samples. The choice of $B = 50$ (resulting in $B^2 = 2500$ bins in total) is motivated by Figure 6a, which shows the accuracy as a function of the discretization level for the case of 50 training samples per class.

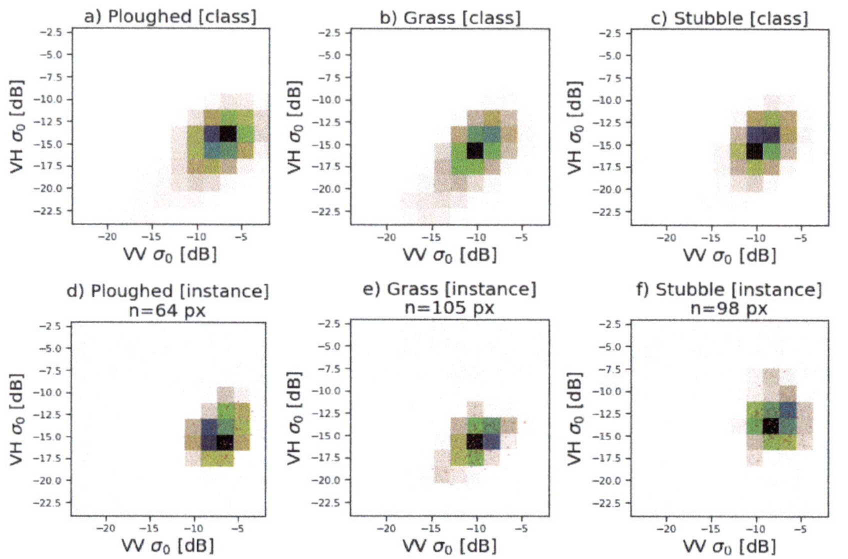

Figure 4. Gaussian kernel density estimate representations of polarimetric SAR intensity measurements for ploughed, grass- and stubble-covered fields. (**a–c**): Class-level density estimates from a large random sample of all pixels of all field parcels of a class. (**d–f**): Density estimates for single parcels of each class. The small red dots indicate individual pixels, with small jitter so multiple pixels with identical values are also visible.

Figure 5 reports the accuracy for varying sizes of training data for three different classifiers. The main observations for our MSI dataset are: (i) All forms of density estimates outperform naive summary statistics. The baseline of using an aggregate summary of all pixels, the median of NDVI and NDTI values, barely beats the random baseline of 33% classification accuracy, whereas all density estimates achieve accuracies between 40% and 60% depending on the case; (ii) direct multivariate estimates are at least as good as histograms and for some cases (SVC) better; (iii) GKDE performs as well as the multivariate histogram and sometimes (SVC and MLP for some training set sizes) marginally better. In summary, the results show that proper density estimators were preferable over both multivariate and marginal histograms as general representations. Even though there was no clear difference for one of the classifiers (RF), there were no cases where using GKDE would hurt.

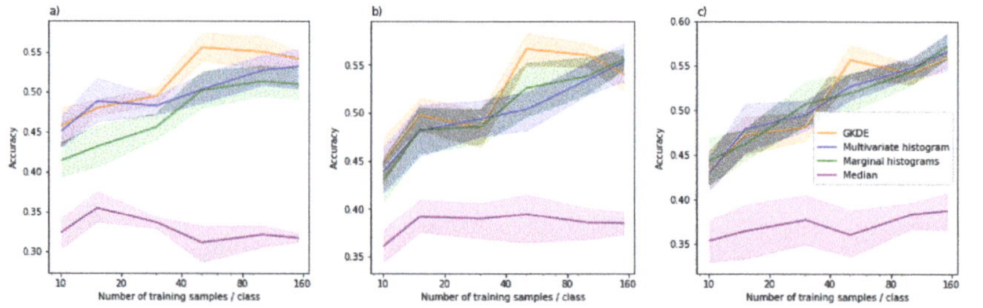

Figure 5. Density estimates, histograms and the median as representations for multispectral NDVI/NDTI data across three different classifiers. (**a**) SVC (**b**) MLP (**c**) RF. Each color corresponds to a representation, the line indicates the average over five random training sets evaluated on a single test set, and the shaded areas represent 95% bootstrapping-based confidence intervals.

3.1.2. Effects of Object Size

Next, we detail the performance of density-based representations under challenging training conditions with very few training instances, highly varying object size, or both. We do this on SAR data, using $B = 12$ bins over the range $-24 \ldots 0$ dB, to keep computational complexity manageable for extensive experimentation on all estimators.

We compare three proper density estimators, GKDE and LGPDE, with two inference algorithms (MCMC and Laplace approximation) and restrict to a single choice of the classifier to streamline the results; the observations are similar for the other classifiers. Figure 7 shows the accuracies for these estimators as function of the size of the training data for three scenarios: *small parcels* that only uses parcels of $20 \ldots 30$ pixels for training and evaluation, *large parcels* that only uses parcels of $90 \ldots 100$ pixels for training and evalution, and *variable parcels* that uses both small and larger parcels (range of $20 \ldots 100$ px). The main results are: (i) The problem is considerably easier if the objects are larger but already for the small parcels of only tens of pixels we comfortably beat the random baseline; (ii) The accuracy naturally improves when we get more training instances, but already relatively small number of approximately 30 parcels per class is enough for good accuracy; (iii) The representations are robust over parcels of varying size, shown by relatively high accuracy for the case that contains both small and large parcels; (iv) There are no clear differences between the three density estimators in terms of accuracy.

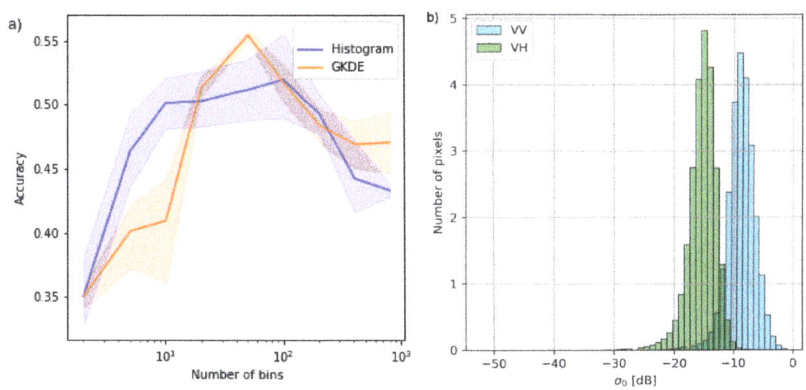

Figure 6. Choice of the discretization bins. (**a**) Accuracy as function of the number of bins for S2 MSI data. (**b**) S1 SAR intensity (sigma nought, σ_0) in crop field pixels concentrate within bounds of $-30 \ldots 0$ dB.

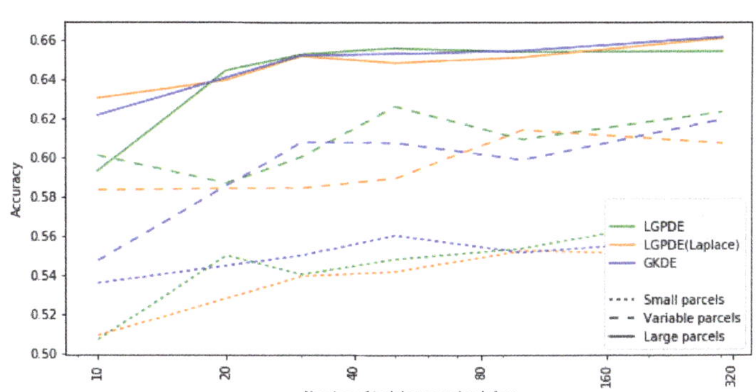

Figure 7. Effect of field parcel size (line style) on MLP accuracy for different estimators (line color). Confidence intervals omitted for visual clarity.

Even though we did not observe a direct improvement in classification accuracy for the more advanced density estimator LGPDE, it has the advantage of explicitly modeling the uncertainty of the estimate and we can propagate it through the classification process for any classifier as explained in Section 2.2.2. To demonstrate this, Figure 8 shows the classification accuracy for the three different classifiers for a dataset of small parcels (20 . . . 30 px), for both PE-C and PP-C. We observe that the PP-C approach that models the uncertainty offers a small but consistent improvement. Figure 9 shows that the resulting class probability distributions behave as expected—for small fields the uncertainty is better captured in the final class distributions.

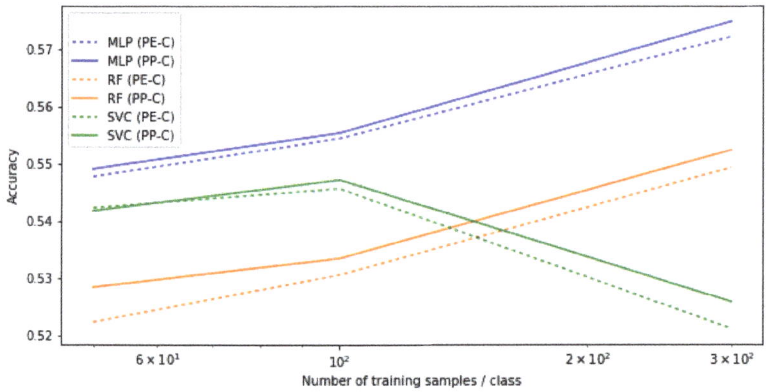

Figure 8. Accuracy for smaller parcels increases using posterior predictive LGPDE classification.

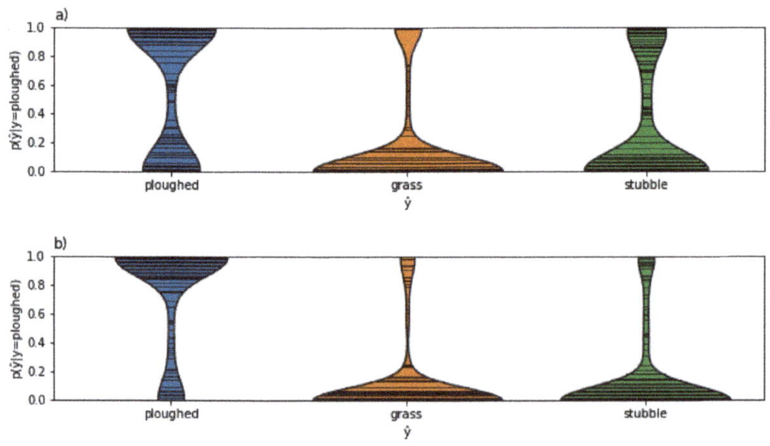

Figure 9. Confidence of small (**a**) vs. large (**b**) ploughed fields being classified as *ploughed* from a probabilistic perspective, with higher uncertainty for small fields, as expected.

3.1.3. Data Fusion

By learning separate representations for each image modality (capture method or sensor) A_i, we can perform easy data integration by simply concatenating the representations $\mathbf{f}_i$. In experiments Sections 3.1.1 and 3.1.2 we showed that both MSI and SAR are valuable sources of information for this task, and Figure 10 shows that by further combining them we get a significant improvement in overall accuracy: The combined solution outperforms MSI, which has the higher accuracy of the single-source capturing methods, on average by approximately 8 percentage points. We show the results on 1500 test parcels for the MLP classifier; the other classifiers followed a similar pattern.

We also evaluated the final accuracy of the data fusion solution for even larger training data to provide a baseline with ample data. With 6500 training parcels per class we reached an accuracy of 82%, validating that the accuracy can be further improved by utilising more data, as expected. However, the improvement over the 78% accuracy obtained already with 160 parcels per class is only modest. On one hand, this implies that the method can be reliably estimated already from small data and does not require access to thousands of or tens of thousands of training instances. On the other hand, it suggests the problem itself is challenging; as shown in Figure 4 and discussed in the next section, some of the classes are highly similar in appearance, which sets natural upper bounds on classification accuracy.

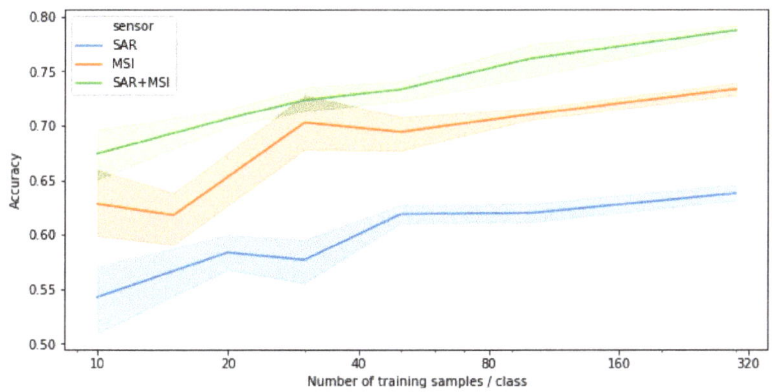

Figure 10. Data integration vs single-source classification on a Random Forest classifier. The integrated solution clearly outperforms both MSI and SAR alone for all training set sizes.

3.2. Soil Tillage Detection

Based on the technical validations above, we made the following choices for solving the soil tillage classification problem: (a) We use both SAR and MSI images; (b) we use RF as the classifier observed to be the most robust one; and (c) we use GKDE as computationally efficient and accurate representation. We use $B = 50$ for MSI and $B = 30$ for SAR within the range of $-30 \ldots 0$ dB in alignment with [48–50]. Motivation for these choices is illustrated in Figure 6.

We now use all six classes described in Section 2.1.1: *ploughed, conservation tillage, autumn crop, grass, stubble* and *stubble with companion crop*. We train the model in total on 43,299 parcels with the number of samples per class ranging from 169 to 15,885, and evaluate the accuracy on 10,666 parcels not used for training. Together these form the full set of parcels we find at the intersecting area of the SAR and MSI images in our data. The overall classification accuracy, evaluated on the test parcels, was 70% and Figure 11 shows the confusion matrix for the test parcels. The largest classes *ploughed, grass* and *stubble* are classified with high accuracy, whereas the smaller classes *conservation tillage, autumn crop* and *stubble with companion crop* are more difficult to classify correctly.

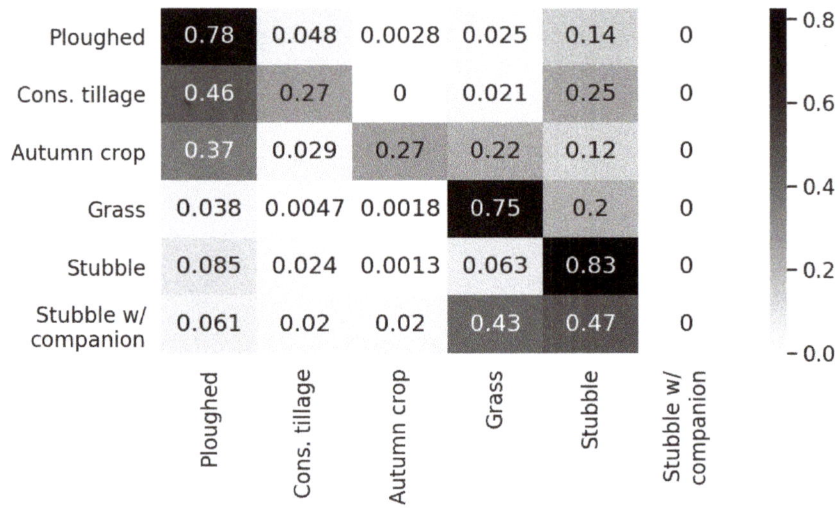

Figure 11. Normalized confusion matrix for classification of fused SAR + MSI image objects for the full set of annotated classes.

4. Discussion

4.1. Soil Tillage Detection

Our main goal was detecting autumn tillage and vegetation cover from earth observations for large-scale agricultural monitoring. Several previous studies such as [26,28,36] on tillage detection with SAR imagery alone or fusion of SAR and MSI have concentrated on tillage intensity classification. However, few studies have detected off-season land cover classes on broader scale including also vegetation covered land cover types [37,51,52]. Shortage of studies on higher granularity of winter-time land cover classes indicates that the task is difficult.

We observed significant and consistent improvement in classification accuracy by combining SAR and MSI data. The result is well in line with those obtained both in crop tillage classification [36,37] as well as in other EO tasks [34,48,53–55]. Since data fusion is easy with the proposed object representations, only requiring georeferencing and simple early fusion, we strongly recommend routinely using both sources for this task. When using a single image capture method, MSI was here clearly more accurate than SAR, but this observation needs to be interpreted with care because our experiment was carried out on images with at most 10% occlusion. During a normal year, the time window for making observations on tillage operations is short and typically cloudy across Southern Finland, and MSI alone could not be trusted.

Somewhat low classification accuracy for the classes *conservation tillage*, *autumn crop* and *stubble with companion crop* is explained by three main reasons: (a) the amount of data for these classes is smaller compared to the other three, (b) under certain conditions some of the classes are virtually indistinguishable, and (c) the ground truth data is imperfect due to mislabeling. Regarding the difficulty of the problem, the *autumn crop*, *stubble* and *stubble with companion crop* all have variable amounts of plant growth that makes the classes highly similar in terms of all EO sources. Also, *ploughed* and *conservation tillage* may resemble each other after snow melt in April on certain soil types, especially where stalks have been highly decomposed.

Regarding mislabeling, the reference data were prepared with automatic rule-based labeling, which is inherently error-prone. Whereas contradictory examples (duplicates) can be removed, mislabeling remains a practical challenge due to inevitable simplifications when building the rules. Imperfections in the underlying information on agricultural practices imply that each class membership has different reliability. For example, planting of autumn crop is explicitly declared by the farmer, thus having really high reliability, whereas ploughing is merely inferred by applying a long classifying set of rules to the information. *Conservation tillage* is an example of low reliability. Farmers explicitly declare to apply conservation tillage in October as it is subsidized, but if weather conditions are not suitable for tilling after the declaration date, fields may remain covered by vegetation. As vegetation cover is considered the more sustainable option, "no-till" is not subject to a penalty. As a result, probability of *stubble* field samples mislabeled as *conservation tillage* is high.

For improving the quality of the reference data, one could consider unsupervised clustering techniques as in [56] to discover structure and compare with supervised techniques and the assumed labels. During data exploration we performed an initial trial with spectral and K-nearest neighbor clustering on the density representation of the objects, the results of which did suggest some internal structure within the given classes of the dataset. Additionally, specific geospatial properties such as latitude can be significant in Finland with varying microclimates affecting vegetation cover and could be used as additional features to improve the accuracy.

4.2. Modelling Aspects

The proposed computational method is applicable also for other agricultural monitoring tasks besides the specific task of tillage and vegetation cover classification, such as crop yield prediction. Furthermore, it can be applied to object-based remote sensing tasks also beyond agricultural monitoring. Hence, we also provide a brief discussion of the method itself. All forms of multivariate density estimates were observed to outperform simple object representations of aggregate summaries and marginal histograms for supervised classification of small and variably sized objects, even though the latter are easier to estimate. Proper density estimators outperformed multivariate histograms in some cases, but not in all and the difference was in general unexpectedly small. We believe this is primarily because evaluation is extremely noisy for the scenarios (the smallest datasets with the smallest objects) that would most benefit from smoothing and uncertainty quantification; more direct measures of representation quality could be considered for stronger conclusions.

Regarding the representations, GKDE [57,58] has only negligible computational overhead compared to histograms and no additional tuning parameters (due to the automatic rule for selecting the bandwidths h_d), and hence works as general plug-in replacement for histograms—we did not observe any reasons to prefer using histograms over GKDE. LGPDE [22] was demonstrated to further slightly improve accuracy while facilitating uncertainty propagation for arbitrary classifiers, but this comes with a significant computational overhead, even when using the more efficient Laplace approximation. Our results indicate that there is value in explicitly modeling the uncertainty of the density estimate itself but we do not yet provide a practical approach for arbitrary problems; to proceed towards computationally efficient but still accurate LGPDE, one could use sparse variational approximations [59]. Besides LGPDE, we could also consider other Bayesian density estimators, for instance, Dirichlet process mixtures [60].

In this work we only considered non-parametric density estimators as representations, since they are generally applicable for all imaging modalities. For SAR specifically, an alternative would be to consider *parametric* distribution estimates [61]. For instance, a Gamma distribution model can be used for pixel intensities [62], and complex-valued SAR backscatter data can modeled using a complex Wishart distribution [63–65]. However, actual observed signal can display behaviors that require increasingly sophisticated distributions to decrease model bias [61].

Finally, we make three generalizing notes on the method. First, it can be applied directly to representing time series of the observations, either by promoting time to an additional feature dimension of the density or by concatenating the representations for the individual time points. Second, the raster images A_i may represent multiple sources with different spatial resolutions, multiple bands and features. Third, an object's pixel set X can be conventionally defined by a geospatial vector polygon, but does not necessarily need to be contiguous or of any regular shape. For instance, it could be a scattered set of individual pixels occluded by atmospheric haze in a cloud detection application.

5. Conclusions

Remote sensing tasks related to agricultural land use frequently involve delineated areas of crop fields, for example, field parcels, as bounded objects of interest that have similarly distributed pixel content with varying degrees of texture. We provided a practical computational pipeline for large-scale agricultural monitoring tasks, combining robust distributional representations computed for individual parcels with standard classifiers. The approach is compatible with arbitrary remote sensing images. We demonstrated the approach here on Sentinel data, using VH and VV polarities of SAR and for NDVI and NDTI spectral indices of MSI, but the computational pipeline is compatible with other EO data sources and indices. Importantly, our approach is amenable for easy data fusion as each source can be processed independently and in parallel. We described and evaluated alternative density estimators for forming the representation, ranging from simple histograms to a non-parametric Bayesian density estimator of LGPDE, and showed that both provide robust and reliable representations. The advantage of using proper estimators is bigger for small training sets consisting of small and varying-sized objects, but we also observed standard multivariate histograms to perform well in most cases. A simple parametric multivariate density estimator GKDE was found to provide the best compromise between computational complexity and accuracy, but for end-to-end uncertainty quantification the LGPDE may offer further advantages.

The approach was demonstrated in the task of off-season soil tillage classification in Southern Finland for the purpose of administrative monitoring. We used a collection of 127,757 field parcels monitored in April 2018 and annotated to six tillage method and vegetation cover classes. The task is challenging due to the small size of many of the individual parcels, unequal distribution of classes, and in particular because of highly similar classes and mislabeling of both training and evaluation instances. By combining MSI and SAR data using the representations that can be estimated already from small parcels, we reached 70% accuracy with six classes and 82% accuracy for a simplified problem considering only the three most important classes. This is already sufficient for partial automation of large-scale tillage monitoring. Furthermore, we showed that, for the three-class problem, we can reach 78% accuracy already on a very small training set of less than 500 parcels. The proposed computational method is applicable also for other agricultural monitoring tasks, such as crop yield prediction. We expect the proposed method to generalize from polygonal annotations of crop fields to other formats of segment annotation and types of human-regulated land use.

Supplementary Materials: The following supporting information can be downloaded at: https://www.mdpi.com/article/10.3390/app12020679/s1, S1. Sentinel-1 data. S2. Sentinel-2 data.

Author Contributions: Conceptualization, M.L., M.Y.-H. and A.K.; methodology, M.L., M.Y.-H. and A.K.; software, M.L. and M.Y.-H.; data curation, M.Y.-H.; writing, M.L., M.Y.-H. and A.K.; supervision, A.K. All authors have read and agreed to the published version of the manuscript.

Funding: This work was supported by the European Union (grant 101033957) and Academy of Finland Flagship programme: Finnish Center for Artificial Intelligence, FCAI.

Data Availability Statement: The parcel delineation data used for the study is not currently publicly available and the authors do not have permission to publish any identifying details. However, we publish a high-level preprocessed and anonymized dataset that does not reveal geometry or geographical location of individual parcels to protect the individual private small farmers that own them. The software for the data flow, the computational methods, the data and instructions for easy execution as a public Docker container are available at: https://github.com/luotsi/vegcover-manuscript-12_2021.

Acknowledgments: Data from Sentinel-1 and Sentinel-2 originates from the European Copernicus Sentinel Program. We thank Mikko Strahlendorff of the Finnish Meteorological Institute for processing the Sentinel-1 mosaics and his valuable comments concerning environmental monitoring with Sentinel-1 and also would like to acknowledge the CSC – IT Center for Science, Finland, for computational resources and user support. Open access funding provided by University of Helsinki.

Conflicts of Interest: The authors declare no conflict of interest. The funders had no role in the design of the study; in the collection, analyses, or interpretation of data; in the writing of the manuscript, or in the decision to publish the results.

References

1. Orynbaikyzy, A.; Gessner, U.; Conrad, C. Crop type classification using a combination of optical and radar remote sensing data: A review. *Int. J. Remote Sens.* **2019**, *40*, 6553–6595. [CrossRef]
2. Felegari, S.; Sharifi, A.; Moravej, K.; Amin, M.; Golchin, A.; Muzirafuti, A.; Tariq, A.; Zhao, N. Integration of Sentinel 1 and Sentinel 2 Satellite Images for Crop Mapping. *Appl. Sci.* **2021**, *11*, 10104. [CrossRef]
3. Garioud, A.; Valero, S.; Giordano, S.; Mallet, C. Recurrent-based regression of Sentinel time series for continuous vegetation monitoring. *Remote Sens. Environ.* **2021**, *263*, 112419. [CrossRef]
4. Mateo-García, G.; Gómez-Chova, L.; Camps-Valls, G. Convolutional neural networks for multispectral image cloud masking. In Proceedings of the 2017 IEEE International Geoscience and Remote Sensing Symposium, Fort Worth, TX, USA, 23–28 July 2017; pp. 2255–2258.
5. Kussul, N.; Lavreniuk, M.; Skakun, S.; Shelestov, A. Deep learning classification of land cover and crop types using remote sensing data. *IEEE Geosci. Remote Sens. Lett.* **2017**, *14*, 778–782. [CrossRef]
6. Luotamo, M.; Metsämäki, S.; Klami, A. Multiscale Cloud Detection in Remote Sensing Images Using a Dual Convolutional Neural Network. *IEEE Trans. Geosci. Remote Sens.* **2021**, *59*, 4972–4983. [CrossRef]
7. Fu, G.; Liu, C.; Zhou, R.; Sun, T.; Zhang, Q. Classification for high resolution remote sensing imagery using a fully convolutional network. *Remote Sens.* **2017**, *9*, 498. [CrossRef]
8. Salehi, B.; Daneshfar, B.; Davidson, A.M. Accurate crop-type classification using multi-temporal optical and multi-polarization SAR data in an object-based image analysis framework. *Int. J. Remote Sens.* **2017**, *38*, 4130–4155. [CrossRef]
9. McNairn, H.; Ellis, J.; Van Der Sanden, J.; Hirose, T.; Brown, R. Providing crop information using RADARSAT-1 and satellite optical imagery. *Int. J. Remote Sens.* **2002**, *23*, 851–870. [CrossRef]
10. Rußwurm, M.; Körner, M. Self-attention for raw optical Satellite Time Series Classification. *ISPRS J. Photogramm. Remote Sens.* **2020**, *169*, 421–435. [CrossRef]
11. Voormansik, K.; Zalite, K.; Sünter, I.; Tamm, T.; Koppel, K.; Verro, T.; Brauns, A.; Jakovels, D.; Praks, J. Separability of Mowing and Ploughing Events on Short Temporal Baseline Sentinel-1 Coherence Time Series. *Remote Sens.* **2020**, *12*, 3784. [CrossRef]
12. De Vroey, M.; Radoux, J.; Defourny, P. Grassland Mowing Detection Using Sentinel-1 Time Series: Potential and Limitations. *Remote Sens.* **2021**, *13*, 348. [CrossRef]
13. Planque, C.; Lucas, R.; Punalekar, S.; Chognard, S.; Hurford, C.; Owers, C.; Horton, C.; Guest, P.; King, S.; Williams, S.; et al. National Crop Mapping Using Sentinel-1 Time Series: A Knowledge-Based Descriptive Algorithm. *Remote Sens.* **2021**, *13*, 846. [CrossRef]
14. Swain, M.J.; Ballard, D.H. Color indexing. *Int. J. Comput. Vis.* **1991**, *7*, 11–32. [CrossRef]
15. Schiele, B.; Crowley, J.L. Probabilistic object recognition using multidimensional receptive field histograms. In Proceedings of the IEEE 13th International Conference on Pattern Recognition, Washington, DC, USA, 25–29 August 1996; Volume 2, pp. 50–54.
16. Barla, A.; Odone, F.; Verri, A. Histogram intersection kernel for image classification. In Proceedings of the IEEE 2003 International Conference on Image Processing, Barcelona, Spain, 14–18 September 2003; Volume 3, pp. 3–513.
17. Zhang, G.; Jia, X.; Kwok, N.M. Super pixel based remote sensing image classification with histogram descriptors on spectral and spatial data. In Proceedings of the 2012 IEEE International Geoscience and Remote Sensing Symposium, Munich, Germany, 22–27 July 2012; pp. 4335–4338.
18. Yang, W.; Hou, K.; Liu, B.; Yu, F.; Lin, L. Two-stage clustering technique based on the neighboring union histogram for hyperspectral remote sensing images. *IEEE Access* **2017**, *5*, 5640–5647. [CrossRef]
19. Demir, B.; Bruzzone, L. Histogram-based attribute profiles for classification of very high resolution remote sensing images. *IEEE Trans. Geosci. Remote Sens.* **2015**, *54*, 2096–2107. [CrossRef]
20. Simonoff, J.S. *Smoothing Methods in Statistics*; Springer Science & Business Media: Berlin/Heidelberg, Germany, 2012.
21. Scott, D.W. *Multivariate Density Estimation: Theory, Practice, and Visualization*; John Wiley & Sons: Hoboken, NJ, USA, 2015.
22. Riihimäki, J.; Vehtari, A. Laplace approximation for logistic Gaussian process density estimation and regression. *Bayesian Anal.* **2014**, *9*, 425–448. [CrossRef]
23. Baker, J.; Laflen, J. Water quality consequences of conservation tillage: New technology is needed to improve the water quality advantages of conservation tillage. *J. Soil Water Conserv.* **1983**, *38*, 186–193.
24. Bechmann, M.E.; Bøe, F. Soil tillage and crop growth effects on surface and subsurface runoff, loss of soil, phosphorus and nitrogen in a cold climate. *Land* **2021**, *10*, 77. [CrossRef]

25. Daughtry, C. Discriminating Crop Residues from Soil by Shortwave Infrared Reflectance. *Agron. J.* **2001**, *93*, 125–131. [CrossRef]
26. Daughtry, C.; Doraiswamy, P.; Hunt, E.; Stern, A.; McMurtrey, J.; Prueger, J. Remote sensing of crop residue cover and soil tillage intensity. *Soil Tillage Res.* **2006**, *91*, 101–108. [CrossRef]
27. Quemada, M.; Daughtry, C.S.T. Spectral Indices to Improve Crop Residue Cover Estimation under Varying Moisture Conditions. *Remote Sens.* **2016**, *8*, 660. [CrossRef]
28. McNairn, H.; Boisvert, J.; Major, D.; Gwyn, Q.; Brown, R.; Smith, A. Identification of Agricultural Tillage Practices from C-Band Radar Backscatter. *Can. J. Remote Sens.* **1996**, *22*, 154–162. [CrossRef]
29. McNairn, H.; Duguay, C.; Boisvert, J.; Huffman, E.; Brisco, B. Defining the Sensitivity of Multi-Frequency and Multi-Polarized Radar Backscatter to Post-Harvest Crop Residue. *Can. J. Remote Sens.* **2001**, *27*, 247–263. [CrossRef]
30. McNairn, H.; Duguay, C.; Brisco, B.; Pultz, T. The effect of soil and crop residue characteristics on polarimetric radar response. *Remote Sens. Environ.* **2002**, *80*, 308–320. [CrossRef]
31. Zhang, J. Multi-source remote sensing data fusion: Status and trends. *Int. J. Image Data Fusion* **2010**, *1*, 5–24. [CrossRef]
32. Haas, J.; Ban, Y. Sentinel-1A SAR and Sentinel-2A MSI data fusion for urban ecosystem service mapping. *Remote Sens. Appl. Soc. Environ.* **2017**, *8*, 41–53. [CrossRef]
33. Ban, Y.; Jacob, A. Object-based fusion of multitemporal multiangle ENVISAT ASAR and HJ-1B multispectral data for urban land-cover mapping. *IEEE Trans. Geosci. Remote Sens.* **2013**, *51*, 1998–2006. [CrossRef]
34. Veloso, A.; Mermoz, S.; Bouvet, A.; Le Toan, T.; Planells, M.; Dejoux, J.F.; Ceschia, E. Understanding the temporal behavior of crops using Sentinel-1 and Sentinel-2-like data for agricultural applications. *Remote Sens. Environ.* **2017**, *199*, 415–426. [CrossRef]
35. Orynbaikyzy, A.; Gessner, U.; Mack, B.; Conrad, C. Crop type classification using fusion of Sentinel-1 and Sentinel-2 data: Assessing the impact of feature selection, optical data availability, and parcel sizes on the accuracies. *Remote Sens.* **2020**, *12*, 2779. [CrossRef]
36. Azzari, G.; Grassini, P.; Edreira, J.I.R.; Conley, S.; Mourtzinis, S.; Lobell, D.B. Satellite mapping of tillage practices in the North Central US region from 2005 to 2016. *Remote Sens. Environ.* **2019**, *221*, 417–429. [CrossRef]
37. Denize, J.; Hubert-Moy, L.; Betbeder, J.; Corgne, S.; Baudry, J.; Pottier, E. Evaluation of Using Sentinel-1 and -2 Time-Series to Identify Winter Land Use in Agricultural Landscapes. *Remote Sens.* **2019**, *11*, 37. [CrossRef]
38. Finnish Meteorological Institute. Sentinel-1 SAR-Image Mosaic (S1sar). Available online: https://ckan.ymparisto.fi/dataset/sentinel-1-sar-image-mosaic-s1sar-sentinel-1-sar-kuvamosaiikki-s1sar (accessed on 8 February 2021).
39. European Space Agency. Sentinel-1 SAR Interferometric Wide Swath. Available online: https://sentinels.copernicus.eu/web/sentinel/user-guides/sentinel-1-sar/acquisition-modes/interferometric-wide-swath (accessed on 8 February 2021).
40. European Space Agency. Level-2A Algorithm Overview/NDVI. Available online: https://sentinels.copernicus.eu/web/sentinel/technical-guides/sentinel-2-msi/level-2a/algorithm (accessed on 26 March 2021).
41. Van Deventer, A.; Ward, A.; Gowda, P.; Lyon, J. Using thematic mapper data to identify contrasting soil plains and tillage practices. *Photogramm. Eng. Remote Sens.* **1997**, *63*, 87–93.
42. Zhang, H.; Kang, J.; Xu, X.; Zhang, L. Accessing the temporal and spectral features in crop type mapping using multi-temporal Sentinel-2 imagery: A case study of Yi'an County, Heilongjiang province, China. *Comput. Electron. Agric.* **2020**, *176*, 105618. [CrossRef]
43. Parzen, E. On estimation of a probability density function and mode. *Ann. Math. Stat.* **1962**, *33*, 1065–1076. [CrossRef]
44. O'Brien, T.A.; Kashinath, K.; Cavanaugh, N.R.; Collins, W.D.; O'Brien, J.P. A fast and objective multidimensional kernel density estimation method: FastKDE. *Comput. Stat. Data Anal.* **2016**, *101*, 148–160. [CrossRef]
45. Leonard, T. Density estimation, stochastic processes and prior information. *J. R. Stat. Soc. Ser. B (Methodol.)* **1978**, *40*, 113–132. [CrossRef]
46. Tokdar, S.T. Towards a faster implementation of density estimation with logistic Gaussian process priors. *J. Comput. Graph. Stat.* **2007**, *16*, 633–655. [CrossRef]
47. Carpenter, B.; Gelman, A.; Hoffman, M.D.; Lee, D.; Goodrich, B.; Betancourt, M.; Brubaker, M.A.; Guo, J.; Li, P.; Riddell, A. Stan: A probabilistic programming language. *J. Stat. Softw.* **2017**, *76*, 1–32. [CrossRef]
48. Van Tricht, K.; Gobin, A.; Gilliams, S.; Piccard, I. Synergistic Use of Radar Sentinel-1 and Optical Sentinel-2 Imagery for Crop Mapping: A Case Study for Belgium. *Remote Sens.* **2018**, *10*, 1642. [CrossRef]
49. Vreugdenhil, M.; Wagner, W.; Bauer-Marschallinger, B.; Pfeil, I.; Teubner, I.; Rüdiger, C.; Strauss, P. Sensitivity of Sentinel-1 Backscatter to Vegetation Dynamics: An Austrian Case Study. *Remote Sens.* **2018**, *10*, 1396. [CrossRef]
50. Vreugdenhil, M.; Navacchi, C.; Bauer-Marschallinger, B.; Hahn, S.; Steele-Dunne, S.; Pfeil, I.; Dorigo, W.; Wagner, W. Sentinel-1 Cross Ratio and Vegetation Optical Depth: A Comparison over Europe. *Remote Sens.* **2020**, *12*, 3404. [CrossRef]
51. Ho Tong Minh, D.; Ienco, D.; Gaetano, R.; Lalande, N.; Ndikumana, E.; Osman, F.; Maurel, P. Deep Recurrent Neural Networks for Winter Vegetation Quality Mapping via Multitemporal SAR Sentinel-1. *IEEE Geosci. Remote Sens. Lett.* **2018**, *15*, 464–468. [CrossRef]
52. Denize, J.; Hubert-Moy, L.; Pottier, E. Polarimetric SAR Time-Series for Identification of Winter Land Use. *Sensors* **2019**, *19*, 5574. [CrossRef]
53. McNairn, H.; Champagne, C.; Shang, J.; Holmstrom, D.; Reichert, G. Integration of optical and Synthetic Aperture Radar (SAR) imagery for delivering operational annual crop inventories. *ISPRS J. Photogramm. Remote Sens.* **2009**, *64*, 434–449. [CrossRef]

54. Inglada, J.; Vincent, A.; Arias, M.; Marais-Sicre, C. Improved Early Crop Type Identification By Joint Use of High Temporal Resolution SAR And Optical Image Time Series. *Remote Sens.* **2016**, *8*, 362. [CrossRef]
55. Torbick, N.; Huang, X.; Ziniti, B.; Johnson, D.; Masek, J.; Reba, M. Fusion of Moderate Resolution Earth Observations for Operational Crop Type Mapping. *Remote Sens.* **2018**, *10*, 1058. [CrossRef]
56. Wang, S.; Azzari, G.; Lobell, D.B. Crop type mapping without field-level labels: Random forest transfer and unsupervised clustering techniques. *Remote Sens. Environ.* **2019**, *222*, 303–317. [CrossRef]
57. John, G.H.; Langley, P. Estimating continuous distributions in Bayesian classifiers (orig. 1995). *arXiv* **2013**, arXiv:1302.4964.
58. Gramacki, A. *Nonparametric Kernel Density Estimation and Its Computational Aspects;* Springer: Berlin/Heidelberg, Germany, 2018.
59. Titsias, M.K. Variational Learning of Inducing Variables in Sparse Gaussian Processes. In Proceedings of the 12th International Conference on Artificial Intelligence and Statistics, Clearwater Beach, FL, USA, 16–18 April 2009; pp. 567–574.
60. Escobar, M.D.; West, M. Bayesian density estimation and inference using mixtures. *J. Am. Stat. Assoc.* **1995**, *90*, 577–588. [CrossRef]
61. Deng, X.; López-Martínez, C.; Chen, J.; Han, P. Statistical Modeling of Polarimetric SAR Data: A Survey and Challenges. *Remote Sens.* **2017**, *9*, 348. [CrossRef]
62. Shuai, Y.; Sun, H.; Xu, G. SAR image segmentation based on level set with stationary global minimum. *IEEE Geosci. Remote Sens. Lett.* **2008**, *5*, 644–648. [CrossRef]
63. Nielsen, A.A.; Skriver, H.; Conradsen, K. Complex Wishart Distribution Based Analysis of Polarimetric Synthetic Aperture Radar Data. In Proceedings of the IEEE 2007 International Workshop on the Analysis of Multi-temporal Remote Sensing Images, Leuven, Belgium, 18–20 July 2007; pp. 1–6.
64. Sánchez, S.; Marpu, P.R.; Plaza, A.; Paz-Gallardo, A. Parallel implementation of polarimetric synthetic aperture radar data processing for unsupervised classification using the complex Wishart classifier. *IEEE J. Sel. Top. Appl. Earth Obs. Remote Sens.* **2015**, *8*, 5376–5387. [CrossRef]
65. Goumehei, E.; Tolpekin, V.; Stein, A.; Yan, W. Surface water body detection in polarimetric SAR data using contextual complex Wishart classification. *Water Resour. Res.* **2019**, *55*, 7047–7059. [CrossRef]

applied sciences

MDPI

Article

Novel Vision Transformer–Based Bi-LSTM Model for LU/LC Prediction—Javadi Hills, India

Sam Navin Mohanrajan and Agilandeeswari Loganathan *

School of Information Technology and Engineering, Vellore Institute of Technology, Vellore 632014, Tamil Nadu, India; samnavin.m@vit.ac.in
* Correspondence: agila.l@vit.ac.in

Abstract: Continuous monitoring and observing of the earth's environment has become interactive research in the field of remote sensing. Many researchers have provided the Land Use/Land Cover information for the past, present, and future for their study areas around the world. This research work builds the Novel Vision Transformer–based Bidirectional long-short term memory model for predicting the Land Use/Land Cover Changes by using the LISS-III and Landsat bands for the forest- and non-forest-covered regions of Javadi Hills, India. The proposed Vision Transformer model achieves a good classification accuracy, with an average of 98.76%. The impact of the Land Surface Temperature map and the Land Use/Land Cover classification map provides good validation results, with an average accuracy of 98.38%, during the process of bidirectional long short-term memory–based prediction analysis. The authors also introduced an application-based explanation of the predicted results through the Google Earth Engine platform of Google Cloud so that the predicted results will be more informative and trustworthy to the urban planners and forest department to take proper actions in the protection of the environment.

Keywords: Land Use/Land Cover; LISS-III; Landsat; Vision Transformer; Bidirectional long-short term memory; Google Earth Engine; Explainable Artificial Intelligence

Citation: Mohanrajan, S.N.; Loganathan, A. Novel Vision Transformer–Based Bi-LSTM Model for LU/LC Prediction—Javadi Hills, India. *Appl. Sci.* **2022**, *12*, 6387. https://doi.org/10.3390/app12136387

Academic Editor: Stefania Pindozzi

Received: 6 May 2022
Accepted: 17 June 2022
Published: 23 June 2022

1. Introduction

The Land Use/Land Cover (LU/LC) prediction is one of the most significant applications of remote sensing and GIS technology. The main causes of LU/LC changes are agricultural/crop damage, wetland change, deforestation, urban expansion, and vegetation loss. Several researchers working in this application area for many years had different findings for their study areas around the world. The importance of this LU/LC prediction research is to provide information about the landscape changes of the specific study area to the government officials, forest department, urban planners, and social workers for the protection of the LU/LC environment [1–3]. Remote sensing technology provides information about the satellite data and helps in performing the LU/LC prediction research effectively. Researchers have used different remote sensing satellite systems for acquiring the data, and some of the satellite system databases are Advanced Land Imager (ALI), Hyperion data, Linear Imaging Self-Scanning Sensor III (LISS-III), Linear Imaging Self-Scanning Sensor IV (LISS-IV), Landsat Series, Sentinel-2A and -2B, Moderate Resolution Imaging Spectroradiometer (MODIS), Rapid Eye Earth Imaging System (REIS), and ASTER Global DEM (Digital Elevation Model). Other data acquisition for performing the LU/LC prediction research can be made through aerial photographs, Google Earth images, government, and field or ground survey data. The advantage of the satellite and airborne data has been used in many applications areas such as oceanography, landscape monitoring, weather forecasting, biodiversity conservation, forestry, cartography, surveillance, and warfare [4–10]. The different band in the multispectral data has been widely used in monitoring the LU/LC changes around the world. The visible (red–blue–green), near infrared (NIR), short-wave infrared (SWIR), and TIRS (thermal infrared sensor) bands were used for calculating the most important LU/LC indices, such as the Land Surface Temperature (LST), Normalized Difference Vegetation Index (NDVI), Normalized Difference Moisture Index (NDMI), Normalized Difference Water Index (NDWI), Normalized Difference Built-Up Index (NBBI), and Normalized Difference Salinity Index (NDSI) [11,12].

The primary processing for correcting the noise and cloud effects in the satellite and airborne data has been achieved through preprocessing. The multispectral satellite data have been used for performing effective research on LU/LC analysis. The noise, atmospheric, geometric, topographic, and radiometric errors in the raw multispectral satellite data are corrected by using the primary process of image preprocessing. Different methods have been used for correcting the satellite image errors, and some of them are Image Registration, Independent Component Analysis (ICA), Linear Discriminant Analysis (LDA), Discrete Wavelet Transform (DWT), Resampling, Quick Atmospheric Correction (QUAC) module, Minimum Noise Fraction (MNF), Dark Object Subtraction (DOS) module, Orthorectification, Rescaling, Principal Component Analysis (PCA), F-mask method, FLAASH (Fast Line-of-Sight Atmospheric Analysis of Hypercubes) module, ASCII Coordinate Conversion, Apparent Reflectance Model (ARM), Georeferencing, Image De-striping, and Lookup Table (LUT) Stretch and Point Spread Convolution methods [13–17]. LU/LC classification has been performed by using different classification algorithms for finding the LU/LC types of a particular location. Some of the LU/LC classification algorithms used by researchers are Maximum Likelihood Classification (MLC), Support Vector Machine (SVM) Classification, k-Nearest Neighbor Classification (kNN), K-Means Clustering, Mahalanobis Distance Classification (MDC), Classification and Regression Tree (CART), Logistic Regression Model (LRM), Artificial Neural Network (ANN) Classification, Random Forest Classification (RFC), Spectral Angle Mapper (SAM) Classification, Minimum Distance to Mean Classification (MDM), Parallelepiped Classification (PLC), Multivariate Adaptive Regression Spline (MARS), Fuzzy C Means (FCM), and Iterative Self-Organizing Data Analysis (ISODATA) clustering. The different LU/LC class types classified are built-up areas, water bodies, forest-cover areas, wetlands, and vegetation areas. The accuracy assessment was performed by comparing the LU/LC classified map with the ground truth data. Based on the accuracy assessment, the performance of the classification method has been measured. The LU/LC change detection has been performed between the LU/LC time-series classified map [18–22].

The LU/LC prediction was performed by calibrating the dependent and independent variables. The LU/LC change map is considered the dependent variable, and the factors associated with the LU/LC change are considered as the independent variables. The factors associated with LU/LC change include slope, elevation, aspect, climatic variables, distance variables (distance from road, forest edge, agricultural land, water bodies, and urban areas), and census data. LU/LC prediction has been performed by using different algorithms for finding the future LU/LC changes in a particular location. Some of the algorithms used by researchers are based on the Markov Chain (MC), Cellular Automata (CA), Conventional Neural Network (CNN), Recurrent Neural Network (RNN), Gated Recurrent Unit (GRU), and Long Short-Term Memory Neural Network (LSTM) [23–30]. In recent technologies, transformer-based models are widely used and processed in image-processing applications. The transformer-based deep-learning model is considered the state-of-the-art model in image recognition, as it focuses on the confident part of inputs to get more efficient results [31,32]. Many researchers have worked on the transformer-based model in the field of natural language processing (NLP) [33,34]. Researchers also performed the transformer-based models in image-recognition problems through remote sensing analysis. The Vision Transformers have been used widely for remote sensing applications. The advantage of using the Vision Transformers for remote-sensing applications provides better classification accuracy than the standard algorithms [35–37].

Explainable artificial intelligence (XAI) is a process of allowing the users to understand and trust the outputs produced by the machine-learning and deep-learning models.

XAI conveys the importance of transparency (presents the significant way of reaching the goal), justification (clarifying why the results provided by the prediction model are acceptable), informativeness (providing new information to researchers), and uncertainty estimation (computing how trustworthy a prediction model is) [38,39]. The few XAI tools for explaining the results of machine-learning and deep-learning models include LIME (Local Interpretable Model-Agnostic Explanations), DeepLIFT (Deep Learning Important Features), SHAP (Shapley Additive explanations), LRP (Layer-Wise Relevance Propagation), Saliency Maps, CIU (Contextual Importance and Utility), DALEX (Model Agnostic Language for Exploration and Explanation), Skater, Occlusion Analysis, and Integrated Gradients/SmoothGrad. The usage of XAI tools varies for every application area of machine- and deep-learning models [40–43].

In the field of remote sensing, we observed that researchers had used the supervised and unsupervised machine-learning models for performing the LU/LC classification and prediction analysis. The supervised-learning models (MLC, SVM, KNN, MDC, CART, LR, ANN, RFC, SAM, MDM, PLC, MARS, MC, CA, CNN, RNN, and LSTM) are considered to be more accurate than the unsupervised-learning models (KM, FCM, and ISODATA). The unsupervised learning is performed with no prior information about the data, and there are no training data available for training the unsupervised algorithms. It performs the LU/LC classification by learning the data without any class labels. The advantage of unsupervised algorithms helps in finding the unknown patterns in the image, which are more difficult to find by using the normal method. The results of the unsupervised classification algorithms were used as the input training data for the supervised algorithms. The advantage of using the unsupervised methods (KM, FCM, and ISODATA) is that they help in separating the similar and dissimilar pixels into clusters through the distance functions. The disadvantage of the unsupervised-learning model is the high computational time when the data are unstructured. The main disadvantage is that unsupervised algorithms are not used during the process of LU/LC prediction analysis since it requires both past and present training data. The supervised learning depends on the user-defined training data for classifying the LU/LC classes. The MLC, SVM, KNN, MDC, SAM, MDM, CART, MARS, and PLC techniques were widely used for classifying the LU/LC classes. The models based on LR, ANN, MC, CA, CNN, RNN, GRU, and LSTM were widely used during the process of LU/LC prediction analysis. The supervised classifiers help in providing the results by using previous experiences. The real-world computation problems were solved by using supervised-learning methods. It performs the classification and prediction with the knowledge of class labels. The supervised-learning models were used during the process of LU/LC prediction analysis. The past and present training data have been analyzed and processed in supervised learning. The accuracy results of the standard classification and prediction algorithms differ for each study area. The results mainly depend on the training parameters and the complexity of the input data. In terms of LU/LC analysis, the misclassification rate has been observed due to the overlapping of pixels in the satellite image. In all the neural network models, the time taken for training and validation is more for massive datasets. The disadvantage of the standard LU/LC machine-learning model is lacking knowledge of the predicted map, resulting in difficulties for the urban planners when further processing the data [44–46].

The rest of the paper is given as follows: Section 2 explains the motivation and contribution of this work. Section 3 explains the proposed methodology of this research work. Section 4 explains the materials and methods proposed in our research work, Section 5 provides the training parameters and validation results of each method used in this research work, Section 6 explains the comparative analysis of our LU/LC prediction model, and Section 7 delivers the conclusion of this research work.

2. Motivations and Contributions

The main contribution of researchers around the world is to provide new innovative information to society, government, and different educational sectors in their respective domains. Many researchers had motivated and contributed to the significant problem of LU/LC prediction analysis. The LU/LC change detection for past, present, and future analysis has been a key research topic to understand the environmental change on the earth's surface. Hence, LU/LC feature extraction has emerged as an essential research aspect, and therefore, the standard and accurate methodology for LU/LC classification and prediction should be made. By use of satellite system technology, we can perform our research on LU/LC change analysis. The main need of this research is to assist the land-resource management, government officials, forest department, and urban planners to take action to protect the earth's environment. From the brief survey on different classification and prediction algorithms, we have found that the sustainable growth of the LU/LC environment for the time-series data requires an accurate classification and prediction map, which was considered the strong motivation for our study. The main contributions of our work are as follows:

- The novel Vision Transformer–based Bidirectional long-short term memory (Bi-LSTM) model is proposed for predicting the LU/LC changes of Javadi Hills, India.
- The use of the LST map with the Vision Transformer–based LU/LC classification map provides the main advantage in achieving good validation accuracy with less computational time during the process of LU/LC prediction analysis through the Bi-LSTM model.
- The impacts of the Multi-Satellite System (LISS-III multispectral with the Landsat TIRS, RED, and NIR bands) on the proposed LU/LC prediction model for Javadi Hills, India, are analyzed.
- Explainable Artificial Intelligence (XAI), an application-based explanation, is also introduced for validating the predicted results through the Google Earth Engine platform of Google Cloud so that the predicted results will be more informative and trustworthy to the urban planners and forest department to take appropriate measures in the protection of the environment.

3. Materials and Methods

This section elaborates the various stages of our proposed prediction model: (i) the study area and data acquisition, (ii) proposed Vision Transformer–based LULC classification, (iii) description of expression for calculating and analyzing the LST map, (iv) Bi-LSTM model for LULC prediction, and (v) description of explainable AI and its importance.

3.1. Study Area and Data Acquisition

The study area in our research work is the forest- and non-forest-covered area of Javadi Hills with the geographic coordinates falling between 78.75 E 12.5 N and 79.0 E 12.75 N. Our study area is located across the Eastern Ghats of Vellore and Tiruvannamalai district, Tamil Nadu, India. The UTM (Universal Transverse Mercator) GCS (geographic coordinate system)/WGS (World Geodetic System) 1984 (44 N) projection system was processed for the extracted satellite data. The location of the Javadi Hills map was extracted from Google Earth Engine (https://www.google.com/earth/ (accessed on 10 November 2021)). The map view of our study area was prepared by using ArcGIS (Version 10.1 developed by ESRI (http://www.esri.com/software/arcgis)) geospatial software, and it is shown in Figure 1.

The multispectral LISS-III satellite images for the years 2012 and 2015 were collected from the Bhuvan Indian Geo-Platform of ISRO (www.bhuvan.com (accessed on 9 December 2019)). The extracted LISS-III multispectral data of Javadi Hills were used for the LU/LC classification process. The TIRS, RED, and NIR bands of Landsat 8 (Band 10) and Landsat 7 (Band 6) were collected from the United States Geological Survey (USGS), United States (https://earthexplorer.usgs.gov (accessed on 16 December 2019)) and were used for the estimation of LST. There was no TIRS Band in the LISS-III sensor, so we extracted the TIRS image from the Landsat Satellite data for our study area. The importance of the TIRS band used in our paper provides the impact of LST on Javadi Hills for the years 2012 and 2015. Table 1 shows the source and characteristics of the remotely sensed satellite images. In our research work, the atmospheric corrections were made to provide good visibility to the extracted LISS-III multispectral satellite image of Javadi Hills. The scan-line error correction was made for filling the gaps in the extracted Landsat TIRS image of Javadi Hills. The geometric correction was made to extract the Region of Interest (ROI) coordinates in the forest- and non-forest-covered area of Javadi Hills that falls between 78.80 E 12.56 N and 78.85 E 12.60 N. Figure 2 represents the preprocessed image of multispectral LISS-III data of Javadi Hills for the years 2012 and 2015. Figures 3–5 represent the preprocessed Landsat TIRS, RED, and NIR bands of the Javadi Hills for the years 2012 and 2015.

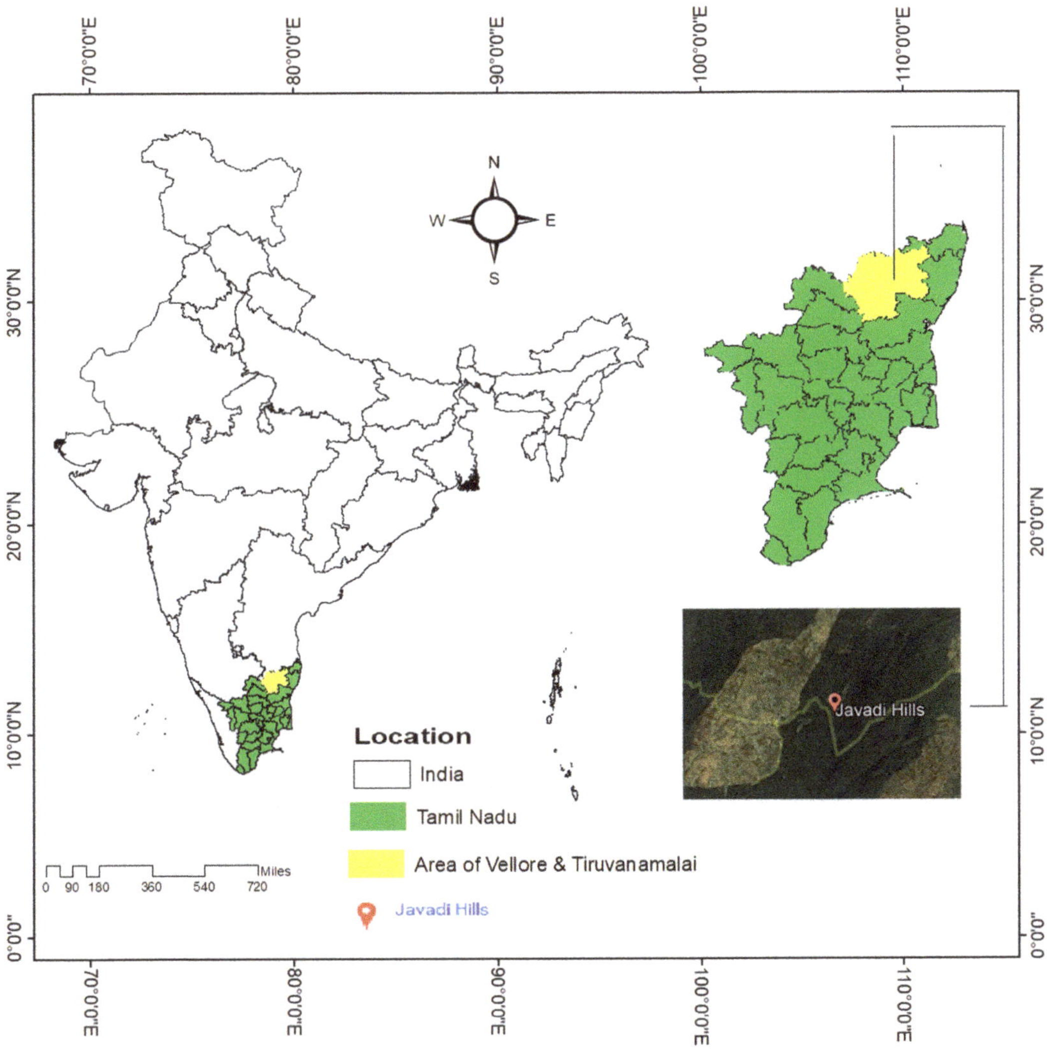

Figure 1. Study Area—Javadi Hills, India.

Table 1. Characteristics and sources of the satellite images.

Satellite	Path	Sensor	Year	Source
Resourcesat-1/2	101/064	LISS-III	18 February 2012 22 March 2015	Bhuvan Indian Geo-Platform of ISRO (www.bhuvan.com (accessed on 9 December 2019))
Landsat 8 OLI/TI and Landsat 7 (ETM+)	143/51	Operational Land Imager (OLI) and the Thermal Infrared (TI) Sensor	27 March 2015	United States Geological Survey (https://earthexplorer.usgs.gov (accessed on 16 December 2019))
		Enhanced Thematic Mapper Plus (ETM+)	26 March 2012	

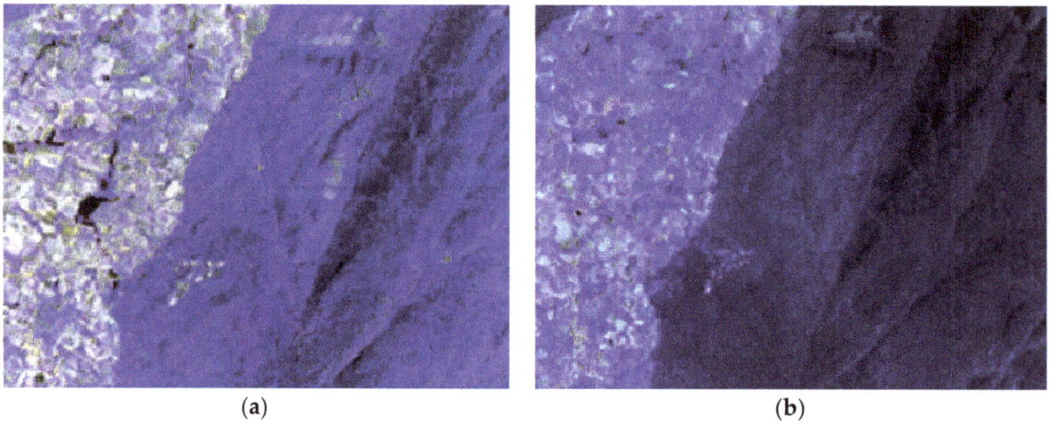

Figure 2. Preprocessed LISS-III multispectral image of Javadi Hill for the years (**a**) 2012 and (**b**) 2015.

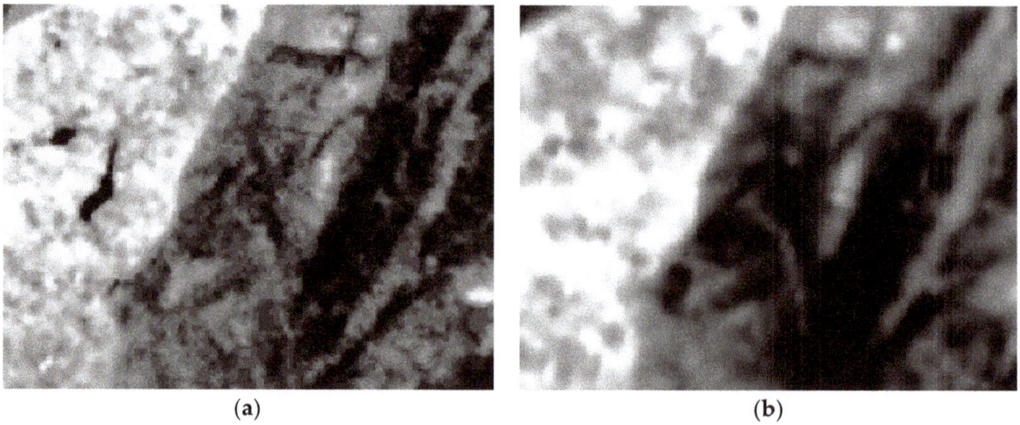

Figure 3. Preprocessed Landsat TIRS bands of Javadi Hill for the years (**a**) 2012 and (**b**) 2015.

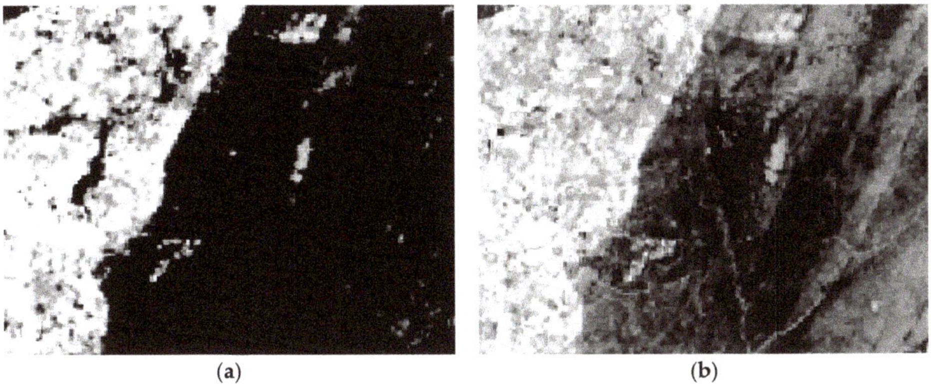

Figure 4. Preprocessed Landsat RED bands of Javadi Hill for the years (a) 2012 and (b) 2015.

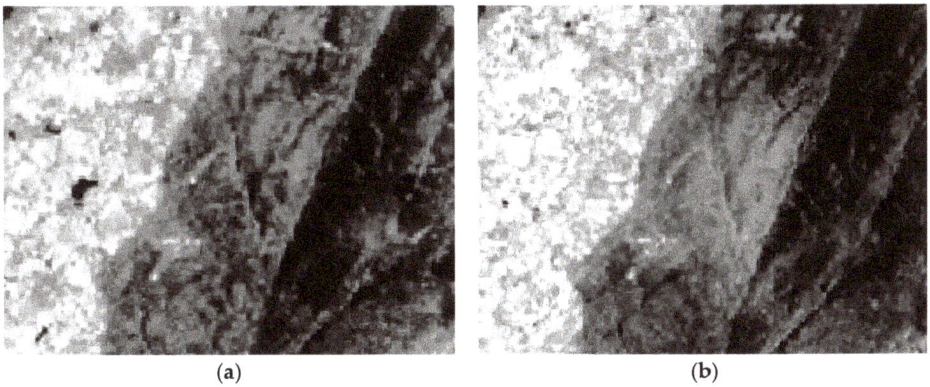

Figure 5. Preprocessed Landsat NIR bands of Javadi Hill for the years (a) 2012 and (b) 2015.

3.2. Proposed Vision Transformer Model for LU/LC Classification

A transformer is a deep-learning model that has emerged through the self-attention mechanism. The transformer follows the encoder–decoder architecture by processing the sequential data parallelly without depending on any recurrent network. It has been widely used in the scientific fields of NLP and computer vision. The Vision Transformer architecture has attracted an interesting view from researchers in recent years by showing good performance in the area of machine- and deep-learning applications. The Vision Transformer has been used in the area of image classification for providing state-of-the-art performance and to outperform the standard classification models. The Vision Transformer develops the encoder module of the transformer for performing the image classification by representing the sequence of image patches to the classified label. The attention mechanism of the Vision Transformer goes through all areas of the image and integrates the information into the full-sized image [47–51]. The end-to-end Vision Transformer model for the classification of satellite images is shown in Figure 6. The Vision Transformer classification model has experimented with the preprocessed LISS-III satellite image of Javadi Hills for the years 2012 and 2015. The Vision Transformer architecture is composed of an embedding, encoder, and classifier layer. Equations (1) and (2) represent the first step of analyzing and dividing the training images into a sequence of patches.

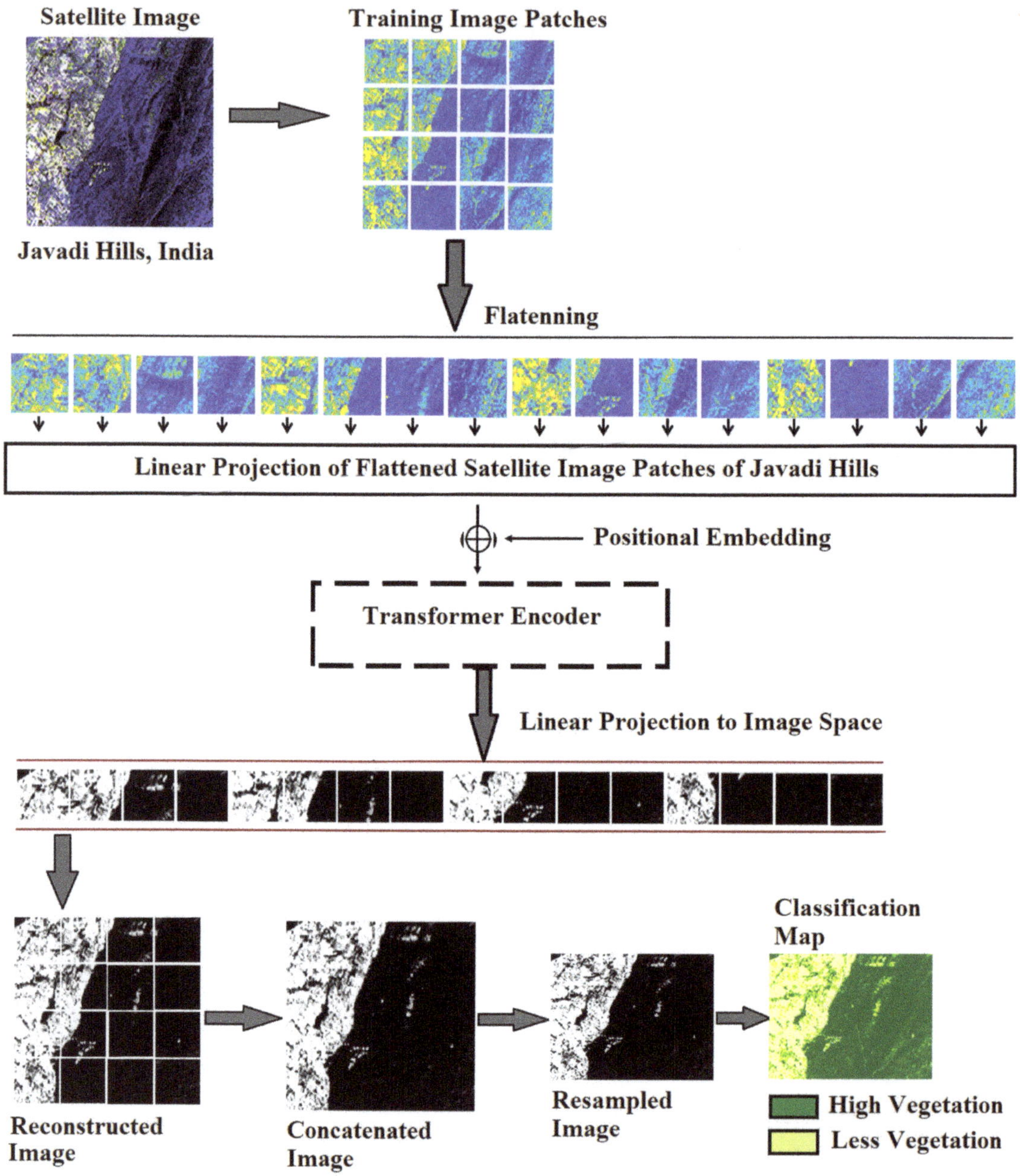

Figure 6. Proposed Vision Transformer model for LU/LC classification.

Let S_i represent a set of training satellite images, r, where X_i is a satellite image; y_i represents the class labels $\{y_i \in 1, 2, \ldots\ldots, m\}$ associated with the X_i, and m denotes the number of defined LU/LC classes for that set.

$$S_i = \{X_i, y_i\}_{i=1}^{r} \tag{1}$$

In the first step of the Vision Transformer model, an image X_i from the training, the set is divided into non-overlapping patches of fixed size. Each patch is observed by the Vision Transformer as an individual token. Thus, from the size $h * w * c$ (where h is the height, c is the number of channels, and w is the width) of an image X_i, we extracted the patches of dimension $c * p * p$ (p is the patch size) from it. The extracted patches are converted to a sequence of images $(x_1, x_2, x_3, \ldots\ldots\ldots, x_n)$ of length n through flattening.

$$n = hw/p^2 \tag{2}$$

The image patches are linearly projected into a vector setup of model dimension, d, using the known embedding matrix, E. The concatenation of embedded representations is processed along with the trained classification token $vclass$ for performing the classification task. The positional information, E_{pos}, is programmed and attached to the patch representation. The spatial arrangements of the trained image patches were processed through positional embedding. The resulting sequence of image patches from positional embedding with token z_0 is given in Equation (3).

$$z_0 = [vclass; x1E; x2E; \ldots\ldots\ldots, xnE] + E_{pos},\ E \in \mathbb{R}^{(p^2c)\,*\,d},\ E_{pos} \in \mathbb{R}^{(n+1)\,*\,d} \tag{3}$$

The resulting sequence of embedded image patches, z_0, is passed into the transformer encoder with L identical layers. It has a multi-head self-attention block (MSA) and fully connected feed-forward MLP (Multilayer Perceptron) block with the GeLU activation function between them. The two subcomponents of the encoder work with the residual skip connections through the normalization layer (LN). The representation of the two main components of the encoder is given in Equations (4) and (5). The last layer of the encoder, the first element in the sequence z_L^0, is passed into the head classifier for attaining the LU/LC classified classes.

$$z_l^1 = MSA\ (LN(z_{l-1})) + z_{l-1},\ l = 1\ldots..L \tag{4}$$

$$z_l = MLP\ \left(LN\left(z_l^1\right)\right) + z_l^1,\ l = 1\ldots..L \tag{5}$$

$$y_i = LN\ \left(z_L^0\right) \tag{6}$$

The transformer block for the classification model is shown in Figure 7. The MSA block of the encoder is considered the central component of the transformer. The MSA block determines the importance of a single patch embedding with the other embeddings in the sequence. There are four layers in the MSA block: the linear layer, the self-attention layer, the concatenation layer, and a final linear layer. The attention weight is computed by calculating the weighted sum of all values in the sequence. The query-key-value scaling dot product is computed by the self-attention (SA) head through the attention weights The Q (query), K (key), and V (value) were generated by multiplying the element against three learned matrices U_{QKV} (Equation (7)). For determining the significance of the elements on the sequence, the dot product is used between the Q vectors of one element with the K vectors of the other elements. The results show the importance of the image patches in the sequence. The outcomes of the dot product were scaled and passed into a Softmax (Equation (8)).

$$[\mathrm{Q},\ \mathrm{K},\ \mathrm{V}] = zU_{QKV},\ U_{QKV} \in \mathbb{R}^{d\,*\,3D_k} \tag{7}$$

$$A = softmax\ \left(\frac{QK^T}{\sqrt{D_K}}\right),\ A \in \mathbb{R}^{n\,*\,n} \tag{8}$$

$$SA\ (z) = A.V \tag{9}$$

$$MSA\ (z) = Concat(SA_1(z);\ SA_2(z);\ldots.\ SA_h(z))W,\ \ W \in \mathbb{R}^{h.D_K\,*\,D} \tag{10}$$

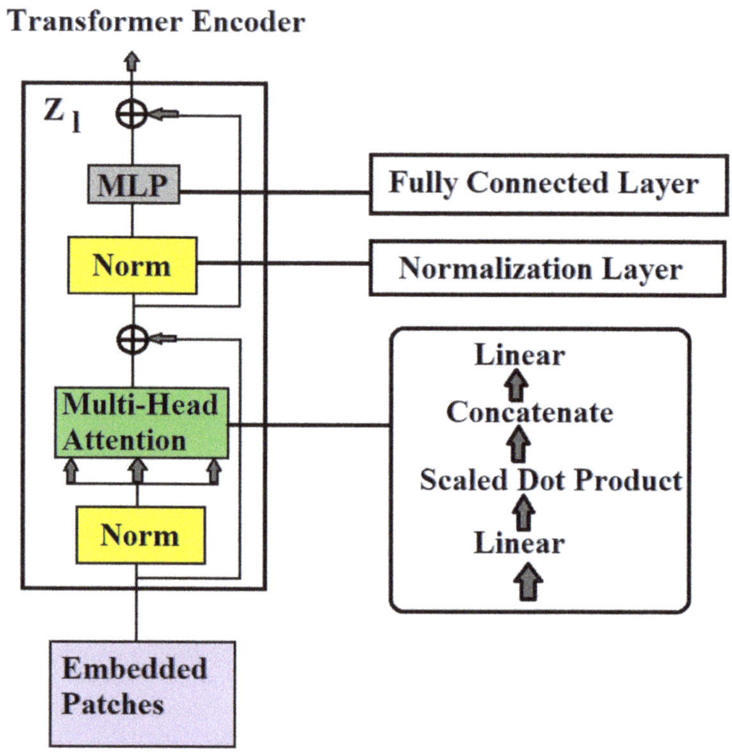

Figure 7. Transformer block for the Vision Transformer classification model.

The scaling-dot-product process achieved by the SA block is related to the standard dot product, but it includes the dimension of the key D_K as a scaling factor. The patches with the high attention scores (Equation (8)) are processed by multiplying the outputs of Softmax with the values of each patch embedding vector. The results of all the attention heads are concatenated and provided to the MLP classifier for attaining the pixel value representation of the feature map (Equation (10)). The resampling was performed for adjusting the size of the feature map so that the output classified image would be represented in the standardized form during the time of accuracy assessment. The training data with different parameters that define the Vision Transformer classification model of our research work are presented in Section 5.1. The LU/LC classification map for the years 2012 and 2015 is shown in Figure 8. The accuracy assessment for the feature-extraction-based classification model is shown in Section 5.2. The evaluation of the LU/LC classification map was achieved through the accuracy assessment. The percentage of the LU/LC change between the years 2012 and 2015 for our study area was calculated. Based on the good accuracy results, the LU/LC change classification map was processed for further findings of the LU/LC prediction map.

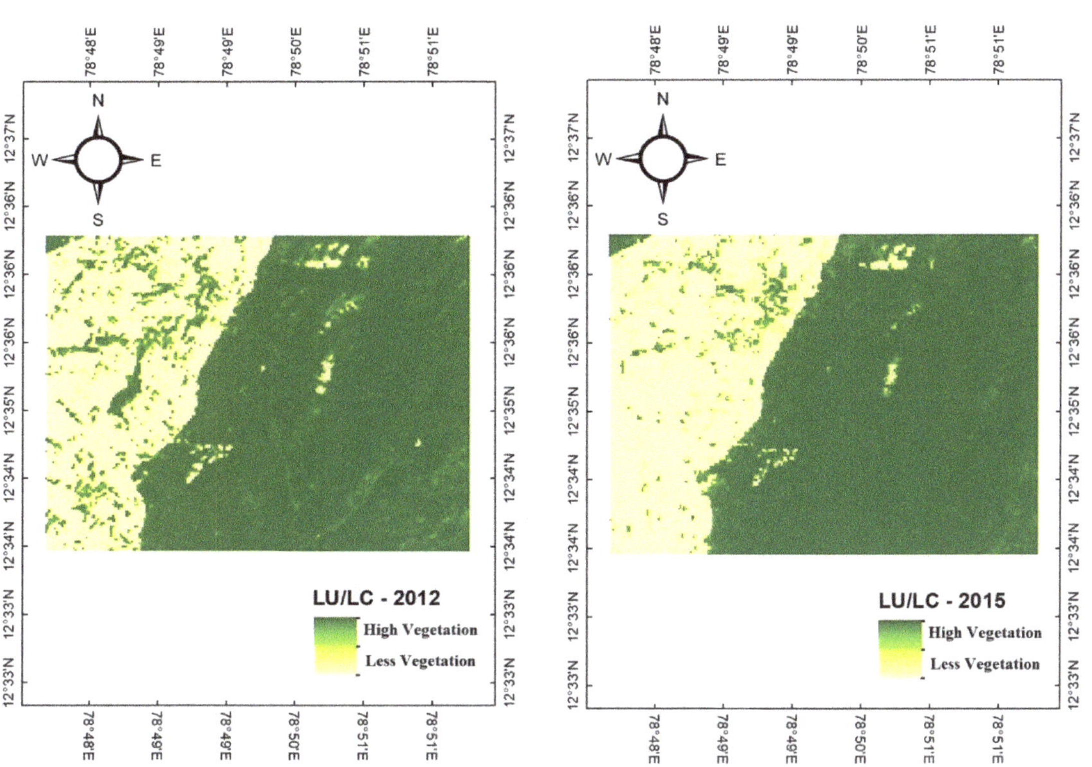

(a) (b)

Figure 8. LU/LC classification map of Javadi Hills for the years (**a**) 2012 and (**b**) 2015.

3.3. Land Surface Temperature

The LST measures the skin temperature of the spatial data in the field of remote sensing. It displays the cold and hot temperature of the earth's surface through the radiant energy reflected within the surface. The thermal-infrared remote-sensing data are used for measuring the LST. The TIRS data help in recognizing the mixture of bare soil and vegetation temperatures through LST [52–54]. In our research work, we estimated the LST for the TIRS bands of Landsat 8. Equations (11)–(13) represent the estimation of LST for TIRS image 7. The conversion of the Digital Number (DN) value to the radiance of the TIRS image is calculated by using Equation (11). The conversion of radiance into the brightness temperature is shown in Equation (12). The degree conversion from Kelvin (K) to Celsius © is shown in Equation (13).

$$L_\lambda = \left(\frac{LMAX_\lambda - LMIN_\lambda}{QCALMAX - QCALMIN}\right) * (QCAL - QCALMIN) + LMIN_\lambda \quad (11)$$

where L_λ represents the spectral radiance in $(Watts/(m^2 * sr^2 * \mu m))$, $QCAL$ represents the quantized calibrated pixel value, $QCALMAX$ represents the maximum quantized calibrated pixel value, $QCALMIN$ represents the minimum quantized calibrated pixel value, $LMAX_\lambda$ represents the spectral radiance scaled to $QCALMAX$, and $LMIN_\lambda$ represents the spectral radiance scaled to $QCALMIN$.

$$T_K = \frac{K2}{\ln\left(\frac{K1}{L_\lambda} + 1\right)} \quad (12)$$

$$C = T_K - 273.15 \tag{13}$$

where T_K represents the effectiveness at the satellite temperature in Kelvin, and $K1$ and $K2$ represent the calibration constants 1 and 2 in $(Watts/(m^2 * sr^2 * \mu m))$, respectively. For Landsat 7, the calibration constant value of $K1$ and $K2$ is 666.09 and 1282.71, respectively.

Equations (14)–(20) represent the estimation of LST for the TIRS image of Landsat 8. By using the radiance rescaling factor, the conversion of Top of Atmosphere (TOA) spectral radiance is shown in Equation (14). By using the thermal infrared constant values in the metadata file of the satellite image, the spectral radiance data are converted to the TOA brightness temperature, and the expression is shown in Equation (15). The *NDVI* is calculated for differentiating the near-infrared and visible reflectance of the vegetation cover of the satellite data. The expression for *NDVI* is shown in Equation (16). The Land Surface Emissivity (LSE) is derived from *NDVI* values for displaying the average emissivity of the earth's surface. The expressions are shown in Equations (17) and (18). By using the results of TOA brightness temperature, emitted radiance wavelength, and LSE, the LST was calculated and is shown in Equation (19).

$$TL_\lambda = ML * QCAL + AL - O_i \tag{14}$$

where TL_λ represents the TOA spectral radiance in $(Watts/(m^2 * sr^2 * \mu m))$, ML represents the radiance multiplicative band rescaling factor of the TIRS image, $QCAL$ represents the quantized calibrated pixel value, AL represents the radiance additive band rescaling factor of TIRS image, and O_i represents the correction value of the TIRS band of Landsat 8.

$$BT_P = \frac{K2}{\ln\left(\frac{K1}{TL_\lambda} + 1\right)} - 273.15 \tag{15}$$

where BT_P represents TOA brightness temperature in Celsius, and $K1$ and $K2$ represent the calibration constant 1 and 2 in $(Watts/(m^2 * sr^2 * \mu m))$, respectively. For Landsat 8, the calibration constant value of $K1$ and $K2$ is 774.8853 and 1321.0789, respectively.

$$NDVI = \frac{(NIR - RED)}{(NIR + RED)} \tag{16}$$

where $NDVI$ represents the Normalized Difference Vegetation Index, NIR represents the reflectance values of the near-infrared band, and RED represents the reflectance values of the red band.

$$PV = ((NDVI - NDVI_{min})/(NDVI_{max} - NDVI_{min}))^2 \tag{17}$$

$$E = 0.004 * PV + 0.986 \tag{18}$$

where E represents the Land Surface Emissivity, PV represents the Proportion of Vegetation, $NDVI$ represents the reflectance values of the $NDVI$ image, $NDVI_{max}$ represents the maximum reflectance value of the $NDVI$ image, and $NDVI_{min}$ represents the minimum reflectance value of the $NDVI$ image.

$$LST = \frac{BT_P}{\left(1 + \left(\frac{\lambda * BT_P}{c2}\right) * \ln(E)\right)} \tag{19}$$

$$c2 = \frac{pk * vl}{bc} \tag{20}$$

where LST represents Land Surface Temperature, BT_P represents the TOA brightness temperature in Celsius ©, λ represents the wavelength of the emitted radiance, pk represents the Planck's constant value of $6.626 * 10^{-34}$ J s, vl represents the velocity of the light value of $2.998 * 108$ m/s, and bc represents the Boltzmann constant value of $1.38 * 10^{-34}$ JK. The statistical modeling of TIRS bands present in the Landsat satellite image was used for analyzing the LU/LC surface temperature of Javadi Hills, and it helps in improving the performance of the LU/LC prediction model. The LST map of Javadi Hills during the years 2012 and 2015 was analyzed by using the TIRS bands of Landsat 7 and 8 for the area of Javadi Hills. The flow of the calculation of LST for our area of Javadi Hills is shown in Figure 9. The LST map for the years 2012 and 2015 is shown in Figure 10. In this research work, we used the spatial features of the LST map and the LU/LC change classification map for evaluating the LU/LC prediction map for Javadi Hills. The LST map shows the features of the high- and low-temperature values of the earth's surface. The high-temperature values indicate less vegetation, and the low-temperature value indicates a high-vegetation area. The impact of the LST map over the LU/LC change classification map provides good accuracy during the process of LU/LC prediction. The relationship between the values of the LST and LU/LC map is shown in Section 5.1.

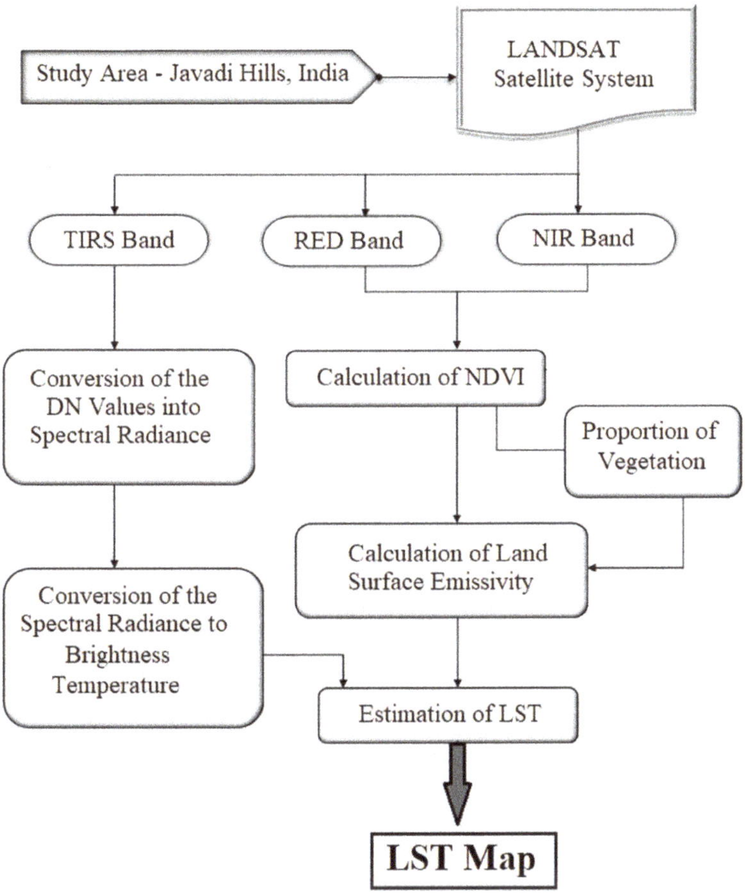

Figure 9. The flow of Land Surface Temperature estimation for the area of Javadi Hills, India.

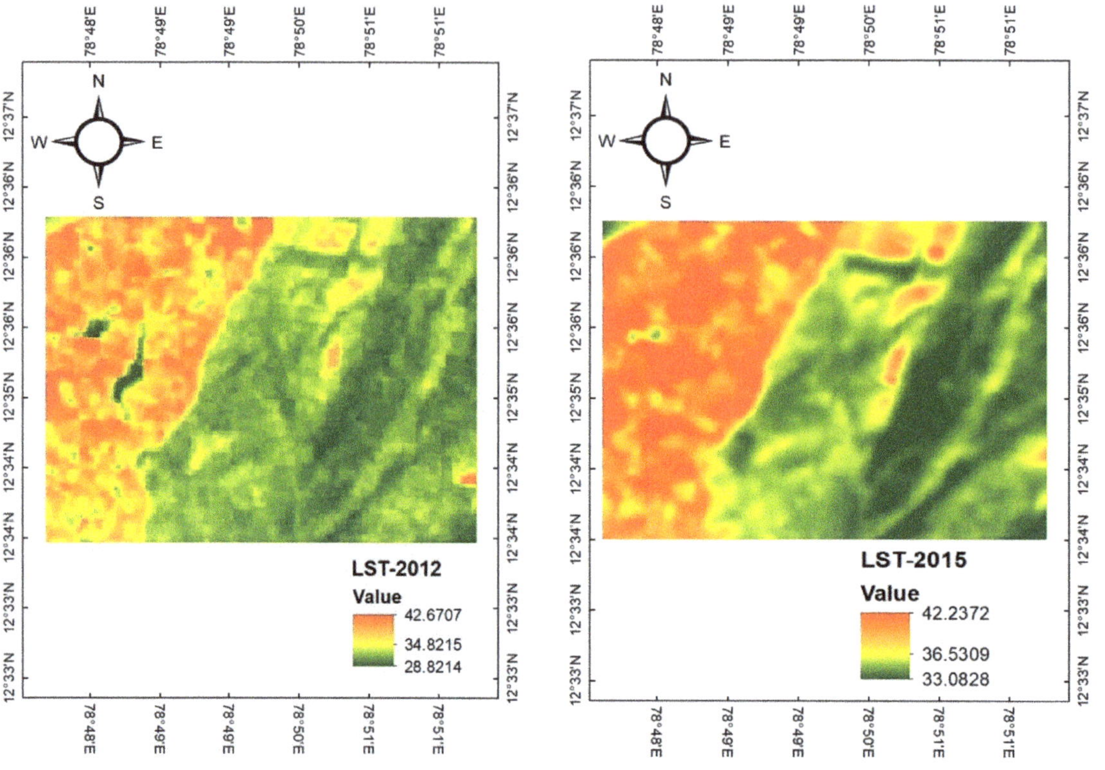

(a) (b)

Figure 10. LST map for the area of Javadi Hills for the years (**a**) 2012 and (**b**) 2015.

3.4. Bidirectional Long Short-Term Memory Model for LU/LC Prediction

The LSTM model is considered the advanced model of RNN, where the long-term dependencies can be learned for the sequence prediction problems. The long-term vanishing-gradient problems are prevented by using the LSTM models. The key elements of the LSTM model are input, forget, and output gate [55–57]. Figure 11 displays the working principle of the LSTM model. In Figure 11, the vector operations represent the element-wise multiplication $(*)$, and element-wise summation $(+)$ respectively. The time step (t) indicates the length of the input sequence in all the Equations (21)–(26). Equation (21) shows the mathematical expression of the forget gate, where f_t represents the memory gate's output at time t, σ represents the sigmoid function ($0 < \sigma < 1$), W_f represents the weight value of ANN, h_{t-1} is the output value of the previous cell, x_t represents the input values, and b_f denotes the bias weight values of the ANN. At the output of the equation, the value 1 will keep the information and the value 0 will forget the information

$$f_t = \sigma\left(W_f \,*\, [h_{t-1}, x_t] + b_f\right) \tag{21}$$

$$I_t = \sigma\left(W_i \,*\, [h_{t-1}, x_t] + b_i\right) \tag{22}$$

$$\widetilde{c_t} = tanh(W_c \,*\, [h_{t-1}, x_t] + b_c) \tag{23}$$

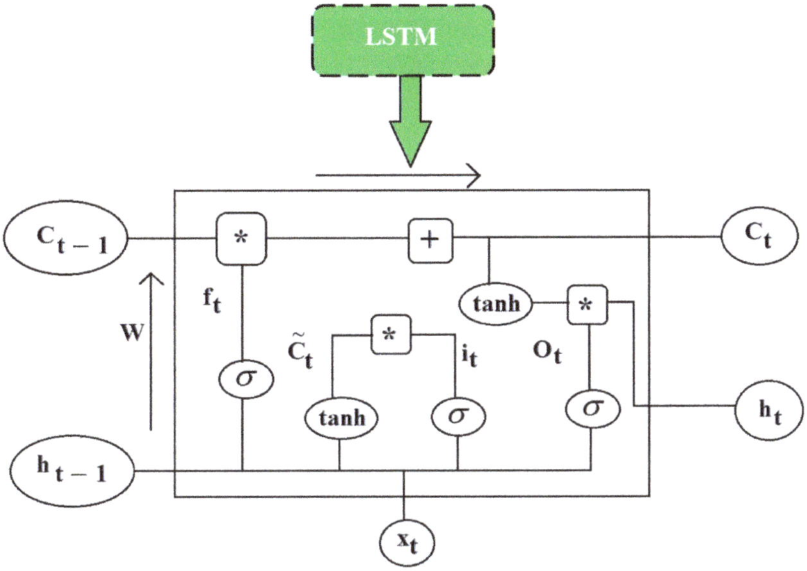

Figure 11. LSTM model.

In Equation (22), I_t represents the output of the input gate, σ represents the sigmoid function, W_i represents the weight values stored in the memory of ANN, h_{t-1} is the output value of the previous cell, x_t represents the input values, and b_i denotes the bias weight values of the ANN.

In Equation (23), $\widetilde{c_t}$ represents the output of ANN with the normalized *tanh* function that outputs the value between −1 and +1, W_c represents the weight values stored in the memory of ANN, h_{t-1} is the output value of the previous cell, x_t represents the input values, and b_c denotes the bias weight values of the ANN.

$$C_t = C_{t-1} * f_t + i_t * \widetilde{c_t} \tag{24}$$

$$O_t = \sigma\left(W_O * [h_{t-1}, x_t] + b_O\right) \tag{25}$$

$$h_t = O_t * tanh\left(C_t\right) \tag{26}$$

Equation (24) shows the mathematical expression of the updated gate, where the memory is updated. The ANN learns the stored or forgotten information from the memory and then updates the newly added information from Equations (21)–(23). Equation (25) shows the mathematical expression of the output gate, where W_O represents the weight values stored in the memory of ANN, h_{t-1} is the output value of the previous cell, x_t represents the input values, and b_O denotes the bias weight values of the ANN. The output value, h_t, was calculated in Equation (26).

The uniform LU/LC classes were generated through the Vision Transformer classification model, and the features of the LST map were extracted for the years 2012 and 2015. In this research work, we used the spatial features of the LST map and the LU/LC change classification map for evaluating the LU/LC prediction map, using the Bi-LSTM model. The idea of Bi-LSTM is to process the sequence data in both forward and backward directions. The Bi-LSTM algorithm was used in our research for extracting the spatial and temporal features of the fifteen-year time-series data from 2012 to 2027 for the area of Javadi Hills. Figure 12 displays the working principle of the Bi-LSTM prediction model.

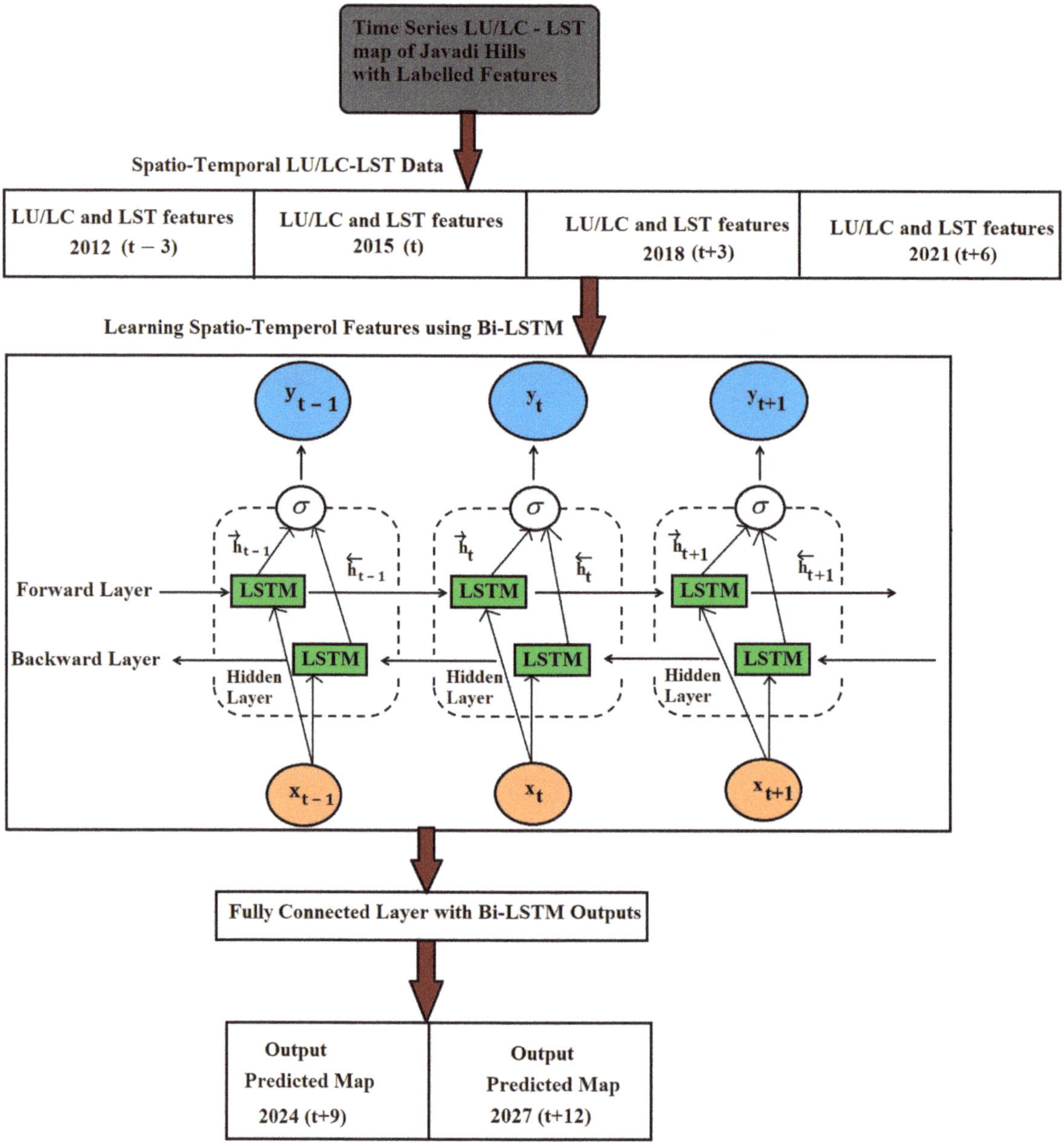

Figure 12. Bi-LSTM model for LU/LC prediction.

The inputs of the Bi-LSTM are given as the 3D vectors (samples, time steps, and features) for producing both spatial and temporal information. The samples define the number of the input LU/LC map ($L\ (j_{m,n})$) of size ($m\ *\ n$) with defined labels (j) for training and validation. With the LU/LC and LST features for the years 2012 and 2015, we have predicted and simulated the LU/LC map for the years 2018 and 2021. With the inputs of 2012 ($t-3$), 2015 (t), 2018 ($t+3$), and 2021 ($t+6$), the Bi-LSTM was processed in forward and backward directions for analyzing the features of time-series data and to project the predicted maps for the years 2021 $(t+9)$ and 2024 $(t+12)$ successfully. The features $(LC(j_{m,n}))$ define the LU/LC classes with the LST temperature values for each time step at defined coordinates. The input set of combined features of the LU/LC and LST map from the Javadi Hills was split by the ratio of 8:2 for the training and validation of the model. The parameters were adjusted through a trial-and-error approach for acquiring good prediction accuracy. The tanh activation function was used for the Bi-LSTM layers, whereas the Softmax activation functions were used for the last layer to calculate the probabilities between the LU/LC classes of Javadi Hills. Through repeated forward and back-propagation processes, the parameters are adjusted until the cost function is minimized. The validation method is part of training the prediction model and adjusting the parameters, which uses a small portion of data to validate and update the model parameters for each training epoch. The significant approach is to ensure that the prediction model is learning from data correctly by minimizing the cost function during the training and validation process. The training data with the parameters that run the Bi-LSTM prediction model for our research work are presented in Section 5.1. The LU/LC prediction map for the years 2018, 2021, 2024, and 2027 is shown in Figures 13 and 14. The validation results of the LU/LC prediction model are shown in Section 5.2. Our proposed model provides good validation accuracy, and the growth patterns of the LU/LC results are shown in Section 5.3.

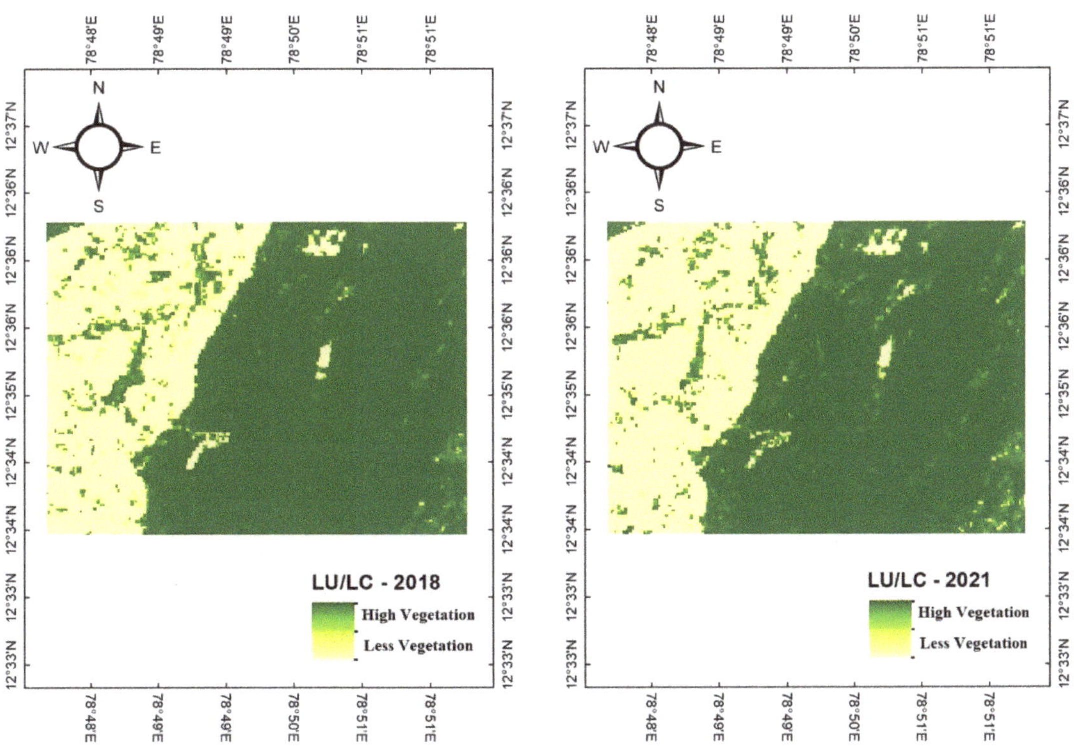

(a) (b)

Figure 13. LU/LC prediction map of Javadi Hills for the years (**a**) 2018 and (**b**) 2021.

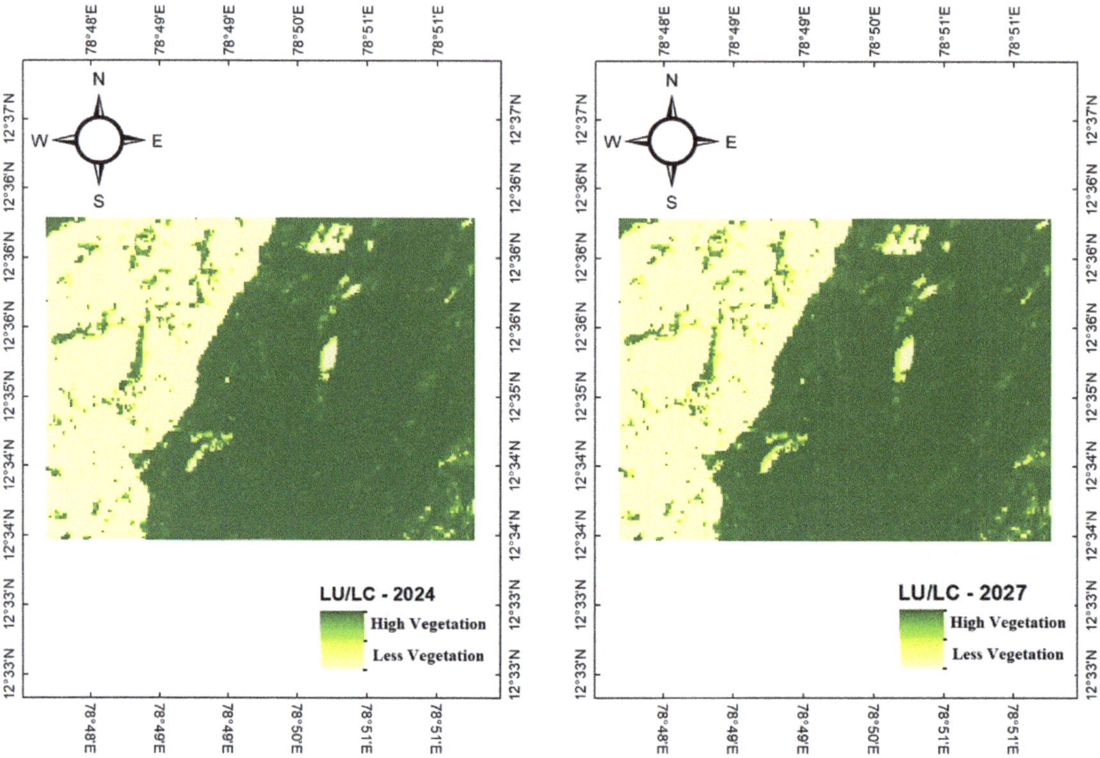

(a) (b)

Figure 14. LU/LC prediction map of Javadi Hills for the years (**a**) 2024 and (**b**) 2027.

3.5. Application-Based Explainable Artificial Intelligence and Its Importance

The XAI provides knowledge to humans about the outcomes achieved by machine- or deep-learning models. The XAI has been used for providing knowledge on the extracted time-series LU/LC information to the urban planners, forest department, and government officials. XAI improves the user's understanding and trust in the products or services. There are many ways of explaining the model through XAI, and the techniques of explaining the model differ for each application area around the world [58–60]. In our research work, we used application-based XAI, and it was observed to be the easiest and fastest way of obtaining knowledge with finite compute resources. The knowledge about the outcomes of the prediction model can be accessed through online applications. Technically, the application-based XAI can be understood by the end-users through third-party applications. In our prediction model, we used the Google Earth Engine (https://www.google.com/earth/ (accessed on 10 November 2021)) platform for explaining our results to urban planners, forest departments, and government officials. The LU/LC predicted results for the years 2018 and 2021 were tested through the Google Earth Engine time-series image. We achieved good testing accuracy for our prediction model. Through the XAI of the Google Earth Engine platform, the end-users can also access and check the LU/LC information. We have shown the model structure of XAI through the Google Earth Engine platform for our research work in Figure 15. The XAI on Google Earth will convey the LU/LC information to the government, forest department, and urban planners to take action in regard to protecting the LU/LC area.

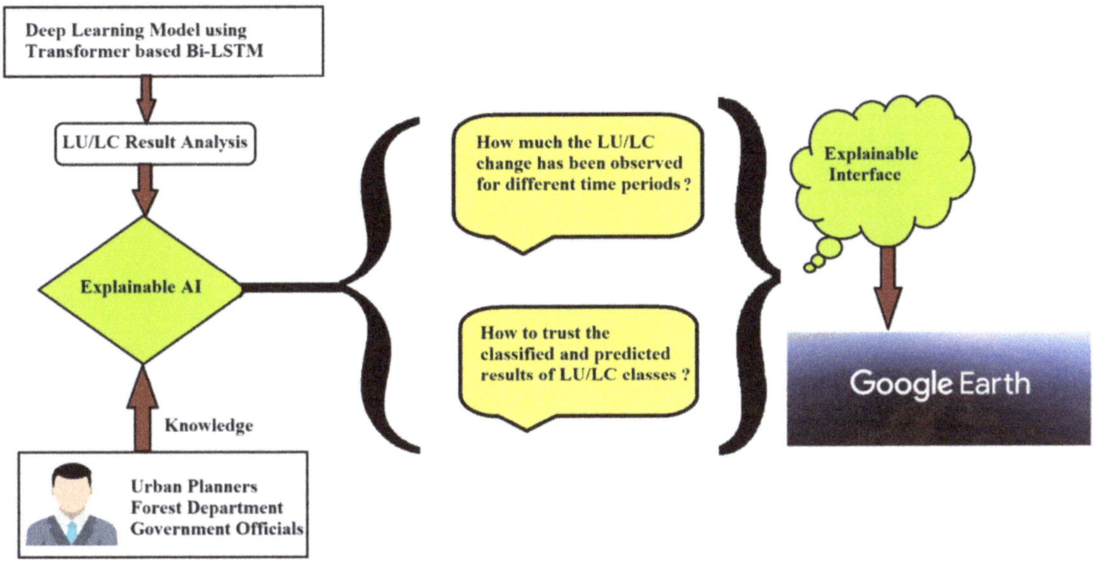

Figure 15. Explainable AI interface through Google Earth Engine platform.

4. Proposed LU/LC Prediction Using Vision Transformer–Based Bi-LSTM Model

This research work aimed to identify the LU/LC changes in the forest-covered (high vegetation) and non-forest-covered (less vegetation) regions of the proposed study area. The flow of LU/LC change for our study area is shown in Figure 16. The proposed flow of this work is described in the following steps,

- The LISS III satellite images for the years 2012 and 2015 of Javadi Hills, India, were collected from Bhuvan-Thematic Services of the National Remote Sensing Centre (NRSC), Indian Space Research Organization (ISRO).
- The Landsat satellite images for the years 2012 and 2015 of Javadi Hills, India, were collected from the United States Geological Survey (USGS), United States.
- Atmospheric, geometric, and radiometric corrections were performed to provide better visibility in the acquired LISS-III and Landsat images.
- The proposed Vision Transformers for classifying LU/LC classes were successfully performed for the years 2012 and 2015 of the LISS-III image.
- An LST map was calculated for the years 2012 and 2015 from Landsat TIRS images for extracting the spatial features.
- The relationship between the spatial features of the LST map with the LU/LC classification map were used to provide good validation results during the prediction process.
- The Bi-LSTM model was successfully applied to forecast the future LU/LC changes of Javadi Hills for the years 2018, 2021, 2024, and 2027.
- The LU/LC changes that occurred in our study area will assist the urban planners and forest department to take proper actions in the protection of the environment through XAI.

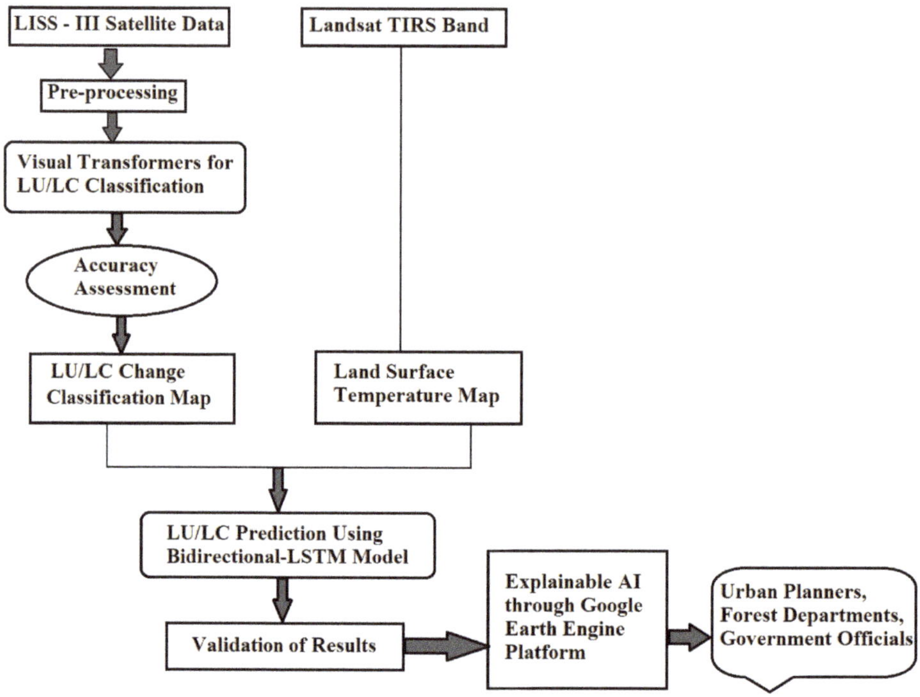

Figure 16. Proposed flow of LU/LC prediction using Vision Transformer–based Bi-LSTM model.

Algorithm to Construct the Vision Transformer–Based Bi-LSTM Model for LU/LC Prediction

Our research is based on the Vision Transformer–based Bi-LSTM model for LU/LC Prediction of Javadi Hills, India. From the brief analysis and validation, we found that the impact of the TIRS LST map with the LU/LC classified provides a good percentage of results with a lower misclassification rate. The detailed steps of our proposed model are presented in Algorithm 1. Each process in our proposed algorithm provides the different aspects of LU/LC information of Javadi Hills. A brief explanation of the input data, training data, parameter settings, and accuracy assessment of our proposed model is explained in Section 5.

Algorithm 1: To Construct the Vision Transformer–Based Bi-LSTM Prediction Model.

Inputs (I_P): The LISS-III multispectral satellite images for the years 2012 and 2015 (I_1, I_2), and Landsat bands for the years 2012 and 2015 ($IR_1,\ IR_2$)
Output (O_P): Predicted LU/LC images 2018, 2021, 2024, and 2027 ($PR_1,\ PR_2,\ PR_3,\ PR_4$)

```
Begin
    Input data (I_P):
    Initialize the input data
        Extract LISS-III multispectral image (M = I_1, I_2)
        Extract Landsat bands (T = IR_1, IR_2)
    Return input data (I_P)

    Preprocessed data (PR_I):
    Initialize the input data for performing the preprocessing for the input data I_P of M and T
    For each initialized input image of M and T
```

Algorithm 1: Continued.

        Calculate the geometric coordinates of the study area G_I (georeferencing)
        Reduce the atmospheric (haze) effects A_I of the georeferenced image
        Correct the radiometric errors R_I for the haze-reduced image
    End for
    Return preprocessed data (PR_I)

    LU/LC classification (LU_I):
    Perform the Vision Transformer–based LU/LC classification by using the preprocessed image PR_I
    For each input image of PR_I
        Load the training data *Ti* and initialize the parameters
        Split an image into patches of fixed size
        Flatten the image patches
        Perform the linear projection from the flattened patches
        Include the positional embeddings
        Feed the sequences as an input to the transformer encoder
        Fine-tune the multi-head self-attention block in the encoder
        Concatenate all the outputs of attention heads and provide the MLP classifier for attaining the pixel value representation of the feature map.
        Generate the LU/LC classification map
    End for
    LU/LC classification (LU_I)

    Accuracy assessment (AA_I):
    Perform the accuracy assessment for the feature extraction–based LU/LC classification map LU_I
    For each classified map of LU_I
        Compare the labels of each classified data LU_I with the Google Earth data
        Build the confusion matrix
        Calculate overall accuracy, precision, recall, and F1-Score
        Summarize the performance of the classified map LU_I
    End for
    Return accuracy assessment (AA_I)

    Change detection (CD_I):
    Perform the LU/LC change detection by using the time-series LU/LC change classification map (LU_I)
    For each classified map of LU_I
    Calculate the percentage of change between the time-series classified map of LU_I
    End For
    Return change detection (CD_I)

    Extracting LST map (LST)
    Initialize the I_P of T
    For each preprocessed image of T
        Calculate Land Surface Temperature using the Landsat bands (TIRS, RED, and NIR)
        Extract the spatial features
    End for
    Return LST (LST_I)

    LU/LC prediction (LP_I):
    Perform the Bi-LSTM prediction model by using the time-series LU/LC classification map of 2012 (LU_1) and 2015 (LU_2) and the spatial features of the LST map of 2012 (LST_1) and 2015 (LST_2)
    For each time-series, LU/LC classified map of LU_I: $\{LU_1, LU_2\}$ and LST map LST_I: $\{LST_1, LST_2\}$
    Perform LU/LC prediction (LP_I) using Bi-LSTM model

Algorithm 1: Continued.

```
        Initialize the inputs for LU/LC prediction
        Input (IP) = {L1, LST1, L2, LST2, ...}
        Combine the information of the time-series LU/LC classified map LU_I with the LST map LST_I
        Load the 3D input vectors {samples, time steps, features}
        Initialize the Bi-LSTM parameters
        Apply tanh activation function for each Bi-LSTM layer
        The output layer is decided by using the Softmax activation function
        Update the parameters until the loss function is minimized
        The output of the predicted time-series data is obtained
        Validate the results
    End for
    Return LU/LC prediction map LP_I {PR1, PR2, ...}
    Analyze the growth patterns of the LU/LC prediction maps

    Explain predicted results to the urban planners, forest department, and government officials, using application-based XAI
End
```

5. Results and Discussion

The problematic study on LU/LC prediction in Javadi was presented in this research work. The LISS-III multispectral, Landsat TIRS, RED, and NIR satellite images were used for predicting the vegetation in the forest- and non-forest-covered regions of the Javadi Hills. All the research experiments were processed on the Intel Xeon processor 2.90 GHz CPU, along with 128 GB RAM in Windows 10 (64-bit) environment. The needed libraries and packages of Python of version 3.10.2 developed by Python Software Foundation (https://www.python.org/) were installed for implementing the proposed model of our research. The backend geospatial software such as QGIS of version 3.6.1 developed by QGIS Development Team (https://qgis.org/en/site/), ArcGIS of version 10.1 developed by ESRI (http://www.esri.com/software/arcgis) and Google Earth Engine developed by Google (https://www.google.com/earth/) was used for preparing and analyzing the satellite data.

5.1. Training Data and Parameter Settings

For appropriate mapping of the input features to the output features using machine-learning or a deep-learning model, the training data and its parameters were used and tuned. Algorithm 1 shows the detailed procedure of our research on LU/LC prediction. The multispectral input map (M) of our study area Javadi Hills for the year 2012 and 2015 was considered as (I_1, I_2). The preprocessed multispectral image was processed for the further processing of our model.

The training samples of an image are divided into patches. The 16 patches (size = 64 × 4) were extracted from the input training image (256 × 256), of which each patch contains the trained LU/LC classes (high and less vegetation). The training samples for the area of Javadi Hills were generated through the latitude and longitudinal coordinates of Javadi Hills manually through Google Earth image. For the input image of Javadi Hills for the years 2012 and 2015, the LU/LC classification was performed through the Vision Transformer model. The working process of the Vision Transformer model was explained in Section 3.2. For a better understanding of our training samples in the patched image, we show the trained patches of 1 and 16 in Figure 17. The hyper-parameters used during the training process of the Vision Transformer model are shown in Table 2. The output extracted at the end of the fully connected layer was used as the LU/LC classified map for further processing.

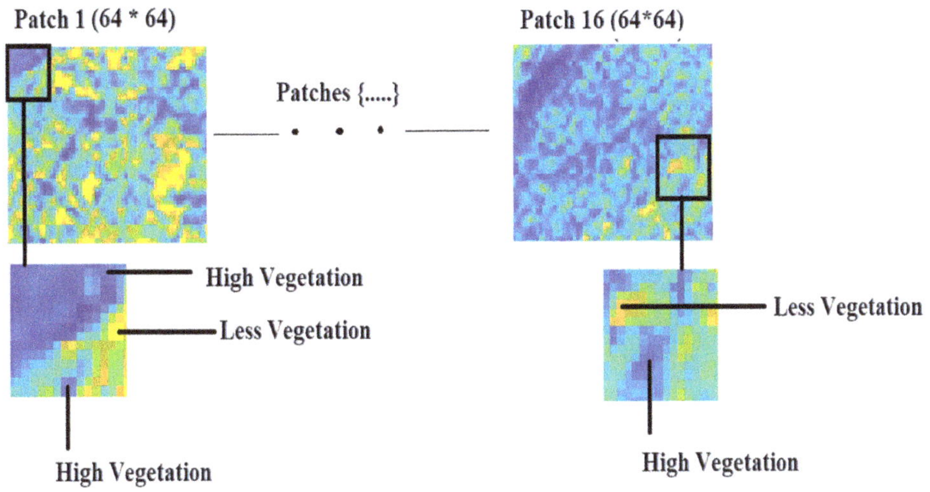

Figure 17. Trained patches for the area of Javadi Hills.

Table 2. Hyperparameters of the Vision Transformer model.

Hyperparameters	Value
Learning Rate	0.001
Weight Decay	0.0001
Batch Size	10
Number of epochs	100
Image size	256 × 256
Patch size	64
Patches per image	16
Number of heads	4
Transformer Layers	8
Activation Function	GeLU
Optimizer	Adam

After the classification, each classified sample was tested through the referenced data of Google Earth images. The LU/LC classified image (LU_I) was tested through the referenced Google Earth image. Each reference datum was labeled according to the respective LU/LC classes of the Javadi Hills through Google Earth images. The LU/LC class considered in our research work includes the high- and less-vegetation regions of the forest- and non-forest-covered regions of Javadi Hills. For better understanding, we have shown the validation of the point shape file with the Google Earth images in Figure 18, and the class values associated with each coordinate of the trained image are shown in Table 3. The accuracy assessment was calculated for the Vision Transformer model, and the results are shown in Section 5.2.

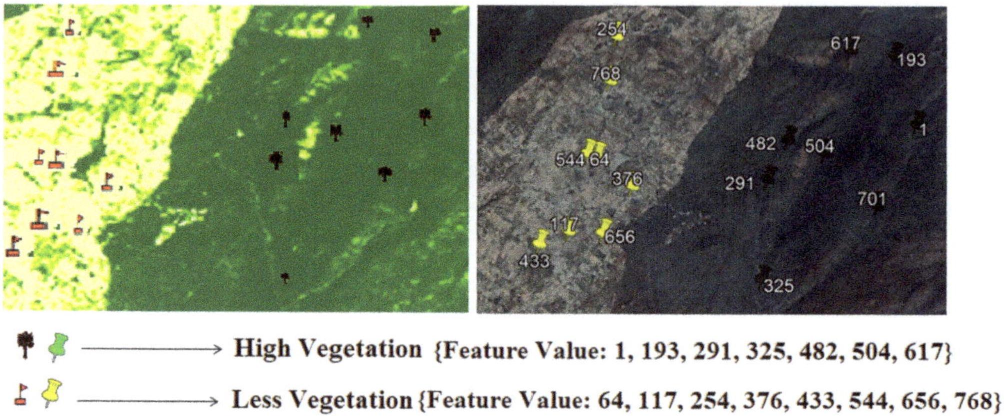

Figure 18. Validation of LU/LC classified map for the area of Javadi Hills.

Table 3. Training data values for the area of Javadi Hills.

Training Feature Value	Longitude	Latitude	Class Value	Class Label
1	78.829746	12.581815	1	High Vegetation
241	78.81025	12.58796	2	Less Vegetation
1785	78.818244	12.580221	2	Less Vegetation
733	78.849159	12.576782	1	High Vegetation
6640	78.81107	12.57028	2	Less Vegetation
6277	78.83463	12.576789	1	High Vegetation
12,354	78.851079	12.59151	1	High Vegetation
12,179	78.80721	12.58024	2	Less Vegetation
20,163	78.81167	12.5669	2	Less Vegetation
30,759	78.841932	12.59148	1	High Vegetation
24,465	78.840458	12.591477	1	High Vegetation
28,861	78.805977	12.580232	2	Less Vegetation
35,655	78.836129	12.591499	1	High Vegetation
33,638	78.812464	12.580187	2	Less Vegetation
63	78.81674	12.60167	1	Less Vegetation
39,388	78.81276	12.58634	2	Less Vegetation

The percentage of LU/LC change detection was calculated for the LU/LC classified image, and the results are shown in Section 5.3. Based on the good accuracy, the LU/LC classification map was processed for further findings of the LU/LC prediction map. The LST map for the years 2012 and 2015 was calculated to extract the spatial features of Javadi Hills. The estimation of the LST map was explained in Section 3.3. The LST map shows the features of the high- and low-temperature values of the earth's surface of Javadi Hills. The high-temperature values indicate less vegetation, and the low-temperature value indicates a high-vegetation area. The LST (LST_I) and the LU/LC (LU_I) classification map was used as an input for predicting the LU/LC map of Javadi Hills. We combined the time-series features of LST and the LU/LC map of Javadi Hills. The impact of LST on the LU/LC map provides good results during the prediction process. For a better understanding, we show the impact of a few *LST* and LU/LC features in Figure 19, and we show the values in Table 4. The impact on the LST and LU/LC map strengthens our proposed predicted model with good validation results.

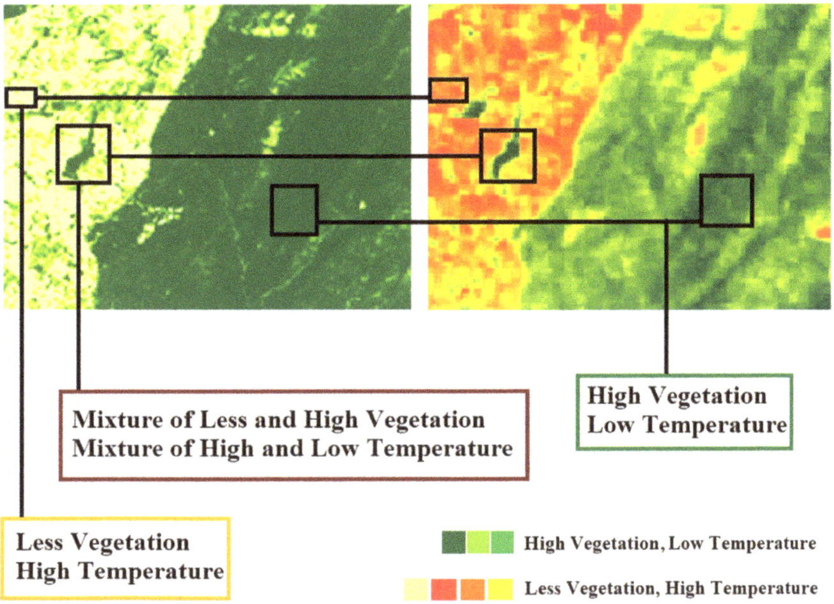

Figure 19. Impact of LST features with the LU/LC classes for Javadi Hills, India.

Table 4. LST and LU/LC values for the area of Javadi Hills.

Feature Value	Longitude	Latitude	Class Value	Temperature Value	Class Label
1	78.82975	12.58182	1	32.183754	High Vegetation
241	78.81025	12.58796	2	37.755061	Less Vegetation
1785	78.81824	12.58022	2	37.755061	Less Vegetation
733	78.84916	12.57678	1	31.708773	High Vegetation
6640	78.81107	12.57028	2	34.998298	Less Vegetation
6277	78.83463	12.57679	1	31.708773	High Vegetation
12,354	78.85108	12.59151	1	30.273344	High Vegetation
12,179	78.80721	12.58024	2	38.20916	Less Vegetation
20,163	78.81167	12.5669	2	34.998298	Less Vegetation
30,759	78.84193	12.59148	1	32.607521	High Vegetation
24,465	78.84046	12.59148	1	32.183754	High Vegetation
28,861	78.80598	12.58023	2	38.20916	Less Vegetation
35,655	78.83613	12.5915	1	31.708773	High Vegetation
33,638	78.81246	12.58019	2	34.533323	Less Vegetation
63	78.81674	12.60167	2	36.842331	Less Vegetation
39,388	78.81276	12.58634	2	38.20916	Less Vegetation

From the input LU/LC and LST features of 2012 and 2015, we predicted the LU/LC map of 2018 by using the Bi-LSTM model with the tuning of different parameters. The validated result provides good accuracy for our proposed model. We used the inputs of the LU/LC map of 2012 and 2015, along with the predicted LU/LC map of 2018 for predicting the LU/LC map for the year 2021. The short-term prediction was performed till the year 2027 for our study area. The working process of the Bi-LSTM model was explained in Section 3.4. The parameter used during the training process of the Bi-LSTM model is shown in Table 5.

Table 5. Hyperparameters for the Bi-LSTM model.

Parameter	Value
Input Image Format	Raster
Number of Training Samples	51,200
Activation Function	*tanh*, Softmax
Dropout	0.1, 0.25
Learning Rate	0.001
Optimizer	Adam
Loss Function	Categorical Cross Entropy
Hidden layers	20
Number of epochs	100
Batch Size	32

The combined features of the LU/LC and LST map were used as the training features during the process of the Bi-LSTM training. Each pixel value was identified through the latitude and longitudinal coordinates of Javadi Hills manually through the combined features of the LU/LC and LST map. Each pixel holds either high or less vegetation for its defined coordinate system. The few combined values were shown in Table 4. For better understanding, we show the combined features map in Figure 20. The accuracy results for the prediction model are shown in Section 5.2. The results were also cross-verified with the time-series Google Earth Engine for acquiring the validation accuracy of our model. With the impact of the LST map with the LU/LC map, good validation accuracy was obtained with a lower misclassification rate.

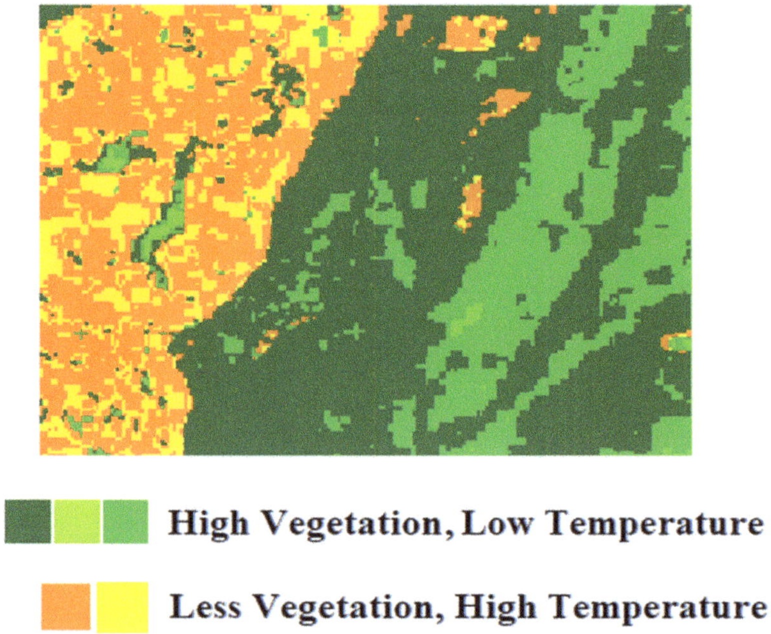

Figure 20. Training LU/LC–LST feature map for Bi-LSTM prediction model—Javadi Hills, India.

5.2. Validation of Vision Transformer–Based Bi-LSTM Model

The Google Earth images with the LU/LC classified images were evaluated for the examination of accuracy assessment. By using the time-series images of the Google Earth Engine, the accuracy assessment was calculated for the LU/LC classified image of Javadi Hills. All the pixel values of the LU/LC classified image were validated with the Google

Earth images. A total of 1008 random training samples were loaded, and the confusion matrix was obtained during the process of accuracy assessment. Table 6 represents the confusion matrix for the years 2012 and 2015. The results of the accuracy assessment for the year 2012 are 0.9891, and for 2015, it is 0.9861. Table 7 represents the LU/LC accuracy assessment for the years 2012 and 2015.

Table 6. LU/LC confusion matrix.

LU/LC Classification	Class	Reference Class			
		2012		2015	
		High Vegetation	Less Vegetation	High Vegetation	Less Vegetation
Actual Class	High Vegetation	694	4	689	6
	Less Vegetation	7	303	8	305

Table 7. LU/LC accuracy assessment for the proposed Vision Transformer model.

LU/LC Classification	2012	2015
Overall Accuracy	0.9891	0.9861
Precision	0.9901	0.9885
Recall	0.9942	0.9913
F1-Score	0.9921	0.9898

The LU/LC prediction was performed, and the results were analyzed and processed. The total number of pixel values was sliced into training and validation sets in an 8:2 proportion. The accuracy values of the prediction method look good for the LU/LC map of 2018 and 2021. The result of the validation accuracy for the year 2018 is 0.9865, and for 2021, it is 0.9811. The results were also cross-verified with the time-series Google Earth Engine image of Javadi Hills for the years 2018 and 2021 for acquiring the testing accuracy of our model. The results of the testing accuracy for our model also provide good results for 2018 and 2021. The results of the testing accuracy for the year 2018 is 0.9696, and for 2021, it is 0.9673. The results of the testing and validation accuracy of the predicted map are presented in Table 8. The validation accuracy refers to the results of the non-trained datasets of the model. The testing accuracy refers to the results of the complete model. We used the inputs of the LU/LC map of 2012 and 2015, along with the predicted LU/LC map of 2018 and 2021 for predicting the LU/LC map for the years 2024 and 2027. The short-term prediction was performed till the year 2027 for our study area. As the Google Earth Engine provides the time-series image till the current date, the validation and testing accuracy for the predicted LU/LC map of 2024 and 2027 was not calculated. With the results of the good validation accuracy for all the LU/LC predicted maps of Javadi Hills, our prediction model provides a lower misclassification rate.

$$Average\ Model\ Accuracy = \left(\frac{A_{Y1} + A_{Y2} + \ldots + A_{Yn}}{\mathrm{T}} \right) * 100 \tag{27}$$

where A_Y represents the accuracy value of years $\{1 \ldots .n\}$, and T represents the total number of years. The importance of providing the performance of the model depends on the average classification and prediction results. The average classification and prediction accuracy for the time series LU/LC data have been calculated by using Equation (27). The accuracy results for the years 2012 (0.9891) and 2015 (0.9861) were used for providing the performance of the calculation model through the average model accuracy. The average classification accuracy that was obtained was 98.76% for the proposed Vision Transformer model. The validation and testing results of our prediction model for the year 2018 are 0.9865 and 0.9696, respectively. The validation and testing results of our prediction model for the year 2021 are 0.9811 and 0.9673, respectively. The average validation accuracy is

98.38%, and the testing accuracy is 96.84% for our prediction model. We infer that the impact of the LST spatial variable from TIRS bands with the classified LU/LC map provides a good percentage of results.

Table 8. Validation and testing process of the proposed Vision Transformer–based Bi-LSTM Prediction Model.

Input Map (Year)	Training Feature Map (256 × 200 Pixels)	Train—Validation Split (8:2) Data	Test Data (Google Earth Image)	Predicted Map	Validation Accuracy	Testing Accuracy
LU/LC Classification—LST Map (2012, and 2015)	51,200	40,960–10,240	51,200	2018	0.9865	0.9696
	51,200	40,960–10,240	51,200	2021	0.9811	0.9673

The computational complexity defines the total time taken by the computer for running an algorithm. The computational complexity of the Vision Transformer model is $O\ (nC)$, where n is the size of input, and C is the number of classified LU/LC classes. The computational complexity of the Bi-LSTM prediction model is $O\ (nkC + 1)$, where k is the size of the spatial maps (LST) associated with input data n. Hence, the total computational time of our proposed algorithm C_c is the arithmetic sum of the classification and prediction model, which is given in Equation (28).

$$C_c = O\ (nC) + O\ (nkC + 1) \tag{28}$$

Although the proposed Vision Transformer–based Bi-LSTM prediction model shows significant performance, its training phase requires the determination of class values associated with spatial maps for each pixel in the n images, and this is computationally expensive.

5.3. Growth Pattern of the LU/LC Area of Javadi Hills

The growth patterns of LU/LC change in the area of Javadi Hills were performed between the years 2012 to 2027, and the results are shown in Table 9. In 2012, the LU/LC multispectral classified map was found to be 1651.04 ha (hectare) of the high vegetation and 736.85 ha of less vegetation. In 2015, the LU/LC multispectral classified map was found to be 1601.22 ha of vegetation and 786.67 ha of less vegetation. In 2018, the LU/LC predicted map was found to be 1621.18 ha of high vegetation and 766.71 ha of less vegetation. In 2021, the LU/LC predicted map was found to be 1596.04 ha of high vegetation and 791.85 ha of less vegetation. In 2024, the LU/LC predicted map was found to be 1568.23 ha of high vegetation and 819.66 ha of less vegetation. In 2027, the LU/LC predicted map was found to be 1553.17 ha of high vegetation and 834.72 ha of less vegetation. It was observed that the LU/LC changes have been frequently happening every three years in the area of Javadi Hills. The results of the LU/LC change that occurred between the years 2012 to 2027 are shown in Table 10. The comparison chart of LU/LC area statistics for the time-series data from 2012 to 2027 is shown in Figure 21.

Table 9. LU/LC area statistics for LU/LC Map (2012–2027).

LU/LC Class	Area (ha)					
	Year					
	2012	2015	2018	2021	2024	2027
High Vegetation	1651.04	1601.22	1621.18	1596.04	1568.23	1553.17
Less Vegetation	736.85	786.67	766.71	791.85	819.66	834.72
Total (ha)	2387.89	2387.89	2387.89	2387.89	2387.89	2387.89

Table 10. Percentage of LU/LC change for the area of Javadi Hills during 2012–2027.

LU/LC Class	**Area (%)**				
	Year				
	2012–2015	**2015–2018**	**2018–2021**	**2021–2024**	**2024–2027**
High Vegetation	−3.01	1.24	−1.55	−1.74	−0.96
Less Vegetation	6.76	−2.53	3.27	3.51	1.83

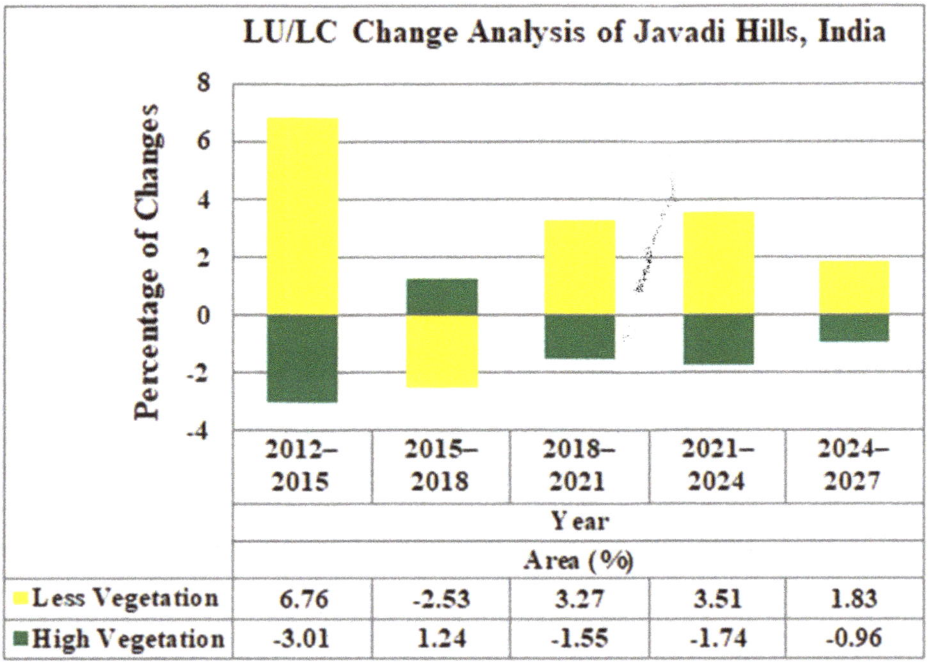

Figure 21. LU/LC change analysis of the Javadi Hills, India (2012–2027).

6. Comparative Analysis

In this research work, we have proposed the Vision Transformer–based Bi-LSTM prediction model for analyzing the past, present, and future changes of Javadi Hills, India. We also infer that the LU/LC prediction accuracy of our model provides a lower error rate, i.e., below 5%. From the thorough analysis, we infer that the use of the LST map has a high impact on the LU/LC environment, and it was considered an important spatial feature for the prediction of the LU/LC vegetation map.

We have compared our model with CNN, DWT, and standard LU/LC classification and prediction techniques for the area of Javadi Hills. Our model outperforms the other standard classification and prediction algorithms in terms of accuracy and computational efficiency. We have executed the standard LU/LC algorithms (DWT [22], CNN [27], SVM [1], MLC [2], and RFC [25]) and provided a comparative analysis of the Vision Transformer model for our study area of Javadi Hills in Table 11. We have also presented the comparative accuracy of the classification model in Figure 22. We have also shown the comparative analysis of our prediction model with the hybrid machine-learning models [7] for the area of Javadi Hills in Table 12.

Table 11. Comparative analysis of the proposed Vision Transformer model with other algorithms for the area of Javadi Hills, India.

Algorithms	Average Accuracy (%)
Ours	98.76
CNN [27]	96.42
DWT [22]	94.21
SVM [1]	97.71
MLC [2]	94.4
RFC [25]	95.6

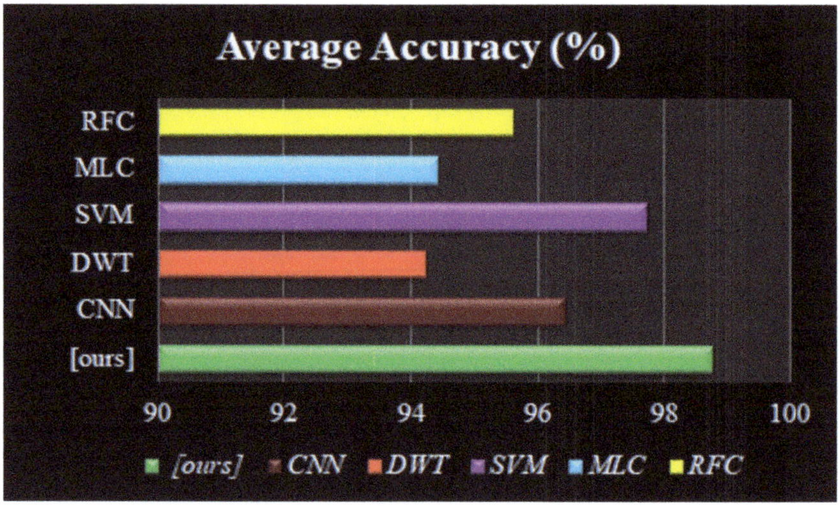

Figure 22. Performance analysis of LU/LC classification model—Javadi Hills, India.

Table 12. Comparative analysis of LU/LC prediction models for the area of Javadi Hills, India.

Study Area	Algorithm	Prediction Accuracy (%)
Javadi Hills, India	Vision Transformer–based Bi-LSTM Model (ours)	98.38%
	RFC-based MC–ANN–CA Model [7]	93.41%

Our model outperforms the hybrid machine-learning models [7] and provides good prediction accuracy. We have validated the use of the LST map with other spatial maps that include a slope, aspect, and distances from the road map [7] for our prediction model. From the thorough analysis, we infer that the use of the LST map has a high impact on the LU/LC environment, and it has been considered an important spatial feature for the prediction of the LU/LC vegetation map. We have shown a few comparisons of the validation results of the LU/LC prediction methods by using LST, slope, aspect, and distance from the road map for the area of Javadi Hills in Table 13.

Table 13. Testing of the Vision Transformer–based Bi-LSTM model using the various combinations of Input Spatial Data for Javadi Hills, India.

Study Area	Input Data	Prediction Accuracy (%)
Javadi Hills, India	LU/LC Classification—LST Map	98.38
	LU/LC Classification—Slope Map	92.33
	LU/LC Classification—Distance from Road Map	91.64
	LU/LC Classification—Slope, Distance from road map	92.52
	LU/LC Classification—Slope, LST map	93.45
	LU/LC Classification—Distance from Road, LST map	93.17
	LU/LC Classification—Slope, Distance from Road, LST map	94.2

We also show a few comparative analyses of overall prediction models for a few different study areas in Table 14. We observed that there is a performance variation in the prediction results for each study area around the world. This variation of the LU/LC classification and prediction results was due to the selection of study area, satellite data, environmental data, and its LU/LC classes. A variation of results was observed for our study area with the assessment of multi-satellite datasets through the proposed algorithm. We delivered a clear view of the importance of Vision Transformer–based LU/LC classification and Bi-LSTM-based prediction for forecasting the time series LU/LC vegetation map. The advantage of our proposed work lies in using only the LST map as the spatial data for predicting the LU/LC vegetation map. We also achieved a good prediction accuracy of 98.38%. Our proposed algorithm can be applied to other study areas around the world in predicting the LU/LC vegetation map. Moreover, our proposed model has been efficient for urban planners, forest departments, and government officials in analyzing the LU/LC information through XAI and taking necessary actions in the protection of the LU/LC environment.

Table 14. Comparative analysis of LU/LC prediction models for different study areas.

Study Area	Algorithm	Prediction Accuracy (%)
Javadi Hills, India (our study)	Vision Transformer–Based Bi-LSTM Model	98.38
Wuhan, China [28]	Self-Adaptive Cellular-Based Deep-Learning LSTM Model	93.1
Guangdong province of South China [23]	Deep Learning (RNN–CNN) and CA–MC Model	95.86
Western Fansu Province, China [26]	CNN–GRU Model	93.46
Dhaka, Bangladesh [29]	CA–MC and ANN Model	90.21
University of Nebraska–Lincoln [24]	CNN–Bi-LSTM Model	91.73
City of Surrey, British Columbia [56]	RNN–ConvLSTM Model	88.0
Klingenberg, Germany [55]	Fully CNN–LSTM Model	87.0
Awadh, Lucknow [6]	CA–MC and LR model	84.0

7. Conclusions

The LU/LC prediction modeling was considered important research in the area of remote sensing. In this research work, the multispectral LISS-III and Landsat satellite image of Javadi Hills for the periods 2012 and 2015 were downloaded and performed for analyzing the LU/LC prediction for the years 2018, 2021, 2024, and 2027. The Vision Transformer model for performing the LU/LC classification was proposed, and the accuracy assessment was performed by using Google Earth Images. The average classification accuracy obtained for our Vision Transformer model was 98.76%. The spatial features from the LST map and LU/LC classified map were used as input for predicting the LU/LC changes in Javadi Hills. For predicting the future LU/LC changes of Javadi Hills, the Bi-LSTM model was successfully applied. We infer that the impact of the LST spatial features with the LU/LC classified map provides a good percentage of results with 98.38%. The predicted results provide the variation in the high- and less-vegetation regions of Javadi Hills from 2012 to 2027. Our Vision Transformer–based Bi-LSTM model has produced good validation results when compared with other standardized models. Our research on LU/LC prediction provides information to the forest departments, urban planners, and government officials to take necessary action in the protection of the LU/LC environment through application-based XAI. In the future, we plan to focus more on using the TIRS bands of hyperspectral data to obtain the temperature values associated with each pixel and to classify the hyperspectral data in real-time scenarios.

Author Contributions: A.L. conceived the study, created the literature review, and designed the flow of the proposed model; S.N.M. contributed to the satellite data acquisition, algorithm development, and writing of the manuscript; A.L. has contributed to testing the performance of the algorithm, and the internal review of the manuscript. All authors have read and agreed to the published version of the manuscript.

Funding: This research received no external funding.

Institutional Review Board Statement: Not applicable.

Informed Consent Statement: Not applicable.

Data Availability Statement: The input satellite images used in our research work can be accessed freely from different online sources. The availability of the LISS-III satellite images for our study area was checked, and the images can be downloaded freely in the open data archive from the website of the Bhuvan Indian Geo-Platform of ISRO (https://bhuvan.nrsc.gov.in/ (accessed on 9 December 2019)). The availability of the Landsat satellite images for our study area was checked, and the images can be downloaded freely from the website of the United States Geological Survey (USGS), United States (https://earthexplorer.usgs.gov (accessed on 16 December 2019)). The images available on the Google Earth Engine platform (https://www.google.com/earth/ (accessed on 10 November 2021)) were used as the reference data during the accuracy assessment for different periods from 1984 to the current date.

Acknowledgments: The authors wish to thank the United States Geological Survey (USGS), United States, for providing Landsat TIRS, RED, and NIR bands. We are thankful to Bhuvan Indian Geo-Platform of ISRO, India, for providing the LISS-III multispectral data. The authors also wish to thank the developers of the Google Earth Engine platform for providing the time-series data with less image resolution. We are thankful to the Vellore Institute of Technology for providing the VIT SEED GRANT for carrying out this work and the CDMM (Centre for Disaster Mitigation and Management) for providing a good lab facility.

Conflicts of Interest: The authors declare no conflict of interest.

References

1. Baig, M.F.; Mustafa, M.R.U.; Baig, I.; Takaijudin, H.B.; Zeshan, M.T. Assessment of land use land cover changes and future predictions using CA-ANN simulation for selangor, Malaysia. *Water* **2022**, *14*, 402. [CrossRef]
2. Imran, M.; Aqsa, M. Analysis and mapping of present and future drivers of local urban climate using remote sensing: A case of Lahore, Pakistan. *Arab. J. Geosci.* **2020**, *13*, 1–14. [CrossRef]
3. Heidarlou, H.B.; Shafiei, A.B.; Erfanian, M.; Tayyebi, A.; Alijanpour, A. Effects of preservation policy on land use changes in Iranian Northern Zagros forests. *Land Use Policy* **2019**, *81*, 76–90. [CrossRef]
4. Chaves, M.E.D.; Michelle, C.A.P.; Ieda, D.S. Recent applications of Landsat 8/OLI and Sentinel-2/MSI for land use and land cover mapping: A systematic review. *Remote Sens.* **2020**, *12*, 3062. [CrossRef]

5. Priyadarshini, K.N.; Kumar, M.; Rahaman, S.A.; Nitheshnirmal, S. A comparative study of advanced land use/land cover classification algorithms using Sentinel-2 data. *Int. Arch. Photogramm. Remote Sens. Spat. Inf. Sci.* **2018**, *42*, 20–23. [CrossRef]
6. Siddiqui, A.; Siddiqui, A.; Maithani, S.; Jha, A.K.; Kumar, P.; Srivastav, S.K. Urban growth dynamics of an Indian metropolitan using CA markov and logistic regression. *Egypt. J. Remote Sens. Space Sci.* **2018**, *21*, 229–236. [CrossRef]
7. Mohan, R.; Sam, N.; Agilandeeswari, L. Modelling spatial drivers for LU/LC change prediction using hybrid machine learning methods in Javadi Hills, Tamil Nadu, India. *J. Indian Soc. Remote Sens.* **2021**, *49*, 913–934. [CrossRef]
8. Sandamali, S.P.I.; Lakshmi, N.K.; Sundaramoorthy, S. Remote sensing data and SLEUTH urban growth model: As decision support tools for urban planning. *Chin. Geogr. Sci.* **2018**, *28*, 274–286. [CrossRef]
9. Hegde, G.; Ahamed, J.M.; Hebbar, R.; Raj, U. Urban land cover classification using hyperspectral data. *Int. Arch. Photogramm. Remote Sens. Spat. Inf. Sci.* **2014**, *8*, 751–754. [CrossRef]
10. Elmore, A.J.; John, F.M. Precision and accuracy of EO-1 Advanced Land Imager (ALI) data for semiarid vegetation studies. *IEEE Trans. Geosci. Remote Sens.* **2003**, *41*, 1311–1320. [CrossRef]
11. Elbeih, S.F.; El-Zeiny Ahmed, M. Qualitative assessment of groundwater quality based on land use spectral retrieved indices: Case study Sohag Governorate, Egypt. *Remote Sens. Appl. Soc. Environ.* **2018**, *10*, 82–92. [CrossRef]
12. Nagy, A.; El-Zeiny, A.; Sowilem, M.; Atwa, W.; Elshaier, M. Mapping mosquito larval densities and assessing area vulnerable to diseases transmission in Nile valley of Giza, Egypt. *Egypt. J. Remote Sens. Space Sci.* **2022**, *25*, 63–71. [CrossRef]
13. Etemadi, H.; Smoak, J.M.; Karami, J. Land use change assessment in coastal mangrove forests of Iran utilizing satellite imagery and CA–Markov algorithms to monitor and predict future change. *Environ. Earth Sci.* **2018**, *77*, 1–13. [CrossRef]
14. Goodarzi Mehr, S.; Ahadnejad, V.; Abbaspour, R.A.; Hamzeh, M. Using the mixture-tuned matched filtering method for lithological mapping with Landsat TM5 images. *Int. J. Remote Sens.* **2013**, *34*, 8803–8816. [CrossRef]
15. Vijayan, D.; Shankar, G.R.; Shankar, T.R. Hyperspectral data for land use/land cover classification. *Int. Arch. Photogramm. Remote Sens. Spat. Inf. Sci.* **2014**, *40*, 991. [CrossRef]
16. Myint, S.W.; Okin, G.S. Modelling land-cover types using multiple endmember spectral mixture analysis in a desert city. *Int. J. Remote Sens.* **2009**, *30*, 2237–2257. [CrossRef]
17. Amigo, J.M.; Carolina, S. Preprocessing of hyperspectral and multispectral images. In *Data Handling in Science and Technology*; Elsevier: Amsterdam, The Netherlands, 2020; Volume 32, pp. 37–53. [CrossRef]
18. HongLei, Y.; JunHuan, P.; BaiRu, X.; DingXuan, Z. Remote sensing classification using fuzzy C-means clustering with spatial constraints based on Markov random field. *Eur. J. Remote Sens.* **2013**, *46*, 305–316. [CrossRef]
19. Navin, M.S.; Agilandeeswari, L. Comprehensive review on land use/land cover change classification in remote sensing. *J. Spectr. Imaging* **2020**, *9*, a8. [CrossRef]
20. Ganasri, B.P.; Dwarakish, G.S. Study of land use/land cover dynamics through classification algorithms for Harangi catchment area, Karnataka State, India. *Aquat. Procedia* **2015**, *4*, 1413–1420. [CrossRef]
21. Sharma, J.; Prasad, R.; Mishra, V.N.; Yadav, V.P.; Bala, R. Land use and land cover classification of multispectral LANDSAT-8 satellite imagery using discrete wavelet transform. *Int. Arch. Photogramm. Remote Sens. Spat. Inf. Sci.* **2018**, *42*, 703–706. [CrossRef]
22. Kabisch, N.; Selsam, P.; Kirsten, T.; Lausch, A.; Bumberger, J. A multi-sensor and multi-temporal remote sensing approach to detect land cover change dynamics in heterogeneous urban landscapes. *Ecol. Indic.* **2019**, *99*, 273–282. [CrossRef]
23. Xing, W.; Qian, Y.; Guan, X.; Yang, T.; Wu, H. A novel cellular automata model integrated with deep learning for dynamic spatio-temporal land use change simulation. *Comput. Geosci.* **2020**, *137*, 104430. [CrossRef]
24. Ali, A.S.A.; Ebrahimi, S.; Ashiq, M.M.; Alasta, M.S.; Azari, B. CNN-Bi LSTM neural network for simulating groundwater level. *Environ. Eng.* **2022**, *8*, 1–7. [CrossRef]
25. Floreano, I.X.; de Moraes, L.A.F. Land use/land cover (LULC) analysis (2009–2019) with google earth engine and 2030 prediction using Markov-CA in the Rondônia State, Brazil. *Environ. Monit. Assess.* **2021**, *193*, 1–17. [CrossRef]
26. Xiao, B.; Liu, J.; Jiao, J.; Li, Y.; Liu, X.; Zhu, W. Modeling dynamic land use changes in the eastern portion of the hexi corridor, China by cnn-gru hybrid model. *GIScience Remote Sens.* **2022**, *59*, 501–519. [CrossRef]
27. Gaetano, R.; Ienco, D.; Ose, K.; Cresson, R. A two-branch CNN architecture for land cover classification of PAN and MS imagery. *Remote Sens.* **2018**, *10*, 1746. [CrossRef]
28. Mu, L.; Wang, L.; Wang, Y.; Chen, X.; Han, W. Urban land use and land cover change prediction via self-adaptive cellular based deep learning with multisourced data. *IEEE J. Sel. Top. Appl. Earth Obs. Remote Sens.* **2019**, *12*, 5233–5247. [CrossRef]
29. Al Kafy, A.; Dey, N.N.; Al Rakib, A.; Rahaman, Z.A.; Nasher, N.R.; Bhatt, A. Modeling the relationship between land use/land cover and land surface temperature in Dhaka, Bangladesh using CA-ANN algorithm. *Environ. Chall.* **2021**, *4*, 100190. [CrossRef]
30. MohanRajan, S.N.; Loganathan, A.; Manoharan, P. Survey on land use/land cover (LU/LC) change analysis in remote sensing and GIS environment: Techniques and challenges. *Environ. Sci. Pollut. Res.* **2020**, *27*, 29900–29926. [CrossRef]
31. Dosovitskiy, A.; Beyer, L.; Kolesnikov, A.; Weissenborn, D.; Zhai, X.; Unterthiner, T.; Dehghani, M.; Minderer, M.; Heigold, G.; Gelly, S.; et al. An image is worth 16x16 words: Transformers for image recognition at scale. *arXiv* **2020**, arXiv:2010.11929. [CrossRef]
32. Hong, D.; Han, Z.; Yao, J.; Gao, L.; Zhang, B.; Plaza, A.; Chanussot, J. SpectralFormer: Rethinking hyperspectral image classification with transformers. *IEEE Trans. Geosci. Remote Sens.* **2021**, *60*, 1–15. [CrossRef]
33. Wen, Q.; Zhou, T.; Zhang, C.; Chen, W.; Ma, Z.; Yan, J.; Sun, L. Transformers in time series: A survey. *arXiv* **2022**, arXiv:2202.07125. [CrossRef]

34. Khan, S.; Naseer, M.; Hayat, M.; Zamir, S.W.; Khan, F.S.; Shah, M. Transformers in vision: A survey. *ACM Comput. Surv. (CSUR)* **2021**, 1–38. [CrossRef]
35. Bazi, Y.; Bashmal, L.; Rahhal, M.M.A.; Dayil, R.A.; Ajlan, N.A. Vision transformers for remote sensing image classification. *Remote Sens.* **2021**, *13*, 516. [CrossRef]
36. Xu, K.; Deng, P.; Huang, H. Vision transformer: An excellent teacher for guiding small networks in remote sensing image scene classification. *IEEE Trans. Geosci. Remote Sens.* **2022**, *60*, 1–15. [CrossRef]
37. Sha, Z.; Li, J. MITformer: A multi-instance vision transformer for remote sensing scene classification. *IEEE Geosci. Remote Sens. Lett.* **2022**, *19*, 1–5. [CrossRef]
38. Xu, F.; Uszkoreit, H.; Du, Y.; Fan, W.; Zhao, D.; Zhu, J. Explainable AI: A brief survey on history, research areas, approaches and challenges. In *CCF International Conference on Natural Language Processing and Chinese Computing*; Tang, J., Kan, M.Y., Zhao, D., Li, S., Zan, H., Eds.; Springer: Cham, Switzerland, 2019. [CrossRef]
39. Schlegel, U.; Arnout, H.; El-Assady, M.; Oelke, D.; Keim, D.A. Towards a rigorous evaluation of XAI methods on time series. In Proceedings of the 2019 IEEE/CVF International Conference on Computer Vision Workshop (ICCVW), Seoul, Korea, 27–28 October 2019; IEEE: Manhattan, NY, USA, 2020. [CrossRef]
40. Saeed, W.; Christian, O. Explainable AI (XAI): A systematic meta-survey of current challenges and future opportunities. *arXiv* **2021**, arXiv:2111.06420. [CrossRef]
41. Ribeiro, J.; Silva, R.; Cardoso, L.; Alves, R. Does dataset complexity matters for model explainers? In Proceedings of the 2021 IEEE International Conference on Big Data (Big Data), Orlando, FL, USA, 15–18 December 2021; IEEE: Manhattan, NY, USA, 2022. [CrossRef]
42. Samek, W.; Montavon, G.; Lapuschkin, S.; Anders, C.J.; Müller, K.R. Explaining deep neural networks and beyond: A review of methods and applications. *IEEE* **2021**, *109*, 247–278. [CrossRef]
43. LinarLinardatos, P.; Papastefanopoulos, V.; Kotsiantis, S. Explainable AI: A review of machine learning interpretability methods. *Entropy* **2020**, *23*, 18. [CrossRef]
44. Navin, M.S.; Agilandeeswari, L. Multispectral and hyperspectral images based land use/land cover change prediction analysis: An extensive review. *Multimed. Tools Appl.* **2020**, *79*, 29751–29774. [CrossRef]
45. Ma, L.; Liu, Y.; Zhang, X.; Ye, Y.; Yin, G.; Johnson, B.A. Deep learning in remote sensing applications: A meta-analysis and review. *ISPRS J. Photogramm. Remote Sens.* **2019**, *152*, 166–177. [CrossRef]
46. Thyagharajan, K.K.; Vignesh, T. Soft computing techniques for land use and land cover monitoring with multispectral remote sensing images: A review. *Arch. Comput. Methods Eng.* **2019**, *26*, 275–301. [CrossRef]
47. Kaselimi, M.; Voulodimos, A.; Daskalopoulos, I.; Doulamis, N.; Doulamis, A. A Vision transformer model for convolution-free multilabel classification of satellite imagery in deforestation monitoring. *IEEE Trans. Neural Netw. Learn. Syst.* **2022**, 1–9. [CrossRef] [PubMed]
48. Liu, X.; Wu, Y.; Liang, W.; Cao, Y.; Li, M. High resolution SAR image classification using global-local network structure based on vision transformer and CNN. *IEEE Geosci. Remote Sens. Lett.* **2022**, *19*, 1–5. [CrossRef]
49. Ghali, R.; Akhloufi, M.A.; Jmal, M.; Souidene Mseddi, W.; Attia, R. Wildfire segmentation using deep vision transformers. *Remote Sensing* **2021**, *13*, 3527. [CrossRef]
50. Chen, X.; Qiu, C.; Guo, W.; Yu, A.; Tong, X.; Schmitt, M. Multiscale feature learning by transformer for building extraction from satellite images. *IEEE Geosci. Remote Sens. Lett.* **2022**, *19*, 1–5. [CrossRef]
51. Meng, X.; Wang, N.; Shao, F.; Li, S. Vision transformer for pansharpening. *IEEE Trans. Geosci. Remote Sens.* **2022**, *60*, 1–11. [CrossRef]
52. Mukherjee, F.; Singh, D. Assessing land use–land cover change and its impact on land surface temperature using LANDSAT data: A comparison of two urban areas in India. *Earth Syst. Environ.* **2020**, *4*, 385–407. [CrossRef]
53. Sekertekin, A.; Stefania, B. Land surface temperature retrieval from Landsat 5, 7, and 8 over rural areas: Assessment of different retrieval algorithms and emissivity models and toolbox implementation. *Remote Sens.* **2020**, *12*, 294. [CrossRef]
54. Nasir, M.J.; Ahmad, W.; Iqbal, J.; Ahmad, B.; Abdo, H.G.; Hamdi, R.; Bateni, S.M. Effect of the urban land use dynamics on land surface temperature: A case study of kohat city in Pakistan for the period 1998–2018. *Earth Syst. Environ.* **2022**, *6*, 237–248. [CrossRef]
55. Sefrin, O.; Riese, F.M.; Keller, S. Deep learning for land cover change detection. *Remote Sens.* **2020**, *13*, 78. [CrossRef]
56. Cao, C.; Dragićević, S.; Li, S. Short-term forecasting of land use change using recurrent neural network models. *Sustainability* **2019**, *11*, 5376. [CrossRef]
57. Wang, H.; Zhao, X.; Zhang, X.; Wu, D.; Du, X. Long time series land cover classification in China from 1982 to 2015 based on Bi-LSTM deep learning. *Remote Sens.* **2019**, *11*, 1639. [CrossRef]
58. Adadi, A.; Mohammed, B. Peeking inside the black-box: A survey on explainable artificial intelligence (XAI). *IEEE Access* **2018**, *6*, 52138–52160. [CrossRef]

59. Fulton, L.B.; Lee, J.Y.; Wang, Q.; Yuan, Z.; Hammer, J.; Perer, A. Getting playful with explainable AI: Games with a purpose to improve human understanding of AI. In Proceedings of the Extended Abstracts of the 2020 CHI Conference on Human Factors in Computing Systems, Honolulu, HI, USA, 25–30 April 2020; Association for Computing Machinery: New York, NY, USA, 2020. [CrossRef]
60. Kute, D.V.; Pradhan, B.; Shukla, N.; Alamri, A. Deep learning and explainable artificial intelligence techniques applied for detecting money laundering–a critical review. *IEEE Access* **2021**, *9*, 82300–82317. [CrossRef]

applied sciences

MDPI

Article

Using Artificial Neural Network Algorithm and Remote Sensing Vegetation Index Improves the Accuracy of the Penman-Monteith Equation to Estimate Cropland Evapotranspiration

Yan Liu [1], Sha Zhang [1], Jiahua Zhang [2], Lili Tang [1] and Yun Bai [1,*]

[1] Centre for Remote Sensing and Digital Earth, College of Computer Science and Technology, Qingdao University, Qingdao 266071, China; liuyan20210409@163.com (Y.L.); zhangsha@qdu.edu.cn (S.Z.); tanglili5208@163.com (L.T.)

[2] Aerospace Information Research Institute, Chinese Academy of Sciences, Beijing 100094, China; zhangjh@radi.ac.cn

* Correspondence: baiyun@qdu.edu.cn; Tel.: +86-137-1880-5053

Abstract: Accurate estimation of evapotranspiration (ET) can provide useful information for water management and sustainable agricultural development. However, most of the existing studies used physical models, which are not accurate enough due to our limited ability to represent the ET process accurately or rarely focused on cropland. In this study, we trained two models of estimating croplands ET. The first is Medlyn-Penman-Monteith (Medlyn-PM) model. It uses artificial neural network (ANN)-derived gross primary production along with Medlyn's stomatal conductance to compute surface conductance (*Gs*), and the computed *Gs* is used to estimate ET using the PM equation. The second model, termed ANN-PM, directly uses ANN to construct *Gs* and simulate ET using the PM equation. The results showed that the two models can reasonably reproduce ET with ANN-PM showing a better performance, as indicated by the lower error and higher determination coefficients. The results also showed that the performances of ANN-PM without the facilitation of any remote sensing (RS) factors degraded significantly compared to the versions that used RS factors. We also evidenced that ANN-PM can reasonably characterize the time-series changes of ET at sites having a dry climate. The ANN-PM method can reasonably estimate the ET of croplands under different environmental conditions.

Keywords: evapotranspiration; penman-monteith equation; artificial neural network; canopy conductance

Citation: Liu, Y.; Zhang, S.; Zhang, J.; Tang, L.; Bai, Y. Using Artificial Neural Network Algorithm and Remote Sensing Vegetation Index Improves the Accuracy of the Penman-Monteith Equation to Estimate Cropland Evapotranspiration. *Appl. Sci.* **2021**, *11*, 8649. https://doi.org/10.3390/app11188649

Academic Editors: Anselme Muzirafuti and Dimitrios S. Paraforos

Received: 27 July 2021
Accepted: 15 September 2021
Published: 17 September 2021

1. Introduction

Evapotranspiration (ET) is the process by which vegetation and groundwater transport water vapor to the atmosphere, mainly including plant transpiration and soil evaporation [1], with transpiration being dominant on a global scale [2]. Estimation of ET is an important basis for reasonable irrigation over croplands at a regional scale [3]; at the same time, as an important part of energy balance and the water cycle, ET also affects atmospheric circulation and plays an important role in regulating climate. Cropland is an important ecosystem on the land surface. Thus, the accurate estimation of cropland ET is of great significance for the rational irrigation of crops and the study of material and energy balance under the background of climate change [4].

The Penman–Monteith (PM) equation is the most commonly used framework for estimating regional or global ET. The regional-scale modeling process based on the PM equation is a simulation of surface conductance (*Gs*), and this parameter accounts for the largest source of uncertainty in ET modeling based on the PM equation on a regional scale. Cleugh et al. [5] tested two models of estimating land surface evaporation, the surface energy balance model and PM-based approach using remote sensing (RS)-derived leaf area index (LAI), to estimate *Gs* at two Australian flux stations, and the PM-based method proved better. Mu et al. [6] found that the surface conductivity model of Cleugh et al. [5] was unreliable when used to estimate the global ET of 19 AmeriFlux sites due to the oversimplified estimates of surface conductance. Therefore, the canopy conductance and ET algorithms based on the PM method of Cleugh et al. [5] were improved by using the RS and global meteorological data. The algorithm of Mu et al. [6] considered the surface energy partitioning process and the environmental constraints of ET, but the performance of Mu et al. [6] still remains uncertain. Mu et al. [7] further improved the global terrestrial ET algorithm and showed the improved algorithm performed better compared to the original. Based on Cleugh et al. [5] and Mu et al. [6], Leuning et al. [8] developed a biophysical model to estimate *Gs* and introduced a simpler soil evaporation algorithm than the MOD16 algorithm [6] to calculate daily average evaporation. The results showed that the PM equation, incorporated with the RS leaf area index, could more reliably estimate the evaporation rate. However, the performances of the model degraded if a fixed value of maximum stomatal conductance (g_{sx}) was used to estimate the surface conductance across a wide range of vegetation categories [8]. Zhang et al. [9] further developed the *Gs* formula and calculated the land surface ET at a spatial resolution of 0.05 ° using the PM equation. Yebra et al. [10] reversed the PM equation to obtain the *Gs* of the plant canopy, and then the estimated *Gs* was used to retrieve actual ET using the parameterized PM equation. Kitao et al. [11] also applied a semi-empirical model dependent on photosynthesis [12] to estimate canopy *Gs*. Because the method of Ball et al. [12] restricted the applicability of the model, Yan et al. [13] used a simple biophysical model to calculate *Gs*, and then the computed *Gs* was used to calculated global ET based on the PM equation. Mallick et al. [14] estimated *Gs* by integrating the radiometric surface temperature into a combined structure of the PM model and the Shuttleworth–Wallace model and used the simplified surface energy balance model to estimate ET. The method of Yan et al. [13] used the leaf area index and surface meteorological data, while Mallick et al. [14] did not use any leaf-scale empirical parameter model to determine *Gs* and ET. However, the method of Mallick et al. [14] had a tendency to overestimate *Gs*. For areas with limited data, the method of Mallick et al. [14] was considered to be further improved. Therefore, Bhattarai et al. [15] used RS and reanalysis data to develop an automatic multi-model to estimate regional ET in important areas.

In order to reduce the uncertainties in ET estimation due to the difficulty in estimating *Gs*, semi-empirical models that use machine learning (ML) to more accurately calculate the *Gs* in the PM equation were proposed [16–18]. For example, Zhang et al. [18] combined ML, in which only temperature (Ta) data was used with the PM equation to estimate crop ET, and showed that the accuracy of the ML-based PM approach was better than the Hargreaves (HARG) method. However, the computational complexity of the model of Zhang et al. [18] is relatively high and requires more storage space. Traore et al. [17] evaluated different ML methods based on only temperature data to calculate ET under the framework of the PM equation. The determination coefficients (R^2) were significantly increased when wind speed data was added to the model of [17]. Thus, only one meteorological input is not enough for reasonably quantifying ET. Multiple data combinations can effectively improve the accuracy of the ET model. Zhao et al. [19] developed a hybrid model to estimate latent heat flux based on various variables (such as soil moisture, carbon dioxide concentration (Ca), etc.), combining ML models with the PM method. The results showed that the hybrid model is more adaptable to extreme environments compared with the pure ML method. Due to a lack of reliable and spatiotemporal continuous soil moisture data sets on a global scale, the model of Zhao et al. [19] is limited to a regional scale and cannot be applied on a global scale. Therefore, using only a single datum or a data set that is difficult to obtain will limit the application of the model on a regional or global scale. Therefore, we use a variety of globally available data combined with ML methods in order to improve the estimates of ET over croplands. The ML approaches can represent the complex and non-linearly relationships between inputs and the target [20], and assess the adaptivity of multiple ET models of different environments [21], with smaller errors under a specific environmental condition.

Nowadays, most of the existing studies on estimating ET use physical models [22–25] or purely rely on ML algorithms [26–31]; these methods are not accurate enough to represent the ET due to the limited ability to understand the ET process. The hybrid ET model that combines the physical framework, namely the PM equation, and ML algorithms has proved to be effective in ET estimates [19,32]. The ML approaches resolved the difficulty of characterizing the complex environmental constraints on ET in the hybrid model, while the PM framework ensures the model's robustness. It is worth noting that the pure ML models may yield comparable or even better performance compared to the hybrid model [19] or individual physical models [26,33]. However, without physical constraints, the reliability of the pure ML models depends on the representativeness of training data [33]. As a result, the pure ML models are vulnerable to extreme environmental conditions [19], while the hybrid models show more robust performances under these conditions [19].

In this study, we aim to improve the estimates of cropland ET by training a hybrid ET model based on an artificial neural network (ANN) and PM equation, investigate whether the use of RS factors can improve the performances of hybrid models, and evaluate the ANN-PM model to simulate ET on a daily scale over flux sites covering a wide range of climate dryness.

2. Material and Methods

The research flow chart of this study is shown in Figure 1. We trained two methods to estimate ET. First, the *Gs* model is constructed using meteorological data and remote sensing data, and subsequently, used to simulate ET under the framework of the PM equation. Secondly, *Gs* is estimated using ANN-derived GPP in conjunction with Medlyn stomatal conductance, and then the computed *Gs* is used to estimate ET using the PM equation.

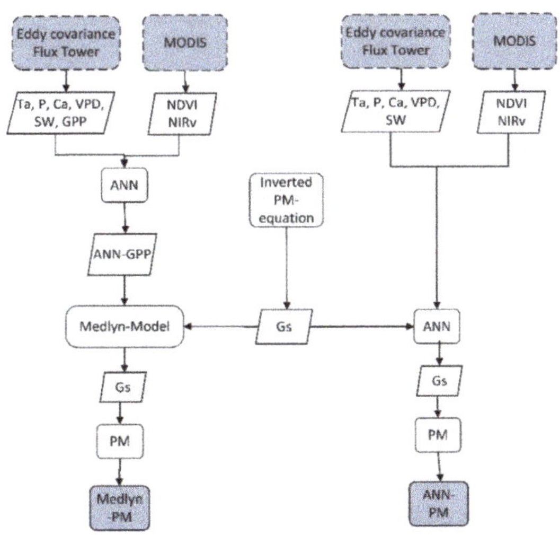

Figure 1. Research flow chart. Ta is temperature, P is precipitation, SW is solar radiation, Ca is carbon dioxide concentration, VPD is vapor pressure deficit, GPP is gross primary production, NDVI is normalized difference vegetation index, NIRv is near-infrared reflectance of vegetation, ANN is artificial neural network, *Gs* is surface conductance, and PM is the Penman–Monteith equation. A white parallelogram denotes a variable, and a white rectangle denotes a method. A gray dotted rectangle denotes the source of the variable, and a gray solid rectangle denotes a model.

2.1. Material

The meteorological data used in this study were retrieved from the meteorological observation data of the eddy covariance flux tower at 17 flux sites. Figure 2 shows the map representation of the 17 flux sites of cropland over the globe.

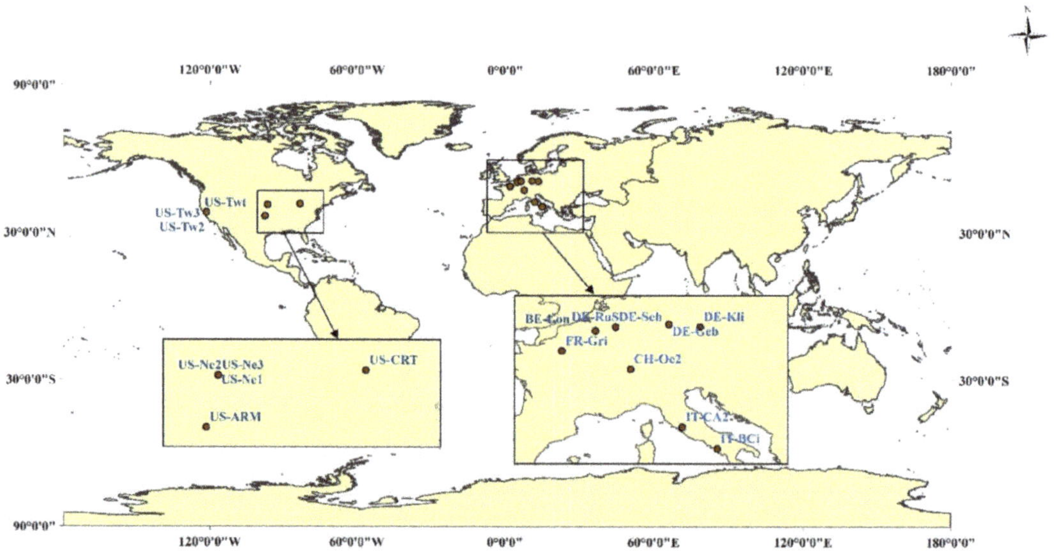

Figure 2. Map representation of 17 eddy covariance flux sites.

The information of the 17 flux sites is shown in Table 1. The 17 flux sites of cropland over the globe were located in different countries (such as Germany, the United States, France, and Italy). DE-Kli and IT-BCi have the lowest (7.77 °C) and highest (17.88 °C) mean annual temperatures, respectively. The annual precipitation of these sites varies from 343.1 (US-Tw3) to 2062.25 mm (CH-Oe2). We divide the flux data set into the training set, validation set, and test set, the ratios of which are 60%, 20%, and 20%, respectively, and the three datasets are used to train, validate, and test the ANN model. The vegetation index and reflectance data were retrieved from MODIS MOD09A1 (https://modis.ornl.gov/data.html, accessed on 27 February 2020), having a spatial resolution of 500 m. These flux data and MODIS data were used to training the two models of estimating ET. The time series of MODIS data were extracted according to the longitude and latitude coordinates of the flux sites. The spectral index was calculated using the MOD43A4 product, following the formulations shown in Table 2. NDVI is usually used to reflect the information of vegetation coverage and growth. In order to obtain information on a larger regional scale, a new vegetation index NIRv is introduced [19], which can reflect the photosynthetic capacity of surface vegetation better. NIRv is the product of the total near-infrared reflectance (NIRt) (MODIS second band) and NDVI. NIRv is a remote sensing measurement of canopy structure, which can more accurately predict photosynthesis [34]. The shortwave infrared band (SWIR) is usually used to reflect water stress and is calculated by using the reflectance data directly.

Table 1. Description of flux sites.

Site Code	Latitude	Longitude	Mean Annual Temperature (°C)	Mean Annual Precipitation (mm)	Years	Reference
BE-Lon	50.55	4.75	11.41	766.50	2004–2014	Moureaux et al. [35]
CH-Oe2	47.29	7.73	9.56	2062.25	2004–2014	Moors et al. [36]
DE-Geb	51.10	10.91	9.67	532.90	2001–2014	Anthoni et al. [37]
DE-Kli	50.89	13.52	7.77	810.30	2004–2014	Brust et al. [38]
DE-RuS	50.86	6.45	10.80	551.15	2011–2014	Eder et al. [39]
DE-Seh	50.87	6.45	10.29	573.05	2007–2010	Korres et al. [40]
FR-Gri	48.84	1.95	10.96	598.60	2004–2014	Loubet et al. [41]
IT-BCi	40.52	14.96	17.88	1197.20	2004–2014	Ranucci et al. [42]
IT-CA2	42.38	12.03	14.84	766.50	2011–2014	
US-ARM	36.60	−97.49	15.27	646.05	2003–2012	Raz-Yaseef et al. [43]
US-CRT	41.63	−83.35	10.85	810.30	2011–2013	Chu et al. [44]
US-Ne1	41.1651	−96.477	10.54	846.80	2001–2013	Verma et al. [45]
US-Ne2	41.1649	−96.470	10.26	876.00	2001–2013	Suyker and Verma [46]
US-Ne3	41.1797	−96.440	10.38	697.15	2001–2013	Suyker and Verma [46]
US-Tw2	38.1047	−121.643	15.23	386.90	2012–2013	Knox et al. [47]
US-Tw3	38.1159	−121.647	16.00	343.10	2013–2014	Baldocchi et al. [48]
US-Twt	38.1087	−121.653	14.75	357.70	2009–2014	Hatala et al. [49]

Table 2. Calculation of vegetation index. *rx* represents the reflectivity of MODIS bands (x = 1–7), NDVI is the normalized difference vegetation index, NIRv is near-infrared reflectance of vegetation.

Index	Formula
NDVI	$\mathrm{NDVI} = \frac{r2-r1}{r2+r1}$
NIRv	$\mathrm{NIRv} = \mathrm{NDVI}*r2$

2.2. Two ET Models Based on ANN

In this study, two models were trained based on the PM equation, and the difference lies in the *Gs* calculation. The following two summaries introduce the two methods in detail. The formula of the PM equation is as follows:

$$\lambda \mathrm{E} = \frac{(Rn - G)\cdot\Delta + \rho\cdot Cp\cdot D\cdot Ga}{\Delta + \gamma(1 + Ga/Gs)} \tag{1}$$

where λE is evapotranspiration, Rn is net radiation, G is soil heat flux, Δ is the gradient of the saturation vapor pressure versus atmospheric temperature, ρ is air density, Cp is the specific heat at constant pressure of air, D is the vapor pressure deficit of the air, Ga is the aerodynamic conductance, and γ is the psychometric constant.

In order to test the effects on the accuracy of using different combinations of input variables, different combinations of input variables in the ANN are shown in Table 3.

Table 3. Different combinations of input variables in the ANN. Ta is temperature, P is precipitation, SW is solar radiation, Ca is carbon dioxide concentration, VPD is vapor pressure deficit, NDVI is normalized difference vegetation index, NIRv is near-infrared reflectance of vegetation, and SWIR is shortwave infrared band.

Number	Input Parameters
1	Ta, P, SW, Ca, VPD
2	Ta, P, SW, Ca, VPD, NDVI
3	Ta, P, SW, Ca, VPD, NDVI, NIRv
4	Ta, P, SW, Ca, VPD, NDVI, NIRv, SWIR

2.2.1. ANN-PM Model

We trained an ANN-PM model based on ANN and PM equations to estimate ET. ANN is a commonly used ML method, which has been widely used in estimating ET. It consists of a large number of nodes, called neurons, which are connected to each other. The typical structure of ANN used to estimate ET is shown in Figure 3.

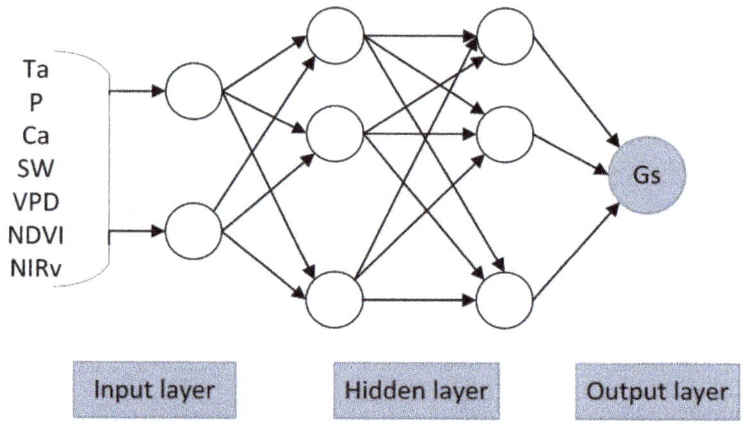

Figure 3. The typical structure of ANN.

ANN contains three layers: the input layer, hidden layer, and output layer. The input layer is responsible for receiving input data, the hidden layer constructs the relationships between the input and output, and the output layer outputs the predicted target values. The variables input to ANN in this study includes Ta, precipitation (P), solar radiation (SW), Ca, vapor pressure deficit (VPD), normalized difference vegetation index (NDVI), and near-infrared reflectance of vegetation (NIRv). In the variables we used, Ta, SW, Ca, and VPD can affect canopy conductivity from different aspects [50]. The consideration of P is mainly to represent the influence of canopy interception on ET. Thus they are selected to model *Gs*. There is an interaction and mutual influence between the transpiration and photosynthetic capacity of plants, and ET is dominated by transpiration. The vegetation index, NIRv, is considered in order to better reflect the impact of the photosynthetic capacity of the surface vegetation on evapotranspiration. NIRv is able to characterize seasonable variations in canopy scale photosynthesis rate without additional environmental factors that are conventionally used to constrain photosynthesis [34]. These variables are used to train ANN to the *Gs* model. Referring to Zhao et al. [19], we used the ANN model to model ln(*Gs*) rather than *Gs* because the logarithmic form can effectively reduce the effect of errors in *Gs* calculated from the observations. Finally, the logarithm of *Gs* obtained by ANN simulation is converted to *Gs*, and then the converted *Gs* is input into the PM equation to calculate ET. Here, *Gs* values used to train the ANN model were calculated from the observed ET along with the inverted PM equation [51]. In order to avoid over-fitting, the network model is repeatedly trained, where the number of hidden layers ranges from 1 to 10, and the number of neurons in each layer increases from 1 to 128, with an interval of 8. Then, we choose the optimal ANN structure as the best model.

2.2.2. Medlyn-PM Model

The Medlyn-PM model uses ANN-derived GPP in conjunction with a theoretical *Gs* model to estimate surface conductance, and then the computed *Gs* is used to estimate ET using the PM equation. Firstly, we use the optimal ANN structure selected above to train the GPP model. Secondly, on the pixel scale, the computed GPP, Ca, and air vapor pressure deficit are used for *Gs* regression analysis to establish the relationship among them and determine the undetermined coefficients g_0 and g_1. Then, we use the above variables and the relationship between them to build the *Gs* model. Finally, the constructed *Gs* is input into the PM equation to calculate ET. The relationship is as follows [52]:

$$Gs = 1.6 * \frac{\text{GPP}}{\text{Ca}} * \left(\frac{g_1}{\sqrt{\text{D}}} + 1 \right) + g_0 \tag{2}$$

where *Gs* is stomatal conductance, GPP is gross primary production, Ca is CO_2 concentration of the air, g_1 and g_0 are undetermined coefficients derived from regression analysis, and D is the vapor pressure deficit of the air. The minimum value of D is fixed to 0.1 KPa.

2.3. ANN Architecture Optimization

The ML method, i.e., ANN, used in the ANN-PM and the Medlyn-PM, considers input variables, including Ta, P, SW, Ca, VPD, NIRv, and NDVI. Usually, in order to reduce over-fitting, the network model is repeatedly trained. Thus, we need to recognize the best ANN structure. In our study, the optimal ANN is determined in terms of mean square error (MSE) while minimizing the number of degrees of freedom based on the Akaike Information Criterion (AIC). AIC is a standard to measure the goodness of fit of the statistical model. AIC encourages the goodness of data fitting but tries to avoid over-fitting. Therefore, the priority model should be the one with the lowest AIC value. Cropland ET is estimated by combining the predictive output of ANN with the PM equation. The calculation formula of the AIC indicator is as follows [53]:

$$\text{AIC} = \log(\text{MSE}) + \frac{2q}{n} \tag{3}$$

where MSE is mean square error, q is the total number of parameters in the network, and n is the number of observations in the training sample.

2.4. Model Evaluation

2.4.1. Model Performance Measurement

The model performance evaluation metrics used in the study include root mean square error (RMSE), mean absolute error (MAE), and determination coefficients (R^2). The calculations of these metrics are shown in Table 4.

Table 4. Calculation formula of evaluation parameters. RMSE is the root mean square error, MAE is the mean absolute error, and R^2 is the determination coefficients. *fi*: Predicted value: $\overline{fi}$ Mean value of the predicted values; *yi*: Experiment value; $\overline{yi}$: Mean value of the observed values; *m*: Total amount of experimental data.

Evaluation Parameters	Formula
RMSE	$\sqrt{\frac{\sum_{i=1}^{m}(fi-yi)^2}{m}}$
MAE	$\frac{\sum_{i=1}^{m}\lvert fi-yi\rvert}{m}$
R^2	$\left(\frac{\sum_{i=1}^{m}(yi-\overline{yi})(fi-\overline{fi})}{\sqrt{\sum_{i=1}^{m}(yi-\overline{yi})^2}\sqrt{\sum_{i=1}^{m}(fi-\overline{fi})^2}}\right)^2$

RMSE is the standard deviation between the predicted and true values, reflecting the degree that the predicted values explain the true values [54]. MAE is the mean error of evaluating a set of predictions and is the average value of the absolute difference between predicted and experimental values on test samples, but MAE is less sensitive to extreme values than RMSE [55]. R^2 is determined by drawing a scatter plot between the observed and predicted value. Lower RMSE, MAE, and higher R^2 correspond to a better performance of the model.

2.4.2. Evaluating the Model Used to Estimate ET under Dry Climate

Modeling ET in dry regions is more challenging than in other regions, especially for croplands. Because the water status of croplands is affected by irrigation, and the information of irrigation on a regional scale is difficult to obtain. On the other hand, in arid areas, most of the precipitation is consumed in the process of ET, and inadequate water supply could substantially limit the growth of crops in these regions. Therefore, accurate estimation of ET plays an important role in the sustainable development of agriculture in arid areas. Research on modeling ET in dry climates can facilitate rational cropland irrigation, maintaining stable crop production in dry regions.

We analyzed the performance of the models we trained in estimating ET under a dry climate. The aridity index (AI) is a means and tool to determine the drought degree and range of a certain period quantitatively, and it is also an indicator of the degree of dry and wet in a region. The calculation formula of the AI is as follows [56]:

$$\mathrm{AI} = \frac{\mathrm{P}}{\mathrm{PET}} \tag{4}$$

where AI is aridity index, PET is potential evapotranspiration, and P is the average precipitation. The AI calculation of each site is limited to the time range covered by the site. Low AI corresponds to a dry climate. We selected the sites with the AI values below 0.5 as arid areas by calculating the AI values of each flux site.

3. Results

3.1. Model Parameter Optimization

The undetermined parameters g_0 and g_1 were required for running Medlyn-PM. They were determined by fitting the analytical *Gs* equation, $Gs = 1.6 * \frac{\mathrm{GPP}}{\mathrm{Ca}} * \left(\frac{g_1}{\sqrt{\mathrm{D}}} + 1\right) + g_0$, and we obtained that g_0 = 0.06 and g_1 = 3.94. The variations in RMSE/MAE/R^2 with the change of the numbers of hidden layers and neurons for the ANN-PM model with training and validation datasets are presented in Figure 4.

The figure shows that the RMSE and MAE of ANN-PM with the training dataset decrease gradually as the number of hidden layers (HL) and the number of neurons increase. The RMSE and MAE of ANN-PM with the validation dataset decrease as the numbers of hidden layers (HL) and neurons increase from 1 (the number of HL) −1 (the number of neurons) to 10–48 but increase after the number of the two parameters become larger than 1–48. As the number of hidden layers and the number of neurons increase to 10–128, the R^2 of the training dataset reaches a maximum value (0.94), and the R^2 of the validation dataset is concentrated around 0.80. Then, considering the AIC values, we identified the best architectures of ANN-PM (AIC = −0.76) and Medlyn-PM (AIC = −0.55) models and the key parameters are shown in Table 5. The ANN-PM model has an ANN structure with two hidden layers and 48 neurons in each layer. The AIC index is also used to select the ANN-based GPP model in Medlyn-PM, and the optimal model has two hidden layers and one neuron in each layer.

Table 5. The key parameters of the two models. ANN is artificial neural network and PM is the Penman-Monteith.

	ANN-PM Model	Medlyn-PM Model
The number of hidden layers	2	2
The number of neurons	48	1

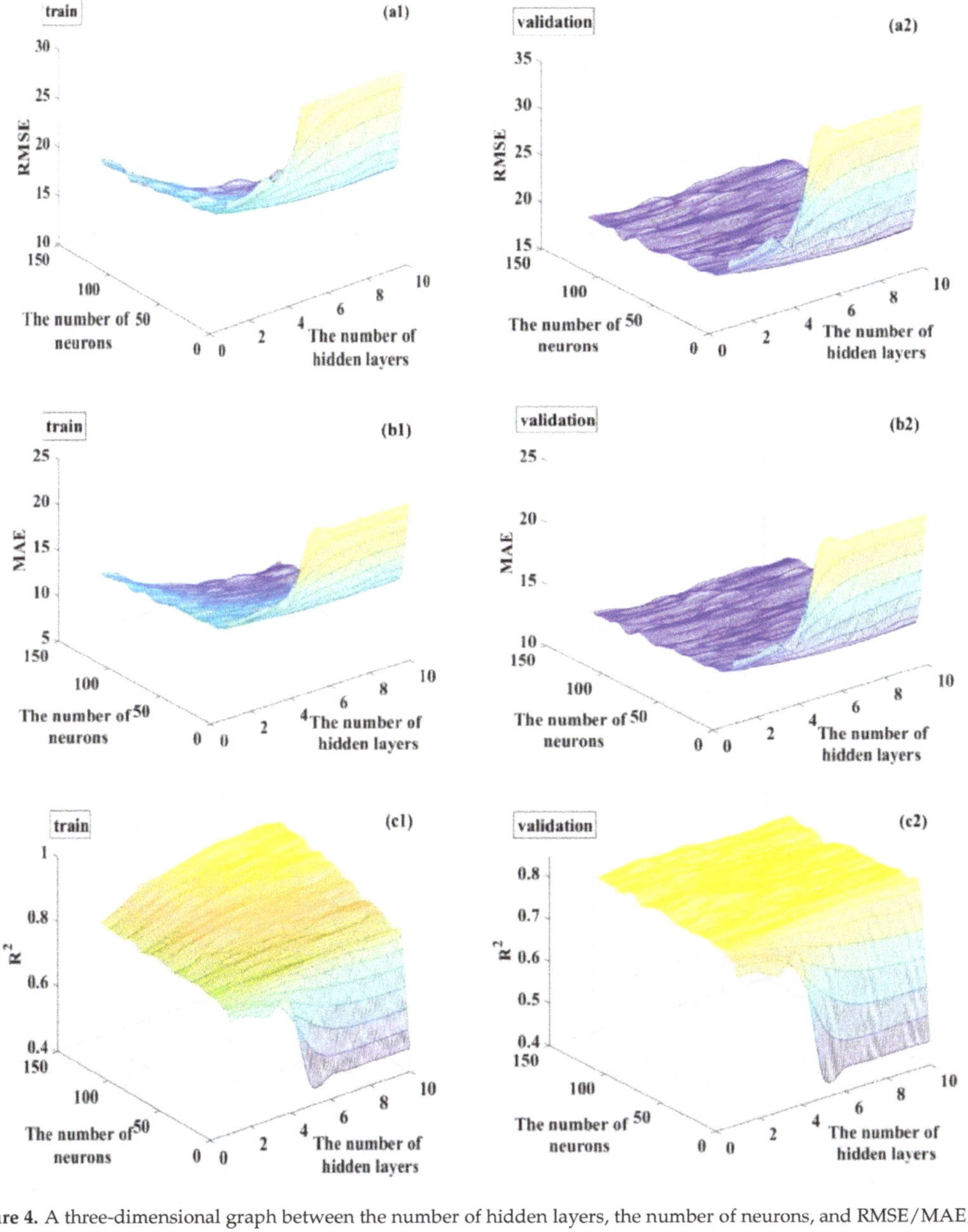

Figure 4. A three-dimensional graph between the number of hidden layers, the number of neurons, and RMSE/MAE/R^2 of the training and validation datasets of the ANN-PM model. (**a1**) is the RMSE of the training, (**a2**) is the RMSE of the validation, (**b1**) is the MAE of the training, (**b2**) is the MAE of the validation, (**c1**) is the R^2 of the training, (**c2**) is the R^2 of the validation. RMSE is the root mean square error, MAE is the mean absolute error, and R^2 is the determination coefficient.

3.2. Comparison of ANN Model with Different Input Data

The input data of ANN in the ANN-PM model includes meteorological data (Ta, P, SW, Ca, and VPD) and remote sensing data (NIRv and NDVI). We investigate the accuracy of estimating ET using the optimized ANN-PM (two hidden layers and 48 neurons in each layer) with several combinations of input data (Table 3). Figure 5 shows the comparisons between the predicted ET values and the measured values of cropland ET in the training, validation, and test datasets across all flux sites.

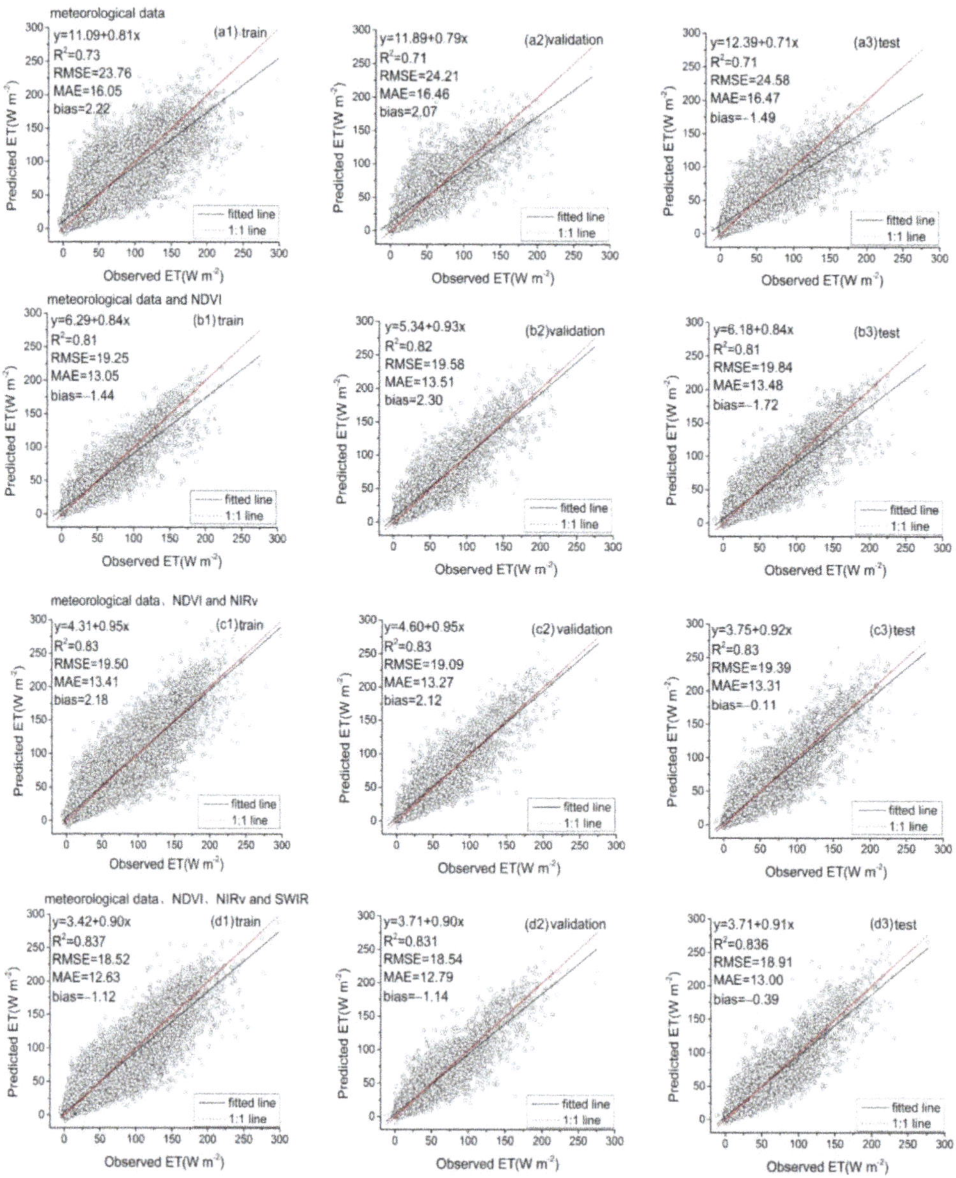

Figure 5. Scatter plots between the predicted ET values and the observed ET values measured from the flux tower in the training, validation, and test datasets of the ANN-PM model. (**a1–a3**) is the scatter plot between the predicted ET values and the observed ET values measured from the flux tower of the ANN-PM model using meteorological data in the three datasets, (**b1–b3**) is the scatter plot using meteorological data and NDVI, (**c1–c3**) is the scatter plot using meteorological data and NDVI and NIRv, (**d1–d3**) is the scatter plot using meteorological data and NDVI and NIRv and SWIR.

As shown in Figure 5, all the employed models provide different accuracies under different input combinations. The accuracy of predicted ET values differs significantly depending on the model types and input combinations. Except for the second input combination, all input combinations show the highest R^2 in the training stage (Figure 5(a1,c1,d1)). The ranks of the input combinations under investigation in terms of prediction accuracy are (the value in parentheses after RMSE indicates the percentage of RMSE relative to the observed value): the fourth input combination (R^2 = 0.831–0.837, RMSE = 18.52–18.91 W m^{-2} (38.42–38.86%), MAE = 12.63–13.00 W m^{-2}), the third input combination (R^2 = 0.83, RMSE = 19.09–19.50 W m^{-2} (39.84–40.46%), MAE = 13.27–13.41 W m^{-2}), the second input combination (R^2 = 0.81–0.82, RMSE = 19.25–19.84 W m^{-2} (39.94–41.05%), MAE = 13.05–13.51 W m^{-2}), and the first input combination (R^2 = 0.71–0.73, RMSE = 23.76–24.58 W m^{-2} (49.29–50.75%), MAE = 16.05–16.47 W m^{-2}). In the testing stage, the models of the third input combination and fourth input combination have identical performance in estimating ET, both of which performed superior to the second input combination and the first input combination in predicting ET. These results confirm that the model using all input variables (meteorological data and three remote sensing data factors (NDVI, NIRv, SWIR)) achieves the best performances (RMSE = 18.52–18.91 W m^{-2} (38.42–38.86%), MAE = 12.63–13.00 W m^{-2}, and R^2 = 0.831–0.837) compared with those using a subset of all the variables. However, the model using meteorological data and two remote sensing data factors (NDVI and NIRv) is also capable of predicting ET with acceptable accuracy, having the RMSE and MAE values of 19.09–19.50 W m^{-2} (39.84–40.46%) and 13.27–13.41 W m^{-2}, respectively. When using only meteorological data, the model shows degraded performance with larger errors (RMSE = 23.76–24.58 W m^{-2} (49.29–50.75%) and MAE = 16.05–16.47 W m^{-2}) and smaller determination coefficients (R^2 = 0.71–0.73). The model using the combination of meteorological data and one remote sensing factor, NDVI, shows intermediate results (RMSE = 19.25–19.84 W m^{-2} (39.94–41.05%), MAE = 13.05–13.51 W m^{-2}, and R^2 = 0.81–0.82). The model using meteorological data and three remote sensing data factors (NDVI, NIRv, and SWIR) showed comparable performance with that using meteorological data and two remote sensing data factors (NDVI, NIRv). Therefore, it can be concluded that remote sensing data in the ANN model facilitated the improvement of the estimates of croplands ET.

3.3. Comparison of ANN-PM and Medlyn-PM

Figure 6 shows the scatter plots of measured ET vs. predicted ET by the Medlyn-PM and the ANN-PM model, respectively. At the site scale, the two models differ substantially in performance from each other. Figure 6 shows good correlations between the observed ET and the predicted ET by the two methods (R^2 = 0.75 and 0.83). Figure 6 also illustrates that the R^2 value of the ANN-PM model is 0.08–0.09 higher than that of the Medlyn-PM model and the RMSE and MAE of ANN-PM are 4.26–4.3 and 3.12–3.34 W m^{-2} smaller than that of the Medlyn-PM model, respectively. Overall, the ANN-PM model shows relatively high accuracy with smaller RMSE and MAE, and larger R^2 (RMSE = 19.09-19.50 W m^{-2} (39.84–40.46%), MAE = 13.27–13.41 W m^{-2}, R^2 = 0.83) in estimating cropland ET compared to the Medlyn-PM model (RMSE = 23.39–23.76 W m^{-2} (49.95–51.14%), MAE = 16.39–16.75 W m^{-2}, and R^2 = 0.74–0.75), indicating a great advantage in estimating cropland ET using the ANN-PM model.

3.4. Accuracy of ANN-PM Model under Dry Climates

In arid areas, most of the precipitation is consumed in the process of ET, and inadequate water supply could substantially limit the growth of crops in these regions. Therefore, accurate estimation of ET plays an important role in the sustainable development of agriculture in arid areas. Hence, we evaluated the ANN-PM model to simulate ET on a daily scale over flux sites covering a wide range of climate dryness, measured using aridity index (AI). The R^2 between simulation and observation is used to measure the model performance. The variations in R^2 of each flux site in relation to site-scale AI are shown in Figure 7, where low AI values correspond to dry climates. The driest site is US-Twt, followed by US-Tw3, US-Tw2, and DE-Rus. The average R^2 of the 16 flux sites is 0.74, and the average R^2 of the driest four flux sites with an AI index lower than 0.5 (DE-Rus = 0.49, US-Tw2 = 0.42, US-Tw3 = 0.30, and US-Twt = 0.26) is 0.77. In terms of R^2, the performances of the ANN-PM model at the dry sites are reasonable and comparable to those at the wet sites (Figure 7).

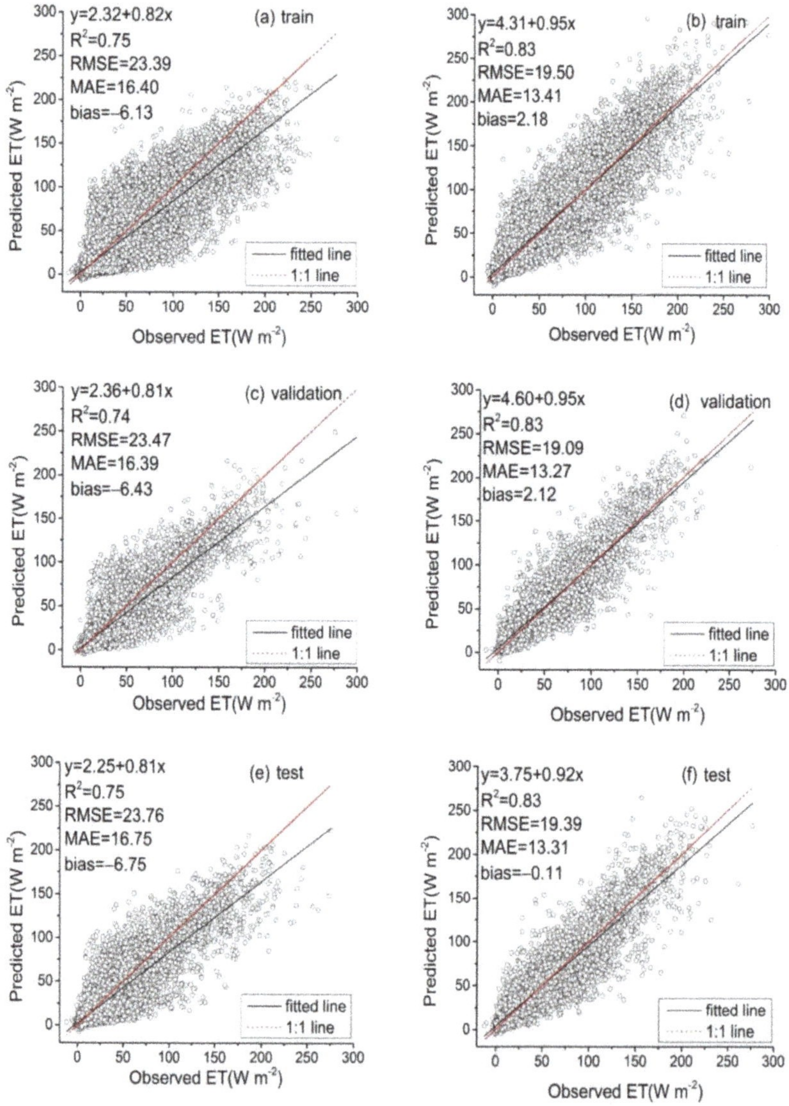

Figure 6. Scatter plots of the observed ET values measured from the flux tower and predicted ET values of the Medlyn-PM (**left**) and the ANN-PM model (**right**) in estimating cropland evapotranspiration. (**a**,**c**,**e**) are the scatter plots of the observed ET values measured from the flux tower and predicted ET values of the Medlyn-PM model in the training, validation, and test datasets, respectively. (**b**,**d**,**f**) are the scatter plots of the observed ET values measured from the flux tower and predicted ET values of the ANN–PM model in the training, validation, and test datasets, respectively.

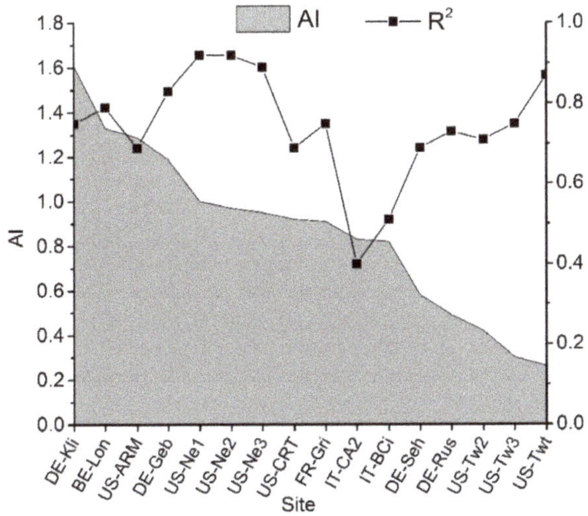

Figure 7. AI and R^2 values of each flux site. AI is aridity index and R^2 is the determination coefficients between simulation and observation.

The ANN-PM model can capture the time-series changes of ET at the dry sites well (Figure 8, four sites with an AI index lower than 0.5). At the driest site, US-Twt, which is a paddy field site, ET predicted by the ANN-PM model agreed well with the observations, indicating that the model can reflect the influence of irrigation on cropland ET under dry conditions. Consequently, the ANN-PM model can simulate cropland ET across a wide range of gradients of climate dryness, showing great potential to estimate cropland ET accurately on a regional scale.

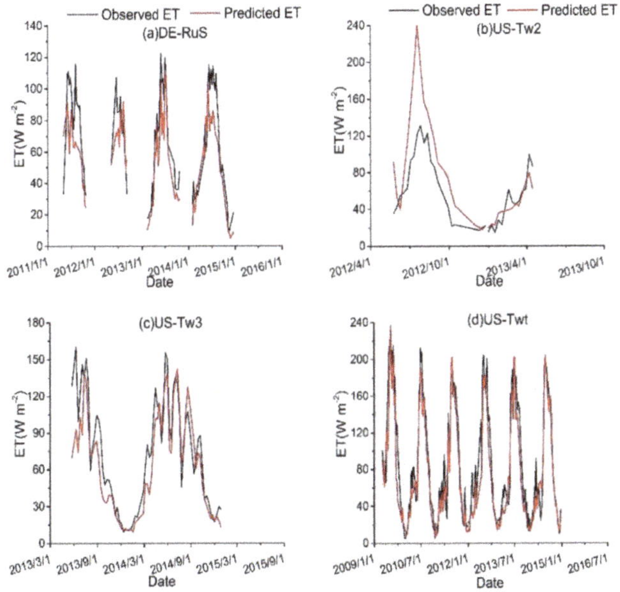

Figure 8. Time-series diagrams of observed ET (**black line**) measured from the flux tower and simulated ET (**red line**) by the ANN-PM model.

4. Discussion

4.1. Discussion of the Number of Sites

We used 17 sites in our study, and the time span of all sites is 2001–2014 (Table 1). The entire dataset contains more than 50,000 samples on a daily scale, which are large enough for establishing the ML-based method. As we know, the size of the sample we used is larger than some existing publications. For example, Zhu et al. [31] used nine stations in the arid region of Northwest China during the period 2002–2016. Yin et al. [57] evaluated ET in the eddy covariance flux observations at 14 Chinese flux tower sites during the period 2003–2017, and each site has at least 3 years of reliable data. Hossein Kazemi et al. [58] only used the daily meteorological records of seven weather stations in Iran for 10 years (2008–2017). Therefore, our data are enough to train a machine learning model. Our study is mainly for cropland. There are currently limited open-access cropland sites, but our sites cover the current main farming areas. These areas cover different climate types. Therefore, our model has wide applicability. There is currently a lack of stations in tropical regions. When applied in this climate region, the model needs to be further tested.

4.2. Comparison between This Research and Existing Research

The ANN-PM model of this study combines ML methods and the PM equation, and the remote sensing data of inputting into ANN contains a recently proposed NIRv index, which can be used to reflect the photosynthetic capacity and water status of the surface vegetation. Combining NIRv with ML and the PM equation shows great advantages in estimating cropland ET. Zhao et al. [19] used an ML method (ANN) and PM equation to estimate ET, but the study used soil moisture data that is difficult to obtain, which limits the application of the model in a large-scale and long-term series. Yamaç and Todorovic [59] combined the PM equation with three ML methods (K nearest neighbor algorithm, ANN, and Adaptive Boosting model) to estimate the ET using available weather input data with four different scenarios (temperature, solar radiation, wind speed, and relative humidity). They showed that using the combination of four data scenarios performs better than any other combinations. The above two studies are based on the theoretical framework of the PM equation and use ML methods. However, the first study uses soil moisture data that is not feasibly accessed on a regional scale, and the second uses only meteorological data, which is only applicable in a limited area. Compared with the above two studies, we combined meteorological data with remote sensing data to estimate ET. The fitting effect is better, and accuracy is improved. The model tested was applicable to a wide range of environmental gradients. He et al. [60] used a process and PM-based ET model, the MOD16 algorithm, to estimate ET for cropland sites (US-Tw2, US-Tw3, and US-Twt). The results showed that the site US-Tw2 has a higher R^2 (0.72) than US-Tw3 and US-Twt. In our study, we evaluated the performance of our ET models at three cropland sites (US-Tw2, US-Tw3, and US-Twt), respectively. Compared with He et al. 's [60] study, our models at the three sites all show higher accuracy (R^2 = 0.74–0.86). Our hybrid model, based on ML and PM, can perform better than the model based on the process and PM equation. Amazirh et al. [61] used the PM equation to estimate ET in semi-arid areas by introducing a simple relationship between surface resistances (r_c) and verified the model at flood and drip irrigation sites. The results showed that the R^2 of these two sites were 0.76 and 0.70, respectively, and the RMSEs were 22 and 23 W m^{-2}, respectively. Feng et al. [62] compared the performance of the PM equation and self-optimizing nearest neighbor algorithm (CCA-k-NN) in estimating ET. The results showed that the performance of CCA-k-NN was comparable with PM (R^2 = 0.8, RMSE = 24.01 W m^{-2}, MAE = 18.06 W m^{-2}). The above studies only used the PM equation to estimate cropland ET. Our study combines ML methods with the PM equation to estimate cropland ET (R^2 = 0.84, RMSE = 17.40 W m^{-2}, MAE = 12.41 W m^{-2}), the estimating accuracy obtained in this study is better, and the physical mechanism of the PM equation can ensure that the simulation result is always within the range of potential evapotranspiration.

4.3. Comparison of the ANN-Based ET Model with Existing ML-Based ET Models

ML algorithms have been more and more widely used to estimate ET on a regional or global scale. In this study, the most widely used ANN algorithm is used to improve the accuracy of the PM equation to estimate cropland ET on a regional scale. There are also many studies that use other ML algorithms to estimate ET, e.g., Abdullah et al. [63], Antonopoulos and Antonopoulos [64], Reis et al. [29], Yamaç and Todorovic [59], Zhu et al. [31], and Ferreira and da Cunha [65]. These studies literally showed different performances of different ML-based ET models. However, it should also be noted that the performance metrics of ET models could vary between different regions, validation data sources, temporal scale of validation, and so on. For example, the ML models estimating the reference ET usually show higher performance metrics than the actual ET models [64,66,67], as reference ET was calculated from only a few meteorological factors. If different data sources are used in modeling ET using the ML algorithm, the efficiency of the ET model can also be different. For example, Fan et al. [67] showed that the performance of the ML algorithm (R^2 = 0.701–0.995, RMSE = 0.106–0.637 mm d^{-1}) in estimating reference ET were significantly different between eight meteorological stations that represented the eight main climate types of China. Zhu et al. [31] showed similar results in modeling reference ET using the ML over nine meteorological stations in the arid region of Northwest China (R^2 = 0.844–0.969, MAE = 0.268–0.635 mm d^{-1}). The ET model focusing on the daily scale also produces different performance metrics from the hourly scale ET model. Ferreira and da Cunha [65] revealed better performances of the deep learning-based models in estimating daily reference ET on a daily scale as compared to the models on an hourly scale, with R^2 increased from 0.78–0.88 to 0.87–0.91, and RMSE decreased from 0.56–0.73 to 0.47–0.60 mm d^{-1}. The above studies show that the performance of the ET models can differ under different temporal scales. The performance metrics of the hybrid model in our study are in line with the range of those ML-based ET models.

4.4. The Reasons for the Low Accuracy of the Medlyn-PM Model and the Lack of the ANN-PM Model

The reason for the degraded performance of Medlyn-PM in estimating cropland ET, as compared to ANN-PM, is that the effect of soil evaporation is not considered in the model. ET includes soil evaporation and plant transpiration, as well as part of the contribution of canopy interception. Soil evaporation cannot be ignored in ET. Yu et al. [68] investigated the contribution of soil evaporation to ET of winter wheat under sprinkler irrigation. Their results showed that soil evaporation was an important part of ET, accounting for 20–28% of ET. Liu et al. [69] used a large-scale weighing permeameter and a micro permeameter to measure the daily evaporation and ET in winter wheat fields, and the study showed that soil evaporation accounted for 30% of the ET. Qin et al. [70] also showed that evaporation accounted for 32% of the total ET during the growth of winter wheat and 65% in the early growth period. These indicated a considerable contribution of soil evaporation in ET. Since ANN-PM used ANN to estimate the bulk surface conductance, which accounts for the effect of both stomatal and soil conductance, it has been found to perform better than the Medlyn-PM model.

The remote sensing information allows ANN-PM to simulate spatiotemporally continuous ET information [71]. However, we did not exhaust all possible RS data in the ANN-PM, which is beyond the scope of this study. In the future, we can evaluate more RS data to improve the accuracy of the ANN-PM model. For example, the development of multi-source RS data and surface parameter inversion products can provide PM models with some basic parameters that promote their application [72], so multi-source remote sensing data and PM models can be combined to estimate cropland ET.

5. Conclusions

The accurate estimation of cropland ET is important for crop irrigation, fertilization, and other management measures. In this study, we proposed an ANN-PM model based on ML and the PM equation to estimate cropland ET. At the same time, we optimized the Medlyn-PM model (uses ANN-derived GPP along with Medlyn's stomatal conductance to compute *Gs*, and the computed *Gs* is used to estimate ET). We compared the two models to get a better method for estimating ET based on the ML approach. Specifically, we used ANN to estimate *Gs* in ANN-PM and GPP that was used to estimate *Gs* in conjunction with Medlyn's *Gs* model in Medlyn-PM. We have the following conclusions.

1. The optimal ANN architecture to estimate *Gs* in ANN-PM consists of two hidden layers with 48 neurons in each layer, and that to estimate GPP in Medlyn-PM, two hidden layers and one neuron in each layer was optimal. The optimized g_0 and g_1 values in Medlyn's *Gs* model are 0.06 and 3.94, respectively.
2. The ANN-PM model can reasonably estimate the ET of cropland (RMSE = 19.09–19.50 W m^{-2}, MAE = 13.27–13.41 W m^{-2}, and R^2 = 0.83 for training, validation, and test datasets) and is proven to perform better than Medlyn-PM with a smaller RMSE and MAE and larger R^2.
3. The ANN approach can represent the water stress impacts on ET well, as ANN-PM can reasonably capture the seasonal variations in ET at the dry sties (AI < 0.5). Additionally, the performances of the ANN-PM model at the dry sites were as good as at the wet sites.

Author Contributions: Conceptualization, Y.L. and Y.B.; Methodology, Y.L. and Y.B.; Software, Y.L. and Y.B.; Validation, Y.L.; Formal analysis, Y.B.; Investigation, Y.L. and Y.B.; Resources, Y.B.; Data Curation, Y.L. and Y.B.; Writing—Original Draft, Y.L.; Writing—Review and Editing, Y.B., S.Z., J.Z., and L.T.; Visualization, Y.L. and Y.B.; Supervision, Y.B.; Project administration, Y.B.; Funding acquisition, Y.B. and J.Z. All authors have read and agreed to the published version of the manuscript.

Funding: This research was funded by the National Natural Science Foundation of China (Grant Nos. 41,901,342, 31,671,585), "Taishan Scholar" Project of Shandong Province, and Key Basic Research Project of Shandong Natural Science Foundation of China (Grant No. ZR2017ZB0422).

Institutional Review Board Statement: Not applicable.

Informed Consent Statement: Not applicable.

Data Availability Statement: The data used in the study can be downloaded through the corresponding link provided in Section 2.4.

Acknowledgments: This work used eddy covariance data acquired and shared by the FLUXNET community, AmeriFlux, AsiaFlux, and European Flux Database Cluster. The FLUXNET also includes these networks: AmeriFlux, AfriFlux, AsiaFlux, CarboAfrica, CarboEuropeIP, Carboltaly, CarboMont, ChinaFlux, Fluxnet-Canada, GreenGrass, ICOS, KoFlux, LBA, NECC, OzFlux-TERN, TCOS-Siberia, and USCCC. The FLUXNET eddy covariance data processing, and harmonization was carried out by the ICOS Ecosystem Thematic Center, AmeriFlux Management Project and Fluxdata project of FLUXNET, with the support of CDIAC, and the OzFlux, ChinaFlux, and AsiaFlux offices.

Conflicts of Interest: The authors declare no conflict of interest.

References

1. Sepaskhah, A.R.; Ilampour, S. Effects of soil moisture stress on evapotranspiration partitioning. *Agric. Water Manag.* **1995**, *28*, 311–323. [CrossRef]
2. Jasechko, S.; Sharp, Z.D.; Gibson, J.J.; Birks, S.J.; Yi, Y.; Fawcett, P.J. Terrestrial water fluxes dominated by transpiration. *Nature* **2013**, *496*, 347–350. [CrossRef]
3. Ding, J.; Peng, S.; Xu, J.; Wei, Z. Evapotranspiration model of water-saving irrigated paddy field based on penman Monteith equation. *J. Agric. Eng.* **2010**, *26*, 31–35.
4. Liu, S.; Liu, Z. Estimation of crop evapotranspiration by Priestley Taylor formula. *Plateau Meteorol.* **1997**, 80–85.

5. Cleugh, H.A.; Leuning, R.; Mu, Q.Z.; Running, S.W. Regional evaporation estimates from flux tower and MODIS satellite data. *Remote Sens. Environ.* **2007**, *106*, 285–304. [CrossRef]
6. Mu, Q.; Heinsch, F.A.; Zhao, M.; Running, S.W. Development of a global evapotranspiration algorithm based on MODIS and global meteorology data. *Remote Sens. Environ.* **2007**, *111*, 519–536. [CrossRef]
7. Mu, Q.; Zhao, M.; Running, S.W. Improvements to a MODIS global terrestrial evapotranspiration algorithm. *Remote Sens. Environ.* **2011**, *115*, 1781–1800. [CrossRef]
8. Leuning, R.; Zhang, Y.Q.; Rajaud, A.; Cleugh, H.; Tu, K. A simple surface conductance model to estimate regional evaporation using MODIS leaf area index and the Penman-Monteith equation. *Water Resour. Res.* **2008**, *44*, 17. [CrossRef]
9. Zhang, Y.; Leuning, R.; Hutley, L.B.; Beringer, J.; McHugh, I.; Walker, J.P. Using long-term water balances to parameterize surface conductances and calculate evaporation at 0.05° spatial resolution. *Water Resour. Res.* **2010**, *46*. [CrossRef]
10. Yebra, M.; Van Dijk, A.; Leuning, R.; Huete, A.; Guerschman, J.P. Evaluation of optical remote sensing to estimate actual evapotranspiration and canopy conductance. *Remote Sens. Environ.* **2013**, *129*, 250–261. [CrossRef]
11. Kitao, M.; Komatsu, M.; Hoshika, Y.; Yazaki, K.; Yoshimura, K.; Fujii, S.; Miyama, T.; Kominami, Y. Seasonal ozone uptake by a warm-temperate mixed deciduous and evergreen broadleaf forest in western Japan estimated by the Penman-Monteith approach combined with a photosynthesis-dependent stomatal model. *Environ. Pollut.* **2014**, *184*, 457–463. [CrossRef]
12. Ball, J.T.; Woodrow, I.E.; Berry, J.A. A model predicting stomatal conductance and its contribution to the control of photosynthesis under different environmental conditions. In *Progress in Photosynthesis Research*; Springer: Dordrecht, The Netherlands, 1987; pp. 221–224. ISBN 978-94-017-0521-9.
13. Yan, H.; Wang, S.Q.; Billesbach, D.; Oechel, W.; Zhang, J.H.; Meyers, T.; Martin, T.A.; Matamala, R.; Baldocchi, D.; Bohrer, G.; et al. Global estimation of evapotranspiration using a leaf area index-based surface energy and water balance model. *Remote Sens. Environ.* **2012**, *124*, 581–595. [CrossRef]
14. Mallick, K.; Trebs, I.; Boegh, E.; Giustarini, L.; Schlerf, M.; Drewry, D.T.; Hoffmann, L.; von Randow, C.; Kruijt, B.; Araùjo, A.; et al. Canopy-scale biophysical controls of transpiration and evaporation in the Amazon Basin. *Hydrol. Earth Syst. Sci.* **2016**, *20*, 4237–4264. [CrossRef]
15. Bhattarai, N.; Mallick, K.; Stuart, J.; Vishwakarma, B.D.; Niraula, R.; Sen, S.; Jain, M. An automated multi-model evapotranspiration mapping framework using remotely sensed and reanalysis data. *Remote Sens. Environ.* **2019**, *229*, 69–92. [CrossRef]
16. Jung, M.; Reichstein, M.; Bondeau, A. Towards global empirical upscaling of FLUXNET eddy covariance observations: Validation of a model tree ensemble approach using a biosphere model. *Biogeosciences* **2009**, *6*, 2001–2013. [CrossRef]
17. Traore, S.; Wang, Y.-M.; Kerh, T. Artificial neural network for modeling reference evapotranspiration complex process in Sudano-Sahelian zone. *Agric. Water Manag.* **2010**, *97*, 707–714. [CrossRef]
18. Zhang, W.; Huo, S.; Jia, Y. Comparison of machine learning models in the calculation of reference crop evapotranspiration in Hebei Province. *Water Sav. Irrig.* **2018**, *4*, 50–58.
19. Zhao, W.L.; Gentine, P.; Reichstein, M.; Zhang, Y.; Zhou, S.; Wen, Y.; Lin, C.; Li, X.; Qiu, G.Y. Physics-constrained machine learning of evapotranspiration. *Geophys. Res. Lett.* **2019**, *46*, 14496–14507. [CrossRef]
20. Du, K.L.; Swamy, M. *Neural Networks and Statistical Learning*; Springer: London, UK, 2014.
21. Bai, Y.; Zhang, S.; Bhattarai, N.; Mallick, K.; Liu, Q.; Tang, L.; Im, J.; Guo, L.; Zhang, J. On the use of machine learning based ensemble approaches to improve evapotranspiration estimates from croplands across a wide environmental gradient. *Agric. For. Meteorol.* **2021**, *298–299*, 108308. [CrossRef]
22. Tabari, H.; Talaee, P.H. Local calibration of the Hargreaves and Priestley-Taylor equations for estimating reference evapotranspiration in arid and cold climates of iran based on the penman-monteith model. *J. Hydrol. Eng.* **2011**, *16*, 837–845. [CrossRef]
23. Yang, Z.; Zhang, Q.; Yang, Y.; Hao, X.; Zhang, H. Evaluation of evapotranspiration models over semi-arid and semi-humid areas of China. *Hydrol. Process.* **2016**, *30*, 4292–4313. [CrossRef]
24. Muhammad, M.; Nashwan, M.; Shahid, S.; Ismail, T.; Song, Y.; Chung, E.-S. Evaluation of empirical reference evapotranspiration models using compromise programming: A case study of Peninsular Malaysia. *Sustainability* **2019**, *11*, 4267. [CrossRef]
25. Peng, L.; Zeng, Z.; Wei, Z.; Chen, A.; Wood, E.F.; Sheffield, J. Determinants of the ratio of actual to potential evapotranspiration. *Glob. Chang. Biol.* **2019**, *25*, 1326–1343. [CrossRef]
26. Liu, M.; Tang, R.; Li, Z.-L.; Yao, Y.; Yan, G. Global land surface evapotranspiration estimation from meteorological and satellite data using the support vector machine and semiempirical algorithm. *IEEE J. Sel. Top. Appl. Earth Obs. Remote Sens.* **2018**, *11*, 513–521. [CrossRef]
27. Tang, D.; Feng, Y.; Gong, D.; Hao, W.; Cui, N. Evaluation of artificial intelligence models for actual crop evapotranspiration modeling in mulched and non-mulched maize croplands. *Comput. Electron. Agric.* **2018**, *152*, 375–384. [CrossRef]
28. Adnan, R.M.; Malik, A.; Kumar, A.; Parmar, K.S.; Kisi, O. Pan evaporation modeling by three different neuro-fuzzy intelligent systems using climatic inputs. *Arab. J. Geosci.* **2019**, *12*, 1–14. [CrossRef]
29. Reis, M.M.; da Silva, A.J.; Zullo Junior, J.; Tuffi Santos, L.D.; Azevedo, A.M.; Lopes, É.M.G. Empirical and learning machine approaches to estimating reference evapotranspiration based on temperature data. *Comput. Electron. Agric.* **2019**, *165*, 104937. [CrossRef]
30. Granata, F.; Gargano, R.; de Marinis, G. Artificial intelligence based approaches to evaluate actual evapotranspiration in wetlands. *Sci. Total Environ.* **2020**, *703*, 135653. [CrossRef]

31. Zhu, B.; Feng, Y.; Gong, D.; Jiang, S.; Zhao, L.; Cui, N. Hybrid particle swarm optimization with extreme learning machine for daily reference evapotranspiration prediction from limited climatic data. *Comput. Electron. Agric.* **2020**, *173*, 105430. [CrossRef]
32. Reichstein, M.; Camps-Valls, G.; Stevens, B.; Jung, M.; Denzler, J.; Carvalhais, N. Deep learning and process understanding for data-driven Earth system science. *Nature* **2019**, *566*, 195–204. [CrossRef] [PubMed]
33. Chen, Y.; Xia, J.; Liang, S.; Feng, J.; Fisher, J.B.; Li, X.; Li, X.; Liu, S.; Ma, Z.; Miyata, A.; et al. Comparison of satellite-based evapotranspiration models over terrestrial ecosystems in China. *Remote Sens. Environ.* **2014**, *140*, 279–293. [CrossRef]
34. Badgley, G.; Anderegg, L.D.L.; Berry, J.A.; Field, C.B. Terrestrial gross primary production: Using NIRV to scale from site to globe. *Glob. Chang. Biol.* **2019**, *25*, 3731–3740. [CrossRef] [PubMed]
35. Moureaux, C.; Debacq, A.; Bodson, B.; Heinesch, B.; Aubinet, M. Annual net ecosystem carbon exchange by a sugar beet crop. *Agric. For. Meteorol.* **2006**, *139*, 25–39. [CrossRef]
36. Moors, E.; Jacobs, C.; Jans, W.; Supit, I.; Kutsch, W.; Bernhofer, C.; Beziat, P.; Buchmann, N.; Carrara, A.; Ceschia, E.; et al. Variability in carbon exchange of European croplands. *Agric. Ecosyst. Environ.* **2010**, *139*, 325–335. [CrossRef]
37. Anthoni, P.M.; Knohl, A.; Rebmann, C.; Freibauer, A.; Mund, M.; Ziegler, W.; Kolle, O.; Schulze, E.D. Forest and agricultural land-use-dependent CO_2 exchange in Thuringia, Germany. *Glob. Chang. Biol.* **2004**, *10*, 2005–2019. [CrossRef]
38. Brust, K.; Hehn, M.; Bernhofer, C. Comparative analysis of matter and energy fluxes determined by Bowen Ratio and Eddy Covariance techniques at a crop site in eastern Germany. In Proceedings of the European Geosciences Union General Assembly, Vienna, Austria, 22–27 April 2012; p. 8006.
39. Eder, F.; Schmidt, M.; Damian, T.; Träumner, K.; Mauder, M. Mesoscale eddies affect near-surface turbulent exchange: Evidence from lidar and tower measurements. *J. Appl. Meteorol. Climatol.* **2015**, *54*, 189–206. [CrossRef]
40. Korres, W.; Reichenau, T.G.; Schneider, K. Patterns and scaling properties of surface soil moisture in an agricultural landscape: An ecohydrological modeling study. *J. Hydrol.* **2013**, *498*, 89–102. [CrossRef]
41. Loubet, B.; Laville, P.; Lehuger, S.; Larmanou, E.; Fléchard, C.; Mascher, N.; Genermont, S.; Roche, R.; Ferrara, R.M.; Stella, P.; et al. Carbon, nitrogen and Greenhouse gases budgets over a four years crop rotation in northern France. *Plant Soil* **2011**, *343*, 109–137. [CrossRef]
42. Ranucci, S.; Bertolini, T.; Vitale, L.; Di Tommasi, P.; Ottaiano, L.; Oliva, M.; Amato, U.; Fierro, A.; Magliulo, V. The influence of management and environmental variables on soil N_2O emissions in a crop system in Southern Italy. *Plant Soil* **2010**, *343*, 83–96. [CrossRef]
43. Raz-Yaseef, N.; Billesbach, D.P.; Fischer, M.L.; Biraud, S.C.; Gunter, S.A.; Bradford, J.A.; Torn, M.S. Vulnerability of crops and native grasses to summer drying in the U.S. Southern Great Plains. *Agric. Ecosyst. Environ.* **2015**, *213*, 209–218. [CrossRef]
44. Chu, H.; Chen, J.; Gottgens, J.F.; Ouyang, Z.; John, R.; Czajkowski, K.; Becker, R. Net ecosystem methane and carbon dioxide exchanges in a Lake Erie coastal marsh and a nearby cropland. *J. Geophys. Res. Biogeosciences* **2014**, *119*, 722–740. [CrossRef]
45. Verma, S.B.; Dobermann, A.; Cassman, K.G.; Walters, D.T.; Knops, J.M.; Arkebauer, T.J.; Suyker, A.E.; Burba, G.G.; Amos, B.; Yang, H.; et al. Annual carbon dioxide exchange in irrigated and rainfed maize-based agroecosystems. *Agric. For. Meteorol.* **2005**, *131*, 77–96. [CrossRef]
46. Suyker, A.E.; Verma, S.B. Gross primary production and ecosystem respiration of irrigated and rainfed maize-soybean cropping systems over 8 years. *Agric. For. Meteorol.* **2012**, *165*, 12–24. [CrossRef]
47. Knox, S.H.; Sturtevant, C.; Matthes, J.H.; Koteen, L.; Verfaillie, J.; Baldocchi, D. Agricultural peatland restoration: Effects of land-use change on greenhouse gas (CO_2 and CH_4) fluxes in the Sacramento-San Joaquin Delta. *Glob. Chang. Biol.* **2015**, *21*, 750–765. [CrossRef]
48. Baldocchi, D.; Sturtevant, C.; Contributors, F. Does day and night sampling reduce spurious correlation between canopy photosynthesis and ecosystem respiration? *Agric. For. Meteorol.* **2015**, *207*, 117–126. [CrossRef]
49. Hatala, J.A.; Detto, M.; Baldocchi, D.D. Gross ecosystem photosynthesis causes a diurnal pattern in methane emission from rice. *Geophys. Res. Lett.* **2012**, *39*. [CrossRef]
50. Jarvis, P.G. The interpretation of the variations in leaf water potential and stomatal conductance found in canopies in the field. *Philos. Trans. R. Soc. B Biol. Sci.* **1976**, *273*, 593–610.
51. Matsumoto, K.; Ohta, T.; Nakai, T.; Kuwada, T.; Daikoku, K.i.; Iida, S.i.; Yabuki, H.; Kononov, A.V.; van der Molen, M.K.; Kodama, Y.; et al. Responses of surface conductance to forest environments in the Far East. *Agric. For. Meteorol.* **2008**, *148*, 1926–1940. [CrossRef]
52. Medlyn, B.E.; Duursma, R.A.; Eamus, D.; Ellsworth, D.S.; Colin Prentice, I.; Barton, C.V.M.; Crous, K.Y.; de Angelis, P.; Freeman, M.; Wingate, L. Reconciling the optimal and empirical approaches to modelling stomatal conductance. *Glob. Chang. Biol.* **2012**, *18*, 3476. [CrossRef]
53. Arifovic, J.; Gençay, R. Using genetic algorithms to select architecture of a feedforward artificial neural network. *Phys. A Stat. Mech. Appl.* **2001**, *289*, 574–594. [CrossRef]
54. Jia, Z.; Liu, S.; Mao, D.; Wang, Z.; Xu, Z.; Zhang, R. Research on verification method of remote sensing monitoring evapotranspiration based on ground observation. *Earth Sci. Prog.* **2010**, *25*, 1248–1260.
55. Fox, D.G. Judging air quality model performance: A summary of the ams workshop on dispersion model performance, woods hole, mass., 8–11 September 1980. *Bull. Am. Meteorol. Soc.* **1981**, *62*, 599–609. [CrossRef]
56. UNESCO. *Map of the World Distribution of Arid Regions*; Explanatory Note; United Nations Educational, Scientific and Cultural Organization: Paris, France, 1979.

57. Yin, L.; Tao, F.; Chen, Y.; Liu, F.; Hu, J. Improving terrestrial evapotranspiration estimation across China during 2000–2018 with machine learning methods. *J. Hydrol.* **2021**, *600*, 126538. [CrossRef]
58. Kazemi, M.H.; Shiri, J.; Marti, P.; Majnooni-Heris, A. Assessing temporal data partitioning scenarios for estimating reference evapotranspiration with machine learning techniques in arid regions. *J. Hydrol.* **2020**, *590*, 125252. [CrossRef]
59. Yamaç, S.S.; Todorovic, M. Estimation of daily potato crop evapotranspiration using three different machine learning algorithms and four scenarios of available meteorological data. *Agric. Water Manag.* **2020**, *228*, 105875. [CrossRef]
60. He, M.; Kimball, J.S.; Yi, Y.; Running, S.W.; Guan, K.; Moreno, A.; Wu, X.; Maneta, M. Satellite data-driven modeling of field scale evapotranspiration in croplands using the MOD16 algorithm framework. *Remote Sens. Environ.* **2019**, *230*, 111201. [CrossRef]
61. Amazirh, A.; Er-Raki, S.; Chehbouni, A.; Rivalland, V.; Diarra, A.; Khabba, S.; Ezzahar, J.; Merlin, O.J.B.E. Modified Penman–Monteith equation for monitoring evapotranspiration of wheat crop: Relationship between the surface resistance and remotely sensed stress index. *Biosyst. Eng.* **2017**, *164*, 68–84. [CrossRef]
62. Feng, K.; Tian, J.; Hong, Y. Self-optimizing nearest neighbor algorithm to estimate potential evapotranspiration in limited meteorological data area. *J. Agric. Eng.* **2019**, *35*, 76–83.
63. Abdullah, S.S.; Malek, M.A.; Abdullah, N.S.; Kisi, O.; Yap, K.S. Extreme Learning Machines: A new approach for prediction of reference evapotranspiration. *J. Hydrol.* **2015**, *527*, 184–195. [CrossRef]
64. Antonopoulos, V.Z.; Antonopoulos, A.V. Daily reference evapotranspiration estimates by artificial neural networks technique and empirical equations using limited input climate variables. *Comput. Electron. Agric.* **2017**, *132*, 86–96. [CrossRef]
65. Ferreira, L.B.; da Cunha, F.F. New approach to estimate daily reference evapotranspiration based on hourly temperature and relative humidity using machine learning and deep learning. *Agric. Water Manag.* **2020**, *234*. [CrossRef]
66. Laaboudi, A.; Mouhouche, B.; Draoui, B. Neural network approach to reference evapotranspiration modeling from limited climatic data in arid regions. *Int. J. Biometeorol.* **2012**, *56*, 831–841. [CrossRef] [PubMed]
67. Fan, J.; Yue, W.; Wu, L.; Zhang, F.; Cai, H.; Wang, X.; Lu, X.; Xiang, Y. Evaluation of SVM, ELM and four tree-based ensemble models for predicting daily reference evapotranspiration using limited meteorological data in different climates of China. *Agric. For. Meteorol.* **2018**, *263*, 225–241. [CrossRef]
68. Yu, L.-P.; Huang, G.-H.; Liu, H.-J.; Wang, X.-P.; Wang, M.-Q. Experimental Investigation of Soil Evaporation and Evapotranspiration of Winter Wheat Under Sprinkler Irrigation. *Agric. Sci. China* **2009**, *8*, 1360–1368. [CrossRef]
69. Liu, C.; Zhang, X.; Zhang, Y. Determination of daily evaporation and evapotranspiration of winter wheat and maize by large-scale weighing lysimeter and micro-lysimeter. *Agric. For. Meteorol.* **2002**, *111*, 109–120. [CrossRef]
70. Qin, F.Y.; Jie, C.H.; Jian, W. Experiment of Soil Evaporation from Winter Wheat Field. *Irrig. Drain.* **2000**, *19*, 2–5.
71. Wang, H.; Ma, M. Estimation of evapotranspiration from different ecosystems in inland river basins based on remote sensing and Penman-Monteith model. *Acta Ecol. Sin.* **2014**, *34*, 5617–5626.
72. Yao, Y.; Cheng, J.; Zhao, S.; Jia, K.; Xie, X.; Sun, L. A review of research on farmland evapotranspiration estimation methods based on thermal infrared remote sensing. *Earth Sci. Prog.* **2012**, *27*, 1308–1318.

applied sciences

Review

The Application of Hyperspectral Remote Sensing Imagery (HRSI) for Weed Detection Analysis in Rice Fields: A Review

Nursyazyla Sulaiman [1], Nik Norasma Che'Ya [1,*], Muhammad Huzaifah Mohd Roslim [2], Abdul Shukor Juraimi [3], Nisfariza Mohd Noor [4] and Wan Fazilah Fazlil Ilahi [1]

1 Department of Agriculture Technology, Faculty of Agriculture, Universiti Putra Malaysia (UPM), Serdang 43400, Malaysia; sya_zyla@yahoo.com (N.S.); wanfazilah@upm.edu.my (W.F.F.I.)
2 Department of Crop Science, Faculty of Agricultural Science and Forestry, Universiti Putra Malaysia Bintulu Campus, Bintulu 97000, Malaysia; muhammadhuzaifah@upm.edu.my
3 Department of Crop Science, Faculty of Agriculture, Universiti Putra Malaysia (UPM), Serdang 43400, Malaysia; ashukur@upm.edu.my
4 Department of Geography, Faculty of Arts and Social Sciences, University of Malaya, Kuala Lumpur 50603, Malaysia; nish@um.edu.my
* Correspondence: niknorasma@upm.edu.my; Tel.: +603-9769-4892

Abstract: Weeds are found on every cropland across the world. Weeds compete for light, water, and nutrients with attractive plants, introduce illnesses or viruses, and attract harmful insects and pests, resulting in yield loss. New weed detection technologies have been developed in recent years to increase weed detection speed and accuracy, resolving the contradiction between the goals of enhancing soil health and achieving sufficient weed control for profitable farming. In recent years, a variety of platforms, such as satellites, airplanes, unmanned aerial vehicles (UAVs), and close-range platforms, have become more commonly available for gathering hyperspectral images with varying spatial, temporal, and spectral resolutions. Plants must be divided into crops and weeds based on their species for successful weed detection. Therefore, hyperspectral image categorization also has become popular since the development of hyperspectral image technology. Unmanned aerial vehicle (UAV) hyperspectral imaging techniques have recently emerged as a valuable tool in agricultural remote sensing, with tremendous promise for weed detection and species separation. Hence, this paper will review the weeds problem in rice fields in Malaysia and focus on the application of hyperspectral remote sensing imagery (HRSI) for weed detection with algorithms and modelling employed for weeds discrimination analysis.

Keywords: rice plant; weed; hyperspectral imagery; remote sensing

Citation: Sulaiman, N.; Che'Ya, N.N.; Mohd Roslim, M.H.; Juraimi, A.S.; Mohd Noor, N.; Fazlil Ilahi, W.F. The Application of Hyperspectral Remote Sensing Imagery (HRSI) for Weed Detection Analysis in Rice Fields: A Review. *Appl. Sci.* **2022**, *12*, 2570. https://doi.org/10.3390/app12052570

Academic Editors: Dimitrios S. Paraforos and Anselme Muzirafuti

Received: 29 September 2021
Accepted: 21 January 2022
Published: 1 March 2022

1. Introduction

The agricultural sector provides significant economic growth by endowing food sources, producing industrial raw materials as well as providing job opportunities for a substantial number of individuals [1,2]. In Malaysia, the agricultural industry has endured as one of the predominant sectors for socio-economic activity, contributing about 8.7% of the annual gross domestic product (GDP) and 11.4% of the total employment [3]. The major agricultural activities in Malaysia are dominated by rubber (*Hevea brasiliensis* (Willd. Ex A. Juss) *Mull. ARg*), oil palm (*Elaeis guineensis*) and rice plant (*Oryza sativa* L.) [4]. The agricultural sector focuses on sustainable food production and proffering consistent, high-quality and safe food products. In line with an increasing population, global food production will need to significantly multiply in the next few years along with limited area expansion [5]. However, there are several issues that have arisen regarding low crop yield production such as uncertain weather conditions, insufficient labour power, unmaintained agricultural instruments, a reduction in soil and seed quality, constraints on the use of new technologies, etc. [4,6].

In the agricultural ecosystem, weeds serve as a competitor with the actual crop for obtaining light source, nutrients, moisture intensity and gaseous exchange which result in a reduction in crop yield and product quality [3,7]. For crop production, the potential of weed-induced deprivation refers to the type of weed, density, emergence time, and duration intrusion including the simultaneous emergence of weeds along with crop-augmented competition towards restricted growth resources that can trigger the risk of critical yield loss [2,8]. This paper will review the weeds problem in rice fields in Malaysia and focus on the application of hyperspectral remote sensing imagery (HRSI) for weed detection analysis. As a result, researchers, particularly in developing nations, can apply their understanding of decreasing weed presence and enhancing yield output. The focus of this work is on weed detection in rice fields utilising a hyperspectral remote sensing platform. However, hyperspectral remote sensing weed detection in other crops is also included in this review.

2. Methodology

This paper is a conventional review paper. Sources of articles and related research papers were browsed and identified from several databases such as Google Scholar, Google Book, Semantic Scholar, UPM EZAccess, MDPI, and ResearchGate. The primary keyword 'hyperspectral remote sensing' and its synonym paired with the secondary keyword 'weed' and the third keyword 'rice plant' was used as the source of content exploration. Each database search made use of these keyword sets. A hand search was also performed to ensure that no related articles were overlooked. The search was carried out in the fourth quarter of 2021.

All search results were filtered using the following criteria: (1) the study must use hyperspectral remote sensing imagery and platform as the primary data input, (2) the study must dispute the application of hyperspectral remote sensing techniques in weed detection analysis, (3) the document must have reported on the research undertaken, (4) the included papers must have been published in the first quarter of 2021, and (5) the articles must be written in English.

The articles were then reviewed by title and abstract to exclude those that did not satisfy the requirements. Finally, the complete text of the remaining articles was scrutinised to determine whether or not they fit the requirements. Finally, data from a number of articles were taken and transferred into a spreadsheet. Citation information, study objectives, hyperspectral remote sensing sensor, crop and weed types, methodologies and techniques employed, accuracy evaluation, study implications, year of publication, and reference data were all included in the details.

This paper is organized into seven sections. The first section describes the weeds problem in agricultural crops. The approach for searching the scientific database for relevant publications is explained in Section 2. Section 3 highlights the significance of weeds presence in Malaysia's rice fields. Section 4 elaborates on the hyperspectral remote sensing system while the literature on several methodologies for handling remote sensing datasets is presented in Section 5. Section 6 explains the weed detection analysis by using hyperspectral remote sensing with the classification by using spectral reflectance and utilization of the modelling and algorithm. Future directions of hyperspectral remote sensing approaches and conclusions are presented in Section 7.

3. Weeds in Malaysia's Rice Field

According to El Pebrian and Ismail [9], rice is one of Malaysia's most widely grown agricultural crops. According to the Department of Agriculture (DoA) Peninsular Malaysia, this crop was grown on 679,239 acres in 2014, making it the third-largest crop in the country after oil palm and rubber. Malaysia produced 2,848,559 metric tonnes of paddy with such a large planted area. Yusof et al. [10] stated that Malaysian farmers produced 70% of the country's rice production while the rice industry's role is not only to contribute to Malaysia's economy but also to ensure the country's food security. The plantation of rice occurs twice a year, for example: (i) Main season (October–March) and off season

(April–September) with two types of methods implemented in rice cultivation known as direct seeding and transplanting. Dilipkumar et al. [4] stated that 90% of the total rice crop in Malaysia was planted using the direct-seeded method and 10% was transplanted.

Most farmers chose the direct seeding method due to labour insufficiency and expensive cost in rice transplanting [11]. Direct-seeded rice systems implement three different principal methods, for example, water seeding, wet seeding and dry seeding [4,12]. However, the substitution of this planting technique presents crucial weeds expansion in rice crops. According to Hossain et al. [13], many studies have reported that the dominance of particular weed species in rice cropping systems is significantly influenced by the crop establishment method. According to Nagarde et al. [14], weeds are a severe danger to rice, with yearly weed yield losses ranging from 15% to 21% worldwide. Due to massive weed infestation, direct-seeded rice yields are predicted to be reduced by 60% and above. Yield reductions of up to 48% in transplanted crops, 53% in direct-seeded crops (flooded conditions) and 74% in direct-seeded crops (dry soils) have been documented. Different types of grasses, broadleaf weed and sedges make up the weed flora in direct-seeded rice (Table 1).

Table 1. Weeds species in Asia's rice field.

Grassy Weeds	Sedges	Broadleaf Weeds
Digitaria setigera	*Cyperus iria*	*Commelina benghalensis*
Digitaria sanguinalis	*Cyperus difformis*	*Caesulia axillaris*
Digitaria ciliaris	*Cyperus rotundus*	*Eclipta prostrata*
Echinochloa colonum	*Fimbristylis miliacea*	*Ipomoea aquatica*
Echinoclhoa crus-galli		*Ludwigia octovalvis*
Eleusine indica		*Ludwigia adscendens*
Ischaemum rugosum		*Monochoria vaginalis*
Leptochloa chinensis		*Sphenoclea zeylanica*
Oryza sativa		
Paspalum		

Source: Nagarde et al. [14]

Direct-seeded rice systems bestow an aerobic environment for weed growth since they are not flooded during the beginning growth stage of the rice plants and it is convenient towards weed expansion [11,15]. The aerobic soil condition in the direct-seeded rice system conserves water, while the weed problem in direct-seeded rice is exacerbated by the lack of stagnant water and the lack of a 'head start' in rice seedlings over sprouting weed seedlings [16]. Toriyama [17] explained that the extensive employment of the direct seeding method with the frequent use of herbicide and a shortage of irrigation supplies accountable on the transference of weed species populations in the rice field ecosystem, for example, the grasses species: *Echinochloa crus-galli, Echinochloa* spp. *(E. oryzicola, E. colona, E. staginina, and E. picta), Leptochloa Chinensis,* and *Ischaemum rugosum*, which were not dominant in Malaysian rice fields, has previously become widespread afterwards (see Table 2). Furthermore, Chauhan et al. [12] found that the density of grassy weeds in zero-tilled direct-seeded rice was higher than in puddled transplanted rice. Sedges and broadleaves, on the other hand, were less abundant. Broadleaves such as *Sagittaria guayanensis* Kunth, *Monochoria vaginalis* (Burm. f.) C. Presl ex Kunth, *Limnocharis flava* (L.) *Buchenau, Ludwigia octovalvis* (Jacq.) P.H. Raven, and *Alternanthera sessilis* (L.) R. Br. Ex DC. and *Ammannia baccifera* L. also had expanded abundance in the puddled transplanted field.

Table 2. Weed shift from transplanting to the direct-seeding method.

Irrigated Transplanting	Extensive Direct Seeding	Intensive Direct Seeding
Grasses		
Isachne globose *Leersia hexandra*	*Echinochloa crus-galli complex* *Lepthochola chinensis* *Ischaemum rugosum* *Oryza sativa* (weedy rice)	*E. crus-galli* *L. chinensis* *I. Rugosum* *O. sativa* (weedy rice)
Broadleaf weeds		
Limnocharis flava *Monochoria vaginalis* *M. minuta* *Sphenoclea zeylanica*	*L. flava* *M. vaginalis* *L. hyssopifolia* *M. minuta*	*L. flava* [a] *S. guyanensis* [b] *Sphenoclea zeylanica* [b] *M. crenata* [b] *Limnophila erecta* [a]
Sedges		
Scirpus grossus	*Cyperus iria* *Fimbristylis miliacea*	*C. iria* *F. milicea* [b] *C. difformis*

[a] Biotypes with herbicide resistance against 2,4-D and ALS-inhibitor herbicides. [b] Species/biotypes with herbicide resistance against 2,4-D. Source: [17].

Weeds are one of the most significant causes of reducing rice productivity, resulting in not only large financial expenditures but also crop quality difficulties. Crops can also be affected by weeds present at any growth stage [18]. In Malaysia, weed-related production losses range from 5% to 85%, depending on the planting method, season, region, major weed flora, weed density, management practices and infestation length [4]. Issues regarding weeds in crops are complex; meanwhile, to reduce their expansion and impacts on the crop, the management strategy chosen must be synchronized in all aspects to make sure that systematic guidance will be assembled to manage the existing weeds as well as to prevent the spreading of new weeds [19]. A particular weed management proposition, for example, mechanical, chemical, manual and biological control strategies were initiated for weeds control in a crop field since these strategies came with certain constraints such as proper climatic circumstances, location of farmers, labour availability and the capability to endure with management expenses [20,21]. Early weed treatment not only reduces the occurrence of pests and diseases but also reduces agricultural yield loss by up to 34%. Chemical and non-chemical weed management strategies have been widely used in rice fields in this scenario. Manual weeding is too time consuming, expensive, and inconvenient as a non-chemical technique. Mechanical weed management is a non-chemical approach [18].

Partel et al. [22] created and constructed a smart sprayer that could distinguish between weeds and non-weed objects using machine vision and artificial intelligence. This targeted approach was combined with a revolutionary precision spraying system that included a state-of-the-art weed detecting technology and a weed mapping system for precise spraying. When compared to traditional broadcast spraying techniques, which often cover the entire field, the results showed that using this system lowered the number of agrochemicals necessary. Huang et al. [23] and Yao and Huang [24] mentioned that agricultural remote sensing has been established and utilised for monitoring crop field conditions such as growth status, soil variability, crops stress from weeds, pests, water and nutrition insufficiency in providing data and information towards the efficient operation. Unmanned aerial vehicle (UAV) technology provides a desirable precision agriculture data gathering platform that is highly flexible and simple to use while collecting high spatial resolution data in a timely way. Due to their geographical and temporal resolution capabilities and cost-effectiveness, UAVs are a better platform for crop monitoring activity [25]. Currently, in conjunction with the evolving transducer technology and sensor, remote sensing approaches were upgraded for weed detection and control particularly with the emergence of hyperspectral sensing and imaging [23].

4. Hyperspectral Remote Sensing: A Brief Overview

According to Weiss et al. [26], agriculture monitoring from remote sensing is a vast subject that has been widely addressed from multiple perspectives, sometimes based on specific applications (e.g., precision farming, yield prediction, irrigation, weed detection), remote sensing platforms (e.g., satellites, unmanned aerial vehicles—UAVs, unmanned ground vehicles—UGVs), or sensors (e.g., active or passive sensing, wavelength domain) or specific locations and climatic contexts (e.g., country or continent, wetlands or drylands). Campbell and Wynne [27] defined remote sensing as the application of acquiring information regarding the Earth's land and water surface by utilising images obtained from an overhead perspective, implementing electromagnetic radiation in one or more regions of the electromagnetic spectrum, reflected or emitted from the Earth's surface. Hyperspectral remote sensing involves extracting information from the objects or scenes that lie on the Earth's surface due to radiance obtained by airborne or spaceborne sensors [28,29].

Generally, hyperspectral imaging is an incorporation of the modern imaging system and traditional spectroscopy technology [30,31]. According to Govender et al. [32], the evolution of airborne and satellite hyperspectral sensor technologies has overcome the restraint of multispectral sensors since hyperspectral sensors assemble several narrow spectral bands from the visible, near-infrared (NIR), mid-infrared, and short-wave infrared portions of the electromagnetic spectrum. The hyperspectral sensor collects about 200 or more spectral bands, each only 10 nm wide [27] which allows the construction of continuous spectral reflectance signatures while the narrow bandwidths element of hyperspectral data enable in-depth examination of Earth surface characteristics which would disappear within the relatively coarse bandwidths acquired with multispectral data. Hyperspectral data are usually assigned as hypercubes (see Figure 1) that contain two spatial dimensions and one spectral dimension, regarding the characteristics of each hyperspectral image, comprising many channels since there were bands—in contrast to grayscale or RGB images—that included only one or three channels, respectively [33].

The hyperspectral data cube in Figure 1 explained that Figure 1a A push-broom sensor on an airborne or spaceborne platform acquire spectral data for a one-dimensional row of cross-track pixels named as scanline; Figure 1b Sequential scan lines including spectra for each row of cross-track pixels are pilled to obtain a three-dimensional hyperspectral data cube which in this illustration the spatial details of a scene are constituted by the x and y dimensions of the cube, while the amplitude spectra of the pixels are projected to the z dimension; Figure 1c the three-dimensional hyperspectral data cube can be analysed as a stack of two-dimensional spatial images whereas each is equivalent to a particular narrow waveband. Usually, hyperspectral data cubes contain hundreds of stacked images; Figure 1d the spectral samples can be marked for each pixel and discrimination of the features in the spectra deliver the primary mechanism for detection and classification in a scene [34,35]. Qian [31] stated that there were about three different methods in obtaining the hyperspectral data regarding the type of imaging spectrometers such as dispersive elements-based approach, spectral filters-based approach and snapshot hyperspectral imaging. In order to collect the hyperspectral images with different spatial and temporal resolutions, the sensors used can, for example, be mounted on different platforms. Unmanned-aerial vehicles (UAVs), airplanes, and close-range platforms [36]. Table 3 shows the comparison of different types of hyperspectral imaging platforms. Kate et al. [37] mentioned that hyperspectral sensors were utilised for providing information such as airborne visible/infrared imaging spectrometer (AVIRIS), Hyperion, Hymap (from HyVista Castle Hill, Australia), and airborne imaging spectroradiometer for applications (AISA). Table 4 below shows different types of hyperspectral sensors used which are usually mounted on the aircraft and satellite [38].

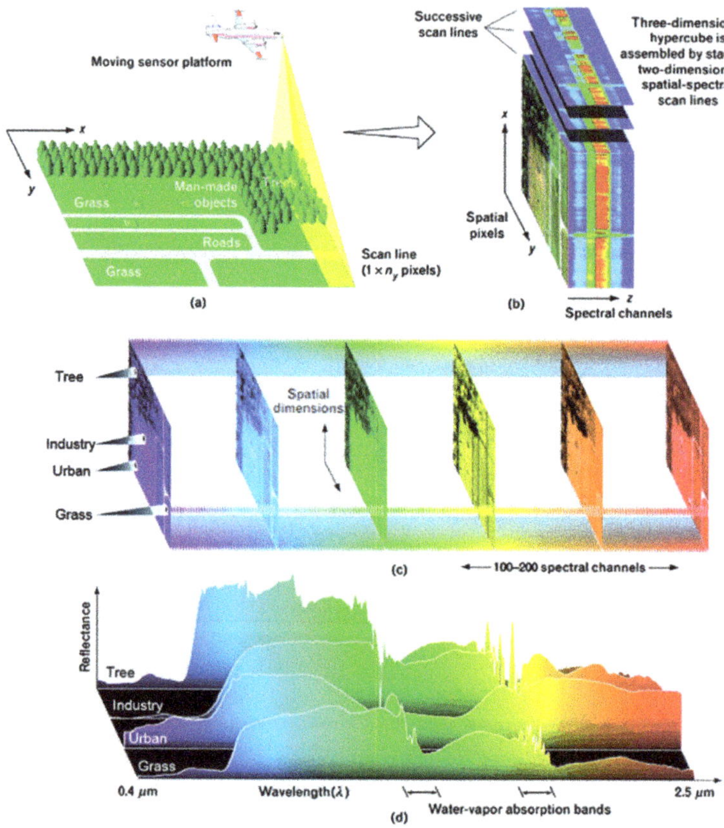

Figure 1. Hyperspectral data cube structure [34,35].

Table 3. Comparison of hyperspectral imaging platforms [36].

Parameters	Satellites	Airplanes	Helicopters	Fixed-Wing UAVs	Multi-Rotor UAVs	Close-Range Platforms
Operational Altitudes	400–700 km	1–20 km	100 m–2 km	<150 m	<150 m	<10 m
Spatial coverage	42 km × 7.7 km	~100 km^2	~10 km^2	~5 km^2	~0.5 km^2	~0.005 km^2
Spatial resolution	20–60 m	1–20 m	0.1–1 m	0.01–0.5 m	0.01–0.5 m	0.0001–0.01 m
Temporal resolution	Days to weeks	Depends on flight operations (hours to days)				
Flexibility	Low (fixed by repeating cycles)	Medium (depend on availability of aviation company)		High		
Operational complexity	Low (provide final data to users)	Medium (depend on users or vendors)		High (operate by users with setting up the hardware and software)		
Applicable scales	Regional–global	Landscape-regional		Canopy–landscape		Leaf–canopy
Major limiting factors	Weathers	Unfavourable flight height/speed, unstable illumination conditions		Short battery endurance, flight regulations		Platform design and operation
Image acquisition cost	Low to medium	High (typically need to hire an aviation company to fly)		Large (due to area coverage)		

Table 4. Type of hyperspectral sensors on aircraft and satellites [38].

Types of Sensors	Producer	Number of Bands	Spectral Image (μm)
Satellite mounted hyperspectral sensors			
FTHSI on MightySat II	Air Force Research (OH, USA)	256	0.35–1.05
Hyperion on EO-	NASA Guddard Space Flight Center (Greenbelt, MA, USA)	242	0.40–250
Aircraft-mounted hyperspectral sensors			
AVIRIS (airborne visible infrared imaging spectrometer)	NASA Jet Propulsion Lab. (Pasadena, CA, USA)	224	0.40–2.50
HYDICE (hyperspectral digital imagery collection experiment)	Naval Research Lab (Washington, DC, USA)	210	0.40–2.50
PROBE-1	Earth Search Sciences Inc. (Kalispell, MT, USA)	128	0.40–2.50
CASI (compact airborne spectrographic imager)	ITRES Research Limited (Calgary, AB, Canada)	Over 22	0.40–1.00
HyMap	Integrated Spectronics	100 la 200	Visible to thermal Infrared
EPS-H (environmental protection system)	GER Corporation	VIS/NIR (76), SWIR1 (32), SWIR2 (32), TIR (12)	VIS/NIR (0.43–1.05) SWIR1 (1.50–1.80) SWIR2 (2.00–2.50) TIR (8–12.50)
DAIS 7915 (digital airborne imaging spectrometer)	GER Corporation (geophysical and environmental research imaging spectrometer)	VIS/NIR (32), SWIR1 (8), SWIR2 (32), MIR (1), TIR (12)	VIS/NIR (0.43–1.05) SWIR1 (1.50–1.80) SWIR2 (2.00–2.50) MIR (3.00–5.00) TIR (8.70–12.30)
DAIS 21115 (digital airborne imaging spectrometer)	GER Corporation	VIS/NIR (76), SWIR1 (64), SWIR2 (64), MIR (1), TIR (6)	VIS/NIR (0.40–1.00) SWIR1 (1.00–1.80) SWIR2 (2.00–2.50) MIR (3.00–5.00) TIR (8.00–12.00)
AISA (airborne imaging spectrometer)	Spectral Imaging	Over 288	0.43–1.00

5. Hyperspectral Remote Sensing Imagery (HRSI) Data Processing and Analysing

5.1. Data Preprocessing

According to Weng and Xiaofei [39], due to the high-dimensional nature of hyperspectral data, as well as the resemblance between the spectra and mixed pixels, hyperspectral image technology still confronts a number of issues, the most pressing of which are the following: (1) Hyperspectral image data have high dimensionality. Because hyperspectral images are created by combining hundreds of bands of spectral reflectance data gathered by airborne or space-borne imaging spectrometers, the spectrum information dimension of hyperspectral images can also be hundreds of dimensions; (2) missing labelled samples. In practical applications, collecting hyperspectral image data is rather simple, but obtaining image-like label information is quite challenging. As a result, the categorization of hyperspectral pictures is sometimes hampered by a shortage of labelled samples; (3) variability in spectral information across space. The spectral information of hyperspectral images changes in the spatial dimension as a result of factors such as atmospheric conditions, sensors, the composition and distribution of ground features, and the surrounding environment, resulting in the ground feature corresponding to each pixel not being single; and lastly (4) image quality which is the interference of noise and background elements during the acquisition of hyperspectral pictures which has a significant impact on the quality of the data collected. The categorization accuracy of hyperspectral images is directly influenced by the image quality.

Hyperspectral images obtained by various platforms and sensors are usually presented in raw format which requires them to be pre-processed (for example, atmospheric, radiometric, and spectral corrections) to rectify detailed information [36]. Assembling hyperspectral data is more intricate than multispectral and RGB sensors because its radiometric and atmospheric calibration workflows are more involuted [40]. Therefore, several steps were required for the hyperspectral imaging processing procedure in order to obtain precise output [33]. The processing of hyperspectral imaging signifies the utilisation of computer algorithms. It includes tasks such as extracting, storing and falsifying information from visible near-infrared (VNIR) or near-infrared (NIR) hyperspectral images. It also provides different information on processing and data mining assignments (for example, analyse, classify, target detection, regression, and pattern identification) [41,42]. Hyperspectral imaging includes extensive data collection stored in pixels while each data particularly correlates to their neighbours [43]. Hyperspectral imaging also comprises the spectral-domain signal as each of the image pixels contains the spectral information; thus, specific tools and approaches have been amplified for processing both spatial and spectral information [42]. This magnitude of data has led to the integration of chemometric and visualisation equipment to competently mine for significant and detailed information [11]. The ordinary hyperspectral image preprocessing procedure is delineated in Figure 2 below [42].

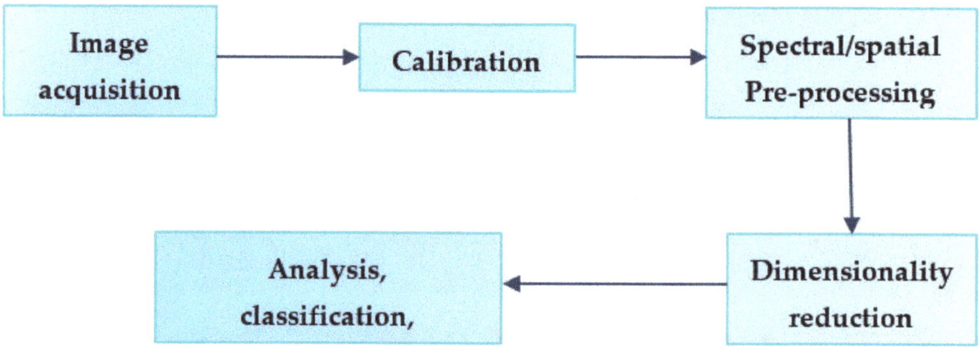

Figure 2. Hyperspectral image preprocessing workflow [42].

According to Burger and Geladi [44], numerous amounts of raw data produced from hyperspectral imaging devices contain lots of errors that can be rectified by calibration. Spatial calibration is one of the steps that correlates each image pixel to known units or features, bestowing information about the spatial dimensions and also rectifying the optical aberrations (smile and keystone effects) [42]. However, three conditions could prevail which invalidate calibration models which are: (1) chemical or physical substitution in samples, (2) change of equipment due to inherent uncertainty or ageing parts and, (3) environment/weather condition, for example, temperature or humidity [14]. Lu et al. [36] mentioned that hundreds of bands are common in hyperspectral photographs, and many of them are highly connected. As a result, dimension reduction is an important step to consider while pre-processing hyperspectral images. Dimensionality reduction is a crucial pre-processing step in hyperspectral image classification that reduces HSI's spectral redundancy, resulting in faster processing and higher classification accuracy. Methods for reducing dimensionality convert high-dimensional data into a low-dimensional space while keeping spectral information [45]. Hence, pre-processing is an important step in increasing the quality of hyperspectral images and preparing them for subsequent analysis.

Basantia et al. [46] stated that hyperspectral imaging generates extensive data collection from a single sample and with thousands of samples that require daily analysis. According to Tamilarasi and Prabu [47], in contrast to other statistical techniques, hyperspectral image analysis uses physical and biological models to absorb light at certain wavelengths. For example, air gases and aerosols could absorb light at specific wavelengths. Dispersion (adding an outside light source to the sensor region of perspective) and absorption are examples of atmospheric diminution (radiance denial). As the outcome, a hyperspectral sensor could not differentiate the radiance recorded with the imaging generated at other times or locations. Hyperspectral image analysis techniques are derived from spectroscopy, which relates to the distinct absorption or patterns of reflection of the context at different wavelengths of a certain material's molecular composition. This image must be subjected to appropriate atmospheric correction techniques in order to compare each pixel's reflection signature to the spectrum of known material; in laboratories and in "library" storage areas, known spectral information of materials include soils, minerals and vegetation types.

5.2. Hyperspectral Image Classification

Hyperspectral imaging (HSI) is classified as supervised, unsupervised, and semi-supervised based on the nature of available training samples. The supervised technique uses ground truth information (labelled data) for classification whereas the unsupervised technique does not require any prior information [48]. According to Wenjing and Xiaofei [39], support vector machines, artificial neural networks, decision trees and maximum likelihood classification methods are examples of commonly used supervised classification methods. The basic process is to first determine the discriminant criteria based on the known sample category and prior knowledge and then calculate the discriminant function. Therefore, in supervised classification, Freitas et al. [49] stated that support vector machines can produce results that are similar to neural networks but at a lower computing cost and faster rate, making them ideal for hyperspectral data analysis.

Unsupervised classification refers to categorization based on hyperspectral data spectral similarity, for example, clustering without prior knowledge. As stated by Wenjing and Xiaofei [39], unsupervised classification can only assume beginning parameters, build clusters through pre-classification processing, and then iterate until the relevant parameters reach the permitted range since no prior knowledge is employed. Examples of unsupervised classification are K-means classification and the iterative self-organizing method (ISODATA). Lastly, is the semi-supervised classification which trains the classifier using both labelled and unlabelled data. The semi-supervised learning paradigm has been successfully utilized beyond hyperspectral imaging [50]. It compensates for the lack of both unsupervised and supervised learning opportunities. On the feature space, this classification approach uses the same type of labelled and unlabelled data. Because a large number of unlabelled examples may better explain the overall properties of the data, the classifier trained using these two samples has superior generalisation. Examples of semi-supervised classification are Laplacian support vector machine (LapSVM) and self-training [39].

Therefore, hyperspectral imaging can be one of the potential techniques for automatic discriminations between crops and weeds. These sensing technologies have been utilized in smart agriculture and made substantial progress by generating large amounts of data from the fields. Machine learning modelling integrating features has also accomplished reasonable accuracy in order to identify whether a plant is a weed or a crop. Table 5 shows the application of hyperspectral imaging for the discrimination of crops from weeds by using machine learning.

Table 5. Hyperspectral imaging for discrimination of crops from weeds using machine learning (adopted from Su. [51]).

No.	Crop	Weed	Model	Optimal Accuracy	Reference
1.	Rice	Barnyard grass, weedy rice	RF, SVM	100%	Zhang et al. (2019)
2.	Maize	Caltrop, curly dock, barnyard grass, ipomoea spp., polymeria spp.	SVM, LDA	>98.35%	Wendel et al. (2016)
3.	Soybean, cotton	Ryegrass	LDA	>90%	Huang et al. (2016)
4.	Wheat	Broadleaf weeds, grass weeds	PLSDA	85%	Hermann et al. (2013)
5.	Broadbean, wheat	Cruciferous weeds	ANN	100%	De Castro et al. (2012)
6.	Sugar beet	Wild buckwheat, Field Horsetail, Green foxtail, Chickweed	LDA	97.3%	Okamoto et al. (2007)
7.	Wheat	Musk thistle	SVM	91%	Mirik et al. (2013)
8.	Maize	*C. arvenis*	RF	>90%	Gao et al. (2018)

RF—random forest; SVM—support vector machines; LDA—linear discriminant analysis; ANN—artificial neural network; PLSDA—partial least square discriminant analysis.

6. HRSI Application in Weed Detection Analysis

6.1. Weed Classification Using the Spectral Reflectance

Weed classification is important in precision farming because weeds are pests to crops and compete for space, nutrients, water, and light, and obstruct the growth of crops in the field [52]. Effective weed management is vital in smart agriculture as weeds can trigger major environmental and economic problems in agriculture [53]. According to Su [51], smart agriculture may utilise intelligent technology to precisely monitor weed dispersion in the field and undertake weed control chores at specific locations, which not only improves pesticide effectiveness but also increases the economic benefits of agricultural products. The most significant aspect of an automatic weed removal system within crop rows is the use of dependable sensing technology to accomplish accurate weed and crop discrimination at specified points in the field. Therefore, the application of remote sensing employed in agricultural research was established for the interaction between electromagnetic radiation and plant materials on the Earth's surface [54,55]. Hyperspectral imaging has been suggested as the most suitable instrument for food quality assessment and safety investigation that has been exerted on an array of spectral imaging modalities, for example, NIR, fluorescence, and Raman hyperspectral imaging [46,56]. Hyperspectral images captured by UAV platforms has lately emerged as a significant tool in agricultural remote sensing, with considerable potential for weed detection and species differentiation [57].

To obtain detailed spectrum information, hyperspectral imaging sensors frequently use more and narrower bands. Hyperspectral images have comprehensive spectrum information in each pixel, which has been used for a number of agricultural applications [37]. According to Pott et al. [58], spectral bands can be utilized for differentiating plants from other non-targets. Plant pigments, such as chlorophyll (chlorophyll a and b), carotenes and xanthophylls, are primarily affected by visible light reflection in plant leaves and canopies [35]. The red-edge band reflectance is affected by a mixture of chlorophyll, intense light scattering and internal cellular plant structure. Internal leaf structure and many leaf layers influence the reflectance qualities of the canopy in the near-infrared (NIR) band [37,58].

Paap [19] stated that plants' spectral reflectance is identified based on the cellular and biochemical leaf structure and leaf canopy. Figure 3 represents a typical spectrum reflectance and transmitted wavelength of green leaf. The contrast of the reflectance and transmittance spectra depend upon absorption which is in the visible spectrum range 400–700 nm, the spectra are controlled by absorption of various pigments and primarily chlorophylls. In near infra-red (NIR), the reflectance spectra are high which is close to 50% and flat while above 1300 nm, the reflectance declines because of the water absorption present in the leaf.

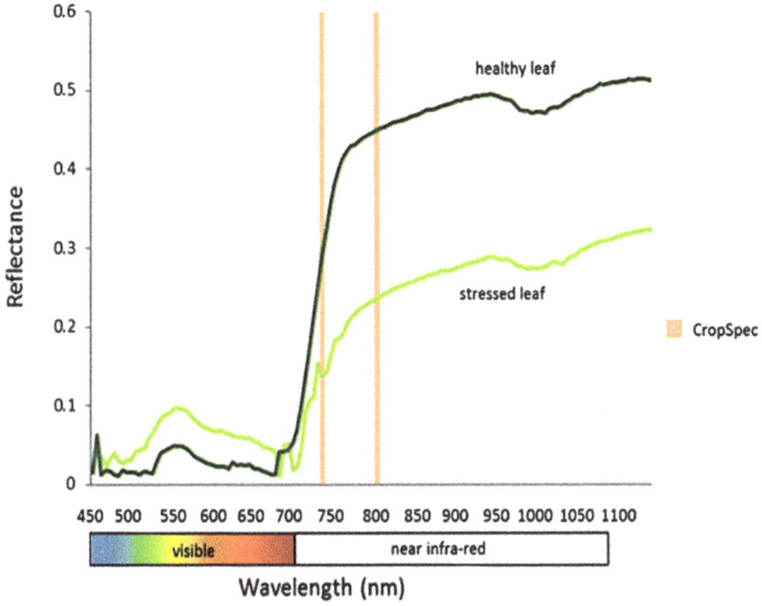

Figure 3. Spectral reflectance for healthy and stressed leaf in visible and NIR wavelength [59].

Further contemplation in the vegetation mapping procedure is the size of the objects to be mapped. Higher spatial resolution imagery is frequently used for mapping narrow vegetation objects which are acquired from airborne sensors [60]. Weeds compete with crops and are difficult to distinguish because of their similar colour, shape, and size [61]. However, a previous study on dispersing reflectance spectra of crop and weed leaves found the potential of weed detection with reflectance measurements. Zhang et al. [62] mentioned that due to the considerable absorption by chlorophylls, a plant leaf typically has a low reflectance in the visible spectral range and a comparatively high reflectance in the near-infrared spectral area due to internal leaf scattering and no absorption. Therefore, according to Thenkabail et al. [63], plant leaf area index and biomass are more sensitive to the red band at roughly 680 nm, while plant moisture status is more sensitive to the NIR near 950 nm. The correlation among the plant pathology has been employed by the remote sensing technique in contemplation to discover the discrete plant characteristics from their spectra reflectance.

The application of hyperspectral remote sensing is comprehensively used for different weed detection analysis studies, for example, weed discrimination in maize [64], discrimination of grassweeds in winter cereal crops [65], in early detection of spotted knapweed (*Centraurea maculosa*) and babysbreath (*Gypsophila paniculate*) with hyperspectral sensor [66], herbicide-resistant weeds classification [67], identification of (*Ranunculus acris* [giant buttercup] and *Cirsium arvense* [Californian thistle]) by Li et al. [53] and spectral features extraction from hyperspectral images to differentiate weedy rice and barnyard grass [68]. Singh et al. [57] stated that hyperspectral imaging application has also been used to identify between crop types, for instance, the utilization of satellite-based hyperspectral sensors to distinguish mustard, potato, sugarcane, sorghum, and wheat in the range of 700–750 nm. The efficacy of hyperspectral sensors for plant species characterization has been documented in a number of different studies, which is included in Table 6 below.

Table 6. Published reports/study on crop and weed species classification using hyperspectral imagery [57].

Crop/Weed Mixture	Reflectance	Wavelength (nm)
Wheat system		
Wheat (*Triticum aestivum*)	0.42–0.45	720–850
Broadleaved weeds	0.6–0.8	720–850
Grass weeds	0.55–0.6	720–850
Wheat stubbles	0.2–0.3	1425–2250
Wheat stubbles heavily grazed	0.3–0.4	1425–2250
Cruciferous weeds	0.65–0.7	750–900
Wheat	0.4–0.5	720–900
Soybean system		
Soybean (*Glycine max*)	0.85–0.90	750–900
Prickly sida (*Sida spinosa*)	0.78–0.82	750–900
Pitted morning glory (*Ipomoea lacunosa*)	0.6–0.65	750–900
Sicklepod (*Cassia obtusifolia*)	0.52–0.55	750–900
Cotton system		
Cotton (*Gossypium hirsutum*)	0.48–0.52	400–600
Cogongrass (*Imperata cylindrica*)	0.5–0.6	400–750
Johnsongrass (*Sorghum halepense*)	0.45–0.52	400–750
Sicklepod (*C. obtusifolia*)	0.21–0.3	400–800
Sorghum system		
Sorghum (*Sorghum bicolor*)	0.55–0.6	720–1000
Common lambsquarters (*Chenopodium album*)	0.39–0.41	720–1000
Pigweed (*Amaranthus* spp.)	0.48–0.52	720–1000
Barnyardgrass (*Echinochloa* spp.)	0.35	72–1000
Mallow (*Malva* spp.)	0.42–0.44	720–1000
Purple nutsedge (*Cyperus rotundus*)	0.28–0.30	720–1000

6.2. Algorithms and Modelling for Weed Detection Analysis

For various agricultural applications, several remote sensing approaches, such as hyperspectral data from airborne, satellite platforms using multispectral and optical imagery have been proposed [69,70]. A Study conducted by Felegari et al. [71] looked into the drawbacks and benefits of using a combination of radar data and optical images to determine the types of crops in the Tarom region (Iran) in which the Sentinel 1 and Sentinel 2 images were utilised to generate a map for the selected research area. Hyperspectral sensing, which measures reflectance from visible to shortwave infrared wavelengths, has allowed vegetation to be classified and mapped at a variety of taxonomic scales, often down to the species level. To reduce the dimensionality of the data to a level suitable for the creation of a classification model, hyperspectral measurements recorded by narrowband spectroradiometers or imaging sensors have typically required some type of spectral feature selection [72]. Therefore, the remote sensing method can detect the existence of non-crop plants between rows, such as the recognition of weeds within rows, whereas segregating weeds from crops and identifying weed species emerging from proximal sensing research has utilized both spectral reflectance and leaf shape analysis for identification [73]. According to Lan et al. [74], to examine the datasets generated by these methodologies, proper and effective sophisticated algorithms as well as high-power computation are required. Genetic programming was utilized by Nguyen et al. [75] to distinguish between rice and other leaf groups. They also employed a scanning window of 20 × 20 pixels on a test image to evaluate the classifier, attaining a 90% accuracy by applying the classifier to each pixel of the window based on a colour threshold.

Therefore, several automatic classification techniques have been employed to classify remote sensing data and plant monitoring procedures, for example, the machine learning method [74]. According to Dadashzadeh et al. [18], machine vision based on image processing has been used to collect data in two different ways: Two-dimensional (2D) vision and three-dimensional (3D) vision. When using 2D cameras, machine vision systems based on two-dimensional (2D) image processing have some drawbacks. First, differences in external illumination have an impact on the quality of images captured by 2D cameras; thus, the camera's field of view must be covered. Second, the overlap of different plant components can make distinguishing weeds from crops difficult.

According to Perez-Ortiz et al. [76], most standard classifiers in machine learning are based on learning a discriminant function from labelled data (i.e., supervised learning). However, obtaining tagged data, as opposed to unlabelled data, can be time consuming and costly. Liakos et al. [77] stated that machine learning (ML) has risen to prominence with big data technology and high-performance computers to open up new avenues for unravelling, quantifying, and understanding data-intensive processes in agricultural operations. ML is characterised as a scientific subject that allows machines to learn without being strictly programmed, among other things. Examples of ML modelling include artificial neural networks (ANNs), Bayesian models (BM), deep learning (DL), dimensionality reduction (DR), decision trees (DT), ensemble learning (EL), instance-based models (IMB) and support vector machines (SVMs).

Dadashzadeh et al. [18] investigated site-specific weed management in the rice field using two metaheuristic algorithms: The bee algorithm (BA) and particle swarm optimisation (PSO), in order to improve the neural network's ability to identify the most effective characteristics and classify different types of weeds. Because of their abundance in the chosen region, this study focused on a rice cultivar (*Tarom Mahali*) and two common types of weeds (narrow-leaf weeds (*Echinochloa crus-Galli, Paspalum distichum*, and *Cyperus difformis*) and wide-leaf weeds (*Alisma plantago-aquatica* and *Eclipta prostrata*) while a stereo camera was used to collect the necessary data in the form of stereo videos, with different channels of each frame extracted. The proposed stereo vision technique, which averaged the related points on various channels and the proposed hybrid ANN-BA classifier for better classification accuracy, proved to have promising capabilities. Zheng et al. [78] created and evaluated a new classification algorithm based on colour indices and support vector data description (SVDD). In the first, second and third years of a three-year case study, overall accuracies of 90.19%, 92.36%, and 93.8%, respectively, were achieved. Kamath et al. [79] looked at how to categorize paddy crops and weeds from digital images utilizing several classifier systems developed with support vector machines (SVM) and random forest classifiers (RFs) in which the dataset included paddy plants and weeds from the seedling stage (1-leaf seedling) to the flowering stage. The results with an accuracy of 91.36% showed that multiple classifier systems were shown to outperform single classifier systems and the extracted features are good for paddy crops and weeds classification.

Li et al. [53] studied weed identification by using hyperspectral data images trained on three classification models, namely partial least squares discriminant analysis, support vector machine and multilayer perceptron (MLP) with an overall accuracy range of about 70–100%. The analysis was run by using the whole plant averaged (Av) spectra and superpixels (Sp) averaged spectra from four different weed samples which comprised two types of grass (*Setaria pumila* [yellow bristle grass] and *Stipa arundinacea* [wind grass]) and two broadleaf weed species (*Ranunculus acris* [giant buttercup] and *Cirsium arvense* [Californian thistle]). Results showed that using both Av and Sp spectra were able to identify the four weeds' species. To solve the challenge of forecasting the pre-planting risk of Stagonospora nodorum blotch (SNB) in winter wheat, Mehra et al. [80] used machine learning approaches such as artificial neural networks (ANNs), category and regression trees, and random forests (RFs). They created risk assessment models that could help with disease control decisions before planting the wheat crop.

Research conducted by Chen et al. [81] combined the application of multi-feature fusion and support vector machine (SVM) to detect corn seedlings and weeds for limiting crop damage with an average recognition accuracy of about 97.50%. The dataset included a small database of corn seedlings and weed and actual field images. The results of the experiments revealed that the fusion feature of rotation invariant local binary pattern (LBP) feature and grey level-gradient co-occurrence matrix based on an SVM classifier accurately detected all types of weeds and corn seedlings. This provides information about weed and crop positions to the spraying herbicide, allowing for exact spraying and fertilising. Chou et al. [82] used a wavelet packet transform paired with a weighted Bayesian distance based on crop texture and leaf data to identify the crop. The dataset needed for this study included field crop images captured with a digital camera with a resolution of 640 × 480 pixels. To discriminate plants, they estimated energy coefficients in multiple frequency bands produced after the change. The crop identification achieved an accuracy of 94.63% by using the decision distance in different climates over three consecutive days of photography.

Bakhshipour and Zareiforoush [83] used integrate decision tree (DT) and fuzzy logic techniques to establish a fuzzy model for differentiating the peanut plant from broadleaf weeds with the overall accuracies on training and testing datasets being, respectively, 92% to 96%. On the input dataset, two feature selection approaches were utilised: Principal component analysis (PCA) and correlation-based feature selection (CFS), and three decision trees (DTs) were used to distinguish between distinct plants: J48, random tree (RT), and reduced error pruning (REP). Another study by Bakhshipour et al. [84] is on texture features recovered from wavelet sub-images to detect and describe four species of weeds in a sugar beet field, while neural networks (NN) were run as a classifier. Images were taken from sugar beet fields with a resolution of 96 × 1280 pixels when the plants had six to eight leaves with significant occlusion and a height of about 80 mm to 160 mm. The research found that even at a stage of beet growth greater than six leaves, the application of wavelets proved to be effective for weed detection. Two-dimensional Gabor filters were employed to extract the features in a study conducted by Tang et al. [85], and an artificial neural network (ANN) was utilized to categorize broadleaf and grass weeds. The seeds of the selected broadleaf weed species were planted and the image was captured four weeks after seeding. The Gabor wavelet/ANN system was created to use texture features to classify weed images into broadleaf and grass categories. Their findings revealed that joint space–frequency texture properties might be used to classify weeds.

Furthermore, in agricultural research, deep learning combined with advancements in computer technology, particularly graphical processing units (GPU) embedded processors, has produced remarkable results for image classification and objection detection [86,87]. According to Alom et al. [88], deep learning (DL) algorithms have many advantages over traditional machine learning approaches for image classification, object detection and localization. To build a feature extractor from raw data, traditional machine learning techniques necessitate extensive domain knowledge [89,90]. The DL approach, on the other hand, employs a representation-learning method in which a machine can automatically discover discriminative features from raw data for classification or object detection problems. DL methods can effectively extract discriminative features of crops and weeds due to their strong feature learning capabilities. Furthermore, as data sets have grown larger, the performance of traditional machine learning approaches has become saturated. When large datasets are used, DL techniques outperform traditional machine learning techniques [88].

Hosseini et al. [91] stated that convolutional neural networks (CNNs) and recurrent neural networks (RNNs) are two commonly used architectures in DL. Although CNNs are used for other types of data, the most common application of CNNs is to analyse and classify images. The term convolution refers to the filtering process. CNN is based on a stack of convolutional layers. Each layer receives input data, transforms or convolves it, and outputs it to the next layer. This convolutional operation eventually simplifies the data so that it can be processed and understood more easily [89]. Mentioned by Bah et al. [92], convolutional neural networks (CNNs) have advanced primarily as a result of their successful use as a method in the ImageNet Large-Scale Vision Recognition Challenge 2012 (ILSCVR12) and the creation of the AlexNet network in 2012, which demonstrated that a large, deep convolutional neural network can achieve record-breaking results on a highly challenging dataset using purely supervised training. Therefore, a deep convolutional neural network (DCNN) system for plant recognition based on plant leaf features and patterns was also documented by Lee et al. [93] based on the leaf's shape, texture, and venation while they presented new hybrid models taking advantage of the correspondence of different contextual information of leaf features.

7. Direction for Future Work and Conclusions

In most situations, removing weeds in agricultural areas requires the use of large amounts of chemical pesticides, which are damaging to the environment regardless of how effective they are at enhancing crop output. Precision spraying might be explored to optimise herbicide application in crop fields, thanks to recent advances in image sensors. In this paper, we have reviewed the situation of weeds in rice crops, the background of hyperspectral imaging and techniques for processing hyperspectral data. This study is interdisciplinary and experts from various disciplines, such as agronomy (weed science), remote sensing, computing, and engineering are collaborating. Hyperspectral remote sensing technology is an important component in precision farming and is being used by a growing number of scientists and agricultural researchers. The capacity to properly and reliably distinguish weeds from crops is a vital step in controlling or eradicating weed infestations in agricultural crops. Due to the abundance of spectral information sensitive to distinct plant biophysical and biochemical properties, hyperspectral imaging offers a lot of potential for applications in agriculture, especially precision agriculture. Hyperspectral remote sensing technology uses the difference in spectral reflectance qualities between weeds and crops to identify weeds in crop stands and aids in the compilation of weed maps in the field, allowing for the application of site-specific and need-based herbicides for weed management.

Hyperspectral imaging data with high spatial resolution along with machine learning algorithms in remote sensing showed good potential in agricultural studies. In the recent decade, sensing technologies and machine learning approaches have grown at a breakneck pace. These advancements are expected to continue to provide more cost-effective and comprehensive datasets, as well as more advanced algorithmic solutions, allowing for better crop and environment status estimates and decision making. For more intricate hyperspectral picture classification, existing theories and algorithms still have some limitations. As a result, future research efforts will focus on developing more tailored hyperspectral image classification systems. Therefore, in order to successfully use the information on weeds and crop monitoring for economic benefit, a state or district level information system based on existing information on diverse crops produced from this hyperspectral remote sensing approach is required. Governments can use hyperspectral remote sensing data to make critical decisions about which policies to pursue and how to address agricultural concerns.

Author Contributions: Conceptualization, N.S. and N.N.C.; methodology, N.S. and M.H.M.R.; resources, N.S.; writing—original draft preparation, N.S.; writing—review and editing, N.S., N.N.C., A.S.J., N.M.N., W.F.F.I. and M.H.M.R.; visualization, N.S. and N.N.C.; supervision, N.N.C., N.M.N., M.H.M.R. and A.S.J.; project administration, A.S.J.; funding acquisition, A.S.J. All authors have read and agreed to the published version of the manuscript.

Funding: This research was funded by Research Project entitled "Pest and Disease Monitoring Using Artificial Intelligent for Risk Management of Rice Under Climate Change" under the Long-term Research Grant Scheme (LRGS), Ministry of Higher Education, Malaysia, LRGS/1/2019/UPM/01/2/5 (vote number: 5545002).

Institutional Review Board Statement: Not applicable.

Informed Consent Statement: Not applicable.

Data Availability Statement: Not applicable.

Acknowledgments: The authors would like to thankfully acknowledge the research project "Pest and Disease Monitoring Using Artificial Intelligent for Risk Management of Rice Under Climate Change" under Long-term Research Grant Scheme (LRGS), Ministry of Higher Education Malaysia for providing financial support, LRGS/1/2019/UPM/01/2/5 (Vote: 5545002).

Conflicts of Interest: The authors declare no conflict of interest.

References

1. Afzal, N.; Ahmad, S. Agricultural input use efficiency in Pakistan: Key issues and reform areas. *Manag. Nat. Resour. Sustain. Future Agric. Res. Brief.* **2009**, *1*, 1–12.
2. Ali, H.H.; Peerzada, A.M.; Hanif, Z.; Hashim, S.; Chauhan, B.S. Weed management using crop competition in Pakistan: A review. *Crop Prot.* **2017**, *95*, 22–30. [CrossRef]
3. Hassan, A.A.G.; Ngah, I.; Applanaidu, S.D. Agricultural Transformation in Malaysia: The Role of Smallholders And Area Development. *Development* **2018**, *15*, 2.
4. Dilipkumar, M.; Chuah, T.S.; Goh, S.S.; Sahid, I. Weed management issues, challenges, and opportunities in Malaysia. *Crop Prot.* **2020**, *134*, 104347. [CrossRef]
5. Chauhan, B.; Johnson, D.E. Row spacing and weed control timing affect yield of aerobic rice. *Field Crop. Res.* **2011**, *121*, 226–231. [CrossRef]
6. Kang, Y.; Khan, S.; Ma, X. Climate change impacts on crop yield, crop water productivity and food security–A review. *Prog. Nat. Sci.* **2009**, *19*, 1665–1674. [CrossRef]
7. Khan, M.A.; Marwat, K.B.; Umm-e-Kalsoom Hussain, Z.; Hashim, S.; Rab, A.; Nawab, K. Weed control effects on the wheat-pea intercropping. *Pak. J. Bot.* **2013**, *45*, 1743–1748.
8. Hussain, S.; Khaliq, A.; Matloob, A.; Fahad, S.; Tanveer, A. Interference and economic threshold level of little seed canary grass in wheat under different sowing times. *Environ. Sci. Pollut. Res.* **2014**, *22*, 441–449. [CrossRef] [PubMed]
9. El Pebrian, D.; Ismail, M.I. The characteristics of the repair and maintenance costs distribution of rice combine harvester in Malaysian paddy fields. *Agric. Eng. Int. CIGR J.* **2019**, *20*, 132–138.
10. Yusof, Z.M.; Misiran, M.; Baharin, N.F.; Yaacob, M.F.; Aziz, N.A.B.A.; Sanan, N.H.B. Projection of paddy production in Kedah Malaysia: A case study. *Asian J. Adv. Agric. Res.* **2019**, *3*, 1–6. [CrossRef]
11. Karim, R.S.; Man, A.B.; Sahid, I.B. Weed problems and their management in rice fields of Malaysia: An overview. *Weed Biol. Manag.* **2004**, *4*, 177–186. [CrossRef]
12. Chauhan, B.S. *Weed Management in Directs Seeded Rice Systems*; International Rice Research Institute: Los Banos, Philippines, 2012; p. 20.
13. Hossain, K.; Timsina, J.; Johnson, D.E.; Gathala, M.K.; Krupnik, T.J. Multi-year weed community dynamics and rice yields as influenced by tillage, crop establishment, and weed control: Implications for rice-maize rotations in the eastern Gangetic plains. *Crop. Prot.* **2020**, *138*, 105334. [CrossRef]
14. Singh, M.; Nagargade, M.; Tyagi, V. Ecologically sustainable integrated weed management in dry and irrigated direct-seeded rice. *Adv. Plants Agric. Res.* **2018**, *8*, 319–331. [CrossRef]
15. Moody, K.; De Datta, S.K. Integration of weed control practices for rice in tropical Asia. In Proceedings of the Workshop on Weed Control in Small Farms, Jakarta, Indonesia, 15–16 July 1977; pp. 37–47.
16. Shekhawat, K.; Rathore, S.S.; Chauhan, B.S. Weed management in dry direct-seeded rice: A review on challenges and opportunities for sustainable rice production. *Agronomy* **2020**, *10*, 1264. [CrossRef]
17. Toriyama, K. (Ed.) *Rice is Life Scientific Perspectives for the 21st Century*; International Rice Research Institute: Los Banos, Philipines; Japan International Research Center for Agricultural Sciences: Tsukuba, Japan, 2005.
18. Dadashzadeh, M.; Abbaspour-Gilandeh, Y.; Mesri-Gundoshmian, T.; Sabzi, S.; Hernández-Hernández, J.L.; Hernández-Hernández, M.; Arribas, J.I. Weed Classification for Site-Specific Weed Management Using an Automated Stereo Computer-Vision Machine-Learning System in Rice Fields. *Plants* **2020**, *9*, 559. [CrossRef] [PubMed]
19. Paap, A.J. Development of an Optical Sensor for Real-Time Weed Detection Using Laser Based Spectroscopy. Ph.D. Dissertation, Edith Cowan University, Perth, Australia. Available online: https://ro.ecu.edu.au/cgi/viewcontent.cgi?referer=https://scholar.google.com/&httpsredir=1&article=2284&context=theses (accessed on 15 May 2021).
20. Ali, H.H.; Tanveer, A.; Naeem, M.; Jamil, M.; Iqbal, M.; Javaid, M.M.; Kashif, M.S. Efficacy of pre-emergence herbicides in controlling Rhynchosia capitata, an emerging summer weed in Pakistan. *Philipp. Agric. Sci.* **2015**, *98*, 301–311.
21. Kandhro, M.N.; Tunio, S.; Rajpar, I.; Chachar, Q. Allelopathic impact of sorghum and sunflower intercropping on weed management and yield enhancement in cotton. *Sarhad J. Agric.* **2014**, *30*, 311–318.
22. Partel, V.; Kakarla, S.C.; Ampatzidis, Y. Development and evaluation of a low-cost and smart technology for precision weed management utilizing artificial intelligence. *Comput. Electron. Agric.* **2019**, *157*, 339–350. [CrossRef]

23. Huang, Y.; Lee, M.A.; Thomson, S.J.; Reddy, K.N. Ground-based hyperspectral remote sensing for weed management in crop production. *Int. J. Agric. Biol.Eng.* **2016**, *9*, 98–109.
24. Yao, H.; Huang, Y. Remote sensing applications to precision farming. *Remote Sens. Nat. Resour.* **2013**, 358–377.
25. Hunter, J.E.; Gannon, T.W.; Richardson, R.J.; Yelverton, F.H.; Leon, R.G. Integration of remote-weed mapping and an autonomous spraying unmanned aerial vehicle for site-specific weed management. *Pest Manag. Sci.* **2019**, *76*, 1386–1392. [CrossRef] [PubMed]
26. Weiss, M.; Jacob, F.; Duveiller, G. Remote sensing for agricultural applications: A meta-review. *Remote Sens. Environ.* **2020**, *236*, 111402. [CrossRef]
27. Campbell, J.B.; Wynne, R.H. *Introduction to Remote Sensing*; Guilford Press: New York, NY, USA, 2011.
28. Bioucas-Dias, J.M.; Plaza, A.; Camps-Valls, G.; Scheunders, P.; Nasrabadi, N.M.; Chanussot, J. Hyperspectral remote sensing data analysis and future challenges. *IEEE Geosci. Remote Sens. Mag.* **2013**, *1*, 6–36. [CrossRef]
29. Camps-Valls, G.; Tuia, D.; Gómez-Chova, L.; Jiménez, S.; Malo, J. Remote sensing image processing. *Synth. Lect. Image Video Multimed. Process.* **2011**, *5*, 1–192. [CrossRef]
30. Qian, S.-E. *Optical Satellite Signal Processing and Enhancement*; SPIE Press: Bellingham, DC, USA; Cardiff, Wales, 2013.
31. Qian, S.E. *Hyperspectral Satellites and System Design*; CRC Press: Boca Raton, FL, USA, 2020.
32. Govender, M.; Chetty, K.; Naiken, V.; Bulcock, H. A comparison of satellite hyperspectral and multispectral remote sensing imagery for improved classification and mapping of vegetation. *Water SA* **2019**, *34*, 147. [CrossRef]
33. Wendel, A. *Hyperspectral Imaging from Ground Based Mobile Platforms and Applications in Precision Agriculture*; School of Aerospace, Mechanical and Mechatronic Engineering, The University of Sydney: Sydney, Australia, 2018.
34. Chen, Y.; Guerschman, J.; Cheng, Z.; Guo, L. Remote sensing for vegetation monitoring in carbon capture storage regions: A review. *Appl. Energy* **2019**, *240*, 312–326. [CrossRef]
35. Shaw, G.A.; Burke, H.K. Spectral imaging for remote sensing. *Linc. Lab. J.* **2003**, *14*, 3–28.
36. Lu, B.; Dao, P.; Liu, J.; He, Y.; Shang, J. Recent Advances of Hyperspectral Imaging Technology and Applications in Agriculture. *Remote Sens.* **2020**, *12*, 2659. [CrossRef]
37. Kate, S.H.; Rocchini, D.; Neteler, M.; Nagendra, H. Benefits of hyperspectral remote sensing for tracking plant invasions. *Divers. Distrib.* **2011**, *17*, 381–392.
38. Vorovencii, I. The hyperspectral sensors used in satellite and aerial remote sensing. *Bull. Transilv. Univ. Brasov. For. Wood Ind. Agric. Food Engineering. Ser. II* **2009**, *2*, 51.
39. Lv, W.; Wang, X. Overview of Hyperspectral Image Classification. *J. Sens.* **2020**, *2020*, 4817234. [CrossRef]
40. Rosle, R.; Che'Ya, N.N.; Ang, Y.; Rahmat, F.; Wayayok, A.; Berahim, Z.; Omar, M.H. Weed Detection in Rice Fields Using Remote Sensing Technique: A Review. *Appl. Sci.* **2021**, *11*, 10701. [CrossRef]
41. Chang, C.I. *Hyperspectral Data Processing: Algorithm Design and Analysis*; WileyInterscience: Hoboken, NJ, USA, 2013.
42. Park, B.; Lu, R. (Eds.) *Hyperspectral Imaging Technology in Food and Agriculture*; Springer: New York, NY, USA, 2015.
43. Vidal, M.; Amigo, J.M. Pre-processing of hyperspectral images. Essential steps before image analysis. *Chemom. Intell. Lab. Syst.* **2012**, *117*, 138–148. [CrossRef]
44. Burger, J.; Geladi, P. Hyperspectral NIR image regression part I: Calibration and correction. *J. Chemom. A J. Chemom. Soc.* **2005**, *19*, 355–363. [CrossRef]
45. Ahmad, M.; Shabbir, S.; Raza, R.A.; Mazzara, M.; Distefano, S.; Khan, A.M. Hyperspectral Image Classification: Artifacts of Dimension Reduction on Hybrid CNN. *arXiv* **2021**, arXiv:2101.10532, preprint.
46. Basantia, N.C.; Nollet, L.M.; Kamruzzaman, M. (Eds.) *Hyperspectral Imaging Analysis and Applications for Food Quality*; CRC Press: Boca Raton, FL, USA, 2018.
47. Tamilarasi, R.; Prabu, S. Application of Machine Learning Techniques for Hyperspectral Image Dimensionality: A Review. *J. Crit. Rev.* **2020**, *7*, 3499–3516.
48. Sawant, S.S.; Prabukumar, M. Semi-supervised techniques based hyper-spectral image classification: A survey. In Proceedings of the 2017 Innovations in Power and Advanced Computing Technologies (i-PACT), Vellore, India, 21–22 April 2017; pp. 1–8.
49. Freitas, S.; Almeida, C.; Silva, H.; Almeida, J.; Silva, E. Supervised classification for hyperspectral imaging in UAV maritime target detection. In Proceedings of the 2018 IEEE International Conference on Autonomous Robot Systems and Competitions (ICARSC), Torres Vedras, Portugal, 25–27 April 2018; pp. 84–90.
50. Romaszewski, M.; Głomb, P.; Cholewa, M. Semi-supervised hyperspectral classification from a small number of training samples using a co-training approach. *ISPRS J. Photogramm. Remote Sens.* **2016**, *121*, 60–76. [CrossRef]
51. Su, W.H. Advanced Machine Learning in Point Spectroscopy, RGB-and hyperspectral-imaging for automatic discriminations of crops and weeds: A review. *Smart Cities* **2020**, *3*, 767–792. [CrossRef]
52. Sarvini, T.; Sneha, T.; Sukanya Gowthami, S.G.; Sushmitha, S.; Kumaraswamy, R. Performance comparison of weed detection algorithms. In Proceedings of the 2019 International Conference on Communication and Signal Processing (ICCSP), Melmaruvathur, India, 4–6 April 2019; pp. 0843–0847.
53. Li, Y.; Al-Sarayreh, M.; Irie, K.; Hackell, D.; Bourdot, G.; Reis, M.M.; Ghamkhar, K. Identification of Weeds Based on Hyperspectral Imaging and Machine Learning. *Front. Plant Sci.* **2021**, *11*, 2324. [CrossRef]
54. Atzberger, C. Advances in Remote Sensing of Agriculture: Context Description, Existing Operational Monitoring Systems and Major Information Needs. *Remote Sens.* **2013**, *5*, 949–981. [CrossRef]

55. Pandey, P.C.; Balzter, H.; Srivastava, P.K.; Petropoulos, G.P.; Bhattacharya, B. Future perspectives and challenges in hyperspectral remote sensing. In *Hyperspectral Remote Sensing*; Elsevier: Amsterdam, The Netherlands, 2020; pp. 429–439.
56. Feng, Y.-Z.; Sun, D.-W. Application of Hyperspectral Imaging in Food Safety Inspection and Control: A Review. *Crit. Rev. Food Sci. Nutr.* **2012**, *52*, 1039–1058. [CrossRef] [PubMed]
57. Singh, V.; Rana, A.; Bishop, M.; Filippi, A.M.; Cope, D.; Rajan, N.; Bagavathiannan, M. Unmanned aircraft systems for precision weed detection and management: Prospects and challenges. *Adv. Agron.* **2020**, *159*, 93–134.
58. Pott, L.P.; Amado, T.J.; Schwalbert, R.A.; Sebem, E.; Jugulam, M.; Ciampitti, I.A. Pre-planting weed detection based on ground field spectral data. *Pest Manag. Sci.* **2020**, *76*, 1173–1182. [CrossRef] [PubMed]
59. Nursyazyla, S.; Syarifah, N.I.S.S.A.; Zaid, R.; Nik, N.C.; Muhammad, H.M.R. A Review on Hyperspectral Remote Sensing for Weed Detection Analysis in Agricultural Crop. Southeast Asian Agricultural Engineering Student Chapter Annual Regional Convention (2021: Sarawak) E-PROCEEDING. 2021. Available online: https://btu.upm.edu.my/upload/dokumen/20211105082434E-Proceeding_ARC2021_eISBN.pdf (accessed on 19 January 2022).
60. Bradter, U.; O'Connell, J.; Kunin, W.E.; Boffey, C.W.; Ellis, R.J.; Benton, T.G. Classifying grass-dominated habitats from remotely sensed data: The influence of spectral resolution, acquisition time and the vegetation classification system on accuracy and thematic resolution. *Sci. Total Environ.* **2019**, *711*, 134584. [CrossRef] [PubMed]
61. Che'Ya, N.; Dunwoody, E.; Gupta, M. Assessment of Weed Classification Using Hyperspectral Reflectance and Optimal Multispectral UAV Imagery. *Agronomy* **2021**, *11*, 1435. [CrossRef]
62. Zhang, G.; Xu, T.; Tian, Y.; Xu, H.; Song, J.; Lan, Y. Assessment of rice leaf blast severity using hyperspectral imaging during late vegetative growth. *Australas. Plant Pathol.* **2020**, *49*, 571–578. [CrossRef]
63. Thenkabail, P.S.; Lyon, J.G.; Huete, A. Hyperspectral remote sensing of vegetation and agricultural crops: Knowledge gain and knowledge gap after 40 years of research. In *Hyperspectral Remote Sensing of Vegetation*; Thenkabail, P.S., Lyon, J.G., Huete, A., Eds.; CRC Press: Boca Raton, FL, USA, 2010; p. 705.
64. Casa, R.; Pascucci, S.; Pignatti, S.; Palombo, A.; Nanni, U.; Harfouche, A.; Fantozzi, P. UAV-based hyperspectral imaging for weed discrimination in maize. In *Precision Agriculture'19*; Wageningen Academic Publishers: Wageningen, The Netherlands, 2019; pp. 24–35.
65. Martin, M.P.; Barreto, L.; Riano, D.; Fernandez-Quintanilla, C.; Vaughan, P. Assessing the potential of hyperspectral remote sensing for the discrimination of grassweeds in winter cereal crops. *Int. J. Remote. Sens.* **2011**, *32*, 49–67. [CrossRef]
66. Lass, L.W.; Prather, T.; Glenn, N.; Weber, K.; Mundt, J.T.; Pettingill, J. A review of remote sensing of invasive weeds and example of the early detection of spotted knapweed (Centaurea maculosa) and babysbreath (Gypsophila paniculata) with a hyperspectral sensor. *Weed Sci.* **2005**, *53*, 242–251. [CrossRef]
67. Scherrer, B.; Sheppard, J.; Jha, P.; Shaw, J.A. Hyperspectral imaging and neural networks to classify herbicide-resistant weeds. *J. Appl. Remote Sens.* **2019**, *13*, 044516. [CrossRef]
68. Zhang, Y.; Gao, J.; Cen, H.; Lu, Y.; Yu, X.; He, Y.; Pieters, J.G. Automated spectral feature extraction from hyperspectral images to differentiate weedy rice and barnyard grass from a rice crop. *Comput. Electron. Agric.* **2019**, *159*, 42–49. [CrossRef]
69. Chaoying, T.A.N.G.; Xianghui, W.E.I.; Biao, W.A.N.G.; Prasad, S. *A CrossBorder Detection Algorithm for Agricultural Spraying UAV*; American Society of Agricultural and Biological Engineers: St. Joseph, MI, USA, 2018.
70. Huang, H.; Deng, J.; Lan, Y.; Yang, A.; Zhang, L.; Wen, S.; Zhang, H.; Zhang, Y.; Deng, Y. Detection of Helminthosporium Leaf Blotch Disease Based on UAV Imagery. *Appl. Sci.* **2019**, *9*, 558. [CrossRef]
71. Felegari, S.; Sharifi, A.; Moravej, K.; Amin, M.; Golchin, A.; Muzirafuti, A.; Tariq, A.; Zhao, N. Integration of Sentinel 1 and Sentinel 2 Satellite Images for Crop Mapping. *Appl. Sci.* **2021**, *11*, 10104. [CrossRef]
72. Hennessy, A.; Clarke, K.; Lewis, M. Hyperspectral classification of plants: A review of waveband selection generalisability. *Remote Sens.* **2020**, *12*, 113. [CrossRef]
73. Shannon, D.K.; Clay, D.E.; Kitchen, N.R. *Precision Agriculture Basics*; John Wiley & Sons: Hoboken, NJ, USA, 2020; Volume 176.
74. Lan, Y.; Huang, Z.; Deng, X.; Zhu, Z.; Huang, H.; Zheng, Z.; Lian, B.; Zeng, G.; Tong, Z. Comparison of machine learning methods for citrus greening detection on UAV multispectral images. *Comput. Electron. Agric.* **2020**, *171*, 105234. [CrossRef]
75. Nguyen, M.L.; Ciesielski, V.; Song, A. Rice leaf detection with genetic programming. In Proceedings of the 2013 IEEE Congress on Evolutionary Computation, Cancun, Mexico, 20–23 June 2013; pp. 1146–1153.
76. Pérez-Ortiz, M.; Gutiérrez, P.A.; Peña, J.M.; Torres-Sánchez, J.; López-Granados, F.; Hervás-Martínez, C. Machine learning paradigms for weed mapping via unmanned aerial vehicles. In Proceedings of the 2016 IEEE Symposium Series on Computational Intelligence (SSCI), Athens, Greece, 6–9 December 2016; pp. 1–8.
77. Liakos, K.G.; Busato, P.; Moshou, D.; Pearson, S.; Bochtis, D. Machine learning in agriculture: A review. *Sensors* **2018**, *18*, 2674. [CrossRef] [PubMed]
78. Zheng, Y.; Zhu, Q.; Huang, M.; Guo, Y.; Qin, J. Maize and weed classification using color indices with support vector data description in outdoor fields. *Comput. Electron. Agric.* **2017**, *141*, 215–222. [CrossRef]
79. Kamath, R.; Balachandra, M.; Prabhu, S. Paddy crop and weed discrimination: A multiple classifier system approach. *Int. J. Agron.* **2020**, *2020*, 6474536. [CrossRef]
80. Mehra, L.; Cowger, C.; Gross, K.; Ojiambo, P.S. Predicting Pre-planting Risk of Stagonospora nodorum blotch in Winter Wheat Using Machine Learning Models. *Front. Plant Sci.* **2016**, *7*, 390. [CrossRef]

81. Chen, Y.; Wu, Z.; Zhao, B.; Fan, C.; Shi, S. Weed and Corn Seedling Detection in Field Based on Multi Feature Fusion and Support Vector Machine. *Sensors* **2020**, *21*, 212. [CrossRef]
82. Chou, J.J.; Chen, C.P.; Yeh, J.T. Crop identification with wavelet packet analysis and weighted Bayesian distance. *Comput. Electron. Agric.* **2007**, *57*, 88–98. [CrossRef]
83. Bakhshipour, A.; Zareiforoush, H. Development of a fuzzy model for differentiating peanut plant from broadleaf weeds using image features. *Plant Methods* **2020**, *16*, 153. [CrossRef]
84. Bakhshipour, A.; Jafari, A.; Nassiri, S.M.; Zare, D. Weed segmentation using texture features extracted from wavelet sub-images. *Biosyst. Eng.* **2017**, *157*, 1–12. [CrossRef]
85. Tang, L.; Tian, L.; Steward, B.L. Classification of broadleaf and grass weeds using Gabor wavelets and an artificial neural network. *Trans. ASAE* **2003**, *46*, 1247. [CrossRef]
86. Gu, J.; Wang, Z.; Kuen, J.; Ma, L.; Shahroudy, A.; Shuai, B.; Chen, T. Recent advances in convolutional neural networks. *Pattern Recognit.* **2018**, *77*, 354–377. [CrossRef]
87. Yu, J.; Sharpe, S.M.; Schumann, A.W.; Boyd, N.S. Deep learning for image-based weed detection in turfgrass. *Eur. J. Agron.* **2019**, *104*, 78–84. [CrossRef]
88. Alom, M.Z.; Taha, T.M.; Yakopcic, C.; Westberg, S.; Sidike, P.; Nasrin, M.S.; Hasan, M.; Van Essen, B.C.; Awwal, A.A.S.; Asari, V.K. A State-of-the-Art Survey on Deep Learning Theory and Architectures. *Electronics* **2019**, *8*, 292. [CrossRef]
89. Hasan, A.S.M.M.; Sohel, F.; Diepeveen, D.; Laga, H.; Jones, M.G. A survey of deep learning techniques for weed detection from images. *Comput. Electron. Agric.* **2021**, *184*, 106067. [CrossRef]
90. LeCun, Y.; Bengio, Y.; Hinton, G. Deep learning. *Nature* **2015**, *521*, 436–444. [CrossRef]
91. Hosseini, M.P.; Lu, S.; Kamaraj, K.; Slowikowski, A.; Venkatesh, H.C. Deep learning architectures. In *Deep Learning: Concepts and Architectures*; Springer: Cham, Switzerland, 2020; pp. 1–24.
92. Bah, M.D.; Hafiane, A.; Canals, R. Deep Learning with Unsupervised Data Labeling for Weed Detection in Line Crops in UAV Images. *Remote. Sens.* **2018**, *10*, 1690. [CrossRef]
93. Lee, S.H.; Chan, C.S.; Mayo, S.J.; Remagnino, P. How deep learning extracts and learns leaf features for plant classification. *Pattern Recognit.* **2017**, *71*, 1–13. [CrossRef]

applied sciences

MDPI

Review

Weed Detection in Rice Fields Using Remote Sensing Technique: A Review

Rhushalshafira Rosle [1], Nik Norasma Che'Ya [1,*], Yuhao Ang [2], Fariq Rahmat [3], Aimrun Wayayok [4], Zulkarami Berahim [5], Wan Fazilah Fazlil Ilahi [1], Mohd Razi Ismail [6] and Mohamad Husni Omar [5]

1. Department of Agriculture Technology, Faculty of Agriculture, Universiti Putra Malaysia (UPM), Serdang 43400, Malaysia; GS54090@student.upm.edu.my (R.R.); wanfazilah@upm.edu.my (W.F.F.I.)
2. Department of Civil Engineering, Faculty of Engineering, Universiti Putra Malaysia (UPM), Serdang 43400, Malaysia; vincentangkhun@gmail.com
3. Department of Electrical and Electronic Engineering, Faculty of Engineering, Universiti Putra Malaysia (UPM), Serdang 43400, Malaysia; fariqrahmat94@gmail.com
4. Department of Biological and Agricultural Engineering, Faculty of Engineering, Universiti Putra Malaysia (UPM), Serdang 43400, Malaysia; aimrun@upm.edu.my
5. Institute of Tropical Agriculture and Food Security (ITAFoS), Universiti Putra Malaysia (UPM), Serdang 43400, Malaysia; zulkerami@upm.edu.my (Z.B.); mohdhusni@upm.edu.my (M.H.O.)
6. Department of Crop Science, Faculty of Agriculture, Universiti Putra Malaysia (UPM), Serdang 43400, Malaysia; razi@upm.edu.my

* Correspondence: niknorasma@upm.edu.my

Abstract: This paper reviewed the weed problems in agriculture and how remote sensing techniques can detect weeds in rice fields. The comparison of weed detection between traditional practices and automated detection using remote sensing platforms is discussed. The ideal stage for controlling weeds in rice fields was highlighted, and the types of weeds usually found in paddy fields were listed. This paper will discuss weed detection using remote sensing techniques, and algorithms commonly used to differentiate them from crops are deliberated. However, weed detection in rice fields using remote sensing platforms is still in its early stages; weed detection in other crops is also discussed. Results show that machine learning (ML) and deep learning (DL) remote sensing techniques have successfully produced a high accuracy map for detecting weeds in crops using RS platforms. Therefore, this technology positively impacts weed management in many aspects, especially in terms of the economic perspective. The implementation of this technology into agricultural development could be extended further.

Keywords: invasive plants; precision agriculture; remote sensing; rice farming; site-specific weed management

Citation: Rosle, R.; Che'Ya, N.N.; Ang, Y.; Rahmat, F.; Wayayok, A.; Berahim, Z.; Fazlil Ilahi, W.F.; Ismail, M.R.; Omar, M.H. Weed Detection in Rice Fields Using Remote Sensing Technique: A Review. *Appl. Sci.* **2021**, *11*, 10701. https://doi.org/10.3390/app112210701

Academic Editors: Dimitrios S. Paraforos and Anselme Muzirafuti

Received: 10 September 2021
Accepted: 1 November 2021
Published: 12 November 2021

1. Introduction

It is undoubtful that weeds, also known as invasive plants, have their roles in the ecosystem. However, their presence in crops such as rice, oil palm, rubber, and other mass plantations influences productivity, causes significant economic consequences, decreases land prices, and reduces company profits [1]. Moreover, the current trend shows that farmers worldwide are strongly dependent on herbicides used to control weeds; other control measures include cultural, physical, biological, and mechanical methods [2].

A statistic released by the Food and Agriculture Organization of the United Nations (FAO) for the years 1990 to 2019 showed that the Asia continent had used approximately 805,412 tonnes of herbicides in controlling the presence of weeds in various types of crops, followed by the Americas (593,619 tonnes), Europe (179,799 tonnes), Oceania (29,309 tonnes), and Africa (21,117 tonnes) [3]. Thus, much money was spent on herbicides to control and manage the presence of weeds in crops. However, too much dependence on herbicides usage to control weeds to maximize yield production has caused herbicide resistance and reduced the choices of herbicides to use [4,5]. Figure 1 illustrates the herbicides usage in controlling weeds for each continent in percentage.

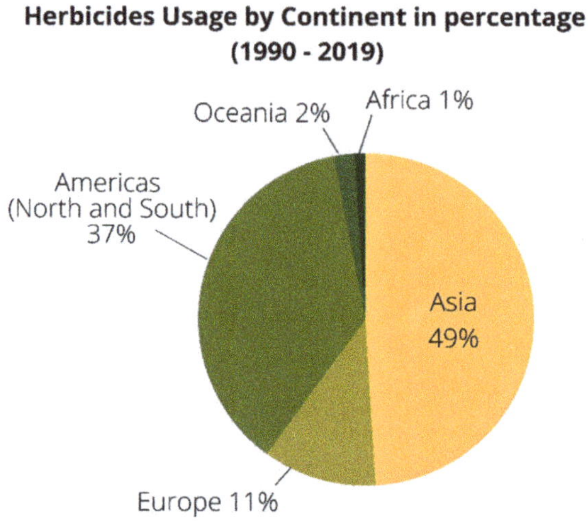

Figure 1. The herbicides usage in controlling weeds by continent from 1990 to 2019.

It is necessary to construct systematic and strategic planning to improve the precision agriculture (PA) sector, especially in weed management, to control and increase yield production, leading to a better economy for the country and farmers. Therefore, remote sensing-based techniques were used to construct and optimize weed management. Remote sensing is a comprehensive framework that monitors and captures earth surface images without direct contact with it. In PA sectors, the data gathered can be used in various applications, such as monitoring rice's morphology [6], yield estimation [7], and mapping irrigated areas for food security and water resource management [8]. However, even though remote sensing has been widely used in weed management, it may not be a permanently adopted by developing countries anytime soon since local farmers still prefer the traditional practices.

Thus, this paper aims to review and discuss the techniques and algorithms used in remote sensing to construct systematic and strategic planning to improve precision agriculture in weed management. As a result, researchers can adapt the knowledge of controlling weed presence and increasing yield production, especially in developing countries. This study's focus was limited to weed detection using a remote sensing platform in the paddy field. However, weed detection in other crops using remote sensing was also included.

This paper is organized into eight sections. Section 1 briefly explains this study's goal in implementing remote sensing techniques into the precision agriculture (PA) sector. Section 2 explains the strategy used to search through the scientific database for relevant publications. Meanwhile, Section 3 discusses the importance of rice and what has been carried out to increase yield. Section 4 highlights the best stage to control weed in paddy, weed types, and traditional farming practices. Section 5 presents the literature covering various types of weed detection using remote sensing techniques. Section 6 reviews the impact of inadequate and good weed management on crops, yield, and economy. Section 7 deliberates the future direction of remote sensing techniques in weed detection. Lastly, in Section 8, the conclusions are presented.

2. Methodology

Articles were searched and identified from nine bibliographic databases: IEEE, Science Direct, MDPI, Web of Science, Scopus, Google Scholar, ProQuest, Springer, and Wiley Online Library. The primary keyword 'remote sensing' and its synonyms were paired with the secondary keyword 'weed' and the third keyword 'detection' and its synonyms, with Boolean operators. These keyword sets were used in each database search. In addition, a hand search was also run to ensure no related articles were missed. The search was conducted in the quarter of 2021.

All search results were filtered based on five criteria: (1) the study must use remote sensing imagery and platform as the primary data input with at least three spectral bands (red, green, and blue), (2) the study must discuss the application of remote sensing techniques in weed detection, (3) the document must have reported the research conducted, (4) the included documents have been published up to the quarter of 2021, and (5) the articles must be in English.

Next, the articles were screened by title and abstract to eliminate articles that did not meet the stated criteria. Finally, the full text of the remaining articles was carefully reviewed to decide whether they met the criteria or not. Lastly, details from selected articles were extracted and compiled into one giant spreadsheet. The details include citation information, study objective, remote sensing sensor, crop and weed types, approaches and technique used, accuracy assessment, study's implications, year of publication, and reference data.

3. The Importance of Rice Productivity

Rice is consumed by around 3.5 billion people worldwide. However, the estimated demand by 2025 is mind-boggling, as rice consumption would grow higher than the population growth in major Asian countries [9]. In general, paddy production had increased globally up to 12% from 1975 to 2008, and nearly 166 million ha of paddy have been harvested in the world [10]. However, in 2020, it was reported that China was the leading country in the world producing paddy (30.5%), followed by India (224.14%), Bangladesh (7.36%), Indonesia (7.14%), Vietnam (5.53%), and Thailand (4.17%) [11].

Numerous research have been conducted to increase the yield of rice production to fulfil consumer demand. Masum et al. [12] had found that the Boterswar variety could help improve the weed-suppressing capacity of rice. The study used five Bangladesh rice varieties named Boterswar, Goria, Biron, Kartiksail, Hashikolmi, and Holoi, and these varieties were planted via a non-weed control method. By using Simpson's diversity ndex (SDI) to measure the infestation rate of weed species, the relative neighbour effect (RNE), and relative competitive intensity (RCI), results showed that Boterswar facilitated the crop–weed interaction compared to the other varieties. This finding will significantly influence methods to control the presence of weeds in paddy fields.

Meanwhile, Yamori et al. [13] found that, to increase plant productivity among various crop species, they must improve the photosynthesis rate at the single-leaf level. To achieve this, they used transgenic rice plants that consist of various amounts of the Rieske FeS protein in the cytochrome (cyt) b6/f complex at between 10 and 100% of wild-type levels. As a result, they decreased the electron transport rates through photosystem II, leading to an increased uptake of carbon dioxide (CO^2) and a successfully increased production yield, up to 40% [14].

Besides improving the photosynthetic activities, improving the irrigation system in paddy is the best practice to increase yield. In Thailand, they practiced an alternate wetting and drying (AWD) method [15]. By setting the threshold at 15 cm of water level below the soil surface for irrigation, this method increased the grain yield by 15% in the wet season and 7% in the dry season, meanwhile improving water usage by 46% and 77% in the wet and dry season, respectively, compared to continuously flooding water into the paddy field. Therefore, the AWD method is a good practice that helps sustain rice production through water-saving. Lahue et al. [16] also obtained the same result, and in addition, their study successfully reduced the total arsenic concentration released by rice grain up to 65%. Meanwhile, Liang et al. [17] managed to reduce the methane emissions into the atmosphere by up to 77.1%.

Climate conditions played a significant role. The potential rice yield will be affected by severe climate conditions due to increased sterility caused by heat and shortening of the growing season [18]. Van Oort [19] implemented a geographical information system (GIS) by producing a map of abiotic stresses in Africa using drought, cold, iron toxicity, and salinity or sodicity information as the input. From the analysis, drought was found as the most critical variable that contributes to stress, where 33% of rice area was potentially affected, followed by iron toxicity (12%), and then cold (7%) and salinity/sodicity (2%). Dossou-Yovo et al. [20] used socio-economic, biophysical, farmer population surveys, and secondary remote sensing data on soil characteristics and demand for water to determine drought input parameters in rice-based inland valley production systems. Their study shows that the average annual standardized precipitation evapotranspiration index and groundwater availability duration were the most critical input to determine drought occurrence in their study area.

It is crucial to find solutions to improve and increase rice yield. However, to achieve rice production sustainably and meet demands, productivity and quality must significantly improve. Therefore, through participatory approaches, it is critical to foster joint working between research, extension, local governments, non-governmental organizations (NGO), and private industry to identify the relevant constraints to high yield, adopt new solutions and technologies, and make systematic decisions to close rice yield gaps.

4. Controlling Weed in Paddy Fields at Different Growth of Stages

In general, rice growth periods can be identified in three stages. They are the vegetative stage, reproductive stage, and maturative or ripening stage [21]. Depending on agricultural and environmental conditions, the whole cycle takes about 120 to 125 days. The International Rice Research Institute (IRRI) splits the growth cycle into five stages [22]. A general idea of the growth cycle is presented in Figure 2, with morphology examples.

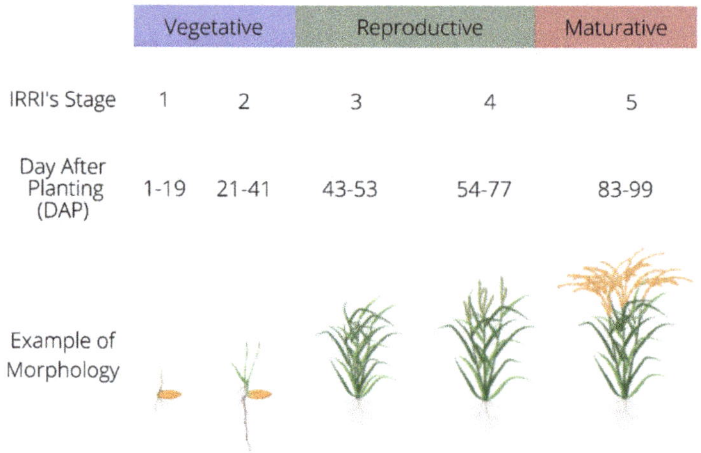

Figure 2. The growth cycle of a rice plant corresponds to the IRRI scale and sample structure.

Rice is generally a weak competitor with weeds. Therefore, the vegetative stage is critical in the paddy growth cycle. Successfully controlling weeds at this stage can deliver a 95% weed-free yield [23]. This is agreed with by Kamath et al. [24] because the effect of weeds in this stage will be at maximum. However, if we fail to prevent weeds from spreading in the vegetative stage, they will dominate the area, leading to a lack of sufficient space, light, and nutrients to grow and develop [25]. As a result, crops will experience uneven flowering and will not mature uniformly for the scheduled harvest [26,27].

Once the tillering reaches its maximum number, the reproductive stage will occur, followed by the maturative or ripening stage. Excess water in the fields is drained, resulting in a drop in the overall biomass due to lower moisture content. The grain is maturing and becoming heavier. At this stage, the presence of weeds will not affect the development of the crop. Nevertheless, we cannot save the yield losses because weeds dominated the paddy plot and the number of paddy crops that survive the competition is nominal. In general, weed in paddy can be classified into three types. They are grasses, sedges and broad leaved weeds [28], and Table 1 shows a compilation of the primary weeds usually found in paddy fields.

Table 1. Type of weeds commonly found in the paddy field.

Family Name	Scientific Name	Common Name
Grasses weeds		
Poaceae	*Oriza sativa complex*	Weedy rice
	Leptochloa chinensis (L.) Nees	Chinese sprangletop
	Chloris barbata Sw.	Swollen fingergrass
	Echinochloa crus-galli (L.) Beauv.	Barnyardgrass
	Echinochloa colana (L.) Link	Jungle rice
	Ischeamum rugosum Salisb	Ribbed murainagrass
	Brachiaria mutica (Forsk.) Stapf	Para grass, buffalo grass
	Cynodon dactylon (L.) Pers.	Bermuda grass
Sedge weeds		
Cyperaceae	*Fimbristylis miliacea* (L.) Vahl.	Fimbry
	Cyperus iria	Rice flat sedge
	Cyperus difformis	Small flower umbrella plant
	Cyperus rotundus	Nut grass, nut sedge
	Eleocharis dulcis (Burm.f) Henschel	Chinese water chestnut
	Fimbristylis globulosa (Retz.) Kunth	Globe fimbry
	Fuirena umbellate Rottb	Yefen, tropical umbrella sedge
	Scirpus grossus L.f.	Tukiu, giant bulrush
	Scirpus juncoides Roxb.	Club-rush, wood club-rush, bulrush
	Scirpus suspinus L.	-
Broad leaved weeds		
Butomaceae	*Limnocharis flava* (L.) Buchenau	Yellow velvet-leaf, sawah lettuce, sawah flower rush
Pontederiaceae	*Monochoria vaginalis* (Burm.f.) C.Presl	Pickerel weed, heartshape false pickerel weed
	Eichhornia crassipes (Mart.) Solms	Floating water-hyacinth
Alismataceae	*Sagittaria guayanensis* Kunth	Arrowhead, swamp potato
Onagraceae	*Ludwigia hyssopifolia* (G.Don) Exell	Seedbox, linear leaf water primrose
Sphenocleaceae	*Sphenoclea zeylanica* Gaertn	Goose weed, wedgewort
Convolvulaceae	*Ipomoea aquatica* Forsk	Kangkong, swamp morning glory, water spinach, swamp cabbage

The environmental relationship between weed and rice is very complicated and complex [29]. The weed management system needs improvement to control the spreading of weeds. The traditional practices that include burning, hand sowing, manual spot spraying, herbicide pre-emergence or post-emergence application, and repetitive blade hoeing are not practical anymore. These practices impacted the non-target species and the ecosystem rather than benefiting production [30].

The traditional weed sampling for practice-oriented management is too costly, and this is not a recent concern. Since 2005, Brown and Noble [31] have developed automated methods for evaluating infestation. Automatic weed sampling provides a way to increase the amount of data obtained in the field (smaller sampling intervals) at lower overall costs of 7–13 USD/ha, and sensor technology is used exclusively for the application of herbicides, resulting in a reduction of herbicide usage of 30–70% [32].

Advanced weed management methods are required to manage weeds effectively. The process may include targeted and site-specific weed control, selection of weed seeds, different herbicide application (depending on weed distribution, spatial arrangement and soil properties), destruction of weed seeds over predation and microbial loss, nano herbicides, and optical spraying techniques. Advanced vision-guided robotics that can be adopted for site-specific weed management (SSWM) are transgenic herbicide-resistant crops, weed control and spraying robots, decision support systems, and pattern recognition modelling [33]. Implementing these technologies will help prevent unwanted species and improve existing weed management systems [34].

5. Weed Detection Using Remote Sensing Technique

Remote sensing technology aims to monitor and capture the earth's information without making direct contact and destroying it. The utilization of the electromagnetic spectrum, ranging from visible to microwave for measuring the earth's properties, is the main idea behind remote sensing technology. Since the target's reactions to various wavelength regions differ, we can exploit them to identify vegetation, water, soil, and other features [35]. Combining the target's reaction with the shape, texture, and pattern information of weeds and crops, we can discriminate them and improve SSWM using remote sensing algorithms.

The image processing workflow to detect weed in paddies can generally be divided into five stages: image data collection, pre-processing, feature extraction and selection, training, image classification and validation [36].

5.1. Image Data Collection

There are multiple platforms available for data gathering for weed detection in crops, such as digital cameras [37], hand-held spectroradiometers [38], polarization spectroscopy [39], and satellites [40]. However, unmanned aerial vehicles (UAV) are the most popular platforms researchers use to identify weeds in crops, due to their availability, high-quality data delivery, and ease of handling [41]. Nevertheless, the data collection differs in the types of sensors attached to UAVs: RGB, multispectral, or hyperspectral.

5.1.1. RGB Sensor

The RGB sensor is the most widely used and widely available commercial camera. Because of their promise in delivering high-quality images and low-cost operational needs, their possible applications have been the focus of most research for many years [42,43]. These sensors are increasingly employed in machine learning algorithms for object recognition, diseases, phenology, and other applications.

These are typical steps to acquire RGB images captured by UAV remote sensing: (1) pre-flight planning, (2) flight and image acquisition and (3) post-processing and indices or dataset extrapolation [44]. However, when preparing the images for machine learning algorithms, the processing steps are different depending on the research's objective [45–47]. The advantage of using this sensor is that radiometric and atmospheric calibration are not required, unlike multispectral and hyperspectral images [41]. Therefore, noises from electromagnetic radiation (EMR) can be ignored.

5.1.2. Multispectral Sensor

The use of the multispectral sensors has become a trend nowadays because it has more than three (RGB) bands installed. Compared to RGB sensors, several vegetation indices that can be investigated are significantly expended. Nevertheless, to obtain accurate indices, radiometric and atmospheric calibration are compulsory. Moreover, unlike RGB sensors, the multispectral sensor is unable to deliver a high-quality spectral resolution image. This drawback can be overcome by using a lower flying height and acceptable percentage of horizontal and vertical image overlap [48].

In general, the typical steps involved in preparing multispectral images captured by UAV remote sensing are: (1) radiometric and atmospheric calibration, (2) locating and avoiding input and output (I/O) errors, missing data, and mission failure, and (3) image rectification, georeferencing, and stacking [41]. In addition, these sensors are increasingly being employed in machine learning algorithms for site-specific weed management (SSWM) [40,49,50].

5.1.3. Hyperspectral Sensor

The hyperspectral sensor analyzes a broad spectrum of light, instead of assigning primary colors (red, green, and blue). These sensors can record hundreds of narrow radiometric spectral bands from visible to infrared, sometimes up to microwave ranges. Its ability in providing narrow radiometric spectral bands can detect specific field concerns. Thus, users can compute narrowband indices, such as the chlorophyll absorption ratio index (CARI), transformed chlorophyll absorption ratio index (TCARI), triangular vegetation index (TVI), and photochemical reflectance index (PRI) [51].

Preparing hyperspectral data is more complicated than RGB and multispectral sensors because its radiometric and atmospheric calibration workflows are more complex. Sensor calibration approaches are generated from the UAV's hyperspectral platforms, which use simulated targets to check data quality, correct radiance, and provide high-quality reflectance information [52]. Therefore, typical steps in acquiring and preparing hyperspectral data captured by UAV remote sensing are: (1) setting up a flight plan, (2) image size and data storage, and (3) quality assessment [41]. Table 2 summarizes the characteristics of each sensor alongside its advantages and disadvantages.

Table 2. Characteristic of RGB, multispectral, and hyperspectral sensors.

Sensors/Details	RGB	Multispectral	Hyperspectral
Resolution (Mpx)	16–42	1.2–3.2	0.0025–2.2
Spectral range (nm)	400–700	400–900	300–2500
Spectral bands	3	3–10	40–660
Weight (approx.) (kg)	0.5–1.5	0.18–0.7	0.032–5
Price (approx.) (USD)	950–1780	3560–20,160	47,434–59,293
Advantages	High-quality images Low-cost operational needs No need for radiometric and atmospheric calibration	Have more than three bands Can generates more vegetation indices than RGB	Hundreds of narrow radiometric bands Can calculate narrowband indices that can target specific concerns.
Disadvantages	Only have three bands A limited number of vegetation indices can be computed	Radiometric and atmospheric calibration is compulsory Unable to deliver a high-quality resolution image	Expensive, heavier, and more extensive compared to the other sensors Complicated system Complex radiometric and atmospheric calibration Unable to deliver a high-quality resolution image

5.2. Image Mosaicking and Calibration

Images acquired from UAVs can be mosaicked using a Pix4D mapper (Pix4D, Prilly, Switzerland), Agisoft Photoscan Pro (Agisoft LLC, 52 St. Petersburg, Russia), and any available commercial software to generate qualitative, high-resolution orthomosaic images. After mosaicking, the process will continue with radiometric calibration and rescale the intensity of the electromagnetic radiation or digital number (DN) into the percentage of reflectance values [53]. Researchers have implemented numerous methods, such as the traditional empirical line correction approach and modern automatic radiometric calibration using available commercial software.

The empirical line correction approach is an atmospheric correction technique that provides a straightforward surface reflectance calibration method, if a set of invariants in the time calibration target measurement is provided. Kelcey and Lucieer [54] implemented this approach to improve six multispectral UAV data quality bands for quantitative data analysis. Similarly, Mafanya et al. [55] applied the same method and obtained a reflectance value of r = 0.997 ($p \leq 0.01$) with an overall root mean square of 0.63. Nevertheless, when dealing with high-quality data, the performance and accuracy must be re-evaluated [56].

In order to improve radiometric calibration accuracy, Xu et al. [57] introduced a spectral angle correction approach, where their method uses all information in each spectral band. Compared to the empirical line correction approach, they successfully improved the mean relative percent error (MRPE) range up to 3% in the visible band and 1% in the near-infrared (NIR) band. This finding will highly benefit the agriculture remote sensing field.

However, the user can also run the radiometric calibration automatically using available commercial software such as Agisoft Photoscan Pro (Agisoft LLC, 52 St. Petersburg, Russia) and Pix4D mapper (Pix4D, Lausanne, Switzerland). The 'reflectance map' tool in Pix4D mapper software is also similar to calibrate 'calibrate reflectance' in Agisoft Photoscan Pro that employs multiple image attributes to determine surface reflectance [58]. In addition, these software packages provide 'color correction/balancing' functions to develop the image information based on a radiometric block correction algorithm. However, the algorithms used in these packages only calculate the homogeneity of the neighbouring image's histogram homogeneity, not the bidirectional reflectance distribution function (BRDF) effect in a single image [59].

5.3. Feature Extraction and Selection

Following the spectral calibration, feature extraction can be extracted or computed for different image processing purposes using various approaches (Table 3). This process will be helpful for the classification and identification of weeds in paddy fields. Feature extraction techniques are beneficial, especially in shape and pattern recognition. As features define the behavior of an image, they show its place in terms of storage taken, classification efficiency, and, obviously, in time consumption [60]. Therefore, optimizing the feature subset is required before feeding it into the machine learning (ML) and deep learning (DL) algorithms for improving the classification process and making it cost and time-efficient [61].

Table 3. An example of features extracted or computed for image classification.

Categories	Feature	Description/Formula	Reference
Vegetation indices	Normalized vegetation index (NDVI) Excess green index (ExG)	(NIR − R)/(NIR + R) $2 \times \frac{G-R-B}{R+G+B}$	[62,63]
Color space transformed features	Hue Saturation Value	A gradation or variety of a color Depth, purity, or shades of the color Brightness intensity of the color tone	[64]
Wavelet transformed coefficients	Wavelet coefficient mean Wavelet coefficient standard deviation	Mean value calculated for a pixel using discrete wavelet transformation Standard deviation calculated for a pixel using discrete wavelet transformation	[65]
Principal components (1)	Principal component 1	Principal component analysis-derived component accounting maximum amount of variance	[66]

5.4. Image Classification and Validation

Many machine learning (ML) and deep learning (ML) algorithms are available for image classification. However, choosing the best one that fits the research's objective is crucial, because different algorithms have different difficulty levels. Therefore, Section 5.6 will further discuss the application of remote sensing algorithms in detecting weeds in crops.

Accuracy assessment is crucial to validate the quality of the classification output that best represents the study area. Overall, the assessment can be carried out by comparing the classified pixels with ground truth pixels using a confusion matrix [67]. The result for weed classification is presented in terms of producer accuracy and overall accuracy. Producer accuracy (Equation (1)) is the probability that a pixel in the classification correctly shows class X. Given the ground truth class is X, producer accuracy can be calculated using

$$Producer\ accuracy = \frac{c_{aa}}{c_{\cdot a}} \times 100\% \tag{1}$$

where:

- c_{aa} = element at a position ath row and ath column.
- $c_{\cdot a}$ = column sums.

Overall accuracy (Equation (2)) is the total percentage of pixels correctly classified, and it can be calculated by using

$$Overall\ accuracy = \frac{\sum_{a=1}^{U} c_{aa}}{Q} \times 100\% \tag{2}$$

where:

- Q = total number of pixels.
- U = total number of classes.

The agreement between variables with ground truth data can be represented by using the kappa coefficient (Equation (3)), and its value can be calculated by using

$$Kappa\ coefficient,\ K = \frac{\sum_{a=1}^{U} \frac{c_{aa}}{Q} - \sum_{a=1}^{U} \frac{c_{a}\cdot c_{a}}{Q^2}}{1 - \sum_{a=1}^{U} \frac{c_{a}\cdot c_{a}}{Q^2}} \times 100\% \tag{3}$$

where:

- c_a = row sums.

However, some limitations occur when dealing with object-based classification, primarily related to the real-world object recognition's thematic and geometrical accuracy [68]. Therefore, to address this concern, De Castro et al. [46] designed Weed detection Accuracy

(WdA), Equation (4). This index analyzes the spatial placement of classified weeds by using the intersection of shapefiles as a spatial relationship rather than the overall overlap.

$$WdA\ (\%) = \frac{Area\ of\ Observed\ Weed\ objects\ intersecting\ Detected\ Weed\ Objects}{Area\ of\ Observed\ Weed} \times 100 \quad (4)$$

The detection of weeds is crucial for successful site-specific weed management (SSWM). However, weed detection is still challenging for automatic weed removal [37]. In addition, low tolerance between the cutting point and the crop location requires an accurate weed classification against the main crop. Therefore, several works have been conducted in the context of remote sensing image processing to detect and improve site-specific management [69–71].

5.5. An Overview of Machine Learning in Agriculture

In recent years, machine learning (ML) has provided a new criterion for agriculture with big data technology and high-performance computing. The development of ML has created new opportunities in agriculture operational management to unravel, measure, and analyze complex data [72]. Generally, the ML framework involves learning from 'experience', known as training data, to execute the classification, regression, or clustering tasks. These training data are usually regarded as a feature described by a set of attributes or variables. The machine learning model works by predicting the pattern and trend of future events in crop monitoring and assessment [73]. The ML model's performance in a particular task is evaluated by performance metrics improved by experience over time. As a result, classification techniques have been a prominent research trend in machine learning for many years, informing various studies. This method seeks to create features from the input data. Furthermore, it is highly field-specific and requires significant human effort, leading to deep learning techniques [36]. Figure 3 shows how machine learning and deep learning techniques work.

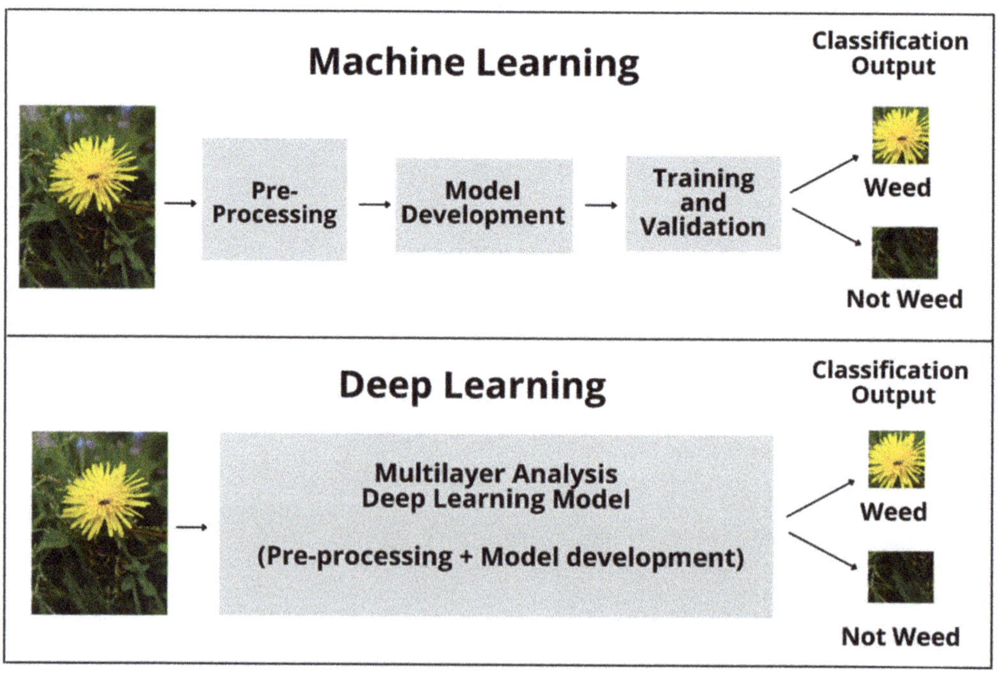

Figure 3. The differences in how deep learning and machine learning techniques work.

Deep learning is a subset of machine learning, but with more complicated image analysis [36], commonly used in agricultural crop monitoring and management. In terms of functionality, machine learning and deep learning share the same purpose: to make intuitive and intelligent decisions using artificial neural networks stacked layer-wise based on what it has learned while being trained [74]. However, in terms of developing an accurate model, machine learning requires a pre-processing stage before the model is developed, trained, and validated. In contrast, deep learning has a 'build in' feature extractor to extract meaningful features from the raw data. It learns features layer by layer, which means that it learns low-level features in the first levels and then progresses up the hierarchy to learn a more abstract representation of the input. [75]. Regardless of which agricultural domain and purpose, it has taken a directive in various crop monitoring purposes such as nutrient disorder, weed detection, plant insects, and disease detection. Many studies on weed detection have utilized deep learning with other remote sensing methods concerning classification or regression performance differences. The outcome has marked high accuracy, outperforming other commonly used image processing techniques [76].

In deep learning (DL), CNN is the most well-known and widely used algorithm [69,70,77]. The fundamental advantage of CNN over the other DL algorithms is that it automatically detects significant elements without the need for human assistance [36]. Comparable to the multi-layer perceptron (MLP), where it consists of three layers known as the input, output, and hidden layer [78], CNN has many convolution layers before sub-sampling (pooling) layers, with fully connected (FC) layers as the last layers. An illustration of the CNN framework for image classification is shown in Figure 4.

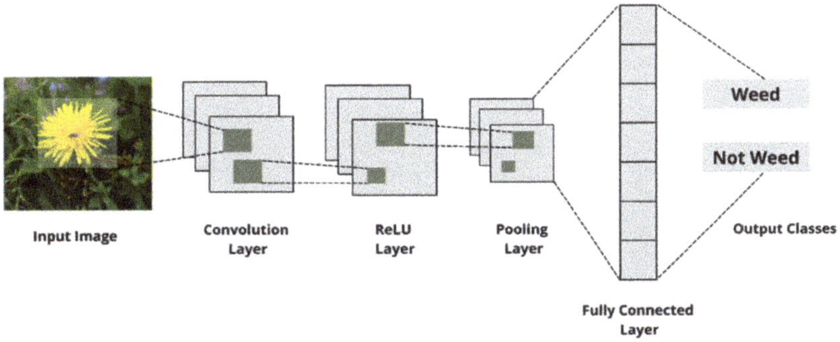

Figure 4. An illustration of the CNN framework for image classification.

A CNN model's input image is structured in three dimensions: height (m), width (m), and depth (r), where height (m) equals the width (m), and the depth (r) is referred to as channel number. For example, the depth (r) of the RGB image in Figure 4 equals three (three bands). The available kernel filters for the convolution layer will be designated by the letter k (n × n × q). However, n must be less than m, and q must be equal to or less than r. The dot product between the input and the weights is calculated by the convolution layer using Equation (5)

$$h^k = f\left(W^k * x + b^k\right) \quad (5)$$

where:

- h^k = feature maps in size (m − n − 1).
- W^k = weightage.
- b^k = bias.

These groundbreaking CNNs were able to achieve such incredible accuracy, partly because of their non-linearity. The rectified linear activation function (ReLU) applies the much-needed nonlinearity to the model. Non-linearity is necessary to produce a non-linear decision boundary, so the output cannot be written as a linear combination of the inputs. If there is no non-linear activation function, the deep CNN architecture will evolve into a single equivalent convolutional layer, and its performance will hardly be so. The ReLU activation function is used explicitly as a non-linear activation function, in contrast to other non-linear functions such as Sigmoid, because it has been observed from experience that the CNN using ReLU trains faster than the corresponding CNN [79]. Furthermore, the ReLU activation function is a one-to-one mathematical operation, as shown in Equation (6).

$$ReLU(x) = max(0, x) \tag{6}$$

It converts the whole values of the input to positive numbers. Thus, lower computational load is the main benefit of *ReLU* over the others. Subsequently, each feature map in the sub-sampling layers is down-sampled, decreasing network parameters, speeding up the learning process, and overcoming the problem related to the overfitting issue. This can be carried out in the pooling layers. The pooling operation (maximum or average) requires selecting a kernel size p × p (p = kernel size) and another two hyperparameters, padding and striding, during architecture design. For example, if max-pooling is used, the operation slides the kernel with the specified stride over the input, while only selecting the most significant value at each kernel slice from the input to yield a value for the output [80].

Padding is an important parameter when the kernel extends beyond the activation map. Padding can save data at the boundary of the activation maps, thereby improving performance, and it can help preserve the size of the input space, allowing architects to build simpler higher-performance networks, while stride indicates how many pixels the kernel should be shifted over at a time. The impact that stride has on a CNN is similar to kernel size. As stride is decreased, more features are learned because more data are extracted [36]. Finally, the fully connected (FC) layers receive the medium and low-level features and generate the high-level generalization, representing the last-stage layers similar to the typical neural network's technique. In other words, it converts a three-dimensional layer into a one-dimensional vector to fit the input of a fully connected layer for classification. Usually, this layer is fitted with a differentiable score function, such as softmax, to provide classification scores. The fundamental purpose of this function is to make sure the CNN outputs the sum to one. Thus, softmax operations are helpful to scale the model output into probabilities [80].

The key benefit of the DL technique is the ability to collect data or generate a data output using prior information. However, the downside of this strategy is that, when the training set lacks samples in a class, the decision boundary may be overstrained. Furthermore, given that it also involves a learning algorithm, DL consumes many data. Nevertheless, DL requires enormous data to build a well-behaved performance model, and as the data grow, the well-behaved performance model can be achieved [36].

5.6. The Application of Remote Sensing and Machine Learning Technique into Weed Detection

Choosing remote sensing (RS) and machine learning algorithms for SSWM can improve precision agriculture (PA). This situation has resulted in integrating remote sensing and machine learning becoming critical, as the need for RGB, multispectral, and hyperspectral processing systems has developed. Numerous researchers who tested the RS technique successfully produced an accurate weed map with promising implications for weed detection and management. Since the weed management using RS technique application in paddy is still in its early stage, Table 4 lists more studies on weed detection and mapping in various crops that apply remote sensing techniques with acceptable accuracy, for further reviews.

Table 4. Weed detection and mapping in various crops that apply remote sensing techniques.

Sensors	Crops	Weed Type	Technique	Accuracy (%)	Implications	Year	Reference
RGB*	Carrots: Autumn King	Grass and broad-leaved	Auto-associative neural network	>75%	Neural network-based allows the system to learn and discriminate between species without predefined plant descriptions	2003	[81]
Hyperspectral images: 72-waveband	Corn	Grass and broad-leaved	Support vector machine (SVM) vs artificial neural network (ANN)	66–76%	The SVM technique outperforms the ANN method	2006	[82]
Multispectral	Winter wheat	Cruciferous weeds	Maximum likelihood classification (MCL)	91.3%	MCL accurately discriminated weed patches field-scale and broad-scale scenarios	2013	[83]
RGB*	Rice	Various types	Overlapping and merging the binary image layers	N/A	RGB images can be used to validate proper growth and discover the irregularities such as weeds in the paddy field	2013	[84]
Multispectral and hyperspectral	Cereals and broad-leaved crops	Grass and broad-leaved	General discriminant analysis (GDA)	87 ± 5.57%	Using GDA, it is feasible to distinguish between crops and weeds	2013	[85]
Hyperspectral 61 bands: 400–1000 nm spectral resolution: 10 nm	Field pea, spring wheat, canola	Sedge and broad-leaved	Artificial neural network (ANN)	94%	ANN successfully discriminates weeds from crops	2014	[86]
Hyperspectral	Soybean	Broad-leaved	Random forest (RF)	>93.8%	Shortwave infrared: best spectrum to differentiate pigweeds from soybean	2016	[38]
RGB*	Rice	N/A	Artificial neural networks (ANN)	99%	ANN can detect weeds in paddy fields with reasonable accuracy, but 50 m above the ground is insufficient for weeds similar to paddy	2016	[45]
RGB*	Sunflower	Broad-leaved	Object-based image analysis (OBIA)	>85%	The OBIA procedure computed multiple data points, allowing herbicide requirements for timely and improved site-specific post-emergence weed seedling management	2016	[87]
RGB*, multispectral	Maize	Grass	Object-based image analysis (OBIA)	86–92%	Successfully produced accurate weed map, reduced spraying herbicides and costs	2016	[88]
Multispectral	Bracken fern	Broad-leaved	Discriminant analysis (DA)	87.80%	WolrdView-2 has the highest overall classification accuracy compared to Landsat 8 OLI, but Landsat 8 OLI* provides valuable information for long term continuous monitoring	2017	[40]
Multispectral camera	Cereals	Broad-leaved	Supervised Kohonen network (SKN), counter-propagation artificial neural network (CP-ANN) and XY-fusion network	>98%	The results demonstrate the feasibility of weed mapping on the multispectral image using hierarchical self-organizing maps	2017	[49]
Multispectral	Cereals	Broad-leaved	Maximum likelihood classification (MCL)	87.04%	The results prove the feasibility of weed mapping using multispectral imaging	2017	[89]
RGB*	Sugarcane	Grass	Artificial neural network (ANN) and random forest (RF)	91.67%	Even though ANN and RF achieved nearly identical accuracy. However, ANN outperform RF classification	2017	[90]
RGB*	Sugar beet	Broad-leaved	Support vector machine SVM vs artificial neural network (ANN)	95.00%	The SVM technique outperformed the ANN method in terms of shape-based weed detection	2018	[37]

Table 4. *Cont.*

Sensors	Crops	Weed Type	Technique	Accuracy (%)	Implications	Year	Reference
RGB*	Rice	Grass and sedge	Pre-trained CNN with the residual framework in an FCN form and transferred to a dataset by fine-tuning.	94.45%	The proposed method produced accurate weed mapping	2018	[42]
RGB*	Rice	Grass and sedge	Fully convolutional neural network (FCN)	93.5%	A fully convolutional network (FCN) outperformed convolutional neural network (CNN)	2018	[43]
RGB*	Sunflower and cotton	Grass and broad-leaved	Object-based image analysis (OBIA) and random forest (RF)	Sunflower (87.9%) and cotton (84%)	The proposed technique allowed short processing time at critical periods, which is critical for preventing yield loss	2018	[46]
Multispectral	Rice	Grass and broad-leaved	ISODATA classification and vegetation indices (VI)	96.5%	SAVI and GSAVI were the best inputs and improved weed classification	2018	[50]
RGB*	Spinach, beet, and bean	N/A	Convolutional neural networks (CNN)	Spinach (81%), beet (93%) and bean (69%)	The proposed method of weed detection was effective in different crop fields	2018	[69]
RGB*	Spinach and bean	N/A	Convolutional neural network (CNN)	94.5%	Best option to replace supervised classification	2018	[70]
RGB*	Rice	Grass and sedge	Fully convolutional neural network (FCN)	>94%	Proposed methods successfully produced prescription and weed maps	2018	[77]
RGB*	N/A	Yellow flag iris	Random forest (RF)	99%	Hybrid image-processing demonstrated good weed classification	2018	[91]
Hyperspectral	Maize	Broad-leaved	Random forest (RF)	C. arvensis (95.9%), Rumex (70.3%) and C. arvense (65.9%,)	RF algorithm successfully discriminated weeds from crops and combination with VIs improved the classification's accuracy	2018	[92]
RGB*	Soybean	Grass and broad-leaved	Joint unsupervised learning of deep representations and image clusters (JULE) and deep clustering for unsupervised learning of visual features (DeepCluster)	97%	Semi-automatic data labelling can reduce the cost of manual data labelling and be easily replicated to different datasets	2019	[71]
RGB* and Multispectral	Wheat	Unwanted crop	Object-based image analysis (OBIA), vegetation index (VIs)	87.48%	30m is the best altitude to detect weed patches within the crop rows and between the crop rows in the wheat field, and VIs successfully extracted green channels and improved weed detection	2019	[93]
RGB*	Upland rice	Grass and broad-leaved	Object-based image analysis (OBIA)	90.4%	Rice and weeds can be distinguished by consumer-grade UAV images using the SLIC-RF algorithm developed in this study with acceptable accuracy	2020	[47]
RGB*	Rice	Grass and sedge	Convolutional neural network (CNN)	80.2%	A fully convolutional network (FCN) outperformed OBIA classification	2020	[94]
RGB*	Barley	Broad-leaved	Linear regression	N/A	Qualitative methods proved to have high-quality classification	2020	[95]
RGB*	Vineyard	Grass	OBIA and combined decision tree (DT–OBIA)	84.03–89.82%	Proposed methods enable winegrowers to apply site-specific weed control while maintaining cover crop-based management systems and their vineyards' benefits.	2020	[96]

Table 4. *Cont.*

Sensors	Crops	Weed Type	Technique	Accuracy (%)	Implications	Year	Reference
RGB*	Cotton	Sedge and broad-leaved	Object-based image analysis (OBIA) and random forest (RF)	83.33% (low density plot), 85.83% (medium density plot) and 89.16% (high density plot)	The findings demonstrated the value of RGB images for weed mapping and density estimation in cotton for precision weed management	2020	[97]
Multispectral and hyperspectral	Sorghum	Grass and broadleaved	OBIA with artificial nearest neighbor (NN) algorithm	92%	The combination of OBIA–ANN demonstrated the feasibility of weed mapping in the sorghum field	2021	[62]

* RGB = red, green, blue; OLI = operational land imager.

Even though numerous platforms for data collection are accessible, a UAV is the best for identifying weeds in paddy because of its availability, high-quality data delivery, and convenience. On the other hand, the review discovered that deep learning (DL) is suitable for classifying grass weeds in paddy and producing high accuracy weed maps. However, when referring to other crops, it might differ for sedge and broad-leaved weeds. Nevertheless, this method necessitates a large amount of training data, resulting in vast agricultural datasets. In the future, to optimize the use of the RS technique, we must know what types of weeds we are dealing with in the paddy fields to choose the best technique for our research. Therefore, to classify weeds, a sophisticated method might not be necessary.

5.6.1. Machine Learning (ML)

Machine learning is a part of artificial intelligence that enables machines to recognize patterns and judge with little or no human input. Back during the early introduction to machine learning, Aitkenhead et al. [81] proposed a simple morphological characteristic measurement of a leaf shape (perimeter^2/area) and a self-organizing neural network to discriminate weeds from carrots using a Nikon Digital Camera E900S. Their proposed method enables the system to learn and differentiate between species with more than 75% accuracy without predefined plant descriptions. Eddy et al. [86] tested an artificial neural network (ANN) to classify weeds (wild oats, redroot pigweed) from crops (field pea, spring wheat, canola) using hyperspectral images. The original data were 61 bands that were reduced to seven bands using principal component analysis (PCA) and stepwise discriminant analysis. A total of 94% overall accuracy was obtained from the ANN classification. Yano et al. [90] also successfully classified weeds from sugarcane using ANN with an overall accuracy of 91.67% with a kappa coefficient of 0.8958.

Barrero et al. [45] investigated the use of artificial neural networks (ANN) to detect weed plants in rice fields using aerial images. To train the algorithm with a flying height of 50 m, they used a gray-level co-occurrence matrix (GCLM) with Haralicks descriptor for texture classification and a normalized difference index (NDI) for color. As a result, they successfully obtained 99% precision for detecting weed on the test data. However, the detection level was low for weeds similar to rice crops, because the image resolution was 50 m above the ground. Later, to evaluate the ANN's performance, Bakhshipour and Jafari [37] used a digital camera to detect weeds using shape features with an improved machine learning algorithm, support vector machine (SVM). Results showed that SVM outperformed the AAN with an overall accuracy of 95.00%, while 93.33% of weeds were correctly classified. Meanwhile, for ANN, its overall accuracy was 92.92%, where 92.50% of weeds were correctly classified.

Doi [84] used ML knowledge to discriminate rice from weeds from paddy fields by overlapping and merging 13 layers of binary images of red-green-blue and other color components (cyan, magenta, yellow, black, and white). These color components were captured using a digital camera (Cyber-shot DSC T-700, Sony) and used as input to specify the pixels with target intensity values based on mean ranges with $\pm 3\times$ standard deviation. The result shows that yellow with 1x standard deviation has the best target intensity values in discriminating paddy from weeds, with improved ratio values from 0.027 to 0.0015.

Shapira et al. [85] used general discriminant analysis (GDA) to detect grasses and broad-leaved weeds among cereal and broad-leaved crops. Using spectral relative reflect-nce values obtained by field spectroscopy as references, total canopy spectral classification by GDA for specific narrow bands was 95 $\pm$ 4.19% for wheat and 94 $\pm$ 5.13% for chickpea. Meanwhile, for vegetation and environmental monitoring on a new micro-satellite (VENµS), total canopy spectral classification was 77 $\pm$ 8.09% for wheat and 88 $\pm$ 6.94% for chickpea, and for the operative satellite advanced land imager (ALI) it was 78 $\pm$ 7.97% for wheat and 82 $\pm$ 8.22% for chickpea. Thus, an overall classification accuracy of 87 $\pm$ 5.57% for >5% vegetation coverage in a wheat field was achieved within the critical timeframe for weed control, thus providing opportunities for herbicide applications to be implemented.

Meanwhile, Rasmussen and Nielsen [95] developed a yield loss due to weed infestation model by combined manual image analysis, automated image analysis, image scoring, field scoring, and weed density data to estimate yield loss by weeds (*Cirsium arvense*) in a barley field on UAV images. With a flying height of 25 m above the ground, they successfully computed the model (Equation (7)) and found that grain moisture increased directly proportional to weed coverage (Equation (8))

$$Y = 100 \cdot (1 - \exp\,(-0.0017 \cdot X) \tag{7}$$

where:

Y = Percentage of crop yield loss.
X = Percentage of weed coverage.

$$M = 0.0310 \cdot X \tag{8}$$

where:

M = Proportional percentage increase in grain moisture.
X = Proportional percentage of weed coverage.

Other than artificial neural networks (ANN), support vector machine (SVM), and simple ML algorithms, other algorithms have been tested to detect and classify weeds from crops. They are maximum likelihood (ML), random forest (RF), vegetation indices (VIs), and discriminant analysis (DA) algorithms. De Castro, López-Granado, and Jurado-Expósito [83] used ML and VIs to classify cruciferous weed patches on a field-scale and broad-scale. Cruciferous weed patches were accurately discriminated against in both scales. However, the ML algorithm has a higher accuracy than VIs, 91.3 % and 89.45%. The same outcome was archived by Tamouridou et al. [89] when they classified *Silybum marianum* (L.) in cereal crops.

Fletcher and Reddy [38] explored the potential of a random forest algorithm in classifying pigweeds in soybean crops using a spectroradiometer (FieldSpec 3, PANalytical Boulder, Boulder, CO, USA) and WorldView-3 satellite data. One nanometer spectral data were grouped into sixteen multispectral bands to match them with the WorldView-3 satellite sensor. The accuracy of weed classifications ranged from 93.8% to 100%, with kappa values ranging from 0.93 to 0.97. The result shows an excellent agreement between the classes predicted by the models and the ground reference data. They also found that the most significant variable in separating pigweeds from soybean is the shortwave infrared (SWIR) band.

Similar to Baron, Hill, and Elmiligi [91] and Gao et al. [92], they used feature selection to train the random forest (RF) algorithm to classify weeds on different platforms: UAV RGB and hyperspectral camera, respectively. Their studies showed that the integration of feature selection with the RF algorithm produced an accurate map. As for Gao et al. [92], their output showed that for *Zea mays*, *Convolvulus arvensis*, *Rumex*, and *Cirsium arvense* weeds, the optimal random forest model with 30 significant spectral features would achieve a mean correct classification rate of 1.0, 0.789, 0.691, and 0.752, respectively. Meanwhile, Matongera et al. [40] tested discriminant analysis (DA) to classify and map invasive plant bracken fern distribution using Landsat 8 OLI. The performance of the classification output was compared with high spatial resolution data, WorldView-2 imagery. Worldview-2 classification outperformed Landsat 8 OLI with overall accuracies of 87.80% and 80.08%, respectively. However, for long term continuous monitoring, Landsat 8 OLI provides valuable information compared to the WorldView-2 commercial sensor.

A few researchers chose object-based image analysis (OBIA) to classify weeds from crops. OBIA is an automatic hierarchal image classification algorithm. It allows numerous image objects to be created and further categorized into user-defined classes [98]. For example, López-Granados et al. [87] used an RGB (red, green, blue) UAV to monitor early-season weeds in a sunflower field using object-based image analysis (OBIA). Their experiment was tested at two different flying heights, 30 m and 60 m, above the surface. They found that both flying heights give satisfactory outputs, with 2.5% to 5% thresholds and an accuracy higher than 85%. The same result was archived by López-Granados et al. [88], Mateen and Zhu [93], and Sapkota et al. [97] when they classified weeds from maize, wheat, and cotton, respectively. Their research helped farmers with rationalization of the herbicide application.

Some of the researchers integrated object-based image analysis (OBIA) with other machine learning algorithms. OBIA's final output can be converted into another GIS format [99], making it flexible to integrate with other algorithms. For example, De Castro et al. [96] successfully classified *Cynodon dactylon* (bermudagrass) in a vineyard by combining OBIA with the decision tree (DT) algorithm. De Castro et al. [46] also managed to produce a weed map of *Convolvulus arvensis* L. (bindweed) in a soybean field by combining OBIA with the RF algorithm. Meanwhile, Che'Ya, Dunwoody, and Gupta [62] successfully generated various types of weed maps in the sorghum's field by integrating OBIA with the artificial nearest neighbor (ANN) algorithm.

Kawamura et al. [47] experimented with the OBIA classification method using the simple linear iterative clustering algorithm–random forest (SLIC–RF). SLIC is a super-pixel method for extracting input feature details for each subject. They used three-color spaces (RGB, hue-saturation-brightness (HSV) and transformation function of RGB images (CIE-L*a*b*)) as the primary input feature and a spatial texture, four VIs (excess green (ExG), excess red (ExR), green–red vegetation index (GRVI), and color index of vegetation extraction (CIVE)), and DSM as the secondary data. The HSV-based SLIC–RF outperformed the other color spaces tested, with an accuracy of 90.4%.

Instead of using an RGB UAV, Stroppiana et al. [50] used UAV multispectral images for early season weed mapping in rice using ISODATA classification. Their input data are spectral indices (normalized different vegetation index (NDVI), soil adjusted vegetation index (SAVI), GSAVI, a simple ratio index related to leaf pigments content and greenness (RGRI), normalized difference red edge (NDRE), and chlorophyll vegetation index (CVI)) and textural metrics. Weed mapping performance was validated by measuring overall accuracy (OA), while for weed class, omission errors (OE) and commission errors (CE) were calculated. The result shows that SAVI and GSAVI gave the best output compared to other indices, with 96.5% and 94.5% overall accuracy. The final production, classification map, weed proportion in the percentage map, weed canopy height measured in meters (m) map, and rice fraction cover map, were successfully produced from SAVI and GSAVI. Pantazi et al. [49] also chose multispectral UAVs to map weeds in cereals.

5.6.2. Deep Learning (DL)

Deep learning has recently become a machine learning component widely utilized in agricultural crop monitoring and management. It has taken a directive in many crop monitoring objectives such as weed detection, nutrient disorder, and disease detection. Huang et al. [43] utilized the fully convolutional network (FCN) method to map weeds in rice using unmanned aerial vehicle red-green-blue (UAV-RGB) imagery. Transfer learning was used to optimize the generalization capacity, and skip architecture was chosen to boost prediction accuracy. The result was then compared with the patch-based convolutional neural networks (CNN) algorithm and the pixel-based CNN method. The findings showed a proposed FCN method that outperformed others, both in efficiency and efficacy in terms of accuracy. The overall accuracy of the FCN method was up to 93.5%, and the accuracy for weed recognition was 88.3%.

Meanwhile, Huang et al. [94] also tested the same algorithm to delineate weeds from rice in multi-rotor UAV images. Using an RGB-UAV with a flying height of 10 m above the surface, they compared the object-based image analysis with the fully convolutional network (FCN). As expected, their finding shows that FCN performs better than OBIA, with an overall accuracy of 80.2% and 66.6%, respectively, which means that this algorithm can produce precise weed cover maps for the evaluated UAV-based RGB imagery.

Bah et al. [69] also tested other deep learning algorithms: convolutional neural networks (CNNs) on other crops, spinach, beet, and bean using UAV images to classify weeds in the crops from a 20 m flying height. The method effectively differentiates weeds from crops with an overall accuracy for beet of 93%, spinach of 81%, and bean of 69%. However, deep learning alone requires a great deal of training data. It is too time-consuming of a process to construct large agricultural datasets with pixel-level identifications by an expert. Therefore, Bah et al. [70] proposed a fully automatic learning method using CNNs with an unsupervised training dataset collection for weed detection from UAV images. The classification started with the identification of inter-row weeds from the automatic detection of crop rows. Then, training datasets from inter-row weeds were made before performing the CNNs to detect crop and weed images. Results obtained were compared with supervised training data, and the difference in accuracy for spinach is 1.5%, and for bean is 6%. The differences between supervised and unsupervised are narrow. This proposed method can be the best option, since supervised labelling is expensive and challenging and requires human expertise.

Dos Santos Ferreira et al. [71] evaluated the unsupervised deep learning performance to discriminate weeds from soybean in UAV images. They tested two unsupervised deep clustering algorithms, joint unsupervised learning of deep representations and image clusters (JULE) and deep clustering for unsupervised learning of visual features (DeepCluster), using two public weed datasets. The first datasets were captured in a soybean plantation in Brazil, and weeds were distinguished between the grass and broad leaf weed. Meanwhile, the second dataset consists of 17,509 labelled images of eight common species originating from Australia. Semi-automatic data labelling in agriculture was used to evaluate the outputs, and the result showed that this method achieved up to 97% accuracy, reduced 100 times in manual annotations.

This study has used the shape, texture, and pattern of weeds and crops trained and classified by remote sensing algorithms. However, more research needs to be carried out to detect and produce an accurate weed coverage map that recognizes weed types: grass, sedge and broad-leaved in the paddy field. This is because different weeds have different characteristics that require other variables to identify them. Nevertheless, based on the previous study, it is not impossible to produce an accurate map that will highly benefit weed management in the paddy field, especially when dealing with herbicide consumption.

5.7. Advantages of Implementation of Remote Sensing in Weed Detection through PA

The usage of herbicides, also known as agrochemicals, to control weeds in paddy fields has caused several impacts on the environment and human health [100]. Therefore, the authorities can consider reducing these inputs to follow an environmentally friendly rice production practice. A study by Jafari, Othman, and Kuhn [101] showed that a 10% reduction in agrochemical grants would reduce agrochemical use. However, it dramatically reduces national welfare and decreases food safety. Nevertheless, we can overcome these issues by implementing remote sensing SSWM techniques into precision agriculture (PA).

Improving weed management can improve our food security. Numerous remote sensing platforms are available to monitor weeds, and unmanned aerial vehicles (UAV) are among the most popular platforms used these days. The excellent part of a UAV is that it can fly low and precisely detect the presence of weeds in the paddy plot. Numerous researchers proved that a UAV could produce an accurate SSWM map with overall accuracy ranging from 66.6% to 99%, depending on the type of weeds found in the plot [49,89,91,94].

The remote sensing technique can be used to locate weed presence in the paddy plot by using multiple approaches such as machine learning [62] and deep learning [57,58] or by combining them both. Previous studies (Table 4) proved that any weeds, grass, sedge, and broad-leaved weeds in crops could be classified using remote sensing techniques. Therefore, this technique can be adopted into paddy field practices. These algorithms were beneficial in detecting weed distribution in the paddy field, with sufficient training data. The weed location will be recorded, and thus, the farmers will know its location and estimate the suitable volume of herbicide needed to control the invasive plant in the plot. Therefore, the over-application of herbicides will not be an issue anymore.

There is no standard method drawn systematically and strategically planned to detect and manage weeds in paddy fields using remote sensing in developing countries. This study is significant for finding the best approach to classify weeds in a paddy plot. Using UAV imagery, Huang et al. [42] chose a semantic labelling approach to generate weed distribution maps in paddy. A residual framework with an ImageNet pre-trained convolutional neural network (CNN) was adapted and transferred into the dataset by a fine-tuning process. A fully connected conditional random field (CRF) was adapted to improve the spatial details. They successfully produced weed distribution maps with an overall accuracy up to 94.45% and kappa coefficient of 0.9128. The newly generated map can guide the sprayer UAV to spray the herbicide only at the weed colony. Thus, the usage of a spraying UAV can minimize the contact between farmers and herbicides and, at the same time, reduce the impact of agriculture on the environment and human health [102].

Different types of weeds need different treatments. Traditional practices are too time-consuming and require many human resources, and they are not effective methods to monitor weed presence. Developing countries' farmers need this technology to improve and increase yield production.

6. Impact of Weeds Management on Crops, Yield and Economy

Weeds cause severe yield losses in agriculture [103] and cause significant damage to the ecosystem and the economy in the territories they enter [104]. For example, a couple of studies have reported that rice production's total yield loss due to weed infestation could be up to 72% [105,106]. This loss happened due to the presence of weeds in crops that compete in nutrient uptake. In addition, uncontrolled chemical products used to control weeds cause farmers health issues and negatively affect the climate, killing livestock and contaminating the air and water [100].

Fertilizer given by the farmers to increase their yield was not 100% absorbed by their crops. For example, in Cambodia, cultivated agricultural land is 3.7 million hectares, of which 76% is planted with lowland rice and 24% with upland crops such as soybean, cassava, vegetables, maize, and sugar cane. At approximately 3 t ha^{-1}, their average rice paddy yield was about 50%, and another 50% of losses were caused by weed competition, which is a significant constraint [107]. Due to weeds, Iranian wheat and chickpea yield losses are more than 25% and 66%, respectively [108].

Weeds are more competitive when moisture is inadequate, and rice seedlings cannot cope well with weeds. Meanwhile, in China, the presence of such invasive species has caused them an economic loss of approximately USD 15 billion [109]. In Pakistan, USD 3 billion is needed annually for a weed management program to increase yield [33]. In England, approximately USD 545 million of gross profit were lost annually, equal to 0.8 million tonnes of yield production, due to the herbicide-resistant weeds [110].

Precision agriculture techniques using high-tech tools can minimize agriculture resources by site-specific application since they can calculate an optimum input to spatial and temporal requirements, reducing greenhouse gas emissions into the atmosphere. In addition, these techniques will positively affect economics and yield productivity with a lower production cost than traditional practices [111].

Malaysian farmers could expect an additional return of rice yield from 0.3 to 0.6 t ha-1 through proper weed management [112]. Meanwhile, in India, improved weed management successfully decreased weed infestation in rice fields from very high intensity (>75%) to a mild (50%) level [113]. Matthews [114] tested herbicide usage using a spraying UAV to demonstrate the impact of technology adaptation into precision farming. The result showed that the study used approximately 200 L of herbicide per hectare than the traditional method, which is 1000 L per hectare. Meanwhile, by applying site-specific treatment maps on a broad scale, Huang et al. [77] successfully saved herbicide consumption from 58.3% to 70.8%. On the other hand, De Castro, López-Granado and Jurado-Expósito [83] saved 61.31% for the no-treatment areas and 13.02% for the low-dose of herbicide practice. The implementation of SSWM into PA proved that it effectively decreased the herbicide cost, optimized weed control, and avoided unnecessary environmental pollutions [108,109,115,116].

7. Future Direction

Machine learning such as deep learning algorithms should be implemented for extracting higher abstract levels of weeds and their relation to the seasonal changes of the paddy for more accurate weed identification. It is challenging to implement remote sensing techniques into paddy. However, when referring to the previous study, De Castro et al. [96] successfully classified Cynodon dactylon (bermudagrass) in a vineyard by integrating OBIA with a decision tree (DT) algorithm. De Castro et al. [46] also managed to produce a weed map of *Convolvulus arvensis* L. (bindweed) in a soybean field. Meanwhile, Huang et al. [94] successfully generated a grass and sedge weed map in a paddy field using a deep learning technique. This study has similarities in shape, texture, and pattern that machine learning and deep learning techniques can classify. In addition, the integration of various platforms, such as ground-based and machine vision technologies, should be considered. Besides, various yield-determining factors, such as climatic or agronomic, should be considered during the developmental stages of paddy. By maintaining the vigorous development of paddy, the existence of weeds can be minimized due to the biological mechanisms of the crops, which can be used to suppress the growth response of weeds towards the crops during the competition process.

8. Conclusions

Traditional practices are too time-consuming and require many human resources. Therefore, adapting automated practices into precision farming (PA) is the best practice to control weeds. Even though various platforms are available for data gathering, UAVs are the best for detecting weeds in paddy due to their availability, high-quality data delivery, and ease of handling. We had complete control over the data collection phase. The review proved that deep learning could convey high accuracy weed maps. However, this method requires a certain number of training data, resulting in massive agricultural databases. Therefore, to decide which algorithm best suits our research, we need to know what types of weeds we are dealing with by observing their types in paddy fields. It is not necessary to use a complicated algorithm to perform weed classification. Although some studies showed that deep learning might not be necessary when dealing with imagery, much simpler algorithms, such as OBIA, can perform adequate image analysis for detecting weeds in paddy fields. When comparing crops and weed types, both algorithms, ML and DL, had successfully generated a high accuracy map ranging from 85% to 99%, depending on the type of weeds and crops. Thus, we can expect the same accuracy in producing weed maps in paddy, regardless of the types of weeds present in the field. More research needs to be carried out, and this review has shown that improved weed management could optimize the usage of herbicides that should be applied on a site-specific basis. Not only did it increase yield production, but it also proved that this technology could control the spreading of weeds. It also effectively maximizes herbicide usage and decreases the budget required to purchase them.

Author Contributions: Conceptualization, R.R., N.N.C. and Y.A.; writing—original draft preparation, R.R.; writing—review and editing, R.R., N.N.C., Y.A., F.R., A.W., M.R.I., Z.B., W.F.F.I. and M.H.O.; visualization, R.R., Y.A. and F.R.; supervision, N.N.C.; funding acquisition, M.R.I. All authors have read and agreed to the published version of the manuscript.

Funding: The authors would like to thank the Ministry of Higher Education Malaysia for providing research funds from Long-term Research Grant Scheme (LRGS/1/2019/ UPM/01/2): Development of climate ready rice for sustaining food security in Malaysia (Vot no. 5545000).

Institutional Review Board Statement: Not applicable.

Informed Consent Statement: Not applicable.

Data Availability Statement: Not applicable.

Acknowledgments: The authors wish to thank Ibrahim Busu for the knowledge he shared in writing this manuscript.

Conflicts of Interest: The authors have no conflict of interest to declare.

References

1. Patel, M.; Jernigan, S.; Richardson, R.; Ferguson, S.; Buckner, G. Autonomous Robotics for Identification and Management of Invasive Aquatic Plant Species. *Appl. Sci.* **2019**, *9*, 2410. [CrossRef]
2. Dilipkumar, M.; Chuah, T.S.; Goh, S.S.; Sahid, I. Weed management issues, challenges, and opportunities in Malaysia. *Crop Prot.* **2020**, *134*, 104347. [CrossRef]
3. Food and Agriculture Organization of the United Nations (F.A.O.). F.A.O.S.T.A.T. 2021. Available online: http://www.fao.org/faostat/en/#data/RP/visualize (accessed on 4 June 2021).
4. Jones, E.A.L.; Owen, M.D.K.; Leon, R.G. Influence of multiple herbicide resistance on growth in *Amaranthus tuberculatus*. *Weed Res.* **2019**, *59*, 235–244. [CrossRef]
5. Löbmann, A.; Christen, O.; Petersen, J. Development of herbicide resistance in weeds in a crop rotation with acetolactate synthase-tolerant sugar beets under varying selection pressure. *Weed Res.* **2019**, *59*, 479–489. [CrossRef]
6. Yuzugullu, O.; Erten, E.; Hajnsek, I. A multi-year study on rice morphological parameter estimation with X-band PolSAR data. *Appl. Sci.* **2017**, *7*, 602. [CrossRef]
7. Shiu, Y.S.; Chuang, Y.C. Yield Estimation of Paddy Rice Based on Satellite Imagery: Comparison of Global and Local Regression Models. *Remote Sens.* **2019**, *11*, 111. [CrossRef]
8. Xiang, K.; Ma, M.; Liu, W.; Dong, J.; Zhu, X.; Yuan, W. Mapping Irrigated Areas of Northeast China in Comparison to Natural Vegetation. *Remote Sens.* **2019**, *11*, 825. [CrossRef]
9. Papademetriou, M.K. Rice production in the Asia-Pacific region: Issues and perspectives. *Bridg. Rice Yield Gap Asia-Pac. Reg.* **2000**, *16*, 5.
10. Pandey, S.; Byerlee, D.; Dawe, D.; Dobermann, A.; Mohanty, S.; Rozelle, S.; Hardy, B. *Rice in the Global Economy*; International Rice Research Institute: Los Banos, Phillipines, 2010.
11. United States Department of Agriculture, Foreign Agricultural Services. World Rice Production, Consumption and Stocks. PSD Reports. Available online: https://apps.fas.usda.gov/psdonline/app/index.html#/app/downloads (accessed on 10 November 2020).
12. Masum, S.M.; Hossain, M.A.; Akamine, H.; Sakagami, J.; Ishii, T.; Nakamura, I.; Asaduzzaman, M.; Bhowmik, P.C. Performance of Bangladesh indigenous rice in a weed infested field and separation of allelopathy from resource competition. *Weed Biol. Manag.* **2019**, *19*, 39–50. [CrossRef]
13. Yamori, W.; Kondo, E.; Sugiura, D.; Terashima, I.; Suzuki, Y.; Makino, A. Enhanced leaf photosynthesis as a target to increase grain yield: Insights from transgenic rice lines with variable Rieske FeS protein content in the cytochrome b6/f complex. *Plant Cell Environ.* **2016**, *39*, 80–87. [CrossRef]
14. Simkin, A.J.; López-Calcagno, P.E.; Raines, C.A. Feeding the world: Improving photosynthetic efficiency for sustainable crop production. *J. Exp. Bot.* **2019**, *70*, 1119–1140. [CrossRef]
15. Maneepitak, S.; Ullah, H.; Paothong, K.; Kachenchart, B.; Datta, A.; Shrestha, R.P. Effect of water and rice straw management practices on yield and water productivity of irrigated lowland rice in the Central Plain of Thailand. *Agric. Water Manag.* **2019**, *211*, 89–97. [CrossRef]
16. LaHue, G.T.; Chaney, R.L.; Adviento-Borbe, M.A.; Linquist, B.A. Alternate wetting and drying in high yielding direct-seeded rice systems accomplishes multiple environmental and agronomic objectives. *Agric. Ecosyst. Environ.* **2016**, *229*, 30–39. [CrossRef]
17. Liang, K.; Zhong, X.; Huang, N.; Lampayan, R.M.; Pan, J.; Tian, K.; Liu, Y. Grain yield, water productivity and CH4 emission of irrigated rice in response to water management in south China. *Agric. Water Manag.* **2016**, *163*, 319–331. [CrossRef]

18. Van Oort, P.A.; Zwart, S.J. Impacts of climate change on rice production in Africa and causes of simulated yield changes. *Glob. Chang. Biol.* **2018**, *24*, 1029–1045. [CrossRef]
19. Van Oort, P.A.J. Mapping abiotic stresses for rice in Africa: Drought, cold, iron toxicity, salinity and sodicity. *Field Crop. Res.* **2018**, *219*, 55–75. [CrossRef]
20. Dossou-Yovo, E.; Zwart, S.; Kouyaté, A.; Ouédraogo, I.; Bakare, O. Predictors of Drought in Inland Valley Landscapes and Enabling Factors for Rice Farmers' Mitigation Measures in the Sudan-Sahel Zone. *Sustainability* **2019**, *11*, 79. [CrossRef]
21. Ariza, A.A. Machine Learning and Big Data Techniques for Satellite-Based Rice Phenology Monitoring. PhD Thesis, The University of Manchester, Manchester, UK, 2019.
22. Global Rice Science Partnership. *Rice Almanac*, 4th ed.; International Rice Research Institute: Los Banos, Philippines, 2013.
23. Anwar, M.P.; Juraimi, A.S.; Samedani, B.; Puteh, A.; Man, A. Critical period of weed control in aerobic rice. *Sci. World J.* **2012**, *2012*, 603043. [CrossRef]
24. Kamath, R.; Mamatha, B.; Srikanth, P. Paddy Crop and Weed Discrimination: A Multiple Classifier System Approach. *Int. J. Agron.* **2020**, *2020*, 6474536. [CrossRef]
25. Chadhar, A.R.; Nadeem, M.A.; Tanveer, A.; Yaseen, M. Weed management boosts yield in fine rice under system of rice intensification. *Planta Daninha* **2014**, *32*, 291–299. [CrossRef]
26. Ahmed, Q.N.; Hussain, P.Z.; Othman, A.S. Comparative study on vegetative and reproductive development between weedy rice morphotypes and commercial rice varieties in Perak, Malaysia. *Trop. Life Sci. Res.* **2012**, *23*, 17.
27. Halip, R.M.; Norasma, N.; Fadzli, W.F.I.; Roslee, R.; Roslin, N.A.; Ismail, M.R.; Berahim, Z.; Omar, M.H. Pemantauan Tanaman Padi Menggunakan Sistem Maklumat Geografi dan Imej Multispektral. *Adv. Agric. Food Res. J.* **2020**, *1*, 1–18.
28. Man, A.; Mohammad Saad, M.; Amzah, B.; Masarudin, M.F.; Jack, A.; Misman, S.N.; Ramachandran, K. Buku Poket Perosak, Penyakit dan Rumpai Padi di Malaysia. In *Cetakan Kelima*; Institut Penyelidikan dan Kemajuan Pertanian Malaysia (MARDI): Kuala Lumpur, Malaysia, 2018.
29. Juraimi, A.S.; Uddin, M.K.; Anwar, M.P.; Mohamed, M.T.M.; Ismail, M.R.; Man, A. Sustainable weed management in direct seeded rice culture: A review. *Aust. J. Crop Sci.* **2013**, *7*, 989.
30. Power, E.F.; Kelly, D.L.; Stout, J.C. The impacts of traditional and novel herbicide application methods on target plants, non-target plants and production in intensive grasslands. *Weed Res.* **2013**, *53*, 131–139. [CrossRef]
31. Brown, R.B.; Noble, S.D. Site-specific weed management: Sensing requirements—what do we need to see? *Weed Sci.* **2005**, *53*, 252–258. [CrossRef]
32. Oebel, H.; Gerhards, R. Site-specific weed control using digital image analysis and georeferenced application maps: On-farm experiences. In *Precision Agriculture '05. Papers presented at the 5th European Conference on Precision Agriculture, Uppsala, Sweden*; Wageningen Academic Publishers: Wageningen, The Nertherlands, 2005; pp. 131–137.
33. Matloob, A.; Safdar, M.E.; Abbas, T.; Aslam, F.; Khaliq, A.; Tanveer, A.; Rehman, A.; Chadhar, A.R. Challenges and prospects for weed management in Pakistan: A review. *Crop Prot.* **2019**, *134*, 104724. [CrossRef]
34. Bajwa, A.A.; Chauhan, B.S.; Farooq, M.; Shabbir, A.; Adkins, S.W. What do we really know about alien plant invasion? A review of the invasion mechanism of one of the world's worst weeds. *Planta* **2016**, *244*, 39–57. [CrossRef] [PubMed]
35. Shanmugapriya, P.; Rathika, S.; Ramesh, T.; Janaki, P. Applications of remote sensing in agriculture—A Review. *Int. J. Curr. Microbiol. Appl. Sci.* **2019**, *8*, 2270–2283. [CrossRef]
36. Alzubaidi, L.; Zhang, J.; Humaidi, A.J.; Al-Dujaili, A.; Duan, Y.; Al-Shamma, O.; Santamaría, J.; Fadhel, M.A.; Al-Amidie, M.; Farhan, L. Review of deep learning: Concepts, CNN architectures, challenges, applications, future directions. *J. Big Data* **2021**, *8*, 1–74. [CrossRef]
37. Bakhshipour, A.; Jafari, A. Evaluation of support vector machine and artificial neural networks in weed detection using shape features. *Comput. Electron. Agric.* **2018**, *145*, 153–160. [CrossRef]
38. Fletcher, R.S.; Reddy, K.N. Random forest and leaf multispectral reflectance data to differentiate three soybean varieties from two pigweeds. *Comput. Electron. Agric.* **2016**, *128*, 199–206. [CrossRef]
39. Fenfang, L.; Dongyan, Z.; Xiu, W.; Taixia, W.; Xinfu, C. Identification of corn and weeds on the leaf scale using polarization spectroscopy. *Infrared Laser Eng.* **2016**, *45*, 1223001. [CrossRef]
40. Matongera, T.N.; Mutanga, O.; Dube, T.; Sibanda, M. Detection and mapping the spatial distribution of bracken fern weeds using the Landsat 8 O.L.I. new generation sensor. *Int. J. Appl. Earth Obs. Geoinf.* **2017**, *57*, 93–103. [CrossRef]
41. Esposito, M.; Crimaldi, M.; Cirillo, V.; Sarghini, F.; Maggio, A. Drone and sensor technology for sustainable weed management: A review. *Chem. Biol. Technol. Agric.* **2021**, *8*, 1–11. [CrossRef]
42. Huang, H.; Lan, Y.; Deng, J.; Yang, A.; Deng, X.; Zhang, L.; Wen, S. A semantic labeling approach for accurate weed mapping of high resolution UAV imagery. *Sensors* **2018**, *18*, 2113. [CrossRef]
43. Huang, H.; Deng, J.; Lan, Y.; Yang, A.; Deng, X.; Zhang, L. A fully convolutional network for weed mapping of unmanned aerial vehicle (UAV) imagery. *PLoS ONE* **2018**, *13*, e0196302. [CrossRef]
44. Manfreda, S.; McCabe, M.F.; Miller, P.E.; Lucas, R.; Madrigal, V.P.; Mallinis, G.; Ben Dor, E.; Helman, D.; Estes, L.; Ciraolo, G.; et al. On the use of unmanned aerial systems for environmental monitoring. *Remote Sens.* **2018**, *10*, 641. [CrossRef]
45. Barrero, O.; Rojas, D.; Gonzalez, C.; Perdomo, S. Weed detection in rice fields using aerial images and neural networks. In Proceedings of the 2016 XXI Symposium on Signal Processing, Images and Artificial Vision (S.T.S.I.V.A.), Bucaramanga, Colombia, 31 August–2 September 2016; pp. 1–4.

46. De Castro, A.; Torres-Sánchez, J.; Peña, J.; Jiménez-Brenes, F.; Csillik, O.; López-Granados, F. An automatic random forest-OBIA algorithm for early weed mapping between and within crop rows using UAV imagery. *Remote Sens.* **2018**, *10*, 285. [CrossRef]
47. Kawamura, K.; Asai, H.; Yasuda, T.; Soisouvanh, P.; Phongchanmixay, S. Discriminating crops/weeds in an upland rice field from UAV images with the SLIC-RF algorithm. *Plant Prod. Sci.* **2020**, *24*, 198–215. [CrossRef]
48. Micasense Inc. Best Practices: Collecting Data with MicaSense Sensors. MicaSense Knowl. Base. 2020. Available online: https://support.micasense.com/hc/en-us/articles/224893167 (accessed on 23 June 2021).
49. Pantazi, X.E.; Tamouridou, A.A.; Alexandridis, T.K.; Lagopodi, A.L.; Kashefi, J.; Moshou, D. Evaluation of hierarchical self-organising maps for weed mapping using U.A.S. multispectral imagery. *Comput. Electron. Agric.* **2017**, *139*, 224–230. [CrossRef]
50. Stroppiana, D.; Villa, P.; Sona, G.; Ronchetti, G.; Candiani, G.; Pepe, M.; Busetto, L.; Migliazzi, M.; Boschetti, M. Early season weed mapping in rice crops using multispectral UAV data. *Int. J. Remote Sens.* **2019**, *39*, 5432–5452. [CrossRef]
51. Adão, T.; Hruška, J.; Pádua, L.; Bessa, J.; Peres, E.; Morais, R.; Sousa, J.J. Hyperspectral imaging: A review on UAV-based sensors, data processing and applications for agriculture and forestry. *Remote Sens.* **2017**, *9*, 1110. [CrossRef]
52. Brook, A.; Ben-Dor, E. Supervised vicarious calibration (S.V.C.) of multi-source hyperspectral remote-sensing data. *Remote Sens.* **2015**, *7*, 6196–6223. [CrossRef]
53. Guo, Y.; Senthilnath, J.; Wu, W.; Zhang, X.; Zeng, Z.; Huang, H. Radiometric calibration for multispectral camera of different imaging conditions mounted on a UAV platform. *Sustainability* **2019**, *11*, 978. [CrossRef]
54. Kelcey, J.; Lucieer, A. Sensor correction and radiometric calibration of a 6-band multispectral imaging sensor for UAV remote sensing. In Proceedings of the 12th Congress of the International Society for Photogrammetry and Remote Sensing, Melbourne, Australia, 25 August–1 September 2019; pp. 393–398.
55. Mafanya, M.; Tsele, P.; Botai, J.O.; Manyama, P.; Chirima, G.J.; Monate, T. Radiometric calibration framework for ultra-high-resolution UAV-derived orthomosaics for large-scale mapping of invasive alien plants in semi-arid woodlands: Harrisia pomanensis as a case study. *Int. J. Remote Sens.* **2018**, *39*, 5119–5140. [CrossRef]
56. Karpouzli, E.; Malthus, T. The empirical line method for the atmospheric correction of IKONOS imagery. *Int. J. Remote Sens.* **2003**, *24*, 1143–1150. [CrossRef]
57. Xu, K.; Gong, Y.; Fang, S.; Wang, K.; Lin, Z.; Wang, F. Radiometric calibration of UAV remote sensing image with spectral angle constraint. *Remote Sens.* **2019**, *11*, 1291. [CrossRef]
58. Parrot. *Application Note: Pixel Value to Irradiance Using the Sensor Calibration Model*; Parrot: Paris, France, 2017; Volume SEQ-AN-01.
59. Tu, Y.H.; Phinn, S.; Johansen, K.; Robson, A. Assessing radiometric correction approaches for multispectral UAS imagery for horticultural applications. *Remote Sens.* **2018**, *10*, 1684. [CrossRef]
60. Kumar, G.; Bhatia, P.K. A detailed review of feature extraction in image processing systems. In Proceedings of the 2014 Fourth International Conference on Advanced Computing and Communication Technologies, Rohtak, India, 8–9 February 2014; pp. 5–12.
61. Xue, B.; Zhang, M.; Browne, W.N. Particle swarm optimization for feature selection in classification: A multi-objective approach. *IEEE Trans. Cybern.* **2013**, *43*, 1656–1671. [CrossRef]
62. Che'Ya, N.N.; Dunwoody, E.; Gupta, M. Assessment of Weed Classification Using Hyperspectral Reflectance and Optimal Multispectral UAV Imagery. *Agronomy* **2021**, *11*, 1435.
63. Woebbecke, D.M.; Meyer, G.E.; Von Bargen, K.; Mortensen, D.A. Color indices for weed identification under various soil, residue, and lighting conditions. *Trans. ASAE* **1995**, *38*, 259–269. [CrossRef]
64. Shapiro, L.; Stockman, G. *Computer Vision*; Prentice Hall Inc.: Upper Saddle River, NJ, USA, 2001.
65. Stanković, R.S.; Falkowski, B.J. The Haar wavelet transform: Its status and achievements. *Comput. Electr. Eng.* **2003**, *29*, 25–44. [CrossRef]
66. Pearson, K. LIII. On lines and planes of closest fit to systems of points in space. *Lond. Edinb. Dublin Philos. Mag. J. Sci.* **1901**, *2*, 559–572. [CrossRef]
67. Taati, A.; Sarmadian, F.; Mousavi, A.; Pour, C.T.H.; Shahir, A.H.E. Land Use Classification using Support Vector Machine and Maximum Likelihood Algorithms by Landsat 5 TM Images. *Walailak J. Sci. Technol.* **2014**, *12*, 681–687.
68. Whitside, T.G.; Maier, S.F.; Boggs, G.S. Area-based and location-based validation of classified image objects. *Int. J. Appl. Earth Obs. Geoinform.* **2014**, *28*, 117–130. [CrossRef]
69. Bah, M.D.; Dericquebourg, E.; Hafiane, A.; Canals, R. Deep learning based classification system for identifying weeds using high-resolution UAV imagery. In *Science and Information Conference*; Springer: Cham, Switzerland, 2018; pp. 176–187.
70. Bah, M.D.; Hafiane, A.; Canals, R. Deep learning with unsupervised data labeling for weed detection in line crops in UAV images. *Remote Sens.* **2018**, *10*, 1690. [CrossRef]
71. Dos Santos Ferreira, A.; Freitas, D.M.; da Silva, G.G.; Pistori, H.; Folhes, M.T. Unsupervised deep learning and semi-automatic data labeling in weed discrimination. *Comput. Electron. Agric.* **2019**, *165*, 104963. [CrossRef]
72. Liakos, K.G.; Busato, P.; Moshou, D.; Pearson, S.; Bochtis, D. Machine learning in agriculture: A review. *Sensors* **2018**, *18*, 2674. [CrossRef]
73. Zhang, J.; Huang, Y.; Pu, R.; Gonzalez-Moreno, P.; Yuan, L.; Wu, K.; Huang, W. Monitoring plant diseases and pests through remote sensing technology: A review. *Comput. Electron. Agric.* **2019**, *165*, 104943. [CrossRef]
74. Janiesch, C.; Zschech, P.; Heinrich, K. Machine learning and deep learning. *Electron. Mark.* **2021**, *31*, 689–695. [CrossRef]
75. Ongsulee, P. Artificial intelligence, machine learning and deep learning. In Proceedings of the 2017 15th International Conference on ICT and Knowledge Engineering (ICT&KE), Bangkok, Thailand, 22–24 November 2017; pp. 1–6.

76. Kamilaris, A.; Prenafeta-Boldú, F.X. Deep learning in agriculture: A survey. *Comput. Electron. Agric.* **2018**, *147*, 70–90. [CrossRef]
77. Huang, H.; Deng, J.; Lan, Y.; Yang, A.; Deng, X.; Wen, S.; Zhang, H.; Zhang, Y. Accurate weed mapping and prescription map generation based on fully convolutional networks using UAV imagery. *Sensors* **2018**, *18*, 3299. [CrossRef] [PubMed]
78. Abirami, S.; Chitra, P. Energy-efficient edge based real-time healthcare support system. In *Advances in Computers*; Elsevier: Amsterdam, The Netherlands, 2020; Volume 117, pp. 339–368.
79. Saha, S.; Nagaraj, N.; Mathur, A.; Yedida, R.; Sneha, H.R. Evolution of novel activation functions in neural network training for astronomy data: Habitability classification of exoplanets. *Eur. Phys. J. Spec. Top.* **2020**, *229*, 2629–2738. [CrossRef] [PubMed]
80. Voulodimos, A.; Doulamis, N.; Doulamis, A.; Protopapadakis, E. Deep learning for computer vision: A brief review. *Comput. Intell. Neurosci.* **2018**, *2018*, 7068349. [CrossRef]
81. Aitkenhead, M.J.; Dalgetty, I.A.; Mullins, C.E.; McDonald, A.J.S.; Strachan, N.J.C. Weed and crop discrimination using image analysis and artificial intelligence methods. *Comput. Electron. Agric.* **2003**, *39*, 157–171. [CrossRef]
82. Karimi, Y.; Prasher, S.O.; Patel, R.M.; Kim, S.H. Application of support vector machine technology for weed and nitrogen stress detection in corn. *Comput. Electron. Agric.* **2006**, *51*, 99–109. [CrossRef]
83. De Castro, A.I.; López-Granados, F.; Jurado-Expósito, M. Broad-scale cruciferous weed patch classification in winter wheat using QuickBird imagery for in-season site-specific control. *Precis. Agric.* **2013**, *14*, 392–413. [CrossRef]
84. Doi, R. Discriminating crop and other canopies by overlapping binary image layers. *Opt. Eng.* **2013**, *52*, 020502. [CrossRef]
85. Shapira, U.; Herrmann, I.; Karnieli, A.; Bonfil, D.J. Field spectroscopy for weed detection in wheat and chickpea fields. *Int. J. Remote Sens.* **2013**, *34*, 6094–6108. [CrossRef]
86. Eddy, P.R.; Smith, A.M.; Hill, B.D.; Peddle, D.R.; Coburn, C.A.; Blackshaw, R.E. Weed and crop discrimination using hyperspectral image data and reduced bandsets. *Can. J. Remote Sens.* **2014**, *39*, 481–490. [CrossRef]
87. López-Granados, F.; Torres-Sánchez, J.; Serrano-Pérez, A.; de Castro, A.I.; Mesas-Carrascosa, F.J.; Pena, J.M. Early season weed mapping in sunflower using UAV technology: Variability of herbicide treatment maps against weed thresholds. *Precis. Agric.* **2016**, *17*, 183–199. [CrossRef]
88. López-Granados, F.; Torres-Sánchez, J.; De Castro, A.I.; Serrano-Pérez, A.; Mesas-Carrascosa, F.J.; Peña, J.M. Object-based early monitoring of a grass weed in a grass crop using high resolution UAV imagery. *Agron. Sustain. Dev.* **2016**, *36*, 67. [CrossRef]
89. Tamouridou, A.A.; Alexandridis, T.K.; Pantazi, X.E.; Lagopodi, A.L.; Kashefi, J.; Moshou, D. Evaluation of UAV imagery for mapping Silybum marianum weed patches. *Int. J. Remote Sens.* **2017**, *38*, 2246–2259. [CrossRef]
90. Yano, I.H.; Santiago, W.E.; Alves, J.R.; Mota, L.T.M.; Teruel, B. Choosing classifier for weed identification in sugarcane fields through images taken by UAV. *Bulg. J. Agric. Sci.* **2017**, *23*, 491–497.
91. Baron, J.; Hill, D.J.; Elmiligi, H. Combining image processing and machine learning to identify invasive plants in high-resolution images. *Int. J. Remote Sens.* **2018**, *39*, 5099–5118. [CrossRef]
92. Gao, J.; Nuyttens, D.; Lootens, P.; He, Y.; Pieters, J.G. Recognising weeds in a maise crop using a random forest machine-learning algorithm and near-infrared snapshot mosaic hyperspectral imagery. *Biosyst. Eng.* **2018**, *170*, 39–50. [CrossRef]
93. Mateen, A.; Zhu, Q. Weed detection in wheat crop using uav for precision agriculture. *Pak. J. Agric. Sci.* **2019**, *56*, 809–817.
94. Huang, H.; Lan, Y.; Yang, A.; Zhang, Y.; Wen, S.; Deng, J. Deep learning versus Object-based Image Analysis (OBIA) in weed mapping of UAV imagery. *Int. J. Remote Sens.* **2020**, *41*, 3446–3479. [CrossRef]
95. Rasmussen, J.; Nielsen, J. A novel approach to estimating the competitive ability of Cirsium arvense in cereals using unmanned aerial vehicle imagery. *Weed Res.* **2020**, *60*, 150–160. [CrossRef]
96. De Castro, A.I.; Peña, J.M.; Torres-Sánchez, J.; Jiménez-Brenes, F.M.; Valencia-Gredilla, F.; Recasens, J.; López-Granados, F. Mapping cynodon dactylon infesting cover crops with an automatic decision tree-OBIA procedure and UAV imagery for precision viticulture. *Remote Sens.* **2020**, *12*, 56. [CrossRef]
97. Sapkota, B.; Singh, V.; Cope, D.; Valasek, J.; Bagavathiannan, M. Mapping and estimating weeds in cotton using unmanned aerial systems-borne imagery. *AgriEngineering* **2020**, *2*, 350–366. [CrossRef]
98. Boukabara, S.A.; Krasnopolsky, V.; Stewart, J.Q.; Maddy, E.S.; Shahroudi, N.; Hoffman, R.N. Leveraging modern artificial intelligence for remote sensing and N.W.P.: Benefits and challenges. *Bull. Am. Meteorol. Soc.* **2019**, *100*, ES473–ES491. [CrossRef]
99. Arroyo, L.A.; Johansen, K.; Phinn, S. Mapping Land Cover Types from Very High Spatial Resolution Imagery: Automatic Application of an Object Based Classification Scheme. In Proceedings of the GEOBIA 2010: Geographic Object-Based Image Analysis, Ghent, Belgium, 29 June–2 July 2010; International Society for Photogrammetry and Remote Sensing: Ghent, Belgium, 2010.
100. Mohamed, Z.; Terano, R.; Shamsudin, M.N.; Abd Latif, I. Paddy farmers' sustainability practices in granary areas in Malaysia. *Resources* **2016**, *5*, 17. [CrossRef]
101. Jafari, Y.; Othman, J.; Kuhn, A. Market and welfare impacts of agri-environmental policy options in the Malaysian rice sector. *Malays. J. Econ. Stud.* **2017**, *54*, 179–201. [CrossRef]
102. Liu, B.; Bruch, R. Weed Detection for Selective Spraying: A Review. *Curr. Robot. Rep.* **2020**, *1*, 19–26. [CrossRef]
103. Hosoya, K.; Sugiyama, S.I. Weed communities and their negative impact on rice yield in no-input paddy fields in the northern part of Japan. *Biol. Agric. Hortic.* **2017**, *33*, 215–224. [CrossRef]
104. Sosa, A.J.; Cardo, M.V.; Julien, M.H. Predicting weed distribution at the regional scale in the native range: Environmental determinants and biocontrol implications of Phyla nodiflora (Verbenaceae). *Weed Res.* **2017**, *57*, 193–203. [CrossRef]

105. Kuan, C.Y.; Ann, L.S.; Ismail, A.A.; Leng, T.; Fee, C.G.; Hashim, K. Crop loss by weeds in Malaysia. In Proceedings of the Third Tropical Weed Science Conference., Kuala Lumpur, Malaysia, 4–6 December 1990; Malaysian Plant Protection Society: Kuala Lumpur, Malaysia, 1990; pp. 1–21.
106. Wayayok, A.; Sooma, M.A.M.; Abdana, K.; Mohammeda, U. Impact of Mulch on Weed Infestation in System of Rice Intensification (S.R.I.) Farming. *Agric. Agric. Sci. Procedia* **2014**, *2*, 353–360.
107. Martin, R.J. Weed research issues, challenges, and opportunities in Cambodia. *Crop Prot.* **2017**, *134*, 104288. [CrossRef]
108. Abdulahi, A.; Nassab, D.M.A.; Nasrolahzadeh, S.; Salmasi, Z.S.; Pourdad, S.S. Evaluation of wheat-chickpea intercrops as influence by nitrogen and weed management. *Am. J. Agric. Biol. Sci.* **2012**, *7*, 447–460.
109. Zhu, J.; Wang, J.; DiTommaso, A.; Zhang, C.; Zheng, G.; Liang, W.; Islam, F.; Yang, C.; Chen, X.; Zhou, W. Weed research status, challenges, and opportunities in China. *Crop Prot.* **2018**, *134*, 104449. [CrossRef]
110. Varah, A.; Ahodo, K.; Coutts, S.R.; Hicks, H.L.; Comont, D.; Crook, L.; Hull, R.; Neve, P.; Childs, D.; Freckleton, R.P.; et al. The costs of human-induced evolution in an agricultural system. *Nat. Sustain.* **2020**, *3*, 63–71. [CrossRef] [PubMed]
111. Balafoutis, A.; Beck, B.; Fountas, S.; Vangeyte, J.; Van Der Wal, T.; Soto, I.; Gómez-Barbero, M.; Barnes, A.; Eory, V. Precision agriculture technologies positively contributing to GHG emissions mitigation, farm productivity and economics. *Sustainability* **2017**, *9*, 1339. [CrossRef]
112. Ruzmi, R.; Ahmad-Hamdani, M.S.; Bakar, B.B. Prevalence of herbicide-resistant weed species in Malaysian rice fields: A review. *Weed Biol. Manag.* **2017**, *17*, 3–16. [CrossRef]
113. Singh, P.K.; Gharde, Y. Adoption level and impact of weed management technologies in rice and wheat: Evidence from farmers of India. *Indian J. Weed Sci.* **2020**, *52*, 64–68. [CrossRef]
114. Matthews, G. Can drones reduce compaction and contamination? *Int. Pest Control* **2018**, *60*, 224–226.
115. Herrera, P.J.; Dorado, J.; Ribeiro, Á. A novel approach for weed type classification based on shape descriptors and a fuzzy decision-making method. *Sensors* **2014**, *14*, 15304–15324. [CrossRef]
116. Gerhards, R.; Oebel, H. Practical experiences with a system for site specific weed control in arable crops using real time image analysis and G.P.S. controlled patch spraying. *Weed Res.* **2006**, *46*, 185–193. [CrossRef]

MDPI
St. Alban-Anlage 66
4052 Basel
Switzerland
Tel. +41 61 683 77 34
Fax +41 61 302 89 18
www.mdpi.com

Applied Sciences Editorial Office
E-mail: applsci@mdpi.com
www.mdpi.com/journal/applsci

www.ingramcontent.com/pod-product-compliance
Lightning Source LLC
Chambersburg PA
CBHW061619220926
51850CB00005B/548
* 9 7 8 3 0 3 6 5 5 3 3 7 5 *